Closer Than Brothers

Closer Than Brothers

Manhood at the Philippine Military Academy

Alfred W. McCoy

Yale University Press
New Haven and London

Published with assistance from the Louis Stern Memorial Fund.

Some portions of Chapter 2 were published as "Same Banana: Hazing and Honor at the Philippine Military Academy," *Journal of Asian Studies* 54, no. 3 (August 1995): 689–726. Several passages in Chapter 5 appeared in "The Myth of the Maharlika," *Veritas* (Manila), 26 January 1986. Preliminary drafts of some sections in Chapter 6 appeared as "The RAM Boys," *National Midweek* (Manila), 21 September, 28 September, 12 October 1988; and "The RAM Boys," *Philippine Daily Inquirer* (Manila), 1–8 January 1990. An earlier and somewhat different version of Chapter 7 was published as "Coup! The Real Story behind the February Revolt," *Veritas* (Manila, special edition), October 1986.

Copyright © 1999 by Yale University.
All rights reserved.
This book may not be reproduced, in whole or in part, including illustrations, in any form (beyond that copying permitted by Sections 107 and 108 of the U.S. Copyright Law and except by reviewers for the public press), without written permission from the publishers.

Set in Adobe Garamond type by The Composing Room of Michigan, Inc.
Printed in the United States of America by Sheridan Books, Chelsea, Michigan.

Library of Congress Cataloging-in-Publication Data
McCoy, Alfred W.
    Closer than brothers : manhood at the Philippine Military Academy
/ by Alfred W. McCoy.
       p.   cm.
    Includes bibliographical references and index.
    ISBN 0-300-07765-3 (cloth : alk. paper)
    1. Philippine Military Academy—History. 2. Philippine Military Academy—Biography. 3. Philippines—History, Military—20th century. I. Title.
U660.P6M33   1999
355′.0071′1599—dc21   99-20239
                                                                                            CIP

A catalogue record for this book is available from the British Library.

The paper in this book meets the guidelines for permanence and durability of the Committee on Production Guidelines for Book Longevity of the Council on Library Resources.

10 9 8 7 6 5 4 3 2 1

Dedicated to my mother
Margarita Piel McCoy

and to my father
Alfred M. McCoy, Jr. (USMA '44)

Contents

Preface, ix

Acknowledgments, xv

Abbreviations, xix

Introduction

1 Class and Corps, 3

Part One: **The Class of 1940**

2 Kaydet Days, 35

3 Baptism by Fire, 74

4 Career Soldiers, 102

5 Myth of the Maharlika, 143

Part Two: **The Class of 1971**

6 Torture, 183

7 Mutiny, 222

8 Coup d'Etat, 259

9 Impunity, 299

10 Reunion, 339

Notes, 353

Index, 409

Preface

Ten years ago when starting work on this book, I hit upon what seemed a promising new way to write military history. Instead of the usual chronicle of battles fought or forces modernized, I would collapse a half-century of war and peace into the story of just two classes who graduated a generation apart at the Philippine Military Academy (PMA). At first glance, this approach seemed enticing. Comparison of the Class of 1940, the academy's first graduating class, with that of 1971, its most controversial, offered a method that was selective yet comprehensive. Lying between the particulars of biography and the generalities of sociology, the history of these two classes, numbering less than a hundred men each, was compact enough to make a coherent narrative. Yet their graduates rose, over time, to commands that make their stories the history of an entire army.

But I still had doubts. Could I find enough information to fill a book about two small classes? Even if I did, could the story of just two hundred officers say much about an entire army that has seen, over the past sixty years, millions of men and women pass through its ranks?

I soon found ample data. Reflecting differences between the two classes, my research developed distinct, yet complementary, methods—sifting for documentary grains on Class '40, winnowing chaff from the massive media coverage of '71. Since most official records for Class '40 were destroyed during World War II, oral interviews and a miscellany of informal sources became the ingredients of their history. By the time I had interviewed thirty of their seventy-nine original graduates, I realized that my concerns about data had been unfounded.

During a quarter-century on active duty, the collective memory of Class '40 had become a reservoir filled deep with details about the armed forces of their day. Once members of the class agreed to interviews, my main challenge was to frame questions that would open the floodgates of memory. Our hour-long meetings stretched out, sometimes for several days, words cascading through my tape recorder into some four thousand pages of transcripted recollections. As officers used to command, they did not hesitate to correct errors and direct the course of our discussions. Sometimes, my greatest gift was the ability to ask the wrong question, prompting some remarkable revelations. Draft chapters came back from classmates demanding changes, duly noted in the text below. They opened their files to share documentary scraps—old yearbooks, six published and unpublished memoirs, two beautifully assembled photo albums, and, above all, the remarkable *Golden Book* compiled for their fiftieth reunion in 1986. The paucity of conventional evidence forced me to read other disciplines for theory and techniques to mine these limited sources.

Over time, my use of the graduating class as an analytic prism seemed validated by the way these sources came together, at times almost of their own accord. A basic narrative rose from my chronological files of newspaper clippings. Quotations from yearbooks and memoirs seemed to slip effortlessly into the text. It appeared that my accidental approach and historical reality, such as I could divine it, had converged.

In dealing with both classes, I struggled constantly to balance the empathy needed for interviewing with the objectivity required for writing. During our interviews in Manila, I came to admire the alumni of Class '40 who entertained me in their homes and invited me to their reunions. As the son of a West Point graduate of their generation, I slipped easily into habits of deference learned as an officer's child on U.S. Army posts of the 1950s. Listening to their stories of fighting on Bataan and starving in Japanese prison camps, I decided that many of Class '40 were, by the standards of their time, heroes. I agreed with their attempts to prevent the coups that made the military the bane of civil society elsewhere. Back at my university office in Madison, studying their statements under a harsh fluorescent light, I became convinced of their basic integrity and decided to let their own voices tell their story. Thus, part 1 on Class '40 seems, at times, a collage of quotations from their interviews, yearbooks, letters, and memoirs. Of course, my narrative hand is there throughout, whether in selecting the quotes or commenting critically on the last years of their careers, which were marked by compromise and conflict.

My relations with the alumni of Class '71, who led the "people power" uprising against President Ferdinand Marcos, reached a different outcome. I first approached them in the heady months after the dictator's downfall in 1986 when they were heroes, idolized by Manila's press and public. Our interviews began on a celebratory note, and I felt privileged to share confidences of officers who had rebelled against a corrupt, brutal dictatorship. But gradually discordant phrases slipped through my microphone and into the transcripts. One senior officer launched, unprovoked, into a frightening reverie of blood and terror. Another let slip his involvement in a coup plot against President Corazon Aquino that began less than six hours after our interview. Back in my university office, reviewing the human rights records of the Marcos military, I found more and more of my erstwhile heroes cited as brutal, even psychopathic torturers. When my first publications proved critical, the leaders of Class '71 tried manipulation, intimidation, then litigation.

For seven years, I followed their descent into coups, terror bombings, and assassinations. The summer of 1993—isolated in my university office studying closely the reports of their tortures to write Chapter 6—was one of the most depressing periods of my life. Afterward, I put this manuscript aside for over a year. Since these events were covered almost daily in the Manila press, part 2 on Class '71 required selection rather than searching. In analyzing this mass of data, I have tried to put distance between the writing and my own feelings about torture and state terror.

Let me not exempt myself from the same biographical examination that I have imposed upon these Filipino officers. In a personal sense, this book did not begin in 1988 or 1986. It began in 1944 when my father graduated from the U.S. Military Academy at West Point, married my mother, and sailed for war in Europe leaving her pregnant with me. After the war, my father rejoined the Field Artillery back in the United States. Its home at Fort Sill, Oklahoma, became my childhood home of memory. There I first encountered history. When I was five, my mother, intrigued by the life of the great Apache warrior, took me to visit Geronimo's grave on the post's eastern firing range. I can still see the wind whipping dust around a low stone monument in the clear sunlight of a prairie afternoon. I can still sense my touch on the rough masonry and my thoughts about the chief below, my first awareness of the past and its inhabitants.

There too I had my first taste of combat, or simulated combat. When I was seven, my father took me to the artillery range in the hills beyond our home for a night firing exercise. He spoke into a telephone and his breath became fire as tanks and trucks on slopes miles away erupted in flames beneath the glare of falling flares.

It was my childhood ambition to follow my father to West Point and into the U.S. Army. Later, in the mid-1960s when I was a student at Columbia University, I became equally determined to avoid service in the U.S. Army, in Vietnam or anywhere else. If a critic were to analyze me as I have these Filipino officers, he or she might say that my admiration for the older graduates of Class '40 is a surrogate reconciliation with my father and my criticism of Class '71 is a rejection of the me that I might have become had I joined the U.S. Army and fought in Vietnam.

I think otherwise. My relationship with my father was a burden and a blessing. Between two wars and work as an electronic engineer, he was often absent or preoccupied. But I watched him closely, learning the meaning of both military honor and posttraumatic stress. For this project, he left me the empathy that every author must bring to each subject. I like to think that I have come, through a mix of childhood admiration and antiwar activism, to a critical appreciation of what a military should and should not be.

My father also influenced this writing in other ways. Midway through the first draft of this book, one of his West Point classmates invited me to collaborate on his long-overdue obituary for an issue of the alumni magazine, *Assembly*, commemorating the fiftieth anniversary of the graduation of Class '44.[1] Negotiating nuances in that text between my mother and my father's mother taught me the sensitivity of biographical details that others, outside a family, might consider inconsequential. As both a son and a historian, I found reading letters from my father's classmates about his career an extraordinary experience. One, a retired lieutenant general, recalled him as "cool" under pressure when their artillery was "bogged down in mud" near the Normandy beachhead during the European campaign.[2] Another of the same rank characterized him after the war at Fort Sill as "quite uptight," a misfit, a malcontent.[3] A third classmate, later the U.S. Army's chief historian, said of those same days at Fort Sill that my father "had a very good reputation with students and faculty alike as a clear thinker and excellent teacher" and "was very engaging socially."[4]

Such sharp disagreement reminded me that informed observers can look closely at the same individual and see things very differently. Indeed, the simultaneity of these two projects, my father's obituary and these Filipino biographies, gave me a new perspective on the balance between empathy and "objectivity." As I composed every sentence that somehow assessed these classes and their cavaliers, I was compelled, like all historians who use oral sources, to weigh each adverb and adjective for a tone that tempered the warm of personal feelings with the cool of professional judgment. If balance, like truth and objectivity, can never be achieved, at least it can be pursued.

Acknowledgments

As work on this book dragged on for more than ten years and pushed me beyond my intellectual and emotional limits, it was sustained, at every stage of its fitful gestation, by the many who shared ideas, helped with the research, and read drafts.

This project began in July 1985 when I was on sabbatical at the U.S. National Archives and came across U.S. Army records that discredited President Ferdinand Marcos's claims to heroism in World War II. Six months later, when Marcos was running for reelection as a war hero, I delivered these documents to the *New York Times*, and stories of his "fraudulent" military record soon made front-page headlines on both sides of the Pacific. In releasing this story, I was helped by reporter Sy Hersh, who vouched for me at the *Times;* correspondent Marian Wilkinson, who negotiated publication in her paper, the *National Times* of Sydney; and Dr. Ed de Jesus, who persuaded the publisher of the Manila weekly *Veritas,* Jimmy Ongpin, to risk Marcos's anger. That story, found in Chapter 5, plunged me into Philippine politics and launched me on a trajectory that led to this book.

In July 1986, the editors of the *National Times* sent me to Manila with two staff reporters, the veteran Marian Wilkinson and the neophyte Gwen Robinson, to write an instant history of the "people power" revolution that had recently toppled Marcos. We were soon caught up in another coup by the colonels of the Reform the Armed Forces Movement, the RAM. In this chaos, both women taught me much about getting the story, particularly Marian, a superb investigative reporter. The editor of the Manila weekly *Veritas,* Melinda de Jesus, published a special edition with our story despite pressure from the RAM leaders. This experience and the contributions of these journalists influenced the writing of Chapter 7.

For another year, I worked on this project as a researcher for Hal McElroy, a Sydney film producer shooting a television miniseries about Marcos's downfall for Home Box Office (HBO). During a 1987 trip to Manila that coincided with RAM's third coup, I studied the reporter's craft ducking shrapnel on Santolan Road with Melinda de Jesus, watching dive-bombers over Camp Aguinaldo with Lin Neumann, and seeing the headquarters of the Armed Forces of the Philippines burn with Seth Mydans. These experiences helped shape Chapter 8.

In 1988–89, a sabbatical in Manila allowed the first of thirty-two interviews with the cavaliers of Class '40 that became part 1 of this book. I am particularly indebted to Colonel Deogracias Caballero (ret.) and Colonel Jose Mendoza (ret.), who served as my intermediaries with their class. The widows of two classmates, Mrs. Fe Navarro and Mrs. Betty Picar, graciously shared their husbands' photographs. In California, Commodore Ramon A. Alcaraz (ret.) introduced me to his classmates in America. Their onetime underclassman, Defense Secretary Rafael Ileto, offered insights from a fifty-year career about the coups then shaking the country's stability.

When I first presented Chapter 6 on RAM's torturers at the Australian Association for Asian Studies at Canberra in 1988, that paper sparked controversy in the pages of the association's journal. I am grateful to its president, Dr. Elaine McKay, for her careful rebuttal of that critique. After RAM's next coup a year later, the *Philippine Daily Inquirer* carried that conference paper on page one, producing a libel suit by RAM's chairman, Colonel Hernani Figueroa. Throughout the litigation, the paper's publisher, Eugenia Apostol, did not waver. As I continued my work on this topic during the next decade, members of the paper's staff—editor Isagani Yambot, senior columnist Amado Doronilla, and librarians Medy Gregorio and Cipring Frias—sustained the project with reflections and research support.

During this research, I met survivors of RAM's safe houses who taught me about torture. In 1988, one of their victims, Maria Elena Ang, called on me in Sydney to share the burden she still bears. At a conference in Canberra, Pete Lacaba spoke with a reporter's detachment about his endless sessions with Lieutenant Colonel Rodolfo Aguinaldo, while ex-priest Luis Jalandoni grew angry as he recalled Aguinaldo's beatings. During the year he worked on his dissertation in Madison, Dr. Temario Rivera listened carefully to my thoughts on this topic without saying a word, not one, about his own torture by Aguinaldo.

When the study required documents, I was helped by Edward Boone of the MacArthur Memorial Archives in Norfolk, Virginia; Oggie Mallilin at the Ateneo de Manila University's Rizal Library; archivists at the University of the Philippines; the gracious staff at the Lopez Memorial Museum in Pasig, Rizal; and employees at the U.S. National Archives in College Park, Maryland.

As my manuscript moved from draft to book, many helped. Dr. Helen Mendoza, a resourceful Manila researcher, responded quickly to my frantic phone calls for elusive materials. Ms. Caroline Diyco of Ateneo University transcribed the interviews carefully. The Honorable Jaime Zobel de Ayala generously provided portraits of the RAM leaders, while photographers Noli Yamsuan and Jose Duran, himself a torture victim, took other images that appear in this text.

Over the past five years, drafts have benefited from comments by Southeast Asian scholars at the University of Wisconsin—Dan Doeppers, Lucy Mathiak, Lauran Schultz, Amy Golden, Sean Kirkpatrick, Vina Lanzona, Carmel Capati, Kevin Whiston, and Janice Newberry. Dr. Mike Cullinane shared reflections on human rights; Professor Katherine Bowie commented on gender; our Southeast Asian librarian, Dr. Carol Mitchell, gathered materials; and David Streckfuss, then a history graduate student, helped sustain me through the close contemplation of torture needed to complete Chapter 6. Elsewhere, Dr. Marguerite Roulet of Madison, Professor Cynthia Enloe of Clark University, and Dr. Joyce E. Canaan of the University of Central England made useful comments about the sections on gender. Other Southeast Asia specialists shared documents and insights, including Professor Patricio Abinales of Ohio University, Dr. John Sidel at the School of Oriental and African Studies, and Dr. Belinda Aquino at the University of Hawaii.

I was fortunate to find skilled editors and readers at every stage. When I submitted an early version of Chapter 2 to the *Journal of Asian Studies,* David Buck edited it judiciously and had it reviewed by Professor Vince Rafael, who made harsh, yet useful, criticisms. At the retirement dinner for my Yale dissertation adviser Hal Conklin in May 1996, Professor Jim Scott indulged my long summary and introduced the project to Yale University Press. There editor-in-chief

Charles Grench identified the work's main fault and suggested revisions. From her cabin in northern Michigan, Jan Opdyke made comments that helped correct that structural flaw. My mother, Professor Emerita Margarita McCoy, read the final chapters with the same sharp attention for clarity and accuracy that she once reserved for urban planning proposals. When all these readers were done, my wife, Mary McCoy, gave the work a final review, identifying lapses in logic and moderating my prose.

Finally, this work of narrative history bears the imprint of my teachers. My mother, a gifted storyteller, introduced me to the form's power to entice and educate. In high school, I had the good fortune to study English for two years with Robert Cluett, later a professor at York University, who gave me the tools to turn story into narrative. At Yale graduate school, Edilberto de Jesus exposed our group of future Southeast Asian historians to the richness of Philippine history. While on sabbatical from Australia in 1981, Rey Ileto opened me to the spiritual dimension of Philippine politics as we drove from church to church around Laguna de Bay and sat for hours before Carlos V. Francisco's vast tableau of Marcos's heroic destiny.

Aside from the media groups that funded my research with their commissions or assignments, this study was supported by grants from the Australian Graduate Research Scheme, the Australian Academy of the Humanities, the Franklin and Eleanor Roosevelt Foundation, the Graduate School of the University of Wisconsin-Madison, the American Philosophical Society in Philadelphia, and the Social Science Research Council of New York City.

In the final phase, Susan Pador did the comprehensive index, Jan Opdyke proofed with her usual care, and Karen Gangel skillfully coordinated these tasks.

I alone, of course, am responsible for all errors of fact or interpretation.

Abbreviations

AFP	Armed Forces of the Philippines
AIB	Allied Intelligence Bureau
ASAC	Anti-Smuggling Action Center
ASSO	Arrest Search and Seizure Order
BCT	Battalion Combat Team
BMA	Bulacan Military Area
CAA	Civil Aviation Administration
CIA	U.S. Central Intelligence Agency
CIC	Counter Intelligence Command
CIS	Constabulary Investigative Service
CRC	Constabulary Revenue-Customs
COMELEC	Commission on Elections
CSU	Constabulary Security Unit
FLAG	Free Legal Assistance Group
HMB	Hukbong Mapagpalaya ng Bayan, or Huk, Hukbalahap
ISAFP	Intelligence Service Armed Forces of the Philippines
KMU	Kilusang Mayo Uno (May First Movement)

Lakas-NUCD	Lakas ng Bansa-National Union of Christian Democrats
LAW	Light Anti-Tank Weapon
METROCOM	Metropolitan Command, Philippine Constabulary
MFP	Movement for a Free Philippines
MISG	Metrocom Intelligence Security Group
NBI	National Bureau of Investigation
NICA	National Intelligence Coordinating Agency
NISA	National Intelligence and Security Agency
NPA	New People's Army
NUC	National Unification Commission
OSP	Off-Shore Patrol
PACC	Presidential Anti-Crime Commission
PAAC	Philippine Army Air Corps
PARGO	Presidential Agency on Reforms and Government Operations
P.C.	Philippine Constabulary
PCA	Philippine Constabulary Academy
PCO	Presidential Commitment Order
PHILCAG	Philippine Civic Action Group
PMA	Philippine Military Academy
PNP	Philippine National Police
PSC	Presidential Security Command
RAM	Reform the Armed Forces Movement; Rebolusyonaryong Alyansang Makabayan
RCT	Rehabilitation and Research Centre for Torture Victims
ROTC	Reserve Officers' Training Corps
SAS	Special Air Service
SEATO	Southeast Asia Treaty Organization
SELDA	Samahan ng mga Ex-Detainee Laban sa Detensyon at para Amnestia
SFP	Soldiers of the Filipino People
SOG	Special Operations Group
TFD	Task Force Detainees
U.N.	United Nations
UNIDO	United Democratic Opposition
U.P.	University of the Philippines
USAFFE	United States Army Forces in the Far East
USMA	United States Military Academy
USNA	United States Naval Academy
YOU	Young Officers' Union

Introduction

Chapter 1 Class and Corps

In August 1967, the superintendent of the Philippine Military Academy, General Reynaldo Mendoza, stood for the last time before the Corps of Cadets on the high, green plateau near Baguio City. Arrayed in rows, over four hundred strong, the cadets were a picture of military perfection in their dress-gray uniforms. In his valedictory, the general celebrated the academy's democratic mission and urged the corps to honor its motto—"integrity, courage, loyalty." Nearly thirty years before, as a young cadet in the academy's first graduating class, he had used those words as a refrain in lyrics that became the school song, "PMA, Oh! Hail to Thee." As his wife wept and the corps rose to applaud, General Mendoza stepped down from the podium, climbed into his blue Volkswagen Beetle sedan, and drove himself into retirement and into history. He went with pride knowing that he and his Class of 1940 had an unblemished record—no scandals, no corruption, and no coups.[1]

Swelling the ranks of the corps that day were the youngest cadets, the 148 "plebes" of the future Class of 1971. As colonels twenty years later, they would lead six abortive coups d'état before retreating into the guerrilla underground for a campaign of robbery, kidnapping, and terror bombings. Now, more than thirty years later, those plebes are assuming command of the military, and their class captain, Gregorio "Gringo" Honasan, has been elected to the Philippine Senate.

In retrospect, this valedictory seems a symbolic moment. Here we can glimpse a basic generational and cultural change in the Armed Forces of the Philippines (AFP). These two classes, 1940 and 1971, span its entire history, from its founding in 1936 to the present. Through their eyes, we gain a unique vantage point on tumultuous events that have shaped the Philippine military over the past half-century. Marked differences in the character of these two classes—one opposing coups and the other plotting them—highlight the military's changing role in Philippine political life.

Although separated by thirty years, both classes faced similar political decisions that make the comparison telling. In the 1953 presidential elections, the opposition, convinced that the incumbent would win by fraud, pressed Class '40 to prepare a coup. They refused. As long as they were in command, the Philippine military, almost alone in Southeast Asia, did not engage in coup plotting. At the highest echelons in the mid-1960s, they were the senior officers whom President Ferdinand Marcos forced out as he prepared for martial-law dictatorship.

In the 1986 presidential elections, the opposition, again certain that the incumbent would cheat, pressed the leaders of Class '71 to plan a coup. They agreed. When President Marcos won by fraud, they led a bungled coup attempt that sparked the famous "people's power" uprising. After the crowds installed their candidate Corazon Aquino in the presidential palace, the colonels of Class '71 made five more coup attempts that greatly troubled the nation's transition to democracy. Their December 1989 coup paralyzed the capital for a week and came close to capturing the palace. Had they won, they planned to rule by terror, with mass slaughter of dissidents and ritual execution of deviant officers.

Other military factions soon joined the scramble for power, producing a total of nine coup attempts in the late 1980s—more than in any other nation. Watching this season of coups by their fellow academy alumni, Class '40, long retired and well into their seventies, condemned the rebel officers, rued the politicization of the military, and urged a wholesale purge of the cadet corps.

CLASSES AND COUPS

Comparison of these two classes allows us to engage the most problematic aspect of the military in the Third World—the remarkable persistence of coup attempts. By forming an army for defense against foreign invasion, newly independent states, the Philippines included, have created the means of their own destruction. As Edmund Burke once noted, "an armed, disciplined body is, in its essence, dangerous to liberty." Indeed, all but two of Latin America's twenty countries have experienced coups since 1945. In Southeast Asia, the military seized power in seven of nine major nations during that period. By 1985, the military ruled 58 of 109 countries that make up the so-called developing world.[2]

Although the dangers that a standing army poses for a democracy should be obvious, Western scholars of civil-military relations have paid insufficient attention to the problem.[3] One notable exception, S. E. Finer, offers a provocative observation: "Instead of asking why the military engage in politics, we ought surely ask why they ever do otherwise. . . . The military possess vastly superior organization. And they possess *arms*." We should not, he argues, ask why the military "rebels against its civilian masters, but why it ever obeys them."[4] The answer to this latter question should be obvious: the military obeys only when it wants to. What is it, then, that makes an army willing to subordinate itself to civil authority?

The study of the Philippines provides some answers. While civilian regimes fell to coups in Burma, Thailand, and Indonesia, the Philippine military was restrained. Until Marcos declared martial law in 1972, the AFP had respected civil authority for a quarter-century. Politicized by fourteen years of authoritarian rule, military factions later attempted a succession of coups, culminating in the bloody 1989 revolt led by over five hundred officers ranking from lieutenant to general.[5] The academy's alumni, though a minority in the AFP, accounted for nearly 40 percent of these rebel officers and provided almost all the coup leaders.[6] But all of these attempts, without exception, failed. By 1990, no other Southeast Asian military had launched so many coups with so little success. How can we explain this contradictory behavior by the Filipino officer corps—supporting rebel factions but in the end resisting their coups?

The answer lies in the socialization of the Filipino officer corps. Unlike its Southeast Asian counterparts, the AFP embraced politics with a certain ambiguity—reaching for power but holding back, serving politicians but feeling tainted by the compromise, or backing a dictator but feeling somehow betrayed in doing so. Setting aside issues of leadership or tactics, these coup attempts collapsed, at one level, because the majority of Filipino officers did not believe in seizing power. Throughout the Marcos years, a belief in military professionalism had somehow survived, slowing the finger on the trigger at critical points in each uprising. Even after the spectacular 1989 coup, military intelligence estimated that only 30 percent of the officer corps had evinced political loyalties of any description. Moreover, a survey of five hundred academy alumni found a remarkable 94 percent felt that "politics and the military do not mix well," indicating a strong, underlying support for an apolitical professionalism.[7]

Such resistance to political involvement among Filipino officers springs, at base, from their socialization into obedience. The contrast between these classes, 1940 and 1971, illuminates the processes of military socialization, both its consolidation and collapse. During their thirty-year careers, there are two formative periods for Filipino officers—the four years of uniform indoctrination at the academy and more diverse, yet defining, experiences after graduation as junior officers. At the Philippine Military Academy, future officers are influenced by several simultaneous processes—a formal indoctrination into the ideal of civil supremacy and a less formal, but lasting, peer bonding to classmates. The official curriculum of drill and classes seeks to infuse an abiding respect for civil authority. As Class '40 wrote in its 1986 *Golden Book,* the PMA imposed "a strict and Spartan-like regimen aimed at instilling . . . that instinctive obedience to the legal orders of their superiors at all times."[8] Through a harsh ritual initiation in the first or "plebe" year, the academy also tried to mold its cadets into a manhood that equated courage with obedience and restraint. For reasons that fill the pages of this book, such socialization proved remarkably effective for Class '40, much less so for '71.

After graduation, young officers test these lessons in their first, formative years of service, which become a second, critical phase in their socialization. Within the hierarchy, junior officers, lieutenants and captains, are still subject to close supervision and intensive training; but with each promotion they gain greater autonomy and authority for choices of consequence. While variations in the PMA experience are usually small, there have been major changes in the demands of active duty. Among the many factors that might explain the different career paths of these two classes, this study focuses on this second phase, finding the interaction between the graduating class and the regime then in power as central, even defining. Each political regime assigns the armed forces a mission, consonant with its character and circumstance, that leaves a strong imprint on the careers, and experiences, of these junior officers. Thus, to understand why Filipino officers have decided to obey or to disobey, we have to understand both the nature of their socialization and the conditions of their service.

The collective persona of every PMA class is shaped by a central paradox: each is identical yet unique. Every class passes lock-step through a four-year progression from neophyte to leadership that has not changed significantly in sixty years. But within this unbending regimen, each enters with a unique mix of individuals and emerges with a distinctive group dynamic. Through the sum of these personalities and their shared experience, every class develops a collective persona that both shapes and interprets their socialization. Despite their strong solidarity, Class '40 remained somehow a collection of individuals who competed with each other for half a century. When I asked class member Pedro Bersola why they never launched a coup, he replied, "In the first place, we didn't have a strong leader." He added, "Of course, if we have a strong leader, maybe you could influence the class to take action."[9] By contrast, Class '71 subordinated itself to a single, dominant leader, Gregorio "Gringo" Honasan, and followed him through the academy and into the Reform the Armed Forces Movement (RAM) and its succession of coup attempts.

After graduation, both classes joined a military that was being cast, or recast, by the regimes that ruled from Malacañang Palace. As junior officers, recent academy graduates are particularly susceptible to the changing character of the state and its military. Whether war, peace, or martial rule, generals keep to their tents while lieutenants form the line and suffer its fate. For nearly half of the twentieth century, the Philippines has been ruled by two leaders with the power to create regimes, and armies, in their image—the pre–World War II president Manuel Quezon and the postwar dictator Ferdinand Marcos. Both saw combat as young lieutenants, were masters of political intrigue, and knew how to make the military an instrument of their will.

The first part of this book, on Class '40, is a study of successful military socialization. During the 1930s, when Class '40 entered the academy, President Quezon ruled a Commonwealth preparing for independence and needed an army led by competent professionals who could defend the nation against invasion. As a leader of uncommon vision, Quezon understood the dangers inherent in his new military and insulated its officers from party politics—a practice that his postwar successors largely continued. Graduating on the eve of war, Class '40 won honors for fighting enemy invaders, were ennobled by privation in Japanese prisoner of war camps, and emerged with their bonds and values stiffened. For the remainder of their careers, this group of seventy surviving officers struggled to make the Philippine military a professional, apolitical service.

Though, on balance, successful, their effort seems, in retrospect, perilously fragile. As soldiers in a society permeated by patronage politics, Class '40 faced incessant pressures to compromise. Their careers required, on a daily basis, mediation of the paradoxical, even contradictory role of the military in a democratic society—subordinated to politicians yet apolitical; armed yet nonviolent;

all-powerful yet powerless. Thus, Class '40 cannot tell a simple tale of honorable service. Instead, their success was the sum of countless failings; their solidarity withstood rivalry, the imperatives of their honor code survived compromises. Their story is one of nuance, not absolutes. Through their biographies, collective and individual, we can gain a sense of the underlying resilience of the core values, ingrained through military socialization, that restrained their reach for power.

By contrast, the book's second part, on Class '71, is a study in the breakdown of military socialization. Marcos, a master of political maneuver, politicized the military hierarchy as legally elected president (1966–72) and then courted its support for martial law in 1972. As dictator, he opened a theater of terror to silence dissent. Just as Class '40 had graduated from the PMA only twenty-one months before Japanese invasion, so '71 left the academy just eighteen months before Marcos's declaration of martial law. Instead of fighting enemy invasion, the young lieutenants of Class '71 were brutalized by combat against Muslims in Mindanao and interrogation of suspected subversives in Manila.

Influenced by radical nationalism and their service in a martial-law military, the eighty-five graduates of Class '71 came to see the armed forces as an instrument of social transformation. Through the experience of torture, they were liberated from the academy's socialization into subordination. They emerged from a decade in the safe houses of the Marcos regime with a superman sense of themselves as creator/destroyers who could seize the state and transform society. So empowered, they rebelled against their military superiors to launch a half-dozen coups d'état. So enamored of self, they believed that the majesty of their violent coups would sweep away all opposition and thus failed, in every attempt, for want of careful planning and disciplined execution.

Despite these contrasts, the two classes are, in significant respects, similar. Most PMA cadets have been recruited from a broad, lower-middle social stratum. During much of the twentieth century, Filipino elites have shown little taste for the military life, making the academy, from its opening, an avenue of social mobility for ambitious young men. As individuals plucked from diverse communities across the archipelago, PMA graduates bear the imprint of their institution, emerging with a strong sense of public service and patriotism. Detached from the usual social moorings of family, class, and region, they are susceptible to ideological extremes in service of the nation. "Our military academy has always been a vehicle for social mobility for children of the lower-middle class or the poor, as my own family was," explained General Jose Almonte (PMA

'56) in a 1994 interview. "Social awareness," he added, "comes naturally to people who grew up in the military."[10]

Over time, moreover, alumni sons and siblings have become a disproportionate share of most classes, increasing the insularity of the officer corps. While only two members of Class '40 were soldiers' sons, eighteen in Class '71 followed fathers or brothers into the academy. Indeed, a survey of PMA classes '51 through '91 found that over half of postwar cadets had relatives in the military. As recruitment became less open and the officer corps more inbred, patriotism and privilege could combine to increase the military's inclination to political action.[11]

Whether actively apolitical like Class '40 or strongly committed to political change like '71, both groups played upon alumni loyalties to realize their aspirations. Just as Class '40 used their identity as a "batch" to discourage any political involvement, so RAM's leaders later tugged at school ties to pull Class '71 in a diametrically opposite direction—into coup attempts and a movement for radical change. For postwar Filipino officers, remaining "apolitical" required as much active political engagement as did plotting a coup d'état. Yet this emphasis on similarity should not be overdrawn. There is, at the level of practical politics, a substantial difference between serving the state and seeking its overthrow.

Comparison of these two classes also offers an effective mechanism for studying other changes in the Philippine armed forces. As a vast, complex organization, the modern military represents a challenge for the historian. How can we possibly compress the doings of millions of men and women over the span of decades between two covers? How can we grasp the paradoxical character of an institution that is immersed in society yet apart, visible yet closed? By focusing on two graduating classes of regular officers, both less than a hundred men, we have groups small enough for close study, yet strategically placed to make the details of their lives significant. Moreover, comparison of these cohorts separated by a generation isolates two slices of time, highlighting key changes in civil-military relations.

Through the collective biography, or prosopography, of these classes, we can perhaps avoid some of the problems inherent in military history. As both elite and mass organizations, armies pull writers toward the poles of biography and sociology. If we focus on the few in command, the usual result is a biography of a general or admiral. I do not wish to add another to this vast literature. If, by contrast, we study an army at peace in all of its complexity, the outcome might well be an organizational chronicle of more men and bigger weapons devoid of narrative or political dynamic. Much of military history focuses on wars and battles, arguably the least important activity for the many of the world's armies that have only rarely, if ever, fired on a foreign enemy. But most importantly, none of these approaches allows us to grasp the routine role of the military in modern society—matters such as budget, factions, seniority, and social control.

The history of any PMA class affords us a privileged view of the military's role within the Philippine polity. Armies have an illusory unity that often deceives analysts of their politics. All their drill and discipline mask countless fault lines and factional divisions. Once an army moves toward power, every officer has a half-dozen lines of alliance—class, service, unit, rank, region, and ethnicity. The deceptive coherence of the chain of command conceals these splits from outsiders until the moment of action, making analysis of military politics difficult. Moreover, the military's habitual secrecy for reasons of national security limits access by outsiders. But the class can take us inside headquarters to observe staff meetings and private conferences. Its collective memory is a vast store of information about the armed forces of its day.

Military academy alumni and their classes are usually important social units in any nation with a standing army. Among the many loyalties that define each officer, the academy ties often remain the strongest and graduates can act with a self-interest that makes them a powerful network. Bonded by their cadet days and bound together by the seniority system, classes can become significant political actors, both within the armed forces and without. Once we understand the role of academy classes as political actors, we gain a new lens that allows us to see significant elements in Philippine politics—the dynamics of coups, the rise of PMA alumni as politicians, and, more specifically, hidden dimensions of major historical events. In 1986, for example, when over a million people massed for the "people power" uprising against Marcos, the outcome at several key points turned on the internal dynamics of just two PMA classes with less than a hundred men each.

Through overlapping anecdotes that spring from these biographies, we gain, for the first time, a sufficient density of detail to write a history of what has remained a closed, sometimes secretive brotherhood—the AFP officer corps. For an institution with few official histories and no archives, biography, individual or collective, is often the only point of access. Finally, by tracing the career cycles of these two classes, from their initiation as plebes through their promotion to generals, we can raise an issue that is curiously absent from the many volumes on military history—the influence of male gender on a distinctly gendered institution.

MILITARY SOCIOLOGY

In considering the issue of militarized masculinity, this study seeks to engender Philippine national history, not to write a history of Filipino men. With a leaf from the text of women's studies, we can see how gender attaches itself to men, influences their collective behavior, and implicates them in the institutions of power.[12]

Instead of trying to extract men from the plot of Philippine history, this study tries to read their gender into it. The story of Class '40, for example, begins with a group of cadets socialized into an all-male institution, the PMA. As classmates rise through the hierarchy, their history becomes, in one sense, a traditional chronicle of presidents, senators, and generals—almost all men. We begin with young men detached from power and end with the usual script of a nation's history acted out largely by powerful men. But there is a difference. Throughout, this narrative is informed by an awareness that its actors are indeed men and does not assume that their actions somehow represent a universal standard of human behavior. Such an approach, hopefully, represents a corrective to the usual kind of national history that has long resisted "problematizing the masculinity of its male subjects."[13]

Surprisingly few scholars of the military have engaged the issues of gender that should arise in the study of any single-sex institution. In the 1980s, by contrast, feminist scholars focused on the military, creating the first gendered studies of modern armies.[14] Through analysis of these academy classes, we can read gender into the history of the Philippine military and, by implication, other modern armies—seeking thereby to understand, in Cynthia Enloe's words, "just how masculinity is created and sustained in the peculiar ways necessary to sustain a military organization."[15] Studies of military or paramilitary units often show how their masculinity is manifested in incidents of brutality. By contrast, the history of Class '40 focuses on a situation, typical of most armies, where masculinity is used to reinforce restraint. Similarly, the study of Class '71 assumes that restraint, not aggression, is the normative standard for military masculinity.[16]

From the outset, we need to distinguish between societies where the military is "highly bounded," or socially isolated, and those where it is "socially pervasive" through mass conscription. In showing how the Philippine Commonwealth constructed a new masculinity at the PMA during the 1930s, we are focused, above all, on questions internal to such a closed, bounded institution. Yet we cannot ignore the impact that this massive mobilization and its supporting propaganda had upon the whole order of gender roles in an emerging nation.[17]

Despite its isolation in the mountains of Baguio, the PMA's training of these young males had lasting implications for the whole of Philippine society. The school served, in effect, as a social laboratory, a crucible for casting a new form of Filipino masculinity. Through hazing, study, and drill, the academy pounded young males into a foreign mold of military manhood. By parading before the masses in Manila and acting in Tagalog films, these prewar PMA cadets became role models for the peasant conscript of the new Philippine Army.

At the core of the complex relationship between masculinity and the military lies the phenomenon of male bonding—the formation of a lasting group identity through shared suffering. Whether the Wehrmacht in World War II or West Point cadets in the 1960s, studies of male group behavior emphasize the influence of this bonding or "binding."[18] One essay on military academies in Britain, France, and America notes that "potent emotional conditioning... has been of greater importance than... academic curriculum."[19] In most armies, American and Philippine included, such solidarity is formed through "hazing"—a ritual passage through ordeal to manhood and acceptance. By breaking down a cadet's civilian identity and then building him back up as a member of a military sodality, the U.S. Coast Guard Academy, for example, created what one study called a "remarkable unity." Similarly, a psychologist who studied the PMA in the 1960s found that "enduring bonds of social solidarity are formed through... the almost overwhelming hazing process."[20]

This history of two academy classes raises some questions about gender that it, ultimately, cannot answer. While a focus on masculinity illuminates aspects of their experience, it lacks sufficient explanatory power, on its own, to supersede more conventional concerns such as civil-military relations. Thus, in the comparative history of these classes, gender, when useful, will intrude to highlight facets of male group behavior. When not, the narrative returns to analysis of narrowly military topics such as training, promotion, and coups.

Studies of the military, this one included, usually pick the officer corps as their Archimedian point for analysis. This leadership echelon shapes the destiny of any military organization, leading it into war, maintaining its discipline in peace, or launching it on a coup. Thus, the history of the Philippine military is nearly synonymous with the formation of an officer corps with its own values, standards, and traditions.

The classic model from American sociology, by Morris Janowitz and Samuel Huntington, has served for decades as the starting point for study of the modern military. In sum, their studies isolate the factors that led Western officers away from an eighteenth-century heroic ideal toward a modern professionalism that fulfills Karl von Clauswitz's dictum: "The subordination of the military point of view to the political is . . . the only thing which is possible."[21] At base, they apply Max Weber on the routinization of charisma to show how, in Janowitz's words, "the heroic leader" has been transformed into the "military manager"—a uniformed civil servant subordinated to civilian authority.[22] Clearly, this model has a universal application in the study of any military, whether in developed or developing nations. But it seems to work best for societies, like the United States, where the change is complete. By their focus on the final product, an apolitical military manager, these classic studies tell us far less about armies where the heroic ideal is still in command.

Applied to the Philippines, their approach highlights an essential contrast in values between the professionalism of Class '40 and the heroic vision of Class '71. Beyond that important insight, this model does not explain why, in many developing countries, the interplay between these ideals follows such a twisted, painful path to this anticipated triumph of the modern military manager. In societies that have suffered the burden of colonization, the imperative of nationalism seems to shift the ground of the debate in a way that privileges the heroic ideal of soldier as national savior.

In a developing society like the Philippines, where the debate over heroism and professionalism is still unresolved, it is necessary to ground these universal theories in the particulars of national history. Rather than study soldiers as uniform and coups as universals, we might begin by restoring the military to the matrix of its society, seeing how it acts and interacts within its country's onward progress. Even if the military models are foreign, as they often are in developing nations, the key issues remain change and cultural adaptation. Like a peasant or religious movement, an army is the sum of its social parts—heroic myths, hybrid ideology, microsocial units, hierarchy, and generational change. Like all human organizations, armies have gender, myths, politics, extraordinary leaders, and particular histories.

Yet, even as we immerse it in its local context, the military remains a unique institution with universal attributes. Although Philippine military cliques share certain similarities with the country's political factions, there is no reason to assume that its colonels and generals will behave like governors or senators, playing the political game by civilian rules. Invested with land and capital, bound by blood and marriage, Filipino elite families can be personalist and parochial, with interests opposed to those of the nation-state. Political families must ultimately try to manipulate, if not subvert, the power of the state, using public resources to build private wealth.[23] By contrast, military factions are ultimately national in character, detached from primordial roots of blood and earth. Officers, socialized and indoctrinated at the academy, are creations of the state and serve, more than any other elite, as the embodiment of its majesty and power. They seek not its subversion but its perfection. Thus, the usual models for Philippine electoral politics can only take us so far in the study of the military.

To study the politicization of the Philippine armed forces over the past thirty years, we need to understand the country's military history—particularly the complex, contradictory burden of its colonial past. All armies have an acutely historical consciousness, carrying banners and regalia emblematic of past triumphs and evocative of identity and ideology. Ruled by Spain from the sixteenth to nineteenth centuries and occupied by both American and Japanese armies in the twentieth, the Philippines has a history that denies its military such simple glory.

COLONIAL ORIGINS

In the Philippines, as in much of Southeast Asia, the origins of the modern military are colonial. But most military histories of the Philippines are written from either a colonial or national perspective, as if peering down the gunsight at enemies brown or white. Colonial chronicles treat the heroism of white officers amid the colored hordes, while nationalist accounts celebrate only those who engaged in "resistance."[24] Instead of such selective approaches, we can better understand the Philippine experience of the military if we trace its history continuously, from colonial past to national present, treating all the armies, Filipino and foreign, that have had an impact upon the peoples of the archipelago, whether as soldiers or subjects.

Through its three centuries of colonial rule, Spain introduced a military tradition that influenced, in significant ways, all succeeding armies. From the outset of its occupation in the sixteenth century, Spain established fortified cities at Manila, Cavite, and Zamboanga for defense against Japanese pirates and rival Dutch navies. By 1600, the Spanish conquistadors had built Manila as a fortress-city ringed with massive stone walls that ran for thirty-five hundred meters around an impressive grid of streets lined with churches, monasteries, barracks, and arsenals. By the early nineteenth century, Manila had become a garrison city for a substantial colonial army. In 1806, a French traveler noted fourteen thousand heavily armed soldiers in a population of only eighty-six thousand. As the army continued to expand and modernize, it served as the vanguard of Spain's expansion into the unsubdued areas of the archipelago, conquering the Sulu Sultanate in 1878 and pacifying the tribes along the mountain spine of northern Luzon.[25]

While the army guarded against foreign invasion, in 1868 Spain formed a paramilitary police, the *Guardia Civil*, to suppress banditry and insure order. To discharge its prime responsibility of "determining the loyalty and disloyalty of individuals," the Guardia Civil could make arrests on mere suspicion and use force to "extort confessions." By the end of Spanish rule, the Guardia Civil had become a visible, and despised, symbol of Spanish authority.[26]

Spain had occupied an archipelago inhabited by Malay warrior societies and would spend the next three centuries trying, with only limited success, to subdue them. Even in their last decades, the Spanish military failed to disarm Filipinos living in nominally pacified provinces. At the margins of the Spanish state in Mountain Province and Muslim Mindanao, local societies preserved a tradition of warfare and a capacity for resistance against any army, colonial or national.[27]

By the end of its era in 1898, Spain had left a mingled legacy. Throughout its rule, the Spanish military, like other colonial armies, was organized on racial lines with European officers in command of native soldiers. Even on the eve of revolution in May 1896, the forty-two hundred Spanish soldiers in the islands were outnumbered by nine thousand native troops.[28] These troops, living with their families in cantonments and serving Spain father to son, established a Filipino tradition of military service. Not only did Spain introduce the idea of a modern military, but it inspired the opposite—a deep-seated popular suspicion of military interference in civil society. The excesses of the Guardia Civil, immortalized in Jose Rizal's novels, created a collective Filipino memory of military abuse that inspired a later commitment to civil supremacy. Spain also left behind a basic division between regular army and paramilitary constabulary that all succeeding armies, colonial and national, would later follow. In the most eloquent testimony to this legacy, when Filipinos launched their revolt against Spain in 1896, they organized their revolutionary army along Spanish lines.

In its first months, the Philippine revolution witnessed a fundamental struggle between two different military ideals. In August 1896, the revolution began when the *Katipunan*, a secret society based in Manila's native districts, launched a poorly planned uprising on the city's outskirts. After Spanish troops scattered the rebels, the group's leader, a visionary named Andres Bonifacio, retreated south to nearby Cavite Province. There he joined the chapters of his secret society led by a landholder named Emilio Aguinaldo. By mobilizing local elites, Aguinaldo quickly assembled an army of fifty-four hundred regulars and fifteen thousand reserves that scored some stunning victories when the Spanish army advanced from Manila.[29]

Simultaneously, however, the revolution began to suffer deep factional splits. After ousting Bonifacio from leadership, Aguinaldo's faction executed him in May 1897—a decision that alienated many loyal members of the Katipunan. While historians have long emphasized these partisan aspects, much of the conflict may have arisen from their very different visions for the revolutionary army.

As a lower-class radical, Bonifacio had a populist ideal of the army as a brotherhood of patriots that made decisions democratically. When the fighting began, he assembled three hundred troops at a camp near Manila and told them to elect a commanding officer and deputy. After a voice vote, Bonifacio nodded his approval and the troops shouted in unison, "Long live the newly elected generals!"[30]

By contrast, his rival, General Aguinaldo, was inspired by the Spanish ideal of the heroic commander to mobilize an army of gentry officers and peasant conscripts. In the revolution's first weeks, he promised local landholders that anyone who "offered his services together with a hundred or so of his tenants . . . would be appointed a lieutenant or captain."[31] From their Spanish rulers, Aguinaldo and his elite officers borrowed the forms of a regular army, including ranks, tactics, and even the concept of guerrilla warfare—building what one historian called a "top-down . . . organization . . . that closely resembled the European army Aguinaldo was fighting against." For these landlord officers, the revolution was a chance for elite males to recover the authority denied them under Spain and its suffocating, emasculating rule.[32]

As the revolution weakened from internal divisions, colonial forces, reinforced from Spain, finally forced Aguinaldo out of Cavite into the rugged Sierra Madre mountains north of Manila. There, in December 1897, he signed a truce and accepted voluntary exile in Hong Kong. Had not the Spanish-American War intervened, Aguinaldo's military career may well have ended in exile.

America's invasion of the Philippines disrupted the continuity of local military traditions. Coming as a conqueror, the United States spent four years breaking the army of the infant Republic and pursuing its guerrilla remnants. Through these years of grueling combat, marked by atrocities and massive civilian casualties, the U.S. Army not only crushed the Filipino forces, but it eclipsed earlier military traditions, both colonial and national.

During its four decades in the archipelago, America formed two organizations that laid foundations for a modern Philippine military. An inexorable imperial logic led the United States to follow the basic Spanish structure: regular regiments of the Philippine Scouts concentrated near Manila to guard against invasion, and a paramilitary constabulary dispersed across the archipelago to enforce colonial order. Reflecting their very different missions, the two services had distinct approaches to "native" officers. The constabulary soon began training large numbers of Filipinos at a local academy, while the Scouts later sent a select few to the U.S. Military Academy at West Point.

America's "native" regiments were born in a brutal war of colonial conquest. From the outset, U.S. occupation of the archipelago was unplanned and its strategy improvised. After sinking the Spanish fleet in Manila Bay and occupying the city in August 1898, American forces found their advance blocked by massive trench works manned by thirty thousand troops loyal to the infant Philippine Republic. In February 1899, the U.S. Army broke through the Filipino fortifications and marched north into the Central Luzon Plain, driving General Aguinaldo's army before it.[33]

By November, the U.S. Army had smashed Philippine defenses and forced Aguinaldo into the mountains, where he ordered his commanders to commence guerrilla warfare. For the next two years, the U.S. Army faced a difficult pacification, finding, like all colonial armies, conventional tactics useless against guerrillas. As frustrations mounted, American commanders brought in seventy thousand troops and adopted a scorched-earth policy.

The U.S. Army soon discovered the utility of native soldiers under the pressures of combat in a harsh tropical terrain. In the war's first months, field commanders, inspired by the Indian scouts of the American West, recruited local Filipinos to serve as guides and interpreters. In July 1901, when Congress ordered the withdrawal of all volunteer regiments, the U.S. administration in Manila established the Philippine Constabulary as a mobile police force. After several months, the U.S. Army inducted its local contingents of Filipino troops into a unified force later called the Philippine Scouts. By the time the war ended in July 1902, there were only 15,500 U.S. Army troops left in the islands, forcing the 11,000 Filipino soldiers in the Scouts and constabulary to shoulder a substantial share of the ongoing pacification.[34]

In little more than two years, the U.S. Army had learned the same colonial lessons that the British and Dutch had distilled from two centuries of using native troops in India and Indonesia. Asian soldiers were, from an imperial point of view, well adapted to withstand the rigors of service in their own country. But only a European had the character required of an officer. As the editor of England's *Statesman* wrote in 1885, educated Indians were "wanting in the courageous and manly behaviour to which we justly attach so high an importance in the culture of our own youth." In European eyes, moreover, not all native races were equally suited for service. Colonials often found dominant lowland groups both "effeminate" and insubordinate. But certain "martial races"—such as the Gurkhas, Ambonese, or Karens—were thought capable of great courage under fire and fierce loyalty to their white officers.[35]

From the outset, the American commander in the islands, General Elwell S. Otis, felt, like most Americans of his day, that elite Filipinos were unfit for command. In an essay for a U.S. military journal in 1900, one American officer dismissed the typical officer in Aguinaldo's army as "a half-breed, a small dealer, a hanger-on of the Spaniards." Thus, the Scout soldiers would all be Filipinos, but their officers were to be white Americans selected from "the line of the Regular Army."[36]

In the last decades of American rule, the Scouts, with sixty-eight hundred men, comprised over half of the ten thousand U.S. Army troops in the Philippines—a strong statement of the War Department's confidence in its native troops. But the U.S. Army still had nagging reservations about Filipino leadership. Only when Filipinos began graduating from West Point and Annapolis in the 1920s did the Scouts gain a significant minority of Filipino officers.[37]

While the Scouts became regular U.S. infantry units, the constabulary remained a hybrid force, deployed in small, localized units like police but armed with rifles like soldiers. So constructed, its troopers pursued bandit and revolutionary remnants with considerable success and brought a colonial order to the countryside.[38]

The constabulary achieved its initial efficiency through a fusion of skilled Filipino soldiering and competent American leadership. Like other colonial armies, the constabulary justified its discrimination by celebrating the heroism of its white officers and their special bond with colored troops. A typical account describes how, in 1908, an American officer, Captain Harold H. Elarth, took a patrol of just ten Filipino troopers to break up a simmering revolt by "a thousand tribesmen, armed and ready for action." When three of their Muslim leaders attacked, "Elarth dropped the first two Moros with skull shots from his pistol, but there was no time to stop the third, who was armed with a spear." The captain was doomed, but his loyal Sergeant Alvarez "leaped forward to take the spear in his chest." The American captain then "blew the Moro's head away with a .45 caliber bullet." From retirement forty years later, Elarth reflected upon these experiences. "By fair dealing, unusual sagacity and confirmed courage," young American officers "pacified and controlled tribes that for three hundred years had continuously warred with the Spaniards." This success, he explained, came from "the psychology of the Malay," which inspired Filipino soldiers to follow their American lieutenants with "adoration."[39]

The constabulary's officer corps thus remained over 90 percent American until war and fiscal crisis allowed Filipino advancement. In 1916, the new Philippine legislature, as part of a general "Filipinization" policy, cut pay for colonial officers and opened an expanded Constabulary Academy at Baguio to train local replacements. During World War I, when over a hundred American officers resigned to join the U.S. Army for the fight in Europe, the colonial government promoted Filipinos and appointed General Rafael Crame, a Spanish-Filipino mestizo, as chief of constabulary. During his decade in command, Crame, a third-generation veteran of the Spanish colonial army, transformed the constabulary into an all-Filipino force with a high standard of service.[40] When the Philippine Army was mobilized in the 1930s, the constabulary, in its last colonial chapter, provided many of the Filipino officers that trained citizen-soldiers for the defense of the new nation.

FILIPINO OFFICERS

In the early years of American rule, Filipino nationalists made the training of native officers a central plank in their campaign for independence. Taking control of the colony's guns was almost as important as occupying its executive offices. By demanding officer training, the all-male nationalist movement challenged colonial assumptions that native men were, by racial character, unsuited for command. Filipino masculinity, disarmed and denied commissions under colonial rule, was affronted by a policy premised upon native effeminacy and inferiority.

For the all-male electorate of the American era, Filipino nationalism meant not only independence but, of equal importance, liberation from colonial emasculation. In the political rhetoric of the day, military drill would advance the nationalist cause by training officers for a future army and stiffening the fiber of the country's youth. To assert their manhood, nationalist leaders seized upon any pretext for military drill, even service under the American flag.

Over time, colonialism produced a nationalist antithesis whose symbolism and social roles were thus marked by an extreme gender dimorphism. When Filipino leaders finally began building a national army in the 1930s, they borrowed a European standard of military masculinity with all its inbuilt biases. By exempting women from conscription and barring them from the Philippine Military Academy, the Commonwealth exaggerated the society's male/female polarities. Once set in 1936, these military regulations and their social influence persisted for decades. It would be nearly thirty years until the military recruited its first women soldiers in 1963, and another thirty years after that before the PMA admitted its first female cadets in 1993.[41] If we accept what one historian has called "the emancipated status of Filipino women in the 19th century," then the nationalist movement, with its rhetoric of militarism and male empowerment, may have skewed the gender balance within the Philippine polity. In a Malay society with a legacy of gender equality—bilateral kinship, matrilocal marriage, and gender-neutral pronouns—this aspect of nationalism seems socially retrogressive.[42] Understandably, postwar historians have overlooked this glorification of masculinity and military valor in their sympathetic studies of Filipino nationalism.

Only a few years after the Philippine-American War, colonials and nationalists began to cooperate in building a Filipino officer corps. Their motivations were, of course, far from identical. American officials were concerned with colonial security, while Filipino leaders felt military training would foster national independence. In 1907, the fledgling Constabulary School at Manila graduated its first Filipino officers from a three-month training course and then moved to permanent quarters in the mountain city of Baguio for a more rigorous six-month curriculum. A year later, the U.S. Congress authorized the admission of Filipinos to the U.S. Military Academy at West Point. While these Filipino cadets would experience the full military curriculum, their commissions were restricted to service in the Philippine Scouts. In 1914, the first Filipino cadet, Vicente P. Lim, graduated with an academic rank of 77 among 107 classmates—an event of such significance that the Philippine resident commissioner, Manual Quezon, made a special trip from Washington, D.C. Though his classmates, with obvious racial overtones, had nicknamed him "Cannibal," his biographer in West Point's yearbook, *The Howitzer*, still praised him as an officer who "has the ability to make good in any arm of the service." Indeed, over the next twenty years, Lim rose steadily through the Scout hierarchy, clearing a career path for younger Filipino officers.[43]

By the outbreak of World War II, a total of twenty-six Filipinos had graduated from West Point and sixteen joined the Scouts. Despite their small numbers, these Filipino alumni from the U.S. service academies became the core cadre that established the PMA in 1936—providing the superintendent, commandant, and four tactical officers. At the newly formed army headquarters, a half-dozen West Point graduates won key postings that gave them influence far beyond their numbers.[44]

Simultaneously, Filipino leaders worked to transform the constabulary, and its officers' school, into the foundation for a national army. When the new bicameral Philippine legislature opened in 1916, Filipino representatives authorized funds to expand the small Constabulary School at Baguio into the Philippine Constabulary Academy as "a sort of West Point" to train officers for a future republic. In 1928, the legislature upgraded this academy, requiring a high school diploma for admission and authorizing a three-year college curriculum.[45] With small classes of only ten to twenty cadets, the new academy could make only a limited contribution to the formation of a future officer corps.

When America entered World War I, the Philippine legislature voted overwhelmingly to raise a Philippine National Guard division and Senate President Quezon crossed the Pacific to lobby personally for Washington's authorization. Even the War Department's determined effort to block its mobilization until 11 November 1918, the very last day of war, could not dampen the Filipino enthusiasm for military service. Over twenty-eight thousand men volunteered. With bands playing and banners flying, the Philippine National Guard drilled for three months until it was disbanded in February 1919.[46]

During the 1920s, the American colonial regime, in a fundamental change of policy, began training Filipinos for command. As head of a presidential inquiry in 1921, General Leonard Wood reported that there was "no adequate organization of the Philippine people for defense of the Islands" and recommended that the United States "should at once take the necessary steps to organize, train and equip such a force." After taking office as governor-general in October, Wood mobilized the resources of the U.S. Army to open officer training programs at Manila's leading colleges.[47]

Drill began at the University of the Philippines (U.P.) in 1922 when its regents funded a Department of Military Science and Tactics, retained an active-duty U.S. Army captain as its chairman, and authorized an armory. Within three months, thanks to Governor Wood, the new Reserve Officers' Training Corps (ROTC) had an arsenal of U.S. Army rifles and a thousand cadets. Indicative of the program's patriotism, in 1925 a former U.P. cadet recalled his ROTC days "with a great degree of pride and satisfaction" and said that he was proud to have prepared "for the greatest service he can ever expect to render to his country." Two years later, U.P. President Rafael Palma, a prominent nationalist, praised the department for establishing "the nucleus of a future national military organization."[48]

As Palma predicted, the ROTC program grew rapidly, adding field artillery in 1929 and machine guns six years later. After passage of the National Defense Act in 1935, the university acquired another two thousand Springfield rifles and doubled its cadet corps to 3,304 trainee officers by 1938. Indicative of the program's prestige, the cadets' Vanguard Fraternity had by then become "the largest, most influential, and richest student organization of the university." Beyond drill and marksmanship, the program indoctrinated its cadets into nationalism. "We need to make . . . our youth . . . so proud of their race and their democracy that they will die fighting for it," President Quezon told the U.P. cadets in 1937. "We have all been trained," wrote the corps' cadet colonel a year later, "with patriotism ever so carefully engraved in our hearts by our military instructors, we are proud to say, as they would have us say, *we are ready*."[49]

While the publicly funded U.P. had the largest cadet program, the elite, Jesuit-run Ateneo de Manila was proud home to the country's top drill corps. In 1921–22, American Jesuits replaced Spanish priests and launched a military program with strong student support. "Boys," the college president explained, "every good citizen should be a soldier ready for any service his country asks of him." Under the command of U.S. Army officers, Ateneo's cadet corps of 450 students began drilling with wooden rifles and soon captured the silver cup at the Manila Carnival. The following year, when the U.S. Army issued rifles to the cadets, "we all rejoiced at the news," one student wrote, "for everybody wanted to carry a real gun, have an army belt around the waist, and a bayonet dangling on the hip."[50]

Over the next four years, the Ateneo corps, under its beloved commandant, the famed Moro fighter Captain Elarth, adopted the West Point uniform, added a crack rifle team, spent summers training in Baguio, and won a string of trophies before thousands of spectators. "To see Corps, at full strength and in immaculate white march with the quick rhythmic step to the stirring music of our unexcelled Bugle Corps," wrote one student in 1927, "was a sight that thrilled the heart of every spectator." After the Philippine Army began training in 1936, Ateneo's corps was transformed from a flashy drill squad into a combat-ready reserve unit. By 1939, its ROTC, now commanded by a Filipino graduate of the Constabulary Academy, had doubled to nine hundred cadets with a machine gun company and a howitzer platoon.[51]

Other Manila universities followed these leads. The 1923 Manila Carnival featured a drill competition by cadets from San Beda, the National University, and, of course, Ateneo and the U.P. Along with basketball and baseball, close-order drill contests would remain a high point of intercollegiate competition until the war. These parades, featuring what one U.P. cadet called "thousands of virile young blood[s] . . . rifles on their shoulders, gallantly marching to the time of their music," drew large crowds and sparked school spirit.[52]

The program was not without its problems. Even after a decade of drilling, the U.S. War Department still refused reserve commissions to its graduates. Among the 1,707 cadets in the U.P. unit in 1931, only 87 were in the advanced course. The majority of cadets in the basic course concentrated on precision, close-order drill for the Carnival competitions of dubious relevance to actual combat. Nonetheless, after taking direct control of all ROTC units in 1939, the Philippine Army reformed the curriculum and extended the program to a total of thirty-three colleges, making them its main source of officers by the outbreak of war.[53]

COMMONWEALTH ARMY

In 1935, national defense suddenly became the most critical issue facing the Filipino people. In Washington, President Franklin Roosevelt approved creation of the Philippine Commonwealth as an autonomous, transitional government with a ten-year timetable to full independence. After his inauguration in November, President Manuel Quezon was suddenly faced with the problem of building a national army. The United States had decided to give up its military bases after independence. Japan was on the march in China. The threat of invasion was very real.

Under the National Defense Act of 1935, President Quezon made mobilization his top priority and committed a quarter of the budget to building a national army that would, by independence in 1945, have ten thousand regular soldiers backed by reserves of four hundred thousand. Starting in 1937, every Filipino male over six was required to attend "preparatory military training." Within three years, over a million schoolboys were marching. While peasant conscripts would have just six months of basic training, their regular officers would be educated for four years in military science at the new Philippine Military Academy (PMA).[54]

Once the mobilization began, the Philippine Army found that formation of an officer corps was its most intractable problem. While arms could be purchased in weeks and troops trained in months, an officer's education would take years. Although the U.S. Military Mission drafted plans for an army led by 930 regular officers, one of its leaders, Major Dwight Eisenhower, was soon complaining to his diary that there was "no officer corps to supervise organization on such a scale, and officers cannot be produced out of thin air." Similarly, General Vicente Lim dismissed the reservists trained at short-term service schools as "not even half-baked officers but a total loss," adding sadly that it would be years before PMA graduates could "lay down the tradition of our future Army."[55]

Despite these reservations about the reservists, the Commonwealth would have to wait four years for the PMA to produce its first graduates. Thus, the army, by necessity, drew its officers from diverse sources: the few Filipino graduates from West Point; the larger number of college reservists; and, most importantly, veterans of the constabulary and Scouts. By early 1941, the Philippine Army had assembled, with surprising speed, an adequate cadre of 770 regular and reserve officers.[56] With war only nine months off, the PMA had already provided 169 officers, most of them lieutenants in command of troops on the line.

Looking back on this mobilization, it is important to reflect on the choices that Quezon made. By the mid-1930s, there were, in effect, two traditions that could have inspired the formation of a national army—the nationalist and the professional. During the revolution of 1896, General Aguinaldo had led an army commanded by landed elites. Inspired by this tradition, students at the University of the Philippines, a bastion of the nationalist movement, had built a large reserve-officer program. Soon after its foundation in 1908, the U.P. dominated professional training and its law graduates moved rapidly through the best Manila firms to powerful positions within the emerging nation-state—its legislature, executive, judiciary, and bureaucracy. With the start of ROTC training in 1922 and army mobilization in 1936, U.P. alumni were well positioned to bring the officer corps within their ambit of influence. By opening the Constabulary Academy in 1905 and sending Filipinos to West Point in 1910, the American regime fostered an alternative form of military professionalism.

In the mid-1930s, Quezon could have built his new army entirely upon the ROTC programs, already well established at every major university. In so doing, he would have made the military part of the control apparatus of the country's landed elites, in effect reviving Aguinaldo's army with its gentry officers. Though once an officer in that army, Quezon instead chose the Western ideal of military professionalism. The United States lent its full support to his decision by sending advisors to Manila and bringing Filipino officers to America for advanced training.

With the help of these advisors, the president opened the new Philippine Military Academy in 1936 with an entering class of 120 cadets and a four-year curriculum modeled, in every detail, upon West Point. Indeed, Quezon made the PMA his "pet project"—selecting its site at Baguio, picking staff, investigating cadet derelictions, and making frequent informal visits. Every summer, when Baguio became the "summer capital" and Manila's elite decamped to the mountain cool, Quezon, reigning over this courtier city, introduced his lower-class cadets to high society. As a careful student of his own culture, Quezon seemed to understand that educational institutions, by forming values and forging social networks, played a major role in shaping the direction of Filipino society. In effect, the president appreciated the influence of what one historian has called the "old school tie."[57]

In establishing the PMA and investing it with his prestige, Quezon, perhaps unwittingly, also created a new elite and an alternative path to power. Unlike their counterparts elsewhere, Filipino political and economic elites have shown little enthusiasm for the military as a career. With few exceptions, they did not send their sons to the PMA or West Point.[58] Consequently, the country's regular officers would be drawn from a lower-middle-class stratum, fostering a social cleavage between senior officers and other Filipino elites. During their years at the PMA, the cadets, starting with the Class of 1940, were indoctrinated into an ideal of military professionalism, and for the next thirty years would resist both corruption and politics. As creations of the state, they accepted its doctrine of civil supremacy and the idea of a subordinate role for the military.

By contrast, their peers who studied law and won reserve commissions at the University of the Philippines would emerge with activist, nationalist values that colored their view of the military. In the 1930s, the U.P. Corps of Cadets trained men who would, thirty years later, seek to politicize the armed forces. Among the forty-five cadets who finished the U.P.'s advanced military courses in 1937, eight later became central figures in the martial-law regime of the 1970s—notably, President Ferdinand Marcos, his chief of staff Romeo Espino, Supreme Court Chief Justice Fred Ruiz Castro, and sugar czar Roberto S. Benedicto. "This swash-buckling major is one of the best if not the best officer in the Corps," the 1937 U.P. yearbook said of Cadet Marcos.[59] As politicians first and reserve officers second, such men had few reservations about using the military to serve both nation and ambition.

On the eve of war, the Commonwealth had thus fostered two competing visions of an officer's role—an academy that, in Janowitz's words, trained the "military manager" and an ROTC program that instilled a more "heroic stance."[60] When Quezon, after supervising the state's step-by-step formation since 1907, rejected the ROTC and decided instead to establish a new institution to train professional officers, he unwittingly broke ground for a long-term struggle between U.P. reservists and PMA regulars. Reflecting this division, President Marcos, preparing for martial law in the late 1960s, would push aside many PMA graduates and, though allied with certain academy classes, fill his military command with reserve officers, many of them fellow alumni from U.P.'s cadet corps.[61]

The history of Quezon's army in World War II could thus be written as the story of a gender and a generation. Under the National Defense Act, he had mobilized an entire generation of men to build an army—conscripting all twenty-year-olds as soldiers and training all college graduates as officers. So indoctrinated and mobilized, this generation held the Japanese invaders at Bataan, led the anti-Japanese guerrilla movement, and then fought the battles of liberation at war's end.

As war threatened in mid-1941, President Franklin Roosevelt ordered the American and Philippine armies integrated into a single command, the United States Army Forces in the Far East (USAFFE). After a strategic withdrawal to the Bataan Peninsula on Manila Bay, the USAFFE defenders, some fifteen thousand Americans and sixty-five thousand Filipinos, fought well against superior Japanese forces. After their surrender in April 1942, the Japanese army ordered the survivors on a harsh, sixty-mile "death march" that killed fourteen thousand en route and ended in prison camps where half the Filipino troops later died of disease. By contrast, USAFFE units on outlying islands eluded capture and, for the duration, mounted a guerrilla resistance that denied the Japanese access to much of the archipelago.

By war's end, the Philippine Army had suffered something akin to an institutional trauma—units broken, regular officers compromised by service in the Japanese-sponsored constabulary, and thousands of reserve officers hardened in a guerrilla warfare that thrust them forward as soldier-politicians. In the immediate postwar period, PMA regulars and reservists fought for both commissions and seniority on the lineal roster. Many reservists, like Ramon Magsaysay and Ferdinand Marcos, used their guerrilla comrades as constituencies to win political office, becoming powerful advocates for the reservists. But the public, recalling the brutality of the guerrillas, seemed to favor the regular officers. By the late 1940s, the armed forces, shorn of their massive prewar reserves, had been rebuilt with U.S. military aid as a compact, professional army with regular officers in command.

PEACETIME ARMY

After the war, the Philippine economy was ravaged and American advisors again played a key role in establishing the Armed Forces of the Philippines (AFP). Without much consultation, Washington decided that the AFP's mission was to maintain internal order and provide land defenses for U.S. air and naval bases.

To fulfill these missions, the U.S. Defense Department determined the AFP's postwar troop strength, set its order of battle, and supplied much of its equipment. In postwar decades, virtually all senior Filipino officers would receive their advanced training in the United States. With this aid, the AFP troop strength rose from thirty-seven thousand in 1947 to fifty-seven thousand five years later—a level that it would maintain for the next twenty years.[62]

By the early 1950s, the Philippines was integrated into an American global system that guaranteed its strategic defense. Clark Field was home to U.S. Air Force fighters with a limited tactical range, while Subic Bay served as the forward base for the U.S. Navy in the Western Pacific. These foreign forces dwarfed local capabilities, thereby denying the AFP ultimate responsibility for its own country's defense.

Such subordination to the American military planted the seeds of an ideological crisis for Filipino officers. As defenders of the nation and heirs to its military traditions, their anger at the unequal alliance simmered just beneath the surface, flaring at any slight. Though Class '40 was more inclined to a sympathetic view of America than younger officers, their nationalist sensibilities were still offended by the shoddy, second-hand equipment the United States transferred under the Military Assistance Agreement. As reserve officers who shared the U.P.'s heroic vision of the nation's history, President Marcos and his circle were inclined to a manipulation of the alliance in the name of nationalism. By contrast, the PMA's Class of 1971, coming of age during the period of radical student politics in the 1960s, saw the U.S. alliance as an affront to national sovereignty. In the turmoil that followed Marcos's downfall, these issues would be debated openly by serving officers, adding to the AFP's internal crisis.

In the immediate postwar years, peasant revolution in Central Luzon threatened the survival of the Philippine state and forced a reorganization of its armed forces. In 1947, the *Hukbalahap*, or *Huks*, rebelled with fifteen thousand peasant guerrillas hardened by wartime combat against the Japanese. This combination of a large mass base and a substantial armed force made their squadrons a formidable enemy for the demoralized, disorganized AFP.

Through close cooperation between Washington and Manila, the situation soon changed. At the height of the crisis in 1950, Ramon Magsaysay, an energetic, charismatic politician, became defense secretary and worked with his American advisors to develop a new doctrine for unconventional warfare. Within fourteen months, in the estimation of his CIA advisor, "the Huks had lost the initiative and were on the run."[63]

By the time he resigned to run for president in 1953, Magsaysay had reorganized and revitalized the armed forces. Through guarding polling places during elections and defeating the Huks, the army had won a reputation for integrity. But his activist style was not without its liabilities. By engaging the military in civic action work and drawing its officers into civil administration, Magsaysay, as both defense secretary and president, also injected an element of political tension within the ranks.

During the next decade, the armed forces remained where Maysaysay had left them. In the early 1960s, the Philippine military was still a small, multiservice force of some fifty-eight thousand. With its victory in the Huk campaign, the AFP had recovered its prewar morale and professional élan. Staffed largely by PMA graduates, the postwar officer corps was indoctrinated into an ideal of civilian supremacy and generally avoided political entanglements. After graduation, regular officers spent thirty years living in military cantonments, physically isolated from the civilian population, an elite apart from the pervasive politics of their society.

In some ways, however, the military did not prosper from its decade at peace. The budget was stagnant and promotion slow. Some officers succumbed to corruption or sought patronage. As the smuggling of American cigarettes, known as "blue seals," grew into a national scandal, Congress charged constabulary officers with taking payoffs. In the 1965 presidential campaign, opposition candidate Ferdinand Marcos tarred the administration with this taint and promised reforms.

MARTIAL LAW

After campaigning as a war hero, President Marcos took office vowing to cleanse a demoralized military. In his first term, he seemed to fulfill this promise. On the pretext of reforming a corrupt constabulary, Marcos ordered a massive shake-up, removing over a third of its provincial commanders and forcing fourteen of the AFP's twenty-five general officers to retire. But his reforms masked an ulterior motive. For, as one study found, every change was calculated "to increase the AFP's responsiveness to the new President." Indeed, in the seven years before martial law, Marcos groomed a hierarchy bound to him by strong personal ties: old classmates from the U.P. cadet corps, blood relatives, and fellow Ilocanos, the northern Luzon ethnic group known for being clannish.[64]

Marcos was also the first president in a decade who increased the military budget, a generosity that muted criticism from the ranks. Reversing the steady decline that blighted careers and sapped morale, he raised the AFP's funding by nearly 50 percent. After dropping from sixty thousand to forty-five thousand troops under his predecessor, the armed forces climbed back to a strength of sixty-three thousand by 1972.[65]

Marcos also manipulated the American alliance to win more resources for his military. When Washington pressed him for soldiers to fight in Vietnam in 1966, Marcos sent just two thousand engineers, the Philippine Civic Action Group (PHILCAG), in exchange for U.S. military aid of $34 million. With these funds, Marcos equipped ten engineering battalions and ordered a massive "civic action" campaign that built roads and dug wells across the countryside. These efforts, heavily publicized, made the AFP seem an activist, populist force. But they also gave it considerable experience in civil administration. Aside from equipment for other AFP units, Marcos also demanded, during the three years the unit served in Vietnam, "special payments" of a million dollars per year—delivered directly to his office by the U.S. Embassy. These negotiable checks, never audited, may have provided Marcos with the black funds for covert units within the AFP.[66]

Toward the end of his first term, veteran observers in the AFP command could see some disturbing signs. Above all, Marcos seemed to be building a "parallel command," promoting loyalists and creating covert-action units outside the formal hierarchy. Indeed, during Marcos's 1969 reelection campaign, a "special force," controlled from AFP headquarters, sent two hundred troops to Marinduque Island and another hundred to the Batanes Islands to coerce votes for pro-administration candidates.[67]

His second administration was a realization of these concerns. To quell demonstrators, Marcos called out the army and soldiers fired on students. As terror bombs went off across Manila, he suspended the writ of habeas corpus and rounded up dissidents. Then, in September 1972, Marcos called a dozen top defense officials for a series of secret meetings—notably, Defense Secretary Juan Ponce Enrile, General Fabian Ver, and General Fidel Ramos. This group included three PMA alumni, all members of Class '51, and all—like Marcos, Ramos, Enrile, and Ver—native Ilocanos. Would they, he asked, support the imposition of martial law? All, save one, said yes.[68]

During the years of dictatorship, the Philippine officer corps lost its professional autonomy. Under authoritarian rule (1972–86), the military served as the bastion of Marcos's "New Society"—enforcing his authority, arresting opponents, and staffing civilian agencies. As the AFP expanded from 62,000 troops in 1972 to 150,000 a decade later, many officers succumbed to political pressures. By 1981, the Marcos regime was an alliance of the first family, a new national oligarchy, and the military hierarchy.

While favoring reservists, the president also courted the regulars. Using his patronage powers, Marcos formed alliances with strategic elements among the active-duty PMA alumni, notably Class '51, whose loyalty won stars for nearly half the class. Like Quezon before him, Marcos lavished attention on the PMA—selecting staff, making frequent campus visits, and speaking at all graduations. Perhaps sensing a source of threat, Marcos ordered a new annual rite called the "testimonial parade," which brought the cadet corps to Manila en masse to swear a personal oath of loyalty before the president. "We did not only Honor the first couple in parades but also in serenades," recalled Class '76 in their yearbook, "that was how dear they are to us." The first family reciprocated, opening the palace to the cadets and hosting a grand ball in their honor.[69] Under martial law, Marcos picked the PMA's superintendents with care—from his first, General Ernesto Gidaya, a member of the ever-loyal Class '51, to his last, General Jose Ma. Zumel ('59), the ultimate loyalist.

In the last years of the Marcos regime, the officer corps thus split into factions and clandestine brotherhoods, fragmenting the armed forces and encouraging coup conspiracies. Many officers would emerge from the Marcos era with a taste for affluence, the habit of political power, and a belief in the efficacy of violence. Just as an earlier subordination to civil authority had promoted apolitical officers, so Marcos's use of the military as the fist of authoritarian rule produced a generation of politicized officers.

After a decade as the instrument of Marcos's will, the Philippine military finally emerged as an independent force in the political crisis of the 1980s. In August 1983, soldiers loyal to Marcos's chief of staff, General Ver, assassinated opposition leader Benigno "Ninoy" Aquino, sparking an eruption of mass protests. Seeking a renewed mandate, Marcos held snap presidential elections in early 1986 with Ninoy's widow, Corazon or "Cory" Aquino, leading a united opposition ticket. But Marcos, with a mix of force and fraud, claimed victory in the February balloting and plunged the nation into crisis.

In the final act of this drama, Marcos's own defense minister played the lead role. Determined to succeed Marcos, Enrile had assembled a strike force of middle-ranking officers, many members of Class '71, led by Colonel Gringo Honasan. He concentrated them in his ministry's security unit, where they formed a faction called the Reform the Armed Forces Movement, or RAM, and plotted a coup.

Events transformed their abortive coup plot into a mass uprising. On the revolt's first day in February 1986, the rebels were saved from defeat by the constabulary chief, General Ramos, who inspired defections of key Marcos commanders. Then millions of ordinary Filipinos massed on the boulevards ringing the camps, facing down Marcos's tanks and forcing him into exile. His successor Cory Aquino, installed in the palace by this mass uprising, lacked electoral legitimacy. In its first year, her government was shaken by four ill-planned coups. This crisis abated, but did not end, in February 1987, when voters overwhelmingly approved a new constitution, giving her a clear electoral mandate.

Against the tide of history, RAM's leaders persisted in their reach for power, even after the president had won her electoral mandate. Over the next four years, they launched three major coups, repudiating the military professionalism taught at the academy and embracing a heroic vision of the officer as national savior—a military tradition embodied in General Aguinaldo and President Marcos.

Often overshadowed by more flamboyant politicians, General Fidel Ramos emerged as a central figure in these tumultuous decades. At every turn, his decisions shaped the course of Philippine politics—his break with Marcos, his loyalty to Aquino during the coups, and, finally, his election as president in 1992. During his six-year term, President Ramos succeeded in restraining and then redirecting the armed forces in ways that had eluded his predecessor. While encouraging the surrender of the Muslim and military rebels, he transformed the armed forces from a bloated instrument of domestic repression into a modernized force for external defense. In his first three years, Ramos reduced troop strength and cut defense spending to 5 percent of the national budget—far below Marcos's peak of 24 percent and Aquino's 14.[70] But in his rise from constabulary chief under Marcos to defense secretary and then president, Ramos, by his very presence, slowed investigation of past human rights abuse in the military and any progress toward fundamental reform.

The three years between RAM's 1989 coup and Ramos's election offered two contrasting, yet closely related, expressions of the military's role in politics. In the heroic manner of a would-be national savior, Colonel Honasan led five hundred officers in a major coup attempt that nearly captured the palace. With the professionalism of a staff officer, General Ramos headed a disciplined presidential campaign and then filled his administration with over a hundred retired officers.[71] Within this contrast, there is, of course, an underlying continuity: by means legal and illegal, the country's officer corps was laying claim to power, finally achieving, in the 1990s, a political influence comparable to that of its peers elsewhere in Southeast Asia. Indeed, though they won office legally, President Ramos and his military cohort proved surprisingly reluctant to give up power.

As his presidency was ending in 1997, Ramos and his security advisor, General Jose Almonte, spent months maneuvering for a constitutional amendment to extend his term beyond its six-year limit, prompting massive protests, a second "people power" mobilization that finally blocked the move. Once Ramos conceded, his defense secretary, General Renato de Villa (PMA '57), then announced his candidacy for the ruling party's nomination and ran a serious, but unsuccessful, campaign in the May 1998 presidential elections.[72]

In its narrative span across half a century, this study of the officer corps opens us to a close examination of topics with global import. Moreover, the Philippine military, in its troubled passage through dictatorship to civilian control, casts light on issues with international implications: the causes of coups d'état, the politics of impunity, and the collective trauma of torture.

Ultimately, however, all of these themes are secondary to this work's focus on military socialization. As long as that training held and officers preserved their professionalism, as they did when Class '40 was in command, the Philippines was spared the trauma of authoritarian rule. But once the president declared martial law and used the military to control civil society, the socialization of younger officers, like those of Class '71, weakened. After a decade of extraordinary power, a generation of officers abandoned its commitment to civil supremacy and a small fraction launched a revolt that did lasting damage to their society. The wide reverberations from this seemingly small rupture indicate the importance of military socialization for any society with a standing army.

Chapter 2 Kaydet Days

On 15 June 1936, 120 fresh cadets boarded a train in Manila for the Philippine Military Academy (PMA), recently established in the mountain city of Baguio. At the end of the rails, the cadets transferred to Benguet Auto Line buses for the five-thousand-foot climb up the famous zigzag road. As the buses pulled into the campus, some forty upperclassmen, all transfers from the superseded Constabulary Academy, were waiting to greet the new arrivals with a ritual called "hazing." Even fifty years later, these cadets would recall their reception as traumatic.

> The drive up the . . . road to Baguio did not take long. Soon they began to feel the change in altitude as their ear drums began to ache. They smelled the scent of pine needles. When they got to the Teachers Camp grounds, a beautiful sight met their eyes: a formation of upper-class cadets, smart in the khaki uniform. . . . "Wow!' O-h-h-h! That's how we will look like soon too. . . ."
>
> Then all hell broke loose! A command was given and the upperclassmen, like a pack of hungry wolves, ran to where the BAL buses were parked and shouted at the incoming cadets to . . . form a straight line in front of table,

pull their necks in, stop rolling their eyes, shoulders back, guts in, etc., etc., etc. The memory of those first few moments will forever be engraved in the minds of the group.[1]

For the next nine months, from 5:50 A.M. to 10:00 P.M., seven days a week, the Class of 1940 shared an ordeal that not only trained them to become career officers but bonded them together for life. As the first class to complete the PMA's four-year curriculum, the arrival of these cadets also marks the start of the modern Philippine military. Indeed, at their graduation four years later, General Vicente Lim, the deputy chief of staff, would describe these seventy-nine graduates as "the beginning of the life of the Philippine Army. When one of these boys becomes the chief of staff in the future then we will have a real army in the Philippines."[2]

The general's prediction proved accurate. Thirty years later, not one but two of these classmates would serve as chief of staff. The Class of 1940 is thus an apt starting point for the study of long-term trends within the Philippine military. Not only were they the PMA's first full class, but their careers spanned a critical period in the country's military history—from the apolitical Philippine Army of the 1930s to the politicized armed forces of the 1960s. Most importantly, their history affords a unique opportunity to grasp how their socialization as cadets shaped their conduct as commanders. As cadets, they idealized the academy's honor code, making it a symbol of their class solidarity. As junior officers fighting enemy invaders, they tested and affirmed its strict maxims as standards for their future conduct. So socialized, they would struggle, for the next quarter century, to reconcile these principles with the reality of military service in an electoral democracy.

Class '40 is both an apt and willing subject for such a study of military socialization. In preparation for their fiftieth reunion in 1986, classmates compiled the *Golden Book,* a 350-page memoir of their military careers. Two years later, when I arrived with my tape recorder for the first of thirty-two interviews with surviving classmates, they were ready for my questions.

While nostalgic about the old days at the PMA and proud of its traditions, the alumni of Class '40 were sharply critical of their alma mater's recent history. During the late 1980s, their four-year cycle of reunions coincided with a series of bloody coup attempts by younger alumni who deployed these ideals and class connections in a bid for power. This juxtaposition of reunion and rebellion transformed reminiscence into reflection. Among themselves, the cavaliers of Class '40 concluded that the younger generation had forgotten the meaning of the academy's values. To us, outside observers with different concerns, they seemed to say that military socialization, so deep and lasting in their own day, had weakened among later generations of graduates.

MILITARY ACADEMY

The opening of the PMA was, as General Lim implied, part of the larger design for a new army to defend the nation after independence in 1946. Under the National Defense Act of 1935, President Manuel Quezon planned for a small cadre of regular officers to train a citizens' army on the Swiss model that would, by 1946, number 400,000 men. Accordingly, in April 1936, some 150,000 Filipino men registered for the country's first draft, and nine months later, 40,000 reported for training. These twenty-year-old draftees would get just six months of marching and marksmanship, but the men who drilled them would be regular officers educated in military science and committed to thirty-year careers.[3]

Forming an officer corps was the most difficult part of this enterprise. As Quezon put it, "the heart of an army is its officers." Along with buying rifles and building camps, the creation of this army required, as the president was well aware, the construction of officers as exemplars for a new image of the Filipino as warrior. To form such leaders, the Defense Act provided for the establishment of a Philippine Military Academy at Baguio for the education of career officers. This academy was, in the words of the Commonwealth's vice president, "the foundation stone of the entire military establishment," providing "the leadership necessary to knit together a scattered and loosely connected citizen army into one whole, living, pulsating, homogenous machine that can fight with courage."[4]

Quezon and his American advisors scrutinized the technical details of their new military, but they did not grasp fully the long-term dangers inherent in their creation. By seeking to defend itself against invasion, the Philippine state had, inadvertently, created the means of its own destruction. Under this legislation, the Commonwealth was investing a self-selecting elite of regular officers with absolute control over the nation's arsenal. While the colonial constabulary had been armed with pistols and rifles, this new army would have aircraft, tanks, and artillery. If, at some future point, even a handful of officers turned only a fraction of this arsenal inward to seize power, civil authorities would be forced to capitulate. But faced with the immediate threat of Japanese invasion, Quezon and his advisors put aside concerns about long-term stability and rushed to mobilize a standing army.

If we examine the logic of the National Defense Act, the Philippine state's only protection against a future coup was the socialization of its officers. Implicit within this sparse legislative language, these restraints rested on two mechanisms: the subordination of the military hierarchy to civil authority; and the indoctrination of its officers, at a military academy, to accept their place in this chain of command. "The surroundings of the Academy, the everyday life of the cadet, and the strict military training to which he is subjected," said Quezon in a 1936 report, "aim to produce a mental attitude which will make him place duty to country over and above any consideration." In sum, the Commonwealth tried to mute the inherent threat of a standing army by indoctrinating its regular officers with an ingrained respect for civil supremacy—thus making the PMA doubly important for the survival of the civil state.[5]

Preoccupied with mobilizing an entire army, Quezon and his legislature provided few guidelines for this new institution, stipulating simply that "there shall be established a military training school to be named the Philippine Military Academy, for the training of selected candidates for permanent commission in the Regular Force." Translation of these broad outlines became the task of Quezon's chief advisor, General Douglas MacArthur, a former superintendent of the U.S. Military Academy at West Point who demonstrated great faith in the capacity of this American curriculum to form a Filipino officer corps. Not only did MacArthur direct that the new academy would be, as he put it, "built on the lines of West Point," but he also recommended its Filipino graduates as the PMA's first superintendent, commandant, and tactical officers.[6]

Transporting the West Point system, with all of its peculiarities, from the bluffs of the Hudson to the mountains of Baguio entailed cultural adaptation. From the perspective of the PMA staff, the new academy would socialize the cadets through its formal curriculum and a four-year progression from neophyte to command. To succeed, however, these formal processes rested upon rituals and symbols that would make the academy's abstractions meaningful to teenaged Filipinos. Drawing upon the country's culture of masculinity, these cadets used rituals of male initiation and group solidarity to reinforce the PMA's institutional imperatives. Entering cadets were, through this process, incorporated into class and corps and developed deep emotional ties to their peers in just a few months. To preserve their membership in this group and survive the relentless four-year grind, cadets had to internalize the academy's values and subject themselves to its rigid discipline. Hence, formal military socialization rested, in large part, upon an informal cadet culture. Through a fusion of the West Point curriculum, faithfully reproduced by the PMA's staff, and informal innovations by these Filipino cadets, an American academy became a viable model for a Philippine institution.[7]

The Philippines, in using this American model, was adopting a particular kind of military academy. In a comparative study of St. Cyr, Sandhurst, and West Point, one scholar found that "at all three academies there are songs, slang, customs and ceremonies that link each annual class together for the rest of their army life." Indeed, he concluded that "this indoctrination, together with drill and discipline," was far more important in academy life than the formal curriculum. Unlike St. Cyr and Sandhurst, however, West Point remains "more Prussian than the Prussians"—that is, more determined in its use of discipline to mold its cadets during their first or "plebe" year.[8]

While homogeneous social elites filled European academies, West Point recruited cadets of diverse backgrounds and used discipline to mold them into a unified cadet corps. In 1906, echoing Social Darwinism and G. Stanley Hall's influential work, a West Point instructor argued that "at the period of adolescence, when character is plastic and impulse wayward . . . control and constraint are the essential forces for impressing permanent form upon young manhood." To this late Victorian pedagogy, West Point added its own particular innovation—a cadet tradition of harsh initiation called hazing. A future general who entered in 1913 recalled that he "was immediately plunged into the ordeal which faces every plebe—the six weeks of rigorous mental, physical and spiritual testing known as 'beast barracks.'" By the 1920s, West Point had developed a distinctive curriculum that tried to discipline by drill, toughen with sport, test by hazing, educate by daily recitation, and refine with drama, debating, and dancing.[9] By adapting this system, the PMA would attempt, through isolation and indoctrination, to sever the primary ties of loyalty that bound up Filipino society and socialize cadets in new values that would make them servants of state power.

Given the Philippines' strong regionalism, this American system of forced integration was, in certain respects, appropriate. Captain Bonner E. Fellers, a U.S. Army officer who advised Quezon in 1936–37, argued that "no other country has more need of an academy like West Point than has the Philippines. As I recall, there are some forty tribes speaking eighty-seven different dialects. . . . Provincial, tribal, and religious feeling runs strong." With students drawn "from every tribe and province," the PMA would, in the captain's view, "lend a solidarity the people never before possessed."[10] Indeed, the same curriculum and culture that had long molded cadets of diverse backgrounds into the West Point cadet corps would also prove effective at the PMA. There were, however, some significant differences.

The historical development of the two schools is, in several respects, a study in contrasts. West Point had evolved slowly with the U.S. Army over the span of a century. The PMA sprang fully grown from the mind of Quezon as he rushed to establish a new nation and its army in less than a decade. Thus, the culture of masculinity, implicit within West Point's traditions, was self-consciously constructed at the PMA.

Moreover, the West Point regimen that inculcated an impersonal loyalty to the state in America, inspired, in the Philippine context, a more personalized identification with classmates. Indeed, the PMA succeeded in its socialization by superseding traditional ties with a new and surprisingly powerful kind of loyalty—not to family or region, but to the "batch" or graduating class. At the PMA, the informal cadet culture recast West Point's rigid system in ways that created an even stronger culture of male bonding.[11]

A uniquely placed Filipino observer, Colonel Felicissimo Castillo, spent a half-century mediating the differences between these two institutions. Castillo had graduated from West Point in 1940 and then joined the Philippine Army, where he became an honorary member of PMA's Class '40. There had been no physical touching of plebes at West Point, but Filipino officers, he discovered, regarded "a lot of hazings" as essential to the PMA experience. West Point's milder hazing produced weaker solidarity when compared to the PMA classes, who were "really, very, very close to each other."[12] If we take Colonel Castillo at his word, the PMA's harsher initiation reinforced a stronger class identity among Filipino officers, bonding them for life into a batch.

Over time, these unplanned ties to class and classmates became the Philippine state's primary defense against the politicization of its armed forces. For over twenty years after independence in 1946, regular officers, largely PMA alumni, used group loyalties to sustain the lessons of their academy socialization, restraining their classmates from the pervasive politics that compromised almost every postwar institution. Numbering nearly three hundred graduates, Class '40 and its underclassmen formed a substantial, even dominant, block of regular officers that defined the values of the postwar Armed Forces of the Philippines (AFP).[13]

Rising from captains to generals after the war, Class '40 infused the military with a professionalism that militated against revolt. After the last of its number retired in 1968, the class could look back on a record free of corruption or coup plotting. Viewed in a regional context, they were stepping down to modest pensions at a time when peers in Bangkok, Jakarta, and Rangoon were clinging to power and prerogatives by force of arms.[14]

MALE INITIATION

Scholars of the Philippine military have often noted, in passing, the role of the PMA experience in bonding its graduates. But none has analyzed, in any detail, the implications of such solidarity for military socialization. After months of hearings on the failed coups of the 1980s, the Davide Commission's report offered only four tantalizing references to these issues in its six hundred-plus pages—remarking, for example, that "intense bonding takes place among cadets, a function presumably of hazing."[15] Similarly, the half-dozen doctoral dissertations on the Philippine military state flatly, without much evidence or elaboration, that cadets form strong ties to classmates during their academy days. Among these, a Chicago psychologist who observed the PMA in the mid-1960s noted that "lifetime bonds" among regular officers are "forged in the crucible of the hazing process."[16]

For many classes, hazing, and the broader experience of plebe initiation, served as a transformative trauma—coloring the subsequent academy experience for individual cadets and uniting a new class through shared suffering. During their first months, plebes were subjected to an unbroken regimen of running, recitations, and drill under nameless, powerful upperclassmen. After the initial "beast barracks," the hazing subsided into a constant low-level harassment that continued for another eight months until the upperclassmen "recognized" them as full members of the corps. Writing in the *Golden Book,* Cesar Montemayor of Class '40 recalled their plebe year as "a one-year initiation period full of rites, rules and requirements" that instilled "desirable manly and military qualities."[17]

What is the meaning of this ritual with its extreme violence? Hazing, seemingly a small issue, has embedded within it larger problems of masculinity central to armies everywhere. In fieldwork around the world, anthropologists have discovered the near universality of male initiation. While a girl "has the option of being absorbed into womanhood without effort," explained anthropologist Michelle Rosaldo, "at some point, the boy must break away from this mother and establish his maleness as a thing apart." Such "fierce warrior peoples" as the Masai of Kenya and the Amhara of Ethiopia practice violent rites to "prepare young boys for the idealized life of the warrior." Nor is male initiation limited to societies that might be called tribal or primitive. By the 1890s, over five million American males, a quarter of the country's total, belonged to seventy thousand fraternal orders organized around elaborate, albeit nonviolent, initiatory rituals to facilitate "young men's troubled passage to manhood in Victorian

America."[18] Around the globe and across time, many societies view manhood as something that must be earned and thus create rituals to test and train their adolescent males.

The most detailed work on male initiation has been done in the rugged, remote Highlands of Papua New Guinea. Observing these rituals, anthropologist Roger Keesing offers a single, simple explanation for the prevalence of harsh male initiation: warfare. Since the Highlands suffer fighting of "a scale and intensity and bloodiness equaled in few parts of the tribal world," survival requires that boys become "brave warriors separated forever from this world of women by . . . commitments to the glory of the battlefield."[19]

At the margins of the modern Philippine state, young men, bonded in peer cohorts, have long been initiated into manhood through ritual testing of their martial valor. Throughout the twentieth century, Muslim groups in the southern islands have formed all-male "minimal alliance groups" to engage in ritualized warfare, while the Ilongot highlanders of northern Luzon require boys to pass "severe tests of manhood" by taking "at least one head" in combat.[20]

From this anthropological perspective, hazing becomes the central rite in a passage from boyhood to manhood, civilian to soldier. "The life of an individual in any society is a series of passages from one age to another," Arnold van Gennep explained in his landmark study of initiation. In his formulation, Filipino plebe and New Guinea adolescent pass through similar initiations to emerge as warriors hardened for battle and bound together for defense of their communities.[21]

There are, however, limitations to such comparisons stripped of history or politics. Anthropologists, until recently, plucked these rituals from their historical contexts and compared their structural essentials through timeless studies of community or folk.[22] By contrast, the study of males strategically placed at a military academy or conscripted into modern armies inclines us, almost axiomatically, to historicize their gender.

Fortuitously, recent historical research has explored the ways that rising European states reconstructed gender roles to support mobilization of modern armies. After Britain's dismal performance in the Crimean War of the 1850s, headmasters at its elite public schools began hardening boys for future command through sport. Indeed, Harrow's headmaster proclaimed that "the esprit de corps, which merit success in cricket or football, are the very qualities which win the day in . . . war." A half-century later in South Africa, British troops faced difficulties subduing Boer farmers, raising questions about the military fitness of ordinary Englishmen. Responding to this perceived crisis, Lord Baden-Powell organized the Boy Scouts in 1908 "to pass as many boys through our character factory as we possibly can."[23]

In his study of the cult of war in nineteenth-century Europe, historian George Mosse asks: "Why did young men in great numbers rush to the colors, eager to face death and acquit themselves in battle?" Simply put, they volunteered because the modern nation-state, through its poets and propagandists, made the passage to manhood synonymous with military service. To become a man in Victoria's England or Bismarck's Germany, a young male had to serve. In the first months of World War I, this cult of war achieved a virtual florescence as young idealists hurled themselves into the slaughter. After 145,000 German soldiers died at Langemarck in 1914, one poet wrote: "Here I stand, proud and all alone, ecstatic that I have become a man." Recalling this battle in *Mein Kampf*, Adolf Hitler said: "Seventeen year old boys now looked like men." Similarly, during World War II, U.S. Army researchers found that American soldiers fought hard to avoid "being branded a 'woman,' a dangerous threat to the contemporary male personality."[24] By marrying anthropologists' universals to Mosse's time-bounded specifics, we can see how European nation-states, by making military service an initiation ritual, primed their males for slaughter on the modern battlefield.

Not only did mass conscription produce soldiers, it also shaped gender roles in the wider society. To prepare every male for military service, European nations constructed a stereotype of men as courageous, honorable, and physically formed on "borrowed Greek standards of male beauty." By the 1920s, their writers had, in Mosse's words, elevated "militarism, male comradery, and heroic youth to a virtual cult." Women were, through this century-long process, "transformed into static immutable symbols in order to command the attention of truly masculine men." Thus, modern warfare, as it developed in Europe, was the mother of a new masculinity propagated globally through colonial armies, boys' schools, and youth movements.[25]

GENDER AND NATIONALISM

As it mobilized its new national army in the 1930s, the Philippines imported this martial masculinity along with the tank, the howitzer, and the pursuit plane. With only a decade to prepare for independence and the burden of defense, the Commonwealth had to fashion a masculinity that would sustain mass conscription. Moreover, the colonial context seemed to redouble Filipino enthusiasm for the project since nationalists, rejecting colonial "emasculation," had long fought for the right to lead troops.

To build popular support for a citizens' army, the neophyte Philippine state deployed a gendered propaganda with men strong, women weak; men the defenders, women the defended. We do not have to read against the grain to tease gender out of the Philippine Army, as if from some recondite cultural text. The key actors—Quezon, army headquarters, and the cadets themselves—were quite self-conscious in their use of such imagery.

By the early 1930s, a decade of reserve-officer training had already encouraged an ideal of military masculinity among cadets at Manila's universities. At the University of the Philippines, trainee officers articulated an ideology that equated masculine strength with national defense. "A nation stands or falls, succeeds or fails, just in proportion to the... manliness of each succeeding generation," wrote a cadet in the 1931 edition of the campus yearbook.[26] Cadet Sergeant Fred Ruiz Castro, a future Supreme Court chief justice, explained that military training helps "engender the proper citizenship"—notably "courtesy to all especially to the old and to the weaker sex." Reinforcing this dimorphism, U.P.'s all-male cadet companies barred women from drill but recruited them as "sponsors" to appear in formal, frilly gowns at full-dress parades. Illustrative of this imbalance, in the late 1920s one of these sponsors gave the corps a "colorful oration" titled "The Woman Behind the Man Behind the Gun."[27]

From its foundation in 1935, the Philippine Commonwealth, through military mobilization, intensified this process of gender reconstruction—encouraging a complementary array of national symbols, militarized masculinity, and domestic roles. Just as the new nation was personified as the feminine "Filipinas" in currency and propaganda, so young men were conscripted to defend her and her defenseless womankind. The government, in this transition to independence, skillfully manipulated public rituals and symbols to make a polarized gender dimorphism central to a new national self-image.

The impact of militarization upon gender representation was most evident at the Manila Carnival—a grand prewar festival that drew vast crowds to celebrate the fecundity of the land and the glories of its people. Like other pre-Lenten festivals across the Hispanic world, Carnival was a mix of the serious and frivolous, celebration and reflection. The sprawling fairgrounds at the heart of Manila held elaborate displays of tropical products and hosted a two-week social whirl that culminated in the crowning of the Carnival queen. With Filipino society on parade, elite actors gained a mass audience for their images of society and nation. In the 1920s, before conscription, the coronation itself had been a lavish, high-society affair with whimsical Roman or Egyptian themes.[28]

With the inauguration of the Commonwealth and its new army just months away, the 1935 Carnival saw whimsy surrender to military symbolism and gender politics. At the grand ball, the constabulary band played a march while Queen Conchita I walked between two files of University of the Philippines cadets with drawn sabers to a throne where the U.S. governor-general placed a crown of diamonds on her head.[29]

This Carnival also launched a national debate on women's rights. Speaking before the Federation of Women's Clubs, Senate President Quezon announced that the Constitutional Convention had just approved conscription and urged the nation's women "to mold the character of . . . youth that we may build up here a citizenry of virile manhood capable of shouldering the burdens of our future independent existence." Federation president Pilar H. Lim, the wife of General Lim, confronted Quezon, demanding that he redress "the injustice" of the convention's refusal to grant women the vote. He assured her that he had "always been in favor of granting this right to women." Indeed, Quezon would keep this promise and two years later a plebiscite on women's suffrage passed by an overwhelming margin.[30]

As military mobilization intensified, succeeding Carnivals expanded this militarized gender dimorphism. In its coverage of the 1936 Carnival, the *Sunday Tribune Magazine* juxtaposed photo-essays of the all-male military review ("platoons marching as one man, the steel helmets . . . glaring in the afternoon sun") and a female fashion revue ("models resplendent in shining silver and satin"). On their night at the Carnival, the U.P. students presented a richly gendered pageant written by Carlos Romulo with a cast of one thousand students (including seven hundred girls) and starring a woman student as "Filipinas," the feminized symbol of the nation. After the "Spirit of History ascends . . . from stage right and writes 'Commonwealth,'" read the libretto, "Filipinas enters from stage left followed by people, including agencies, soldiers, dancers." Then comes: "Invasion—call to arms. Battle." The drama ends as "Filipinas rises from the center of the floor, flag over her. National hymn is sung by all."[31]

During the next two years, as the new army mobilized, the Carnival's military symbolism kept pace. By 1938, the queen's escorts were uniformed officers and the cadet parade became a spectacle of military might. With thousands of spectators packed along the boulevards, armed columns of Philippine Army, Philippine Scouts, and college cadets tramped past the Legislative Building and then formations of bombers and pursuit planes roared overhead.[32]

Outside of Carnival season, gender was on parade at the city's campuses. By 1936, the U.P. cadets had expanded their corps of sponsors to forty coeds such as Miss Eva Estrada, the muse of the Second Artillery Battalion and a future senator. On National Heroes Day, the ROTC cadets staged a mock battle in the city's main park, the Luneta. "Planes sweep down from the clouds to drop their deadly bombs," read the college yearbook, "men shoot, advance, fall . . . beneath the smoke the unseen drama of war with its horrors and victories." As male cadets litter Luneta's smoking battlefield, "the Nurses' Corps recruited from the ranks of the Sponsors rush to the field to give aid to the wounded and the dying."

Among the cadets, appeal to women was deemed an essential attribute of future military leadership. "The girls go for him in a big way (very big way)," said the 1937 U.P. yearbook of Cadet Major Ferdinand Marcos, "so much so that most of the time he has to put up the sign 'Standing Room Only.' Claims his heart is impregnable to feminine allure, and insists on calling guys who fall in love inebriated weaklings." Marcos himself internalized this gendered duality to write, after the war, of sacrificing his manhood to defend a feminized nation he calls Filipinas. "We cursed ourselves . . . for having given up our arms and with them our manhood. . . . " Marcos wrote of their wartime surrender to Japan on Bataan. "Filipinas had welcomed us in spite of the disgrace of our defeat in Bataan. But it seemed that although she had smiled at us through her tears, she would not bind up our wounds."[33]

After its creation in 1936, the Philippine Army deployed a similar dualism to build support for conscription among a people without a tradition of military service. As the date for draft registration approached, the Commonwealth plastered public spaces with recruiting posters. One depicted a statuesque Filipina, neckline cut low and bare arms outstretched for the embrace, calling on "Young Men" to "Heed Your Country's Call!" Another asked, "Which Would You Rather Be . . . this or that?"—and then showed a snappy soldier smiling at two admiring women while a civilian male skulks in the rear, with his hands shoved symbolically in his pockets.[34]

Then, at 8:30 A.M. on 15 May 1936, each provincial governor supervised an elaborate ritual to select the first conscripts for basic training. Before the public, the governor, flanked by military guards, placed the registration cards for all twenty-year-old men in two large jars. "Two young ladies, not over 18 years of age, shall . . . make the drawing," read the army regulations. "These young ladies shall be blind-folded and shall wear dresses with short sleeves—not reaching beyond elbow."[35]

Film played a central role in this process of gender reconstruction. In 1938–39, the Philippine Army authorized two Filipino movie companies to shoot feature films at the PMA with polarized imagery—muscular males as cadets in hypermasculine uniforms and attractive women as their dates in feminine fashions. Press coverage for the premiere of *Punit na Bandila* [The Torn Flag] spoke of the "lovely" Lucita Goyena, the "virile" Fernando Poe, Sr., and the story of "a brilliant cadet of the Philippine Military Academy who gets kicked out . . . for defending the honor of his mother."[36]

The Quezon administration was aggressive in its defense of this heroic masculinity. In mid-1939, MGM Studios began editing *The Real Glory,* a drama set at century's turn with scenes showing Filipino constabulary troopers as cowards under fire. A Filipino extra on the Hollywood set wrote to a Manila magazine complaining of scenes that "thoroughly disparage the character of Filipino soldiers as a whole." So alerted, Quezon ordered his executive secretary to fire off a letter to MGM studio chief Samuel Goldwyn, protesting a scene showing "sixty Filipino recruits, equipped with bayonets and rifles . . . made to cower and . . . then flee in terror at the sight of a Moro fanatic armed with only a 'kris.'" The film, he continued, will "create an impression in the audience of an extreme of cowardice on the part of the Filipino unequaled by any other race in the world." In August, after allowing Quezon's emissary private screenings at the MGM studio, Goldwyn cabled Manila: "I have today ordered certain deletions" and promised that "my picture is a real tribute to the courage of the Filipino as a soldier."[37] On the surface, Quezon had reacted to a national insult. But at a more fundamental level he was defending his new image of Filipino manhood.

As centerpiece in this project of gender reconstruction, the PMA indoctrinated its Filipino cadets into a Euro-American ideal of military manhood. To ensure that its officers would be archetypes of masculine beauty, the academy barred applicants with "any deformity which is repulsive" or any who suffered from "extreme ugliness." Medical examiners had to insure, moreover, that an applicant's face was free from any "lack of symmetrical development" or "unsightly deformities such as large birthmarks, large hairy moles, . . . mutilations due to injuries or surgical operation."[38]

To mold these exemplary males, the PMA became a total institution that would, like West Point, leave a lasting imprint upon every graduate. After selecting the best cadets by national examination, the new academy drilled them daily for discipline, demanded high academic standards, and required absolute integrity.[39] The PMA's 1938 yearbook thus described the tactical department and its drill instructors as "a veritable forging shop in which the raw and crude materials are ... purified of their undesirable qualities." In their song *P.M.A. Forever,* cadets celebrated their academy's capacity to make men:

> Within the walls of old and glorious P.M.A.
> They're molded to the real men that they should be—
> Men who can face the bitter realities of life
> With courage even in the midst of bloody strife.[40]

With its alien curriculum, the PMA, more than any Philippine institution of its era, aspired to effect a cultural transformation, a remaking of its cadets on a European model of masculinity. The academy made its imprint through a program of moral formation that used body movement, incessant supervision, and formal indoctrination. Through constant drill and conditioning, the academy demanded a new kind self-consciousness of the body—its hardening, shaping, and draping. In its own words, the academy taught "soldierly movements to inculcate prompt obedience" in daily marching, "knowledge of ballroom ethics" with weekly waltz lessons, and "self-reliance, poise, initiative, judgment, enthusiasm, and discipline" in gymnastics.[41] Cadets, under constant supervision by peers and superiors, experienced a social immersion devoid of any privacy or individuality. In this isolation from family, community, and women, the PMA taught cadets, through its rhetoric and rituals of solidarity, to prize male camaraderie and assay their own masculinity on a universal standard of military competence.

The academy's formal and informal cadet traditions complemented this curriculum by making initiation a central rite of passage—from civilian to soldier, plebe to cadet. Entering plebes arrived at the academy from communities with their own rituals of male initiation and expectations for manhood.[42] In many lowland villages, adolescent males passed through an initiation, such as circumcision, and had elaborate codes for masculine friendship epitomized in peer groups called *barkada.* In the villages of Central Luzon, for example, Tagalog males who joined tenancy unions during the 1930s were tested in an elaborate midnight ritual in which each was branded on the upper arm with a poker plucked white-hot from a raging bonfire.[43] The PMA's staff had adopted the West Point system verbatim, but cadets reshaped these imported values through their own culture of masculinity.

Such harsh initiation had become part of an emerging military tradition when officer training started at the University of the Philippines in the 1920s. Cadet Sergeant Macario Peralta, Jr., a future defense secretary, noted in the 1932 *Philippinensian* that the corps had faced difficulties in "breaking in the new cadets," but made sure that troublesome plebes receive "sundry other polite attentions."[44] Peralta's yearbook biography, published two years later when he was cadet colonel, revealed the meaning of this euphemism. "One year after the Colonel sprouted in the university campus, he commenced hazing the plebes and beasts with unrelenting inhumanity. He is still at it."[45]

Influenced by this culture of manhood, PMA cadets modified the academy's official system and its masculine archetype—personalizing its sense of honor and accentuating the ideal of loyalty to classmates. From its founding in 1936, the PMA would face recurring problems with these intertwined issues of hazing and honor—partly of its own making and partly an unwanted legacy from the colony's original military school.

ANTECEDENTS FOR AN ACADEMY

Not only was hazing integral to the PMA experience, it also precipitated the dissolution of its predecessor, the Philippine Constabulary Academy. Established in 1916 and expanded twice in the 1920s, this school had enjoyed considerable prestige until 1935, when a hazing scandal prompted its sudden dissolution.

During the Commonwealth inauguration of November 1935, the Constabulary Academy's harsh hazing was finally exposed to public scrutiny. After months of beating and slapping in Baguio, the plebes of Class '38 reached their breaking point when upperclassmen humiliated them before spectators in Manila with duck walks and double-time drills. On a routine inspection of the campground, the school's commandant passed by a tent and overheard the voice of an angry plebe, Eugenio Lara, saying to his fellows, "I am resigning tomorrow. But before tomorrow arrives and somebody touches me with even the tip of his finger, I'm going to break their bones." When the plebes refused to talk, a board of inquiry exposed a serious pattern of abuse, including the stabbing of a plebe by upperclass cadet Pedro Quezon Molina.[46]

Outraged, President Quezon discharged eight cadets, including his own nephew Molina, insisting that the academy "teach subordination . . . but must not permit servility." Only a few months before the PMA was due to open, the president made a surprise inspection of the Constabulary Academy in Baguio, dismissing the American superintendent and meeting with the cadets to insist that hazing end. MacArthur's deputy, Major Dwight D. Eisenhower, noted in his diary of February 1936 that Quezon had "issued such sharp reprimands to the faculty, in the presence of the student body, there was no other recourse than to replace the superintendent and all his assistants." In the aftermath, Eisenhower and his fellow advisors drew up plans for the new school, adopting the U.S. service curriculum and selecting Filipino officers for its staff.[47]

The Constabulary Academy closed after graduating its last class of eight cadets in March 1936. Two months later, the new Philippine Military Academy abandoned the confines of Camp Allen for larger quarters at nearby Teachers' Camp. Under a time-share arrangement with the Bureau of Education, the PMA would decamp to tents on Baguio's Polo Field every summer when schoolteachers arrived for their annual in-service seminars.[48]

Though the PMA inherited the Constabulary Academy's cadets and traditions, there were marked differences. Instead of the old school's genteel round of tea parties and tennis matches hosted by an American superintendent and his gracious lady, the new academy was seized with the seriousness of training the nation's future generals. While the constabulary school had admitted classes of a dozen cadets for a three-year course in police procedure, the new PMA had an entering class of 120 plebes who would learn military science and earn a four-year bachelor of science degree. A small staff of drill instructors gave way to a credentialed faculty divided into four departments—tactics, social arts, mathematics, and engineering.

Each department had a clearly articulated mission within a curriculum designed to mold future commanders. Tactics taught soldiering to underclassmen and the principles of command to seniors. To train cultured, well-rounded officers, social arts ranged from surveys of English literature and Western history to the practicalities of public speaking, economics, and law. But mathematics and engineering dominated the school week, shifting the curriculum toward demanding technical subjects. Second-year cadets, for example, would study 475 hours of math, physics, and surveying; 200 hours of history and English; and only 70 hours of tactics. In the words of the PMA's 1938 yearbook, "this is an age of mechanized warfare, of rapid movements, and of still more rapid communication" that made math-based skills essential for future officers. Finally, in the new academy's additional fourth year, cadets would concentrate on military science. Aside from courses such as Fortifications, Military Engineering, and Gunnery, they also learned broad military theory by studying the "important campaigns from the Ancient Wars up to . . . Frederick the Great" in Military Art and History; the "requisite qualities of a successful leader such as . . . loyalty to subordinate and superior officers" in Principles of Command; and "the aid military forces lend to civil powers" in sixty hours of Government and Military Aid to Civil Power.[49]

With a legacy of scandal, temporary quarters, and a foreign curriculum, the

PMA's origins were not auspicious. Recruited from towns across the archipelago, the first batch of 120 entering cadets, the future Class of 1940, were faced with a challenge of cultural invention. It would be their task to translate an imported institutional culture into a Filipino idiom. It would be their duty to invest these barren, borrowed halls with the mystique of alma mater.

PLEBE YEAR

In my interviews with Class '40, recollections of the academy usually began with the entrance examination. Following the West Point system, the first batch of cadets was selected by a national examination given to over six thousand applicants in April and May of 1936.[50]

In a society where patronage was pervasive, the exams proved an effective mechanism for social mobility. Drawn from every region and social stratum, the Class of 1940 proved a diverse and talented group. Although there were two graduate civil engineers and several Manila socialites, most cadets came from the broad lower-middle stratum of Filipino society. Among the thirty-two alumni interviewed, seven were from poor households, seventeen from a lower-middle stratum of small farmers or minor officials, and only eight from affluent families. While their mothers had stayed at home raising six or seven children, their fathers held jobs ranging from farmer to senator. In a memoir written before his death in 1987, classmate Liberato Picar recalled a village childhood that seems typical:

> The house where I was raised had two rooms, one room which served as living and bedroom for all of us. . . . It was thatch-roofed, with cogon grass, walled with coconut leaves and floored with split bamboo. Surrounding it were a few coconut trees, which father had planted. I was told I was born in this simple house. . . .
> For elementary education I enrolled in the Caba Elementary School, which is . . . about two kilometers from our house. As usual, I had to trek my way to school barefooted for three years. . . . Because our town had no electricity, for all these years I had been studying beside an oil lamp. The study efforts exerted were rewarded when I graduated number three in a class of about fifty.[51]

All of Class '40 had graduated from high school, most were honor students, and some twenty-five had been valedictorian or salutatorian. But these academic achievements had already been devalued by the flood of high school graduates, and most found their advancement blocked by lack of funds for higher education. By the time they sat for the entrance exams, many of the future cadets were unemployed, some were working students, and a few were getting by as lower-echelon employees or day laborers—road workers, stevedores, farmers, or soldiers.[52] Among the thirty-two classmates interviewed, twenty-three cited the need for a free college education as their main reason for applying to the PMA.

Selected by merit rather than connections, the class set athletic and academic records that stood for decades. Cadet Pedro Baban, child of an impoverished Ibaloi tribal family, ran the four hundred-meter dash in fifty-eight seconds, a school record that lasted thirty years. Victor Osias, son of a leading legislator, set a long-jump record of twenty-two feet and four inches that stood fifty years. Class '40 had six "starmen" cited on the superintendent's list for outstanding academic achievement—a number unsurpassed for the next half-century. Among the seventy graduates who survived World War II, fourteen would win promotion to general or an equivalent navy grade—over half the twenty-seven star-ranks then allotted to the armed forces. They would meet the challenge of their commandant, Captain Rufo Romero (USMA '31), who told them, at the start of their first academic year, to become "an aristocracy of brains." Cheating, he said, would be punished by expulsion and "no amount of political influence could keep a failing cadet in the PMA."[53]

The cadets of Class '40 spent their first year as plebes, untested neophytes subjected to endless, arbitrary harassment by upperclassmen. Recalling van Gennep's rites of passage, the plebe-year initiation proceeded through three distinct phases: a "separation" during three weeks of "beast barracks," when testing was most intense; a transitional, or "liminal," period of qualified admission into the corps during the academic year; and, finally, a year-end ritual of full acceptance, or "reintegration," by upperclassmen called recognition. Through incessant command-and-response rituals, the plebes were socialized into a chain of command where officers obeyed superior orders without hesitation.[54]

Instead of the year-long hazing of the old Constabulary Academy, the PMA adopted West Point's beast barracks—an initial period of intense harassment that ended with the start of classes. "Beast barracks was a veritable hell barracks," the historian for Class '40 wrote at the end of their plebe year. "With painful limbs and aching backs, we hated every day that came, some of us even cursed their fate." Yet upperclassmen, chastened by the recent expulsions for hazing, were cautious and officers were protective of the incoming plebes. With a few memorable exceptions, the members of Class '38 were tough but not sadistic.[55]

Still, the three "sadists" among the twenty-seven yearlings of Class '38 inflicted considerable suffering. The school's bantam-weight boxing champion, Pedro Felix, delighted in using plebe bellies as punching bags, occasionally beating his victims badly. Since cadet culture regarded squealing as "conduct unbecoming," Class '40 had to suffer his abuse in silence. Even Class '38 described in their own yearbook how classmate Renato Barretto used "the mosquito bar support" and "came close to strangling our innocent beings one morning."[56]

After weeks of round-the-clock drill, beast barracks ended and classes began. To mark this passage, the staff assembled the plebes for an oath to the Commonwealth that marked their official admission to the cadet corps. Some classmates who had suffered badly in these first weeks were moved. "My, my, my hair stood when I was parading the grounds," recalled Ramon Gelvezon fifty years later, "and I said to myself, 'even if you get dismissed from the academy for academics, you belong to that class where you were incorporated.'"[57]

While the staff may have regarded the class as full members of the corps, the upperclassmen still withheld their recognition. For another eight months, they continued to address the plebes as "dumbguard," "ducrot," and "dumbflicket." But with most of the day now occupied by classes, there was simply less time for harassment.[58]

In a plebe's long day, meals and the minutes before taps were the times that he was most vulnerable to upperclass attentions. "At the mess hall," one classmate wrote of the plebe's life for a Manila magazine, "he occupies one half of his chair and eats with his chin drawn in and shoulders thrown back." As table commanders, first classmen could deny plebes food by seating them under the table or ordering a contorted posture. Classmate Felix Apolinario recalls that a playfully vindictive upperclassman denied him all food between soup and dessert until another intervened, pointing out that the plebe's health was at risk.[59]

Throughout each meal, table commanders quizzed the class in all manner of p.k. or "plebe knowledge"—a borrowed West Point custom of convoluted replies to absurd questions such as "How is the cow?" The answer, verbatim from the banks of the Hudson, went: "Sir, she walks, she talks, she's full of chalk." In the gap between dessert and dismissal, plebes often performed pop songs or amused their table with ad hoc nonsense such as classmate Hospicio Tuazon's fish theory: "After many years of hazardous toil and profound researches, I finally arrived at the conclusion that no matter how long a fish swims, it will never perspire." One such composition, "Why on the Double?" by a member of Class '41, captured the logic of plebe year so succinctly that it gained immediate entry into the p.k. canon: "The force coming from the itinerate glances of the immaculates are so powerful that . . . a double timing ensues which develops an invulnerable machine in the body of the degraded mammal."[60]

In the privacy of their quarters upperclassmen could be capricious or even cruel. Quotations clipped from interviews with Class '40 form a textual collage of this varied experience. Pedro Baban: "We are not supposed to squeal, that is the tradition, no matter what is done to you." Felipe Fetalvero: "Cadet Cleofe, Senen Cleofe, when you are told to report to his room, you'll find that there are already several plebes there.... They are floor deeping, they are being quartered, they are squatting, squatting with their rifle. You know, we really hated him." Reynaldo Mendoza: "That . . . was not hazing. Those are normal rituals of a group of young men trying to . . . initiate somebody."

There were, however, limits. Academy regulations barred any physical abuse and required that upperclassmen ask permission before touching a plebe. When cadet Renato Barretto ('38) hit plebe Ramon Nosce ('40) with his sword "unintentionally or accidentally" during a double-time drill, the upperclassman was "busted" without mercy—stripped of rank and made to march for hours with fixed bayonet.[61] In sum, the plebe's experience of hazing was harsh enough to induce solidarity from shared suffering. But it was not so rough that it moved beyond testing to brutality, leaving lasting scars.

In his memoirs, Edmundo Navarro captured the paradoxical mix of pride and anger that Class '40 took away from the plebe experience. "I was 168 pounds with a 39-inch waistline when I arrived in Baguio City," he recalled. Within minutes of my arrival, I had been slapped, kicked, boxed and subjected to such indignities as I had never experienced before." After months of incessant forced exercise he thought "concocted by some evil upperclassman," Navarro was proudly surprised to find, he wrote, that "my weight had been reduced to 131 pounds and my waistline had shrunk to 29 inches!"[62]

The harshness of plebe year left the class with an enduring sense of solidarity. From their interviews, voices rise in unison to proclaim lifelong bonds. Licurgo Estrada: "We felt really like real brothers as we called each other 'cavaliers' or 'mistah.'" Lucendro Galang: "We are like brothers there, or more than a brother." Jose Mendoza: "Like my blood brothers." Eduardo Soliman: "We were closer than brothers . . . Because when you have passed through hardships like plebe year, you are molded into one."

After nine months of humiliation, Class '40 finally won "recognition" from their senior tormentors during graduation week of 1937. The day before this ritual, however, the first classmen subjected Class '40 to a final round of humiliations that an alumni history called "a series of 'gymnastics,' 'acrobatics,' the verbal abuse reminiscent of the din and confusion" that they had experienced when they first reported to the academy. That night, Class '40 struck back. During the graduation dance, the plebes retaliated, ransacking first-class rooms by "disarranging of lockers, wetting of beds by pouring water between the sheets, razor trimming of shoe laces."[63]

The upperclassmen could not let these insults pass without punishment. At reveille formation the next morning, the cadet first sergeant, a member of class '38, barked: "All plebes, hang yourselves." So Class '40 reached upward and swung from the pergola that shaded the walkway. The sergeant gave a second order: "Upperclassmen, fall out and do your thing." The plebes suffered angry blows to the stomach until they "fell to the pavement semi-conscious." Although bitter about a beating they considered sadistic, Class '40 had no time for further retaliation.[64]

Eight hours later, they were recognized as full members of the corps. At the close of commencement, the band struck up "Auld Lang Syne." Then, the graduates of Class '37 separated themselves from the corps and stood at attention while the cadets passed in review under the command of the new first class. At the command "stack arms," the upperclassmen of Class '38 approached, one by one, to shake hands with the plebes they had harassed for the past nine months. It was an emotional ritual with "much weeping." Instead of "beast," "mammal," or "ducrot," upperclassmen would now address Class '40 respectfully as "cavalier" or "mistah"—a passage from animal, to subhuman, and, finally, human male. "That's a symbol that you passed the test," said classmate David Pelayo. "And you can congratulate yourself for being a man, for having underwent a rigid test of manhood for a year." Years later, many classmates would recall recognition as an emotional high point, a teary-eyed moment of triumph and acceptance.[65]

The pressure of plebe year was relentless and Class '40 had bonded to survive. With faculty distant and upperclassmen speaking only to command, classmates turned to each other for support during these ten months of unbroken confinement. From reveille to taps, Class '40 marched in formation to meals, classes, and assemblies, living their lives in an incessant, syncopated drill. It seems that this constant proximity, the sheer physicality of sharing discipline and isolation, inculcated an unspoken but real intimacy that deepened during the three years that followed. "Living together under one roof, moving together as one body, eating at the same time," recalled classmate Pedro Bartolome, "actually makes you a team, a team."[66]

The supreme statement of this solidarity was the PMA's renowned formal parades. Arrayed by height in columns that created an optical illusion of perfect uniformity, the whole corps fused to become one in precision drills, rigid and sinuous, that seemed to extinguish their individuality. The prewar PMA used the German close-order drill, a demanding form with complicated commands and stiff body movements that produced a heightened image of mechanistic precision accentuated by sharp clicking sounds. "You can put a ruler in any of the group and you can see that . . . from head to each step, the line is there," recalled classmate Reynaldo Mendoza with visible pride. Through daily practice, each plebe had to learn the complex movements to precision: "chin in, breast out, no rolling eyes"; left arm swinging in unison with the eleven others in the file; riflestock on the right shoulder at thirty degrees angle to the body; white-gloved hands aligned downward with the toes and across to the other twenty-two hands "so when you look . . . it's only one."[67]

Within these ranks, each taut body was enveloped in a full-dress uniform that made a dramatic statement. A cadet's physique, shaped to classical male form through constant conditioning, was sheathed, from heel to chin, in a smooth, gray fabric that projected an aura of martial power. The pants rose in tight wrap around groin and buttocks into a high waist where the tunic spread seamlessly across a swelling chest accented with horizontal rows of exoskeletal black piping and a flared "V" of oversized brass buttons. Though cut from woolen fiber, the tunic's focus on the flared torso evoked the steel breastplate of a medieval knight. Yet the uniform's tight cut and rows of twill said here was a soldier whose power was constrained within the discipline of a modern army.

The crowd's reaction to this spectacle was affirming. Toward the end of their plebe year in November, the entire corps was sent to Manila to march in the Commonwealth Day celebration, giving the class their first lesson in civil relations. The new cadets proved an unexpected sensation as they stepped past the presidential reviewing stand. "Suddenly the crowd lining the route of march burst out in waves upon waves of spontaneous applause," the class history recalled, "as platoon after platoon of gray clad cadets . . . marched in perfect cadence. Never had they seen such a thrilling and hair-raising sight!"[68] Individual class members recalled the crowd's reaction to their precision drill as "electric," and felt a certain connection with the masses that lined Manila's boulevards.[69] Since they were over 100 of the 160 cadets on parade, Class '40 knew that this triumph was theirs.

Through these rituals of hazing and recognition, degradation and acceptance, the cadets not only bonded as a class but felt brotherhood with the whole corps, even their upperclass tormentors. Plebe year was, moreover, the first in four that replicated the rhythms of a military career. Through this progress from plebe to command of the corps, the academy gave cadets a cameo of careers that would take them from third lieutenants to generals—thereby training them to accept their place in a chain of command.

UPPERCLASSMEN

Their second year as "yearlings" was a time of transition for Class '40. No longer subject to hazing, they focused on academics, military training, and personal relations, forming close personal ties to classmates impossible under the pressures of plebe year. They gained, moreover, their first experience of leadership as the class responsible for the initiation of the incoming plebes of Class '41.[70]

When the new plebes arrived in April 1937, the newly minted yearlings of Class '40 "gave vent to 'transmit undiminished' all the suffering they had received from the upperclassmen." They ordered plebe Manuel Yan, a future chief of staff, to eat the newspaper clipping he proudly carried announcing his appointment to the academy, relenting only when he "was about to place the . . . bitesize wad into his watering mouth."[71] They gave plebe Roberto Lim, the son of General Lim, the special treatment due the son of a senior officer, forcing him to recite poems in a booming voice and eat his meals under the table. At year's end, the yearlings of Class '40 found it was their turn to suffer slashed shoelaces and ransacked rooms on the eve of plebe recognition. But the next day's ceremony brought what Jose Crisol ('42) later called "firm and manly" handshakes that erased any resentments "in a flood of conciliatory tears."

Instead of brutal physical punishment, Class '40 gave this initiation a less threatening air of psychological pressure and pranks "all in the spirit of clean fun." Indeed, the class was proud to claim that, during their three years in command of the plebes, hazing was strictly controlled. In their own plebe year, classmates had decided to end the abuse that had been synonymous with the old Constabulary Academy. As yearlings, their class president, Victor Osias, recalls advising his classmates, "You should keep them from hating us and . . . remember that they will join us and we will all be in the same uniform."[72]

Two classmates, who had been rather meek as plebes, proved harsh disciplinarians as both yearlings and first classmen. But class leaders warned the violators to shape up. "First of all, the people in the same company . . . would speak to him about it and see if he could tone down his actions," recalled Victor Osias. "And if this persisted, we have to go to the battalion level. Eventually, it would end up at the First Class Club." There leaders confronted the erring classmates. "You would get this righteous indignation, sort of initial reaction," Osias said, "but then on . . . second or third meeting they come around."[73]

During their last two years, Class '40 left the initiation of incoming plebes to the new yearlings but kept close watch to prevent abuses. When several members of Class '42 whipped plebes with the metal T-bar for mosquito nets, the leaders of Class '40 ordered their immediate dismissal. This fragile tradition of humane hazing continued after their graduation. Observing the initiation of Class '44 a month after Class '40 had graduated, General Lim wrote his family that "they have taken away that old Constabulary way of hazing and now they are treating the Plebes humanly [*sic*] with the idea of correcting them for the good of the service."[74]

As yearlings, members of the class were allowed to leave campus for social events in Baguio and holidays at home with family. When Manila socialites escaped to the highland cool for Baguio's summer season, cadets, wearing their snappy dress uniforms and schooled in the social graces, were in demand at dances and parties. President Quezon, the social leader of the summer capital, set the tone, dropping by the campus with his daughters and making it clear to all that the academy was his pet project. Cadets could also attend the academy's weekly "hops" and receive female visitors on weekends, affording them both casual liaisons and serious courtship. Despite their access to Manila society, only a few classmates managed to negotiate Baguio's summer whirl and marry upward on the social scale.[75] More importantly, by filling their dances and visitors' room with admiring women, the cadets affirmed the gender division that inspired the new army—they were the defenders and they had the attentions of the defended.

Midway through second year, the class went to Manila for the Commonwealth Day parade and got a painful lesson in the politics of their position as the nation's military elite. At a formal review at the University of the Philippines, where they were billeted, the PMA cadets, drilling with enviable precision, "saw the biggest crowd and . . . the greatest applause ever recorded." That evening atop the campus roof garden, the PMA cadets in their dashing dress uniforms edged the local cadets off the dance floor and disappeared into the darkness with the U.P. sponsors. "The first classmen could tell you a mouthful," said the academy's 1938 yearbook with a knowing wink, "yes, about the moon!" Classmate Pedro Bartolome recalled that "the UP boys were also jealous in a way because of the admiration of their girl friends for our group."[76] Two days later at the Commonwealth Day parade, the PMA marched down streets "jammed to the fullest . . . amidst a field of comparison supplied by Manila's R.O.T.C. units"—and outshone them all. At parade's end, however, the corps, in the words of its 1938 yearbook, suddenly faced a "severe" crisis which "involved not only the cadets but Alma Mater itself." What had happened? The yearbook says only that: "The worst had come up. . . . But the 'affair' is best forgotten."[77]

Fifty years later, classmate Edmundo Navarro explained that a senior staff officer, Colonel Fidel Segundo (USMA '17), confronted them. "I have never been so embarrassed in all my . . . life," said the colonel, with visible rage. "I have just visited the women's . . . toilets and professors' room in the University of the Philippines and all I saw was a sea of human shit." Without pausing to investigate, he "blamed the academy for . . . the writing of obscene things with excreta on the walls." The corps suffered his insults in silence, knowing that this was retaliation. "Obviously," explained Navarro, "the U.P. cadets resented our bringing the sponsors out for the night."[78]

Such visceral anger in this competition over women was but one manifestation of a deeper political conflict. After a decade as the country's premier military unit, the U.P. cadet corps suddenly found itself relegated to secondary status. With every step on the parade ground and each turn on the dance floor, the PMA cadets had communicated their future dominance. Just as they maneuvered for women that night on the roof garden, so in later years the PMA regulars and U.P. reservists would compete for ultimate control over the nation's arsenal. These trivial events were the first signs of a struggle between regular and reserve officers that would continue, unabated, for the next fifty years.

FIRST CLASSMEN

In their third year at the PMA, Class '40 had the extraordinary experience of rising to a de facto first-class status. The transition from the three-year Constabulary Academy to the four-year PMA produced a missing class in 1939, allowing the class to spend their last two years in command of the corps—an experience that redoubled their lessons of leadership and reinforced an emerging hierarchy.

Through constant competition in academics, sport, and drill, the class had, by its third year, established a ranking of leaders and followers. To prepare future officers, the academy had two contradictory missions: formation of a unified officer corps and identification of the most talented individuals for command. By its third year, this small group of seventy-nine men was arrayed in four distinct hierarchies—academic, athletic, military, and moral. Within every exercise that seemed to instill solidarity, there was a parallel element of individual testing. "There was a lot of camaraderie as well as an awful lot of keen competition," recalled classmate Victor Osias. "There was an awful lot of pride involved in trying to maintain a status and status proficiency that would make you proud of being in a given company or battalion."[79]

When the class took command of the corps in their third year, informal leadership gave way to a formal hierarchy ranging from captain down to corporal or private and distinguished by chevrons, swords, and plumage. Moreover, through recitations in every course every day, the class was forming a precise academic ranking from one to seventy-nine. Courses were divided into sections scaled by difficulty and the top student in each became the "section marcher" who led the columns into the classroom and then supervised their performance through the hour. At the apex of the academic hierarchy were the "starmen"—the five or six top scholars who wore two powerfully symbolic stars on the high collars of their dress uniforms.

Cutting against the grain of their class solidarity, competition left a legacy of personal rivalries that threatened to rupture their unity. As every cadet knew, class rank at graduation would, for the next twenty-six years, fix their position in the seniority roster. Indeed, even fifty years later, almost all could recall their class rank as easily as their birthday.[80] Locked together in this hierarchy for decades, some classmates would, in later years, become bitter with each other in the competition for command.

For the time being, however, hierarchy complemented solidarity. It was the duty of the class leaders, at the academy and afterward, to channel the competition in positive directions that would preserve unity. Despite some strong tensions, these leaders would retain, through coming decades, sufficient informal authority to mobilize the group. In their third year, reflecting their status as moral leaders, the class chose Deogracias Caballero and Cesar Montemayor to serve on the honor committee. And, in the decades after graduation, they would continue to exercise an informal moral leadership.[81]

Class '40 also mediated the colonial and the national to create a distinct institutional culture for their new school—a contradictory task of cultural invention. At one level, cadets took a certain pride in their West Point antecedents. As the editors of the 1938 yearbook noted, the PMA "has grown up under the influence of West Point traditions and customs and out of purely local tendencies." The members of Class '40 had reached maturity in the last decade of American rule, a virtual florescence of colonial culture, and thus accepted, almost without question, a foreign model for their academy and careers. Most were products of the public schools and had ten years of education conducted in English, making them fluent in this foreign language and sympathetic to its ethos. Similarly, the colonial constabulary enjoyed remarkable respect in the 1930s, one reason that many classmates had applied to the PMA.[82]

While the PMA's Americanisms might have seemed curious, even bizarre, in Manila or Cebu, they seemed somehow appropriate in Baguio, the only American-built city in the Philippines. Designed by Chicago's renowned urban planner Daniel K. Burnham as a colonial hill station, Baguio's every detail evoked an exotic North American sensibility—its cool five thousand-foot elevation, pine groves, curving garden-city roadways, California-style bungalows, and formal gardens filled with temperate ornamentals. Within Baguio, the academy occupied the grounds of another laboratory of American social engineering, Teachers' Camp, which the colonial regime had built as a summer training center for its schoolteachers. Its cluster of white, clapboard buildings lying on a pine-covered slope had the feel of an austere Adirondack resort.

By their bearing and intellect, the West Point alumni on the PMA's staff served as military exemplars for the class. Not only had these Filipino officers attended the U.S. Military Academy, which colonialism imbued with a special aura, but several had scored the top marks needed to become a "starman." Throughout the years Class '40 spent at the PMA, the superintendent was Colonel Pastor Martelino (USMA '20), who exercised an aloof but unassailable authority. Their first commandant was Captain Rufo Romero, a starman in West Point's class of 1931, known as a "pugnacious but brilliant man who did not hesitate to 'bust' cadet officers . . . for breaches of discipline." His successor, Captain Jaime Velasquez, another starman in that same class, impressed them with his agile intellect and commanding presence.[83]

Despite their admiration for West Point, some aspects of the institutional imitation seemed inappropriate, even silly. The copy was simply too exact. In their rush to create a complete institution in just three months, the PMA's staff had transported every detail of West Point's particular culture unaltered to the mountains of Baguio—drill, dancing, uniforms, curriculum, songs, and slang. Of the 147 English-language words that Class '40 listed in their yearbook as their distinctive cadet slang, or "slingo," forty-one were borrowed verbatim from West Point. At weekly parades the band played West Point marches, and the PMA's graduation had a romantic "ring hop" just like West Point's.[84]

Feeling demeaned by being mere replicas, Class '40 set about Filipinizing the details to create their own traditions. For a start, they drew upon the rituals of the old Constabulary Academy and retained much of its argot—terms such as "mistah" and "cavalier." Trained as an architect, classmate Washington Sagun was enlisted to design items small and large that comprised, in sum, a distinctive visual iconography for the academy. The class created rules for the new honor committee, composed a new school song, wrote the cadet prayer, and adapted West Point's "cadet slang" to make it "Peemay slingo." A sympathetic tactical officer, Lieutenant T. G. Fajardo (USMA '34), contributed to this project of cultural construction with a throbbing, sentimental waltz titled "My

Kaydet Girl." At the academy's first ring hop during graduation week in 1937, a Class '40 quartet sang the song for the first time while the "dates of graduating cadets slipped Class Rings on the cadets' ring fingers."[85]

Not only did they create a distinct institutional culture, but the class also acted as extras in Filipino films and thus projected their image of masculinity into an emerging national consciousness. Only a year after the PMA opened, a Manila film crew shot a two-reel documentary, titled *The West Point of the Philippines*, which, the cadet yearbook reported, was "being featured at the Ideal Theatre" and was "taking Manila by storm." Inspired by *Flirtation Walk*, the 1934 Hollywood musical starring Dick Powell, the Filipino film *Madaling Araw* (The Dawn) drew a record crowd at its 1938 Manila premiere, filling seven screenings daily for two weeks. A year later, *Punit na Bandila* (Torn Flag) opened with the entire cadet corps moving in mass formation through precision drills that spelled out its title letter by letter. When these films were shown at a Baguio cinema, the corps marched into town, two abreast and a hundred deep. "The cadets were projected as somebody," recalled classmate Manuel Acosta, "something upright and [with] all the desirable qualities of manhood."[86]

Years later, as they looked back from retirement, the class began to reflect critically on their role as cultural mediators. In their PMA applications, all had to vow to "recognize and accept the supreme authority of the United States of America in the Philippines." In effect, to become defenders of the Philippines, these would-be officers had to pass an exam in English, with questions on the Mayflower Compact or Dred Scott, and swear loyalty to America. "Like most of the young men of their time," the class wrote of itself, "they were products of an educational system established by the American victors . . . in the Filipino-American War of 1898–1902." At the academy, they read English literature from Chaucer to Wordsworth. Accordingly, in the "class favorites" section of their original yearbook, they picked American icons: "*Actress*—Bette Davis, *Actor*—Tyrone Power, *Cigarette*—Camel."

But in their *Golden Book* nearly fifty years later, these aging alumni regretted this alienation from popular Filipino culture that made them think Filipino stars "the likes of . . . Rosa del Rosario, Carmen Rosales and others, were for the *bakya* [wooden clog] crowd." Instead of the light cotton uniforms of the Filipino revolutionaries of 1896, cadets wore, the class recalled, American woolens "to make the parody of being the 'West Point of the Philippines' complete." While the class admitted to an unwitting alienation from Filipino culture, they denied being "American boys" and insisted that they were "fiercely 'nationalistic' whenever US-Philippine relations were under discussion."[87]

Though they helped create a distinct institutional ethos that could stand the test of time, they would, after the war, find it far more difficult to defend its West Point ideals, which gained, after independence, the taint of colonialism. In classes on law and politics, the academy's instructors had drummed in the lesson that the military was the instrument of the civil state, not its master. "We were very well impressed with the limits of the military, that the civilian is supreme as far as democracy is concerned," recalled classmate Deogracias Caballero. "The military takes orders from the civilians."[88] Though this language was, under the circumstances, inherently colonial, the ideal of apolitical service was universal to all standing armies. If the Philippine military were to restrain its reach for power, then professionalism, as taught at the PMA, had to prevail. It would become the task of Class '40, and its successors, to adapt values inherited from a colonial military to make them viable for the army of a new nation.

"SAME BANANA"

Although freed from the discipline of upperclassmen in their third year, Class '40 soon found themselves in a bitter confrontation with the academic staff. Ultimately, this conflict would prove defining for Class '40, affirming their collective identity as men of integrity.

To create a four-year college curriculum, the PMA had hired Professor Alejandro Melchor from the University of the Philippines to recruit instructors from civilian universities, including many from his alma mater. The academy integrated these academics into the military hierarchy by awarding them reserve ranks as third lieutenants. "So we had the ludicrous situation," recalled classmate Caballero, "that one day an instructor was teaching the cadet but upon his graduation he [the cadet] automatically outranked his former instructor. This situation rankled among the reserve officers in the academic department."[89] By their second year, Class '40, the first large batch of future regulars, began to sense hostility from these academic reservists. Moreover, this suspicion was reinforced by the winnowing of classmates after each semester's exams—fourteen flunking out in the first and eight in the second.[90]

That was not the end of it. Three months later during the final exams, nine more classmates flunked major subjects and were dismissed, reducing the class from its original 120 to only 80. The class could hardly believe it when Noli Reyes, a good scholar, failed surveying. The commandant suggested that he retake the exam and agree to being turned back to Class '41 if he passed. Reyes refused and resigned, reducing the class to just seventy-nine.[91]

A year later, on the eve of exam week in March 1939, the class faced the threat of more dismissals. After losing a third of their original members, many felt that "some examinations were intended to make cadets fail." There was, moreover, a volatile personal element in this conflict. As unmarried males, both cadets and instructors competed for the attentions of "the cream of Manila's socialites" at academy dances. In the words of one classmate, the "rivalry started from these girls . . . and that rivalry went to the classrooms."[92]

Convinced that their instructors were guilty of academic "sadism," the class began to consider some sort of protest, even though they could be charged with mutiny and dismissed or even sent to jail. Commandant Cepeda was also concerned about the loss of so many promising cadets and advised the class, secretly, on appropriate tactics. At a meeting in his quarters with the cadet officers of Company B, he solved the problem of signaling a strike during the exams: if the questions were obviously biased, he said, have their officer-of-the-day give the class bells a special ring.[93]

At 9:15 P.M. after the first day of exams, tattoo sounded the release from quarters and cadets rushed to the bulletin board. "We were quite surprised to find that only about 80 percent of the first sections got blue [passing] marks," recalled classmate Caballero, "and the rest, including all the second and third sections, got red marks. . . . At the rate the exams were going, we estimated that about one-third would not make it." That night, meeting in their dorms, the class decided to boycott and the underclassmen agreed to go along.[94]

When cadets marched into the exams the next day at 8:00 A.M., the instructors distributed test papers and the class took their usual places. One classmate who was already a certified engineer, Licurgo Estrada, reviewed the questions and found them unreasonable. Then the bell sounded a short buzz, as if a mistake—the signal![95] The cadets mimed proper examination behavior for a few minutes before handing in blank papers and marching out to meetings.

When news of the boycott reached Superintendent Pastor Martelino, he ordered the class to assemble. "You can't do this," he told them. "You've got to take the exams and give it your best try. Because if you don't, this is mutiny and you are all subject to the Articles of War." When a few classmates voiced their anger at the professors, he replied that he was "really disappointed with our class." Then Colonel Martelino, a West Point graduate, broke down and wept. They were moved.[96]

Throughout the rest of the day, there were informal meetings among classmates, who agonized over whether they should continue. When one cadet asked what they should do at the next exam, another offered the "rather incongruous answer . . . 'Same banana!'" With this unlikely battle cry ringing down the corridors, they decided to take the risk. The next day, mindful of the superintendent's warning, the cadets sat through the entire exam period filling up their pages with erroneous answers. When the professors realized that their students had submitted a stack of nonsense, the academy appealed to army headquarters.[97]

Advised of the boycott, Chief of Staff Basilio Valdes motored to Baguio. In a meeting with the staff, the general reportedly ordered new, unbiased examinations. But he offered no compromises as he stood before the cadet corps later that morning: "Gentlemen, I came here on a mission. Are you going back to your classes?" The corps rose "as one solid wave" and shouted with one voice, "Yes!"[98] The strike was over.

The underclassmen resisted the return to exams. "Oh, it's good for you guys because you're more or less assured of 100 percent passing," they told the leaders of Class '40. "But what about us?" Invoking tradition, the class insisted that everyone take the exams. "You mean to say that this . . . institution . . . will be a diploma mill?" they told the underclassmen. "You have to take your casualties." In the words of class historian Jose Mendoza, "we actually forced them to take the exams again."[99]

Despite the anticlimax of their meeting with General Valdes, the class soon realized that they had won. The instructors issued new exams, sparing them another round of academic casualties. The faculty's bid for full recognition as regular officers was quashed when Commandant Cepeda ordered that addressing faculty by military rank "is purely an act of individual courtesy." But it was a victory with a price. Headquarters could not ignore this breach of discipline, and punished those responsible, stripping Cepeda of his post and "busting" the cadet officers of Company B, Ramon Alcaraz and Edmundo Navarro, from lieutenant to private. Still, the class turned even these demotions into a ritual of solidarity. "Before the evening parade," recalled Navarro, "we assembled the plebes of the First Battalion, who sang the Philippine national anthem while a classmate, using a razor blade, cut the stitches of our chevrons from our sleeves. It was very unmilitary to cry but we did when we lost our hard-earned chevrons."[100]

More than any other single incident, the "same banana" boycott inspired a sense of class unity and became a defining moment in their identity. As their *Golden Book* put it, "the cadets realized that, provided the cause was just and they did not engage in any violent or destructive acts, they had and could exercise power as long as THEY WERE ALL UNITED!"[101]

HONOR

Beyond this testing and bonding, Class '40 also went through an intense indoctrination into the honor code. Collective defense of the academy's honor was a source of class pride, just as individual compliance was necessary for membership in the corps. In their 1938 yearbook, the cadets themselves called the honor system "that mighty and enduring code of all time" and claimed that "in no other institution in this country has it been so rigidly enforced as in this institution where gentlemen are made."[102]

Honor was entwined within every aspect of academy life, but its precise definition was implicit and imprecise. There were two parallel disciplinary codes at the PMA: the official, written regulations enforced to the letter by the tactical officers; and an "honor system" whose spirit was guarded by the cadets themselves. For routine infractions of regulations, the official Battalion Board, chaired by a tactical officer, could recommend penalties such as confinement to quarters or reduction in rank. By contrast, the cadet honor committee dealt only with major moral derelictions such as lying and could impose just a few punishments—forced resignation or, if the erring cadet refused, ostracism.[103]

The honor code fused an internalized self-discipline with severe sanctions to become a potent instrument for character formation. In their first two years, Class '40 practiced the code's "intense . . . inhibition from lying, stealing, and cheating," learning both to admit fault and to report themselves. They observed that those called before the honor committee were usually sent out of the academy, even for minor infractions. "We were scared of the honor committee," said classmate Conrado Nano, "because we know their decision is final. No appeal." In their last two years, Class '40 staffed the honor committee and were strict in enforcing its sanctions. When several cadets were found guilty but refused to resign, the class forced them out by imposing "the silence"—a sanction so strong that "nobody will talk to you, even your roommate."[104]

Reflecting the intensity of this indoctrination, the class devoted a full page to the honor committee in their yearbook. "Men of character must necessarily be honorable men," they wrote. "It is the great conscience of the Corps which imposes on all Cadets the obligation to do right." Then, in a statement simultaneously clear and opaque, they added that, "The system is not concerned with the means or mode by which one can live an unblemished life; it is rather interested with only one thing—that each and every Cadet comes up to the standard of honor set and must maintain it thus preserving the self-individuality of the Cadet Corps."[105]

These words are remarkable for both their detail and their ultimate ambiguity. In the end, they fail to answer the key question: what is honor? As articulated by the class, honor was an absolute but its maintenance was a shared responsibility. In essence, the class defined a breach of honor as a rupture in the solidarity of the group rather than an individual's defiance of an immutable code. Thus, the yearbook's maxim that "every Cadet will have the same notion about honor" and its convoluted language about "preserving the self-individuality of the Cadet Corps." The individual behaved to protect the "self-individuality" of all cadets, thus submerging individual will to group identity. The individual merged into corps, and the corps itself became an individual.

Writing in their reunion *Golden Book* fifty years later, a former member of the class honor committee, Deogracias Caballero, explained that a new lieutenant, so trained in the academy's tradition, should know that his oath to uphold the Philippine constitution bars him from any "personal loyalty to the Chief of Staff . . . or even to the President no matter how advantageous it might be."[106]

Cadets did not dwell on ethical issues, but they often sang the school song, "PMA, Oh! Hail to Thee" written by two classmates. After several months of singing West Point's "Alma Mater" as their own, two Class '40 roommates, Quirico Evangelista and Reynaldo Mendoza, tired of this borrowed American song and felt they had to do something. Drawing upon his high school English, Mendoza drafted lyrics that "hailed the PMA as an institution worth worshipping" and Evangelista wrote the musical score, producing a composition that won a contest for the new school song. Over time, it became the academy's official song, one that "has never failed to inspire the Corps whenever and wherever it is sung up to this day."[107]

The lyrics are didactic—an oath, not an anthem, that requires the individual to affirm the academy's ethics and thereby merge with "those solid ranks of gray."[108] The song seems to define the "honor you instill" as a masculine attribute that would make them all "men of integrity, courage and loyalty." Stanza by stanza, the song proclaims a lifelong defense of personal honor as the supreme duty of each cadet.

> At every end of day,
> We hope and fervent pray,
> The honor you instill
> Doth guide our will;
> May thy sons ever be
> Men of INTEGRITY, COURAGE and LOYALTY.
> PMA, Oh! Hail to Thee.

When bells for us are rung
And our last "Taps" is sung,
Let generations see
Our country free;
Oh! lead to righteous way
Those solid ranks of Gray
Thine virtues to display Academy.
Oh! Hail to thee PMA.

Embedded within the honor code was a view of the academy as autonomous—with those "solid ranks of gray" insulated from society and its politics. At the PMA, cadets could "hope and fervent pray" before a vision of honor as a noble abstraction, an absolute, demanding unqualified compliance. By the time they retired a quarter-century later, constant compromises would reduce the code to a few minimal maxims—no corruption, no patronage, no coups.

The academy experience identified the group as an elite, a term that was both inclusive and exclusive, positive and negative. In this bonding through shared rituals and principles, the classmates also came to identify themselves in terms of what they were not—reservists, politicians, or women. By their third year at the PMA, the class, in its own narrative, had strengthened its solidarity in a moral mutiny against the academic reservists. As career officers, the group would defend their status as regular officers against these same reservists. In their rise to command, they extended this same disdain to politicians, whose systematic graft and self-aggrandizement were antithetical to their own values of code and corps. But above all, they were men—men of courage, men of integrity. Even so, their view of women was still more nuanced and less distancing than their attitude toward reservists or politicians—making women both distinctive and defining.

MOTHERS, FEMMES, O.A.O.

Unlike officers elsewhere who often defined their manhood in opposition to the feminine, prewar PMA cadets constructed a military masculinity that incorporated women as affirmation of identity.[109] Their stereotypical self-image of athleticism, camaraderie, and integrity somehow coexisted with a more supple sense of gender roles.

Indeed, Class '40 showed a playful transvestitism by dressing women in their cadet caps for photographs and donning female dress for plays and musicales. At the academy's 1937 "Amateur Frolic," star athlete Vic Osias and "his Hawaiians" danced in combat boots and grass skirts before an approving audience, while the cadet who played Juliet in a Shakespeare recitation could, in the yearbook's words, "have passed for a belledame from that balcony!" In this spirit, one member of Class '40 photographed a "cadet's girl" in full-dress uniform for the cover of the campus literary review.[110] At one level, juxtaposing feminine smiles with the cadet's dress uniform tried to show just how unsuited women were for military service. At another, incorporating women through rituals and military regalia softened the institution's ingrained gender polarities.

But women still remained the ultimate opposite, an alluring, defining antithesis. Confined in the academy's all-male environs for their first year and limited in their social contacts thereafter, the cadets remained, for four full years, obsessed with women. Then, in a series of elaborate rituals culminating in graduation, the class crossed over the boundaries that had separated them from women and civilian society. Symbolic of their legitimating role, women, usually the cadet's mother or sweetheart, officiated in the annual PMA graduation rites by slipping the class ring on each graduate's finger at the "ring hop."[111] In their exit from the PMA, the class thus prepared for reentry into society with rituals that tried to transcend the extremes of their gendered military identity.

In their last months, the class compiled a yearbook, *The Sword of 1940*, that made a strong, albeit unconscious, statement about gender. In short, stylized biographies, the class assessed its members on four criteria—personality, academics, athletics, and women. Among the seventy-nine graduates, fifty biographies commented in some way on the cadets' relations with "femmes," even though they were almost absent from academy life. The yearbook biography of Manuel Acosta, for example, skips over his daily "diligence in studies . . . skill in boxing and fencing" to describe him as a cadet who "began a woman-hater and ended a passionate swain." Similarly, Victor Osias—the class's president, best athlete, and outstanding cadet—is given the ultimate accolade: "A woman (who, by the way?) couldn't ask for more."[112]

The paradoxical role of women as a defining other, nearly invisible yet somehow central, is indicated in the overall structure of the yearbook itself. With an almost military symmetry, a mother stands sentry behind the title page, an all-male chronicle of cadet life fills the volume's center, and future wives patrol the closing pages.

On page two, the yearbook is dedicated, in bold gothic print, "TO THE MOTHERS OF THE CORPS, Whose . . . solicitous care has made possible the consummation of cherished dreams." Above these words, a mature woman, dressed in a traditional *terno* dress with full butterfly sleeves, clutches a sheathed saber to her bosom. At her feet stands a silhouetted officer—drawn curiously small, about the size of a newborn infant—who reaches toward her torso. Is she about to hand her child this sword, arming him at birth? Is the class somehow saying that their mothers gave them life so they could prove their manhood by facing death?

After 133 all-male institutional pages, women dominate the final 37 pages of "features." An "inspirational" page with a large silhouette of a woman's head, filled with snapshots of women and cadets socializing, frames a dedication to "the O.A.O. [One and Only] whose affectionate encouragements, unfaltering faith, endearing devotion and infinite love has provided the brighter aspects of an otherwise rigid, duty-bound, regimented Cadet life."

Two pages on is a whimsical tribute to women in the musical score for "My Kaydet Girl," the romantic ballad composed by the class's tactical officer, Lieutenant Tirso Fajardo (USMA '34). As the song opens ("on her lips was a smile") and its melody rises ("in her eyes are the stars above"), the head of each note is filled with a cameo photo of a Manila beauty or academy "drag," some wearing cadets' dress caps cocked at jaunty, unmilitary angles ("For I love you my Kaydet Girl"). Women thus stand guard at the book's ends—mothers dispatching cadets into the academy's all-male isolation, wives-to-be extracting them for reentry into civil society.

The ultimate exercise in gender inversion came in their last months. Following the West Point tradition, Class '40 marked the countdown to their graduation by presenting a "100th Nite Show" titled "Love Pirates of Hawaii," with classmates forming a chorus of wahines dressed in muumuus that, in the words of the *Golden Book*, "literally stole the show."[113] At the apex of their cadet careers, when the class members were about to take their oaths as officers and step before the nation as models of a new masculinity, they donned the guise of women to sing and dance.

MARCH WEEK

Their last days at the academy gave Class '40 some final lessons in solidarity and socialization. In the seven days of graduation rituals called "March Week," the corps paused to celebrate their collective advance up the chain of command—plebes to yearlings, first classmen to third lieutenants. Everyone moved up and onward, but the corps remained, symbolic of their solidarity and shared values.

The celebrations began on Sunday, 10 March 1940, with a baccalaureate mass for the graduating class beneath the pines at Baguio's Forbes Field. Then followed a week of daily parades and nightly dancing. Monday morning brought the close-order drill competition, with the four cadet companies competing for the Araneta Cup; while afternoon saw the Athletic Review, an open-air ceremony awarding letters and numerals to outstanding athletes. That night the academy hosted the "A-Monogram and Numeral Dance," where a cadet could present the golden "A-pin" to his one-and-only. The next day, Tuesday, featured the academy's first equitation demonstration, with sixteen first classmen going through precision drills on horseback with lances.[114]

On Wednesday at 4:50 P.M., President Quezon, unable to attend graduation later that week, entered the campus to the sound of a twenty-one gun salute and reviewed the corps, massed before him on the parade ground. Flanked by his generals, Valdes and Lim, Quezon stood bolt upright in a topcoat, lines of concern marking his face as he spoke extemporaneously, telling the cadets of their historic mission. Half a world away, Finland's soldiers were retreating before Soviet invasion and Quezon seemed concerned about the fate of his own army. The Class of 1940, he said, "constitutes the first fruit of the government to provide the army with officers fully trained in a military academy." Since they would set the army's standards, theirs was a "task of transcendental importance." Above all, he concluded, "soldiers should never mix in politics" and should remain "loyal to the Philippine government and not to any individual person." At a reception for the graduates, Quezon personally congratulated the class.

That week, the front pages of the Manila press featured a photograph of President Quezon shaking hands with Cadet Licurgo Estrada, winner of the presidential saber for academic excellence, bending at the waist, arm fully extended—a sinuous arc of military perfection from his shined shoes to the chevrons on his sleeve. Among the underclassmen Quezon honored that day for their academic achievement were two—Manuel Yan of Class '41 and Rafael Ileto of '43—who would reach the apex of the military hierarchy.[115]

The emotional peak of March Week came on Thursday afternoon with the full-dress Graduation Parade. While "Auld Lang Syne" played in march tempo, Class '40 executed a "front and center" and stepped out of the line to watch as the rest of the corps passed in review. When the parade finished, the companies held formation for a moment until the command rang out, "front ranks, about face." In this ritual of recognition, the upperclassmen extended their hands to welcome the plebes of Class '43, some weeping, as full members of the corps.[116]

That same day, Class '40 received their service assignments and were outraged at obvious signs of political interference. Army headquarters had assigned the sole Cavalry vacancy to Antonio Perez, an unexceptional horseman. While other services lacked cachet, the Cavalry had become a reserve unit for polo-playing members of the Manila elite, and Cadet Perez belonged to an old Manila family with social connections.[117] In their months of mandatory riding lessons, the class had seen that Florencio Causin was a superb rider and felt their principles violated by this appointment.

In conversations around the dance floor at their Graduation Ball that night, classmates were vocal that the Perez appointment smacked of nepotism. The army's deputy chief of staff, General Lim, a legendary disciplinarian, attended the dance and evidently overheard these remarks. The next morning at the breakfast formation, Cadet Captain Ramon Olbes saluted General Lim and stood aside. The general called out: "Mr. Causin, one step forward! Mr. Perez, one step forward! . . . Mr. Causin, you are in the Cavalry. Mr. Perez, you are in the infantry."[118] Once more, solidarity brought moral victory.

At 10:00 A.M., the class assembled for graduation on a slope overlooking the parade ground. Vice President Osmeña, acting for the president, delivered the commencement address to the class, arrayed beneath the pines in curved rows. Calling the academy "the foundation stone of the entire military establishment," he reminded them of their duties as officers—with words whose full implications no doubt eluded everyone present. "The eyes of the people are upon you today," Osmeña said. "Yours is a responsibility of the greatest magnitude. Before you beckon duties as noble in peace as they will be glorious in war. You cannot fail."[119]

The vice president then handed each graduate his diploma as they were called by academic rank from one to seventy-nine. When the cadet with the lowest average, the "class goat" Ramon Gelvezon, stepped forward, the crowd broke into cheers and applause. Then, Chief of Staff Valdes led the class in their oaths as third lieutenants. Finally, in their last act of perfect union, the Class of 1940 tossed their white caps into the air, with a great, echoing shout. A cloud of white caps formed for a second against the green of Baguio's pines, before falling downward into the hands of each individual owner.[120]

As they packed their bags that afternoon for a journey beyond the academy, Class '40 carried core values, both complementary and contradictory, that would inform their relations with classmates over the next quarter-century—solidarity, hierarchy, and competition. Through this intense academy experience, the class had forged a lasting group identity and lifelong friendships. Five or even six decades after graduation, their reunions, held every three months, would draw most surviving classmates and fill a banquet hall. When nine of these aging alumni, then in their late seventies, were asked to identify a faded class photograph taken sixty years before, all could rattle off the names, unerringly, for forty-five blurred photo-faces smaller than a fingernail—as if a visual image of each classmate had been imprinted in memory.[121] So bonded, the class would try to advance its members while using this group identity to maintain a high standard of integrity.

Within this overarching unity, the class had also developed a sense of hierarchy—strong over weak, smart over slow, aggressive over passive. In the powerful symbolism of their graduation ceremony, Class '40 not only marched in solid ranks of gray but they were arrayed in a precise seniority from one to seventy-nine. While the group maintained its solidarity before outsiders, there was an incessant, internal competition for distinction—classes and athletics at the academy, medals and promotions in the service.

Despite conflicts and contradictions, their military socialization, the sum of all these experiences at the PMA, transformed them from ordinary young males into instruments of state power. Principles of integrity, obedience, and subordination to civil authority—taught in the context of a daily discipline that inculcated class solidarity—laid the emotional foundations for a lasting socialization.

They also left the academy with a strong sense of patriotic mission, ready to assume their role as defenders of the nation. As class historian Jose Mendoza put it: "We realized that . . . if there was going to be anybody . . . to invade our country, it was our duty to repel that invader." As they moved on to training camps across the archipelago, the class sensed "a very visible preparation for war" and soon felt "a high respect" from their people as leaders of this national mobilization.[122]

When war did come twenty months later, the sufferings of combat and the degradation of captivity would affirm the principles they had learned at the academy. These formative experiences as junior officers were the final phase in a military socialization that would be severely tested by the pressures of service in the postwar Republic.

Chapter 3 Baptism by Fire

In their first years on active duty, the Class of 1940 affirmed the academy's lessons of loyalty and honor. Whether training with elite units or drilling the infantry's raw recruits, most classmates found their duties fulfilling. After the discipline and self-denial of the academy, they would look back on these years as lowly third lieutenants with fondness. With war on the horizon and an army to build, they were busy with responsibilities that gave them the respect of conscripts and the community. Even all the misery that followed, the brutality of combat and the degradation of captivity, would only serve to strengthen their comradeship.

After graduation, the class scattered to service assignments—twenty-nine to the infantry, seventeen to the Air Corps, ten to the Off-Shore Patrol, seven to the Coast Artillery, seven to the Corps of Engineers, six to the Field Artillery, two to the Signal Corps, and one to the Cavalry.[1] This diaspora could have fragmented the class, but service life allowed frequent contacts and entry into groups with sympathetic values. After nearly half their thirty air trainees washed out into the infantry and artillery, most of Class '40 wound up in dismal provincial camps drilling peasant conscripts with wooden rifles. Often bored, they sought out fellow officers and kept in touch with classmates.

Those assigned to elite services found an ethos sympathetic to the PMA experience. The Philippine Army Air Corps (PAAC) provided a second rite of passage for the seventeen classmates who won their wings. Under the pressure of forming squadrons that could maneuver in unison, the pilots of Class '40 joined an aviators' fraternity that, to some degree, superseded academy ties. Whether in new groups or old, the PMA's lessons of camaraderie proved apt training for an officer's life.

The Japanese invasion of December 1941 shattered the Philippine Army, forcing classmates to rely on academy ties to survive. Not only did the class suffer a high incidence of injury and illness, but nine of their seventy-nine graduates would die in World War II, leaving survivors with a mingled emotional legacy.[2] Adding to the emotional burden, in June 1941, only fifteen months after their oath to the Commonwealth, the class, along with the rest of the Philippine Army, had to swear loyalty to the United States when they were integrated into the U.S. Army—an incidental act at the time but one that clashed with their heightened nationalism of later years.[3] Combat itself offered a varied experience. Those in the infantry were demoralized almost from the outset by the ill-trained troops and short supplies. But others in elite units, such as the Air Corps and Off-Shore Patrol, experienced heady moments of exhilaration in the first weeks of war.

As the Battle of Bataan ground onward to a painful defeat, most surrendered with their units in April 1942, suffered the Death March, and were imprisoned at the Capas concentration camp. At least fifty-seven of the seventy-nine classmates fought on Bataan, and most of these experienced both the horrors of defeat and the months of demoralizing confinement that followed. To survive these ordeals, classmates relied on each other.[4] About ten of them later joined the Japanese-sponsored Constabulary after their release from captivity and a few others retreated into private life for the duration. But many, about thirty of their number, eventually joined anti-Japanese guerrilla groups after recuperation from the ordeals of combat and confinement. Though scattered across the archipelago, classmates, in groups of two or three, clustered in these disparate underground units and assumed commands beyond their junior ranks.

All of these experiences added up to a second, critical phase in their military socialization. Viewed objectively, garrison duty and warfare translated the academy's abstractions into the reality of military experience, testing and affirming its moral maxims. The degradation of defeat and incarceration seemed, paradoxically, to have ennobled the class, deepening their commitment to shared values of honor and solidarity. By the time they emerged from the war, these young lieutenants had finished their professional formation and internalized principles that would influence the rest of their careers.

OFF-SHORE PATROL

The academy and its first full class had the most immediate impact upon the Philippine Army's newest branch, the Off-Shore Patrol (OSP). While Class '40's third lieutenants disappeared into the swelling ranks of the infantry, its alumni were a distinct, even dominant element in this infant navy. Under MacArthur's defense plan, the Philippines was to have fifty torpedo boats, or "Q-boats," by 1946, but only three were actually commissioned before the war. In their final year at the academy, the OSP chief, Captain Jose Andrada (U.S. Naval Academy '30), picked the cream of the class, including First Captain Ramon Olbes, chess champion Heracleo Alano, and boxing champion Ramon Alcaraz.

After graduation in March, ten members of Class '40 reported to the OSP for training in Manila's Port Area, joining earlier PMA graduates and five Filipino alumni from Annapolis. While the group was completing a year of schooling in navigation and gunnery, this new service took delivery of its first Q-boats. The neophyte officers were soon plunged into sea trials aboard these demanding craft. While the Q-111 and Q-112 were imported fully assembled from England, their commander, Captain Andrada, supervised a team of Filipino and foreign specialists that built the Q-113 at Manila's Engineer Island dockyard. The First Lady christened it *Agusan* in March 1941, and it proved so maneuverable in sea trials that the army soon laid keels for eight more.[5]

Designed by Thornycroft, Ltd., of England and powered by three-thousand horsepower engines, the fifty-five-foot Q-112 could leap from a dead stop to sixty-five miles per hour in just two minutes and then slash ninety degree turns across the water at full speed, while firing three heavy machine guns and launching torpedoes or depth charges. As the boats slammed across the chop of Manila Bay, the crew, wearing ear plugs to muffle the engines' roar, had to stand on deck with knees constantly bent to absorb the jarring, erratic reverberations as the rapier-thin hulls skimmed from wave to wave. "Riding the 'battle bridge' of a wide open Q-boat was a thrilling experience," wrote a *Life* magazine correspondent who rode the Q-112 in October 1941. "The broad, buoyant bow rose, the stern buried itself, and the propellers churned water into whipped cream. Spray became liquid bullets."[6]

After six months of training, the OSP selection board selected crews of two officers and five men for this three-ship navy, honoring Class '40 by picking its alumni for service on each of the Q-boats—Ramon Alcaraz as skipper of *Abra*, and Liberato Picar and Heracleo Alano as executive officers of the other two. As war approached, MacArthur canceled Q-boat construction, relieving Captain Andrada when he protested and leaving a single squadron of just three ships.[7]

While war brought humiliation to classmates in the infantry and artillery, those in the Off-Shore Patrol found glory in its first weeks. By early January, with the USAFFE forces bottled up on the Bataan peninsula, Japanese forces attained overwhelming tactical superiority, making conventional navy operations impossible. With their speed, however, the torpedo boats could still maneuver effectively, allowing them some exciting missions. After declaring Manila an "open city" on New Year's Day 1942, MacArthur called Lieutenant Ramon Alcaraz into his headquarters and ordered him to sink the ships anchored along the Manila waterfront. By 8:00 P.M. that evening, Alcaraz and his crew, slipping quietly through the anchorage on the cruising engine, scuttled and burned fifteen vessels, equivalent to some fifty thousand tons of shipping. Other high-risk missions followed over the next two months—ferrying MacArthur from Corregidor to Bataan, collecting intelligence, and escorting supply craft through enemy waters.[8]

At dawn on January 17, the Q-111 *Luzon* and Q-112 *Abra* were returning from a routine patrol when nine Japanese aircraft attacked. A formation of Japanese Zero-type dive-bombers spotted the boats streaking across Manila Bay about a mile apart and formed into groups of three to strafe. As the Zeros came in only one hundred feet above the water with guns firing, Lieutenant Alcaraz of the Q-112 opened the throttle full to sixty miles per hour and ordered his two machine gunners to fire at will. Each time the Zeros dove to sight their guns dead on the rear of his ship, Alcaraz, keeping his boxer's eye on the approaching aircraft and "scared but not scared stiff," pulled the wheel hard port or starboard, making a sudden ninety degree turn that veered the ship away from the incoming fire. Simultaneously, the Q-111 reacted by prearranged signal and turned hard ninety degrees in the opposite direction, catapulting the two ships away from each other at a parting speed of 120 miles per hour. Then, as the aircraft passed overhead, Alcaraz again turned sharply, looping back to give his men a clear shot at the Zeros, which were now, in his words, "easy targets to the Q-boat gunners."[9]

For about thirty minutes, there was a running battle between nine diving Zeros and two zigzagging Q-boats. With its sharp turns and sudden acceleration, the Q-112 outmaneuvered the aircraft, hitting three Zeros repeatedly. The lead aircraft started wobbling and then two more "started whirling about, trailing black smoke and . . . heading toward Formosa." Ten days later, MacArthur's headquarters issued General Order No. 16, awarding Silver Stars to the crews, including two members of Class '40—Alcaraz of the Q-112 and the executive officer of Q-111, Alano. The general then called Alcaraz to his office and promoted him to first lieutenant on the spot.[10]

Through skillful seamanship, the entire Q-boat squadron remained operational until Bataan fell to the Japanese on 8 April 1942. As enemy artillery began to concentrate on Corregidor Island, the three Q-boats made a run for the open sea but were blocked by Japanese patrols. Under attack by enemy ships and aircraft, the Q-111 *Luzon* headed for the Batangas coast, where its crew scuttled successfully and fled. Rather than surrender, Alcaraz headed back across Manila Bay, entered the Pampanga River delta, and sank his ship in the familiar waters of his home province, Bulacan. Japanese patrols rounded up these crews as they came ashore and shipped them to Capas concentration camp, sparing them the Death March but subjecting them to months of confinement. By the time their ships went to the bottom, over 70 percent of the crewmen had won individual decorations for "heroism and gallantry," a remarkable record. After their release in August 1942, both Alcaraz and Alano eventually joined guerrilla units and fought with them until the war's end.

As they looked back from retirement, neither Alcaraz nor Alano expressed the bitterness that many of their comrades felt about American bungling during the Battle of Bataan. Alano's widow Dolly dictated his biography for the *Golden Book* and spoke of the "gallant crew" of Q-111. Alcaraz has reminisced in his memoirs of the glory of that battle. "I . . . entrusted my fate to the good Lord," he wrote. "After the engagement, I was so humbled by the experience. I realized the value of prayers." In his 1986 memoirs, Alcaraz devoted twenty-one pages to a detailed account of Q-boat combat and less than single page to his confinement.[11]

Not only were these sailors spared the horror that the infantry suffered in the hills of Bataan; they were privileged to command the only USAFFE weaponry that was technically superior to Japanese armaments. Through four years in the classroom and a decade in the ring, Alcaraz had been preparing for this moment. The Q-boats were weapons made for modern gladiators; in Alcaraz's words, they were "the most glamorous naval units then afloat."[12] Two societies, Japan and

America, had trained their best warriors and armed them with their most advanced weaponry. Then, on that day, there was a marriage of man and machine that gave Alcaraz victory in a duel against skilled Japanese aviators at the peak of their powers—victors at Pearl Harbor and masters of the Pacific. Sweeping across Manila Bay at sixty knots, three-thousand horsepower engines roaring and .50 caliber machine guns blazing, Alcaraz and Alano felt mastery, a sense of heroism, that would sustain them through years of imprisonment and guerrilla warfare. It was a feeling that the men of the infantry and Coast Artillery, who fought on the ground at Bataan, could not share.

INFANTRY

For the many classmates in the infantry, the first months of active duty were dreary. Posted to makeshift camps across the islands, most became training officers for the legions of peasants drafted for six months of marksmanship and close-order drill. Almost all the reserves were badly equipped. The heavy, elongated Enfield rifle, manufactured for American soldiers during World War I, dwarfed the average Filipino. Not only was it inaccurate, but its extractor often broke after a few rounds, jamming the magazine and rendering it useless. Instead of steel helmets and combat boots, soldiers were outfitted with short pants, rubber shoes, and coconut hats. For heavy weapons, most units had a few antiquated .30 caliber machine guns and superseded three-inch mortars. Lacking ammunition, units drilled by "simulated marksmanship" and many recruits finished training without firing a shot.[13]

After washing out of flying school, David Pelayo of Class '40 had the good fortune to land as a training officer in the First Regular Division, one of the better-equipped infantry units. Based at Camp Murphy near Manila, the division was organized in 1936 as the nucleus of the new army. Its First Battalion was drawn from the old constabulary's Seventh Regiment, lending its operations a professionalism lacking in the reserves. By mid-1941, moreover, its ranks were stiffened by an influx of PMA graduates: notably, the young lieutenants of Class '40, with Jose Javier in command of Company A, Hospicio Tuazon and Pedro Bersola in B, David Pelayo in Company C, and their underclassman, Fausto Valencia (PMA '41) in D.[14] Reflecting its professionalism, the division's veteran officers mounted prolonged maneuvers that went beyond the usual company-sized drills.

Still, there were serious problems. There was only a fistful of bullets for each trainee, insufficient for target practice. The division drew its recruits from Central Luzon's five major ethnic groups, making the camp a Babel and English a poorly understood medium of command. Even these elementary drills were soon forgotten once the six-month training was done.[15] Since the division's mission was to serve, above all, as "an instructional cadre," its regular troops were maintained at skeleton strength and skilled staff were constantly transferred to stiffen the newer units. At the outbreak of war, only 30 percent of the division's three thousand troops were regulars and the balance reservists still in basic training.[16]

When war came to the Philippines on 8 December 1941, the First Division was, in the words of its official history, "undermanned and ill-equipped, and without any combat training on the regimental level." Compounding these problems, Philippine Army headquarters apparently overlooked the division in the confusion of the war's first days and failed to issue orders for its mobilization. Headquarters did not, however, forget the PMA. Baguio had been bombed on the first day of war, and army command, fearing further Japanese air raids, ordered an immediate evacuation. On the sixth day of war, the entire Corps of Cadets, led by Superintendent Fidel Segundo, arrived in Manila. When he reported to USAFFE headquarters, Segundo, a distinguished West Point graduate (Class '17), received orders to disband the PMA and take command of the First Division with the rank of general.

On December 15, the academy died and was reborn as a division. General Segundo assembled the PMA's cadets at Santo Tomas University in downtown Manila. "The supreme test of a soldier's worth is at hand," he told them. "You, young officers, must be equal to this grim and glorious sacrifice ahead of you." The general proclaimed all upperclassmen, the classes of 1942 and 1943, graduates and swore them in as third lieutenants. Their underclassmen, the classes of 1944 and 1945, were deemed too young for command in battle and sent home.

Within forty-eight hours, the PMA, through its officers and cadets, was reborn as the First Division. The commandant of cadets, Lieutenant Colonel Santiago Guevara (USMA '23), became the division's chief of staff; the assistant commandant, Captain Alfredo Santos, assumed command of the First Regiment; and former PMA instructors took most of the senior staff jobs. The division's new staff then "hand-picked the choicest PMA graduates" to become company commanders.

General Segundo and his officers scarcely slept a wink for the next five days as they transformed a training unit of 3,200 men into a combat division of 6,350 troops. To fill their ranks, officers swore in the "long, unending stream of . . . students, rig drivers, stevedores, taxi drivers, common laborers and military reservists" that appeared at Camp Murphy to volunteer. With no alternative, the division had to turn these raw recruits into soldiers overnight. Indeed, they were so raw that some were killed during drills when they failed to duck during the Japanese air raids.[17]

In this chaos, Captain Santos faced enormous difficulties in activating the division's First Regiment, since its battalions were still at skeletal strength. To assist in this rapid mobilization, the captain picked two members of Class '40 for key staff posts—Lieutenant Job Mayo as his S-1 (personnel) and Lieutenant Alfredo Filart as S-2 (intelligence). In Company C, only Lieutenant David Pelayo and his twenty-eight regular soldiers, the training cadre, were on duty. In just a few days, however, the company swelled to its full strength of 110 men and the battalion reached its complement of 440—a jumbled mix of reservists, ROTC cadets, and volunteers that had never even drilled together. As the troops joined his unit at random, Pelayo, without any screening for training or service, just "listed them, like that, and organized them to squads, platoons, and then the company. That was all!" After only a few days of ragtag mobilization, division headquarters ordered the First Battalion to draw equipment for three rifle companies and a machine-gun company. Quartermasters issued each soldier an Enfield rifle, a bayonet, a bandoleer of ammunition, a pair of rubber shoes, and a stiff coconut-fiber hat. There were no steel helmets.[18] On December 19, the First Division was sworn in as a unit of the U.S. Army. The next day, General Segundo received orders to attach his First Regiment to General George Parker's South Luzon force and defend the beaches of Luzon's long Pacific coast "at all costs."

Within hours, the 440 troops of the First Battalion had commandeered a fleet of civilian buses, loaded their eight machine guns, and headed down the highway to meet a possible Japanese invasion. Driving at night without headlights to avoid air raids, the convoy lurched down the potholed highway toward the town of Mauban on the Pacific coast. Most of the troops dozed or scraped the packing grease out of their rifle barrels with bamboo sticks. Stopping in towns along the way, officers commandeered trenching shovels, malaria pills, and food rations from local shops.[19]

At daybreak on December 21, the battalion's advance units drove through Mauban, a modest fishing town, and parked on the beach. The officers surveyed the exposed stretch of sand fronting the wide arc of Lamon Bay as the sun rose from the Pacific. It was a terrain that favored neither defender nor invader. The sea was usually rough during the monsoon season and coral reefs lay just offshore. But once the enemy landed, there was no natural barrier to block their easy progress up the gently sloping beach, through town, and down the highway that led to Manila. Feeling that the rough sea made an enemy landing at Mauban unlikely, General Segundo had told the battalion's commander, Lieutenant Godofredo Mendoza, "Oh hell! You won't see the eyes of the Japanese for a month!" Engineers had orders to build barricades along the beach, but the troops reached Mauban without beams or barbed wire.[20] If the unexpected happened and the Japanese did land, the First Division's forces would be stretched thin by this static defense strategy. The general had no reserves to spare for Mauban.[21]

With little equipment, less training, and no reinforcements, the difference between mere defeat and outright disaster on this beach would be the leadership of the battalion's young, PMA-trained lieutenants. Their commander, Lieutenant Mendoza, had graduated from the academy in 1938, where he had been a good athlete, a member of the honor committee, and a competent platoon commander known for being "slow and calculating in all his movements." With few alternatives in this open terrain, the lieutenant arrayed his troops along the three kilometers of beachfront and ordered them to dig in: Company B under Lieutenant Javier, a member of Class '40, near the town pier; Company C under Lieutenant Pelayo, his classmate, to the right, straddling the swampy banks of the Maapon River; and Company A under a reserve officer, Lieutenant Jose Ortega, to the far right across the river. The remaining unit, Company D, commanded by Lieutenant Valencia (PMA '41), dispersed its machine-gun squads among the other companies. "We instructed our soldiers to dig foxholes and a ditch," recalled Pelayo. "That was all!"[22]

Orders from their regimental commander, Captain Alfredo Santos, the tactical officer of Class '40 at the academy, were simple—"drive the enemy from the beaches." For the next two days, the troops dug in and sat in their foxholes, staring out to sea. To meet a landing by a thousand battle-hardened Japanese regulars, veterans of the China War, the First Battalion's five hundred untrained conscripts had no mortars, no artillery support, and only sixty rounds of rifle ammunition each. Many had never fired their rifles. "We told them," said Pelayo, "do not shoot until you see the Japanese coming in."[23]

Shortly after midnight on December 24, a moonless night, the battalion spotted five ships emerging from the darkness. On the town's telegraph, the civilian operator tapped out a message for regimental headquarters: "Eight transports dropping anchor Lamon Bay. Definitely enemy ships." At about 1:30 A.M., a thousand soldiers of Japan's Second Battalion, Twentieth Infantry, Sixteenth Division, clambered over the side of the transports into landing craft. Nearing the beach, some Japanese troops began wading ashore in columns, holding rifles high above their helmets. "By the light from the ocean," recalled Pelayo, "you could see their silhouette." Despite their lack of training, the fire discipline of the First Battalion was "amazingly splendid" and there was perfect silence.[24]

Shrouded in the darkness, the Filipino troops took aim, while Lieutenant Mendoza moved among them, reminding them to hold their fire and urging them, while his own hand trembled, to stay calm. As the Japanese neared the beach, Mendoza fired a shot from his .45 caliber pistol. At that signal, the battalion opened up with rifles and machine guns, raking the Japanese and inflicting considerable casualties. Among those killed was Lieutenant Sueo Oe, Japan's star athlete at the 1936 Berlin Olympics. For nearly an hour, the Filipino forces maintained a withering fire until their ammunition boxes were emptied.[25] The machine guns overheated, and many Enfield rifles jammed, misfired, or blew up. In the face of this unexpected resistance, the Japanese commander recalled his landing craft while his navy escorts maneuvered to fire. A military history by Uldarico Baclagon, Pelayo's PMA classmate, summarizes what followed: "Surprised by the effective fire of the defenders, the Japanese suspended the landing operations and subjected the beach to heavy aerial and naval bombardment. The defending force was practically decimated."[26]

These sparse words cannot convey the chaos and terror on that beach. After a thirty-minute silence, the Japanese navy's heavy guns opened up with a deafening salvo that became a sustained barrage. While the shells rained down on the sands, the Filipino soldiers huddled in their foxholes, some crying in fright. A few died from direct hits. At 7:00 A.M., just after sunrise, Japanese aircraft began bombing the Filipino positions. Then, a second wave of landing craft headed for the beach. A thousand Japanese troops, armed with machine guns and grenade launchers, stormed ashore, attacking Company B's trenches near the pier. In the hand-to-hand combat, the untrained Filipino troops fared badly. Of the company's 110 men, 26 were killed. Their commander, Lieutenant Javier, already suffering from measles, was wounded. "I am sick," Javier told his classmate David Pelayo, whose own company had been spared the brunt of the fighting. "I am going to attach to you my platoon, my remaining platoon." Pelayo assured him that the troops would be cared for and ordered his classmate's evacuation to the field surgeon. During the intense attack, the battalion's only reserve officer, Lieutenant Ortega, had fled in panic and most of Company A had followed him into the hills.[27]

Within an hour, the battle for Mauban was over. Pinned down by the naval gunfire and their lines smashed by the landing, the officers of the First Battalion conferred on the beach, deciding to withdraw beyond the range of the warships, where they could be reinforced and resupplied. As the Filipino troops pulled back, the Japanese forces, with some twenty-five dead and forty-five wounded, secured the beachhead.[28]

With Pelayo's riflemen and Valencia's machine guns covering their retreat, the First Battalion's remaining 150 men withdrew eight kilometers along the highway to high ground. There they were joined by reinforcements from the division's Second Battalion and established new defenses. When Japanese infantry advanced up that zigzag road from Mauban, the Filipino defenders "opened up with a withering automatic and small arms fire," inflicting scores of casualties and forcing their withdrawal. By noon, the Japanese had regrouped and launched a powerful attack, backed by air and artillery support. When one bomb scored a direct hit on the Second Battalion, its commander, Captain Honorato Ramos, was blown to bits with a group of officers and men.

As shells rained down, the regiment's Combat Company arrived with machine guns under Lieutenant Arcadio Mayor, a former classmate who had dropped back to Class '41. Their *Golden Book* recalled him as an outstanding wrestler, "husky, dark, barrel-chested with a loud and reverberating voice." One classmate described him as the "true embodiment of an officer and a gentleman."[29] That afternoon, their lines held. At about 3:00 A.M. on Christmas Day, orders for withdrawal arrived. The surviving troops began boarding their buses.

Lieutenant Pelayo urged his classmate to join them. But Mayor refused, saying, "I am going to wait for the Japanese here." With a hundred men and a machine gun, he would try to block the advance of a thousand Japanese troops backed by field artillery and air power. As the Japanese columns closed in, he ordered his soldiers to escape and covered their retreat—alone atop a vantage point, firing a machine gun at the advancing enemy. In the words of the division's official history, "Lieutenant Arcadio Mayor . . . remained to the death with his machine gun until he had fired his last belt of .30 Cal. ammunition."[30]

After suffering heavy losses of men and material in two days of hard fighting, the First Battalion was no longer an effective combat force. For the next five days, its survivors merged with other units of the USAFFE in a slow, strategic withdrawal toward Manila, blowing bridges and staging delaying actions. Once they reached the city's southern approaches, the troops disengaged from the enemy, skirted the capital, and rushed across Central Luzon to join the main USAFFE forces for a last stand on the Bataan peninsula.[31]

Assessments of the First Battalion's beach defense were strongly positive on both sides of the battle lines. After the war, an intelligence officer in Japan's Sixteenth Division, Major Shoji Ohta, recalled that their forces had met "stiff resistance" at Mauban. Indeed, it was the only place on Luzon where the Japanese landings had met a creditable defense. The U.S. Army's official history by Louis Morton has just a few lines for this minor battle, calling the battalion's firing an "an effective crossfire." In his roster of Filipino heroes, historian Baclagon singled out two of his PMA classmates, praising Lieutenant Javier, who "stuck it out with his troops in spite of wounds," and Lieutenant Pelayo, who covered the battalion's retreat under "punishing naval gunfire and air bombardment." The Philippine government later awarded the battalion's commander, Lieutenant Mendoza, the Gold Cross medal for his "bold conduct of defense at Mauban."[32]

After years of training an army to defend the archipelago's coastline, General MacArthur's tactics had their only real test on that beach at Mauban. While his strategy was deeply flawed, its chosen instrument had not failed him. Bound together by ties to class, corps, and nation, the academy's young graduates showed a remarkable willingness to fight and die.

COAST ARTILLERY

The handful of classmates assigned to the Coast Artillery deepened their friendships during garrison duty and supported each other throughout the war. After graduation in March 1940, Jose, or "Joe," Mendoza washed out of flying school at Zablan Field and wound up at Fort Wint as a battery officer in the Coast Artillery Corps. Built to guard the approaches to Subic Bay, Fort Wint was located on Grande Island in the South China Sea and still retained the graceful aura of the old army. Mendoza found service on this isolated island with classmates Reynaldo Mendoza and Francisco Lumen remarkably enjoyable. Though training twenty-year-old peasants to fire heavy artillery was frustrating, they still found the slow pace of beaches, band practice, and nine-to-five routine refreshing after four years at the PMA.

After Fort Wint was bombed in the opening days of war, the Coast Artillery withdrew south to the nearby Bataan battlefield, where these classmates commanded 155 mm guns in artillery duels against Japanese regulars. Joe Mendoza became a battery commander and Lumen served as his executive officer in action at Saysain Point and Bobo Point. Reynaldo Mendoza became an intelligence officer in the First Coast Artillery Battalion.

It is here, at the start of his military career, that Joe Mendoza's 140-page memoir of war begins. Unlike the Alcaraz biography, with its celebration of combat exploits, Mendoza's memoir is a chronicle of defeat, humiliation, indignity, and death. As his account moves beyond the Battle of Bataan to the agonies of captivity, Mendoza, in recollections reconstructed for his children forty-four years later, clings to his classmates for survival. When the chaos of war breaks up regiments into a demoralized rabble, members of Class '40 fall back upon their academy bonds to create an ad hoc social and moral order. As defeat first scatters tens of thousands of soldiers across the Bataan peninsula and then collects them in the dying pens of Capas concentration camp, Mendoza's moral universe collapses to its social fundamentals—classmates, family, and friends. In a narrative with one hundred thousand possible characters, it is only this slender cast who people the plot of Joe Mendoza's memory.

On page 63 of his neat, handwritten notebook, Mendoza describes the events of April 1942—the demoralizing order to surrender, the sudden dissolution of his unit, and the disorganized march out of the battlefield toward the Japanese army. The warmth of his camaraderie with classmates stands out in sharp contrast to his revulsion at the nameless thousands in the mob of filthy, demoralized prisoners.[33]

> Soon we reached Mariveles [Bataan], and a Japanese indicated to us to get down from the truck. We got down, and started mixing with the thousands of soldiers aimlessly wandering about on the beach, facing Manila Bay, with Corregidor silent in the background.
>
> I hadn't eaten anything since breakfast, and I was beginning to be hungry. By some stroke of luck, I bumped across [my classmate] Dodong "Pop Eye" Caballero . . . I was happy to see Dodong, with whom I had lost contact since about October of '41. He informed me that his wife Ching had delivered a baby girl, prematurely, sometime in January. . . .
>
> As I gazed around me, I was appalled at the sight that greeted me: soldiers, both Filipinos and Americans, were moving about or lying down or squatting or sitting, but listlessly, with nobody giving orders, and nobody wanting to assume any responsibility for anything. I saw piles of steel helmets, piles of rifles and cartridge belts, but nobody in charge. . . .
>
> When I woke up the next morning, Dodong was already awake, and was drinking some coffee. He offered me some, and I was happy to sip. . . . When the sun had gone up, we wandered about for a while and met [classmate] Francis Lumen and [fellow officer] Ben "Toots" Tolentino. We agreed to stick together and move out of Mariveles as soon as the afternoon is cool enough. . . .
>
> We were off, bright and early the next day, with Balanga as our objective. It was at Balanga that we soon realized that the Japs were not going to let us go free. . . . This time, Japanese soldiers with bayonets fixed on their rifles stood guard over us. Every

once in a while, they would indicate to a group to form a column of fours, and then would march the column off. . . .

I don't remember when we left Balanga. It must have been about 8 or 9 am. . . . Except for Dodong, Toots, & Francis, with whom I formed a foursome, we didn't know anyone else in our column. . . . We sort of just shuffled along in what must have been an easy gait to the people at the head of the column. . . . The first few hours of the march were not too bad, but by noon, when we had been marching some 3 or 4 hours without rest, some of the men in the column began to get tired or thirsty. . . . Our . . . guards would not allow us to break ranks to get water. . . . But one man, who must have been really thirsty, disregarded the sentries' shouts, and, rushing to the next well, kneeled to drink from the freely flowing water. Our rear guard ran to him, and without any hesitation, gave the guy a butt stroke on the skull. I saw him with his eyes pleading to the guard for mercy, before I heard the crack on his skull. I then looked forward, not wanting to see any further. A few moments later, we heard a gun shot.

Later that day, Mendoza suffered a recurrence of the malaria he had contracted in Bataan. As the chills seized his body, he became delirious and decided to break ranks. His classmates, though weak from thirst and hunger, exhausted themselves to save him—aware that the Japanese would shoot anyone who "fell out."

Soon we crossed the provincial boundary between Bataan and Pampanga, and were on the road towards Lubao. There were no trees along the road, and sun beat down mercilessly on our heads. . . . Pretty soon I began to feel myself get feverish again. . . . By about 3 pm, I felt myself feverish, and started to feel somewhat delirious. I began to weaken, and with the sun beating down on me, I told my companions that I was going to fall out.

Dodong told me to hang on, and he held on to my left side, while Francis held on to my right, and we kept on walking like this for a few miles. I could sense that I was getting heavier and heavier to hold on to, and I felt that I was just being a drag to my companions, so I told them, "It's useless. I'm just being a burden to you. I'll just look for a nice hiding place in the ditch. . . ." But Dodong and Francis won't let me. I was at that stage of the malarial attack where I was chilling.

The group's solidarity was tested after their arrival that night at the town of Lubao, Pampanga. It was the hometown of Toots Tolentino, and a crowd of his lifelong friends offered him escape from possible death. Faced with competing loyalties, Tolentino chose by an act of hesitation. Repulsed by the drifting, defecating mass of prisoners, this foursome of comrades had become its own moral universe.

We reached Lubao just before nightfall, and we found ourselves marched into the compound of the NARIC warehouse. When we got there, the place was already jam-packed with prisoners of war like us.... As soon as we sat down to rest, we practically couldn't move an inch. Everywhere I looked, I saw people lying down, squatting, or seated on their haunches, surrounded by fecal matter teeming with maggots and flies as big as bees. The whole place stank with the smell of human waste, urine and sweat....

I don't know how I lasted that night, but... it soon became light, and people were milling about towards the... open gate... where a Jap soldier was handing out balls of rice to each POW as we emerged from the gate to be formed again into columns of four....

Somehow Dodong, Francis, & Toots & I managed to stay close together so that when we joined the column that was being formed outside the compound, we again formed a foursome.

As we began to form the column, some townspeople began to gather alongside the road, and some people began talking in Pampango to Toots. Lubao was Toots' hometown, and the people were telling him just mix with the crowds and escape! We told Toots to go ahead, but somehow, he hesitated, and the moment to escape passed, when... we were to stand up and resume our march—this time towards San Fernando!

After another two days of marching and a train ride packed into boxcars, their column finally reached the Japanese concentration camp near Capas, Tarlac. Following a cursory inspection of their personal goods, the POWs were assigned to quarters. Mendoza's group still stayed together.

We found ourselves assigned to Group X, and somebody took charge of us, and marched us to a group of buildings. I spotted Ben Alejo, PMA '41, in another column that was halted beside our column.... When he saw me, he said, "Can you give me some water? I'm so thirsty. I just want a sip...." I gave my canteen to Ben and said, "Here take a sip...."

The four of us, Dodong, Ben, Francis, and I, were assigned together with about a dozen enlisted men, to a small building about 6 m. × 6 m.... Dodong took charge of our building.... We found out that most of them were Cebuanos, and since Dodong spoke Cebuano himself, he soon established good rapport with most of them.

As dysentery swept the camp, Joe Mendoza noticed the symptoms of blood and mucus in his stools, but managed, upon medical advice, to treat himself by boiling *duhat* seedlings into an effective herbal tea.[34] Still weakened, he soon found Francisco Lumen, his classmate and comrade, delirious and dying. Again, these bonds held firm even under the desperate conditions that killed half the camp's inmates. When Francis started slipping away toward death, Mendoza tried to inspire a will to survive by invoking their academy days.

> Almost immediately after our arrival in the camp, people started dropping. Weak, undernourished, and sick from malaria, dysentery, and cholera, the casualties soon increased daily. . . . One morning, I woke up to find Francis feverish. Without any medicine, the best we could tell him was to wrap himself in a blanket and sweat the fever out. I didn't know what caused the fever. . . . On the second day, I got alarmed. He was suddenly very weak, and his eyes had turned glassy.
>
> When I placed my palm on his forehead to feel his fever, he looked at me, and said, almost in a whisper, "Cavalier, thanks for taking care of me. Just tell our classmates . . ." He didn't go on.
>
> As I wiped off his perspiration, I started talking to him, saying, "Mistah, you're not going to die. Look, I'll go around the camp, and get all the medicine I can get from our classmates . . . and you'll see. You'll get well. Just trust me."
>
> And so, after I wiped off his sweat, and changed his wet shirt (as I pulled over his shirt from his head, I could feel that he was very light and limp, scaring me even more), I went around the camp calling on our classmates, other PMAers, and former co-officers in Fort Wint, asking them to spare any medicines they could, for Francis who was very sick. By about 2 pm I had collected something like 8 or 9 pills of assorted and dubious quality, but when I gave them to Francis, 2 at a time, every 4 hours or so, by some stroke of a miracle . . . the next day his fever left him, and in a couple of days he was feeling fine.
>
> Francis later on became my *compadre* [ritual brother] . . . and we became quite close after the war. He never forgot how I took care of him in Capas.

Mendoza, along with most of his brother officers, watched thousands die at Capas as weeks dragged on. Many volunteered to attend a Japanese indoctrination course held at another base where conditions were believed better. With his group scattered by the transfer, Mendoza turned to a new batch of PMA classmates for support. When he was finally released and took the train to Manila, it was his mother, whom he calls "Moms," and his classmates, rather than the state, who were there to assure a safe transition back to civilian life.

> Silently, we all walked towards the gates of Tutuban [Station], where the members of our families and friends were waiting for us. Moms saw me before I saw her, and we walked to Azcarraga [Avenue] to get a calesa [horse-drawn carriage]. . . .
>
> When we went up to [classmate] Dodong's house and knocked, it was Dodong [Caballero] who opened the door. We had a happy reunion that day. . . .
>
> A few days after this, Moms came to Manila again, to bring me home for good. . . .

When we were at the dinner table, she told Dodong and Ching, "Dodong, Ching, I would like to invite you to my farm. . . . I live in a very small nipa [thatch] hut, but I can have a small room constructed for you, and you are welcome to live with us for as long as you want, without any obligation. I will treat you as if you were my own children. I cannot help but pity [your baby] Linda, who seems so thin. I have a milking goat in the farm, and she can have all the goat's milk she wants."

Dodong and Ching were somewhat dumb stricken. . . . Early the next morning, Dodong told Moms, "Mrs. Mendoza, we are happy to accept your invitation. . . ."

Moms was elated. She left that morning to get the room for the Caballeros constructed. Within the week, she was back, and one day in October '42 we left Rizal Avenue and took the train to Malasiqui, Pangasinan.

Two more ailing classmates, Ricardo Foronda and Reynaldo Mendoza, coauthor of the PMA song, joined their comrades on the farm at Malasiqui. For nearly a year, the four recuperated in the countryside together with Caballero's wife and child, living off the land and playing basketball on village courts. They mended quickly and in January 1943 their team won the basketball championship at the local town fiesta.[35] After recovering their health, Caballero and Rey Mendoza decided to rejoin their families in remote Cagayan Province, where they later served with distinction as staff officers in the USAFIP–Northern Luzon guerrilla unit.[36]

Well over half the class were confined at Capas and many had similar experiences of suffering and sharing. Most entered the camp at a muscular 135 to 145 pounds, and emerged emaciated at weights like 90, 100, or 110 pounds.[37] Built as a training base outside the town of Capas in Tarlac Province, Camp O'Donnell was a treeless, sprawling cantonment filled with rows of bamboo barracks and few amenities. To convert this camp into a prison, the Japanese strung barbed wire to divide it into separate Filipino and American compounds. They then subdivided the Filipino side into some fifteen "groups," each with its own commander, sick bay, and barracks—single-story structures with nipa-thatch roofs, bamboo walls, no water, and no beds. Into these crude shelters built for forty soldiers were now packed seventy to one hundred prisoners without mosquito nets to block infection in this malaria-infested region. As the trains pulled up at the nearby rail siding, the boxcars discharged some nine thousand American and fifty thousand Filipino prisoners in groups of several hundred. Then the Japanese sentries marched them into the areas where they would remain for the next six months—confined to barracks, subsisting on thin rice gruel, and suffering from disease or malnutrition.[38]

There were only two ways to win a temporary reprieve from this enforced idleness—guava detail or burial detail. Camp doctors ordered everyone to drink a potion made from boiled guava leaves to slow dysentery, and prisoners were marched out of camp under guard to gather guava leaves, allowing them to pass messages and buy food. In mid-May, when deaths from dysentery and malaria reached five hundred a day, the strong formed details to bury their comrades. "Every day, you'll see there was sort of an open platform, a wooden platform filled with dead people," recalled classmate David Pelayo. "We could see these flies in the mouths, in the eyes, in the ears. . . . You could see a terrific sight! Rain, dead people under the rain before they are buried." Every day, after the bodies were cleared out of the barracks and stacked three or four deep on the platforms, the burial details worked overtime to carry them to shallow graves outside the camp. "I tell you, one day we marched about four hundred dead in a blanket with stilts . . . to the cemetery and put those dead there," recalled classmate Horacio Farolan. "And by the time they come home, the guy lifting him is already dying." Three months later, when the Japanese began releasing the prisoners, survivors were "only bones with little skin, you know . . . walking skeletons." Capas was, as classmate Pedro Bersola put it, "like being in the house of death."[39]

Individual classmates adopted survival strategies that ranged from self-confident leadership to somber retreat. "If anybody survives this holocaust," Salvador Piccio told himself, "I'll be one of them." Believing that God would watch over him, Piccio commanded his company of a hundred prisoners with an entrepreneurial verve. Using his pocket money of a hundred pesos as capital, Piccio ran an aggressive buy-and-sell operation—sending his men on work details to buy goods outside the barbed wire and selling food at exorbitant rates inside the camp, then using the profits to purchase medicine for his group. When they were released in September, everyone in his company had survived and Piccio had doubled his money. By contrast, Conrado Nano endured hunger and malaria in isolation. "I was alone. I was alone in one barrack," Nano recalled. "I have no classmate anymore." He withdrew from others and refused work details, saying to himself, "Why should I be spending my energy hiking and doing all that? We have a different problem here . . . : how to stay alive." Despite such striking differences, classmates tried, whenever possible, to reach out to each other for emotional and material support.[40]

During their six to ten months as prisoners of war, Class '40 nursed each other and took risks to honor the few who died. During a visit to another area inside Capas, Hospicio Tuazon found his classmate Ciceron de la Cruz suffering from a serious case of dysentery in a primitive sick bay. "He could not stand any more," recalled Tuazon. "He was bathing in his own filth." Tuazon washed the feces from his classmate's body, and then brought him to his own barracks to nurse him through the crisis. Similarly, during the Death March, Manuel Acosta, though weak from amputation of his left arm, helped Washington Sagun, a classmate weakened by dysentery. Inside the camp, Acosta scrounged for food and passed extra rations to him through the barbed wire. When Sagun finally succumbed on May 16, Tuazon gathered his class ring, cash, and personal effects. Despite the difficulties of preserving property in a concentration camp, he managed, after his release, to deliver these effects to Sagun's sister.[41]

Indeed, the class, almost instinctively, made a ritual of recovering graduation rings from dead classmates and returning them to next of kin. When Manuel Cancio was killed by an accidental pistol discharge during a poker game on Bataan, a classmate, Patrocinio Lapus, carried his PMA ring and personal effects into battle and through the Death March. At war's end in June 1945, Epimaco Orias spotted a PMA ring on the finger of a passing guerrilla officer at Camp Murphy and asked to see it. Inside was inscribed "Escobar," the name of a classmate who had been executed by the Japanese only six months before. The guerrilla had found the ring on the body of a Japanese he had killed in northern Luzon and now surrendered it to Orias, who, in turn, delivered it to his classmate's widow.[42]

These experiences strengthened ties among the members of the class, giving them a quality akin to the familial. When asked in interviews fifty years later about the strength of their bonds, many members of the class cited these incidents of shared suffering as evidence of their solidarity.

AIR CORPS

The seventeen classmates who passed the rigors of pilot training to join the Philippine Army Air Corps had experiences that mingled the technological thrills of the navy with the grind of the infantry. The Q-boat was the most advanced naval craft in its class, but Filipino pilots, by contrast, took to the air in antiquated P-26 pursuit planes whose slow speed and weak guns were no match for the lethal Mitsubishi Zero fighters. Even so, the first days of war brought moments of triumph and heroism that inspired strong esprit among the brotherhood of aviators. On the third day of war, Alberto Aranzaso became one of the first heroes of Class '40. After passing flight training in early 1941, Aranzaso—along with classmates Cabangbang, Caldoza, and Farolan—was among the twenty top pilots picked by the legendary Captain Jesus Villamor for his Sixth Pursuit Squadron.

On December 10, three pilots from the squadron scrambled from Zablan Field to engage a superior Japanese force in the air above Manila. In the words of his Silver Star award, Aranzaso "in the face of heavy fire . . . contributed to the rout of the hostile formation . . . and served as an inspiration to his entire squadron." In a later mission, alone and unarmed over the South China Sea, Aranzaso's aircraft was badly damaged by fire from a Japanese cruiser, but he still managed to make a safe landing. Assigned to the artillery when all their aircraft had been destroyed, he was serving on Corregidor Island when the garrison capitulated in May 1942.

Instead of surrendering, Aranzaso and an American officer tried to flee in a small boat and were paddling far from shore when Japanese fire sank their fragile craft. For the next twenty-four hours, the two clung to wreckage, unable to make for land since the Filipino pilot could not swim. "Aranzaso turned to me and asked if I was getting tired," the American officer, Captain Damon Gause, later wrote. "I told him I was just about finished. . . . He was silent for a few minutes, then he asked me to look back at Corregidor. It was a towering mountain of fire. . . . When I turned back to Aranzaso, he was gone. . . . I realized that he had deliberately slipped from his support to drown himself so that I might have a better chance for life and freedom. . . . He was a gallant officer."[43]

While war scattered Class '40, the fraternity of pilots remained in close contact throughout. Bartolome Cabangbang, a leader in the "same banana" boycott, won his wings after graduation and was selected, along with Aranzaso, for the elite Sixth Pursuit Squadron. Under Captain Villamor, the squadron distinguished itself against Japanese fighters over Manila in the first days of the war. When the pilots were grounded by the loss of their planes, Cabangbang, along with five classmates in the Air Corps, joined the artillery on Corregidor Island. There, on May 2, a direct hit by a Japanese shell sealed the mouth of a tunnel, burying the troops behind a wall of earth. Working under heavy enemy fire, Cabangbang risked his life to lead a rescue crew that saved the men—an act that won him a Silver Star.[44]

Cabangbang described the balance of his experiences in a memorandum written at MacArthur's Australian headquarters sometime in 1943–44.[45] Aside from the rich detail, his chronicle is striking for its vivid portrayal of suffering. Through the bombardment of Corregidor, brutality at Capas, and flight to the guerrillas, the memoir records relentless physical and emotional pain—pain compounded by the disorientation of social disorder. In this sea of suffering and moral chaos, Cabangbang relied on classmates to survive.

The narrative begins in February 1942, when Cabangbang was transferred to

Corregidor Island in Manila Bay with a contingent of Air Corps personnel. Although conditions were at first better on Corregidor than they had been on the Bataan battlefield, the situation changed suddenly on April 29. That day, over one hundred Japanese guns unleashed a devastating artillery barrage on the island that continued until its surrender a week later. When the shelling began, Cabangbang wrote: "We could hardly believe that they were guns firing at us. We heard the explosions of the batteries like beats of drums, and . . . then the burst of shells just like machine gun fire." The troops were packed into the island's deep concrete tunnels, which "really quaked and trembled under the heavy poundings of artillery shells." Under this incessant bombardment from land and air, "Many of us got demoralized so much that soldiers began cursing openly. . . . Day and night everybody was scared to death by the whistles of bombs and shells. We had to go to the tunnels to live even if we had to go hungry." Wounded during shelling on May 2, Cabangbang was sent to the Corregidor hospital, where "there were about 2,500 . . . either without legs, eyes, wounded or burned." The announcement of the garrison's surrender on May 6 "caused most to cry."[46]

Three weeks later, the Japanese shipped the Corrigedor prisoners across Manila Bay and marched them through the city to Tutuban Station, where trains were departing for Capas concentration camp. Cabangbang and his fellow POWs found their landing at Pier 7 and passage through Manila devastating. "We thought that our friends, relatives or the 'Big Shots' would meet us. . . . But we were again disappointed because . . . there was not even a shadow of our friends or relatives who dared to meet us on the pier to say the usual 'Mabuhay' [Welcome]." Instead, they were marched four abreast through city streets lined with Filipino police who stood with their backs turned. "There were civilians too along Quezon Boulevard and Azcarraga," Cabangbang recalled. "We recognized some of those on the streets so we waved our hands to them and shouted at them, almost involuntarily. These people did not answer . . . : they were too scared to show any recognition." Instead of the electric communion with the crowds of prewar parades, the defeated officers were now shunned—heroes under the Commonwealth, pariahs in Japan's new order.

Capas concentration camp was a charnel house. Packed "like sardines in groups of one hundred inside the boxcars" at Manila, the POWs were shipped to Capas Station, Tarlac, and then marched for fifteen kilometers to the concentration camp. There the Japanese guards were, in Cabangbang's words, "scientifically exterminating" the Filipino troops by denying them food and medicine. Fed on a diet of impure river water and short rice rations without vegetables or protein, prisoners soon sickened. "In most cases," he recalled, "we got two or three diseases at one time—malaria, beri-beri, and dysentery usually go together. . . . In fact there were so many diseases that even our doctors didn't know some of them." Speaking with the physicians, he learned that "these diseases were easy to cure." Cabangbang felt that the Japanese were barring medicines from the camp as a part of their systematic slaughter. "The Philippine Red Cross was trying to give us medicines . . . but the Japs wouldn't accept the offer, and we understood the motive, to kill us." The death rate rose "from 100, 200, 300, 400, up to more than 500 dead a day," peaking at 556 in July. "There were so many dead that we had to put twenty-five bodies per grave." Cabangbang estimated that half of the sixty thousand Filipino soldiers in the camp died.

To cope with disease and death, the Filipino POWs, now separated from the Americans, established a camp structure modeled after their army organization. This attempt at survival by revival of the familiar Philippine chain of command ended abruptly in July, when the Japanese transferred two thousand able-bodied officers to the "Stotsenberg Educational Training Center," beginning three months of day and night indoctrination. Cabangbang recalled that the Japanese Propaganda Corps taught "how America spoiled the Filipinos with easy and extravagant life; . . . how great and prosperous . . . all other Japanese occupied territories are; . . . how the Philippines could . . . play an important role in the 'Greater East Asia Family.'"

When the Japanese offered to release anyone who joined their Bureau of Constabulary, six classmates met to discuss the ethics of serving Japan after swearing an oath to America. "By God, this is better than staying here," announced Salvador Piccio. "Here, if we stay any longer, we might be facing sure death. Outside we have a good fighting chance. . . . How do you know we will not organize a guerrilla [unit] by the time we get out? Never mind those oaths and swearing . . . , that's only a piece of paper." Accepting this pragmatic view, these classmates—Alcaraz, Delfin Argao, Mayo, Jose Rodriguez, Piccio, and Cabangbang—decided that taking the Japanese training was the quickest way to escape their control.[47]

On September 1, Cabangbang, his classmates, and forty other officers enrolled for Japanese indoctrination in Academy Number Three at Manila's Torres High School. After four weeks of "Japanese drill, courtesies and customs," they joined the Japanese-sponsored Bureau of Constabulary. Most of the class would negotiate this dangerous terrain between Filipino guerrillas and Japanese army, but Jose Rodriquez was one of the few who slipped. He apparently internalized the indoctrination. Some months later, strolling on the Escolta in downtown Manila in his Japanese uniform, Rodriguez ran into a PMA classmate working as a horse-cart driver, a lowly *cochero,* and demanded, "Why aren't you in uniform?" The classmate shot back pointedly, "Why aren't you a cochero?" Nonetheless, in their *Golden Book,* the class praised the professionalism of Rodriguez's service to Japan, saying, "He died in the line of duty on 8 October '44 in San Felipe, Nueva Ecija, in an encounter with guerrilla forces."[48]

Determined to join the guerrillas, Cabangbang spent the next nine weeks planning his desertion. In this odyssey across an uncharted no-man's-land to Bohol Island and ultimately Australia, Cabangbang found that he could trust only his classmates. With Japanese spies in the cities and guerrillas shooting traitors summarily in the countryside, all but the firmest loyalties were now suspect. Indeed, in the structure of his narrative, classmates appear as beacons of safety. Shipped down to Cebu City with the Japanese constabulary in November 1942, he began to "mix and converse with" two classmates, Pedro Baban and Alberto Acenas. Cabangbang learned that guerrillas were active in the region and planned to escape to Bohol with another classmate, Salvador Piccio. In the end, he slipped away from Cebu on his own, crossing the stormy straits to Bohol Island in a small sailboat. Even here, in his home province, it was classmates who assured his survival. Landing on nearby Santa Rosa Island, he was intercepted by guerrillas and interrogated. "Luckily" he wrote, "the Bolo Battalion Chief was my classmate at the Visayan Institute, Cebu." With his endorsement, Cabangbang was sent under escort to guerrilla headquarters for further investigation. There, he recalled, "I met Maj. Engeniero, C[ommanding] O[fficer] Bohol Force, [and] Capts. Bayron and Alano my classmates at PMA." With their support, Cabangbang was appointed the unit's chief inspector. Here his narrative ends.[49]

There were only four Boholanos in the Class of 1940. In the chaos of defeat, all four had sought refuge on their native island and joined the fledgling resistance. Cabangbang soon rose to a senior staff post and, along with his classmates, infused the local guerrilla organization with professionalism. Most important for his future political career, Cabangbang's guerrilla work may have linked him to the island's dominant politician, Senator Carlos Garcia, who was serving as an advisor to the Bohol resistance.[50]

In January 1943, Major Jesus Villamor, the ace pilot who had joined the Allied Intelligence Bureau (AIB), arrived from Australia by submarine and landed on Negros Island. Needing agents to build his intelligence net, Villamor sent an operative to Bohol to contact Cabangbang. With their commander's consent, Cabangbang and his PMA classmate Urbano Caldoza joined Villamor on Negros, leaving another classmate, Sofio Bayron, as the AIB agent on Bohol. After six of his pilots joined him in the mountains of Negros, Villamor "felt a renewed sense of vitality surrounded by these able and brave men." When his mission finished in October, Villamor radioed headquarters for permission to bring back Cabangbang. Since he "would be of later value to send back to the area in GHQ [penetration] parties," headquarters approved.[51]

On October 20, Villamor and Cabangbang paddled a wooden canoe three miles out to sea and spent five hours "riding the crests of gentle rollers" until "the long length of submarine appeared, water tumbling around its steel parts . . . churning waves that rocked our little outrigger." The hatches opened, bristling with rifle barrels. Villamor gave the password, and the pilots climbed aboard for the seventeen-day voyage to Perth, Australia.[52]

Only days after completing Australia's jungle-warfare course in July 1944, Cabangbang was sent back to the Philippines via submarine in command of a twenty-man team. In the words of MacArthur's orders, they were to "penetrate into Luzon, join Major Anderson's guerrillas, and drive communication and intelligence channels through south central Luzon"—an ambitious mission for a young officer.[53]

After the submarine surfaced at Dingalan Bay on Luzon's Pacific coast, Cabangbang hiked for thirty kilometers across the Sierra Madre mountains to join a guerrilla unit under an American officer, Colonel Bernard Anderson. En route, Cabangbang happened across the encampment of Captain Alejo Santos, a Bataan veteran who had broken with Anderson to form an independent command called the Bulacan Military Area (BMA). Needing a deputy familiar with Central Luzon, Cabangbang noticed that the guerrilla roster listed the name of his PMA classmate David Pelayo, a native of nearby Pampanga. In Pelayo's words: "He [Cabangbang] told . . . Santos that we were classmates, so he got me in his unit."[54] Cabangbang later met another classmate, Ramon Alcaraz, commander of the BMA's Kakarong Regiment, at a guerrilla propaganda rally and the two reminisced until dawn.

Surrounded by comrades and classmates, Cabangbang decided to remain in Captain Santos's camp at Victory Hill, just north of Manila. There Cabangbang established a base to control a network of guerrilla radio posts, collecting intelligence from all radios on Luzon and transmitting it to MacArthur's headquarters in New Guinea. Throughout these dangerous months of active guerrilla warfare, three classmates formed a tight network. While Pelayo and Alcaraz used their resources to extend Cabangbang's intelligence net across the Central Plain, he reciprocated by providing Alcaraz's regiment with two of his demolition experts.[55]

Cabangbang's net grew rapidly to stretch from Pangasinan to Manila, the largest of any AIB agent. Over the next six months, his unit established eighteen satellite radio stations and sent 893 messages to MacArthur's headquarters detailing, with speed and accuracy, Japanese positions and internal communications. Radioing from the mountains of Bulacan on December 18, for example, he reported the precise movements of Japanese gasoline and ammunition dumps around Manila the day before, thus identifying bombing targets. Indeed, Cabangbang's unit supplied much of the intelligence on Japanese troop dispositions that guided the U.S. Army's advance on Manila.[56]

Cabangbang also mediated the murderous rivalry between Santos's resistance and the communist-led *Hukbalahap* guerrillas, forging an ad hoc alliance for the liberation of Manila. After killings on both sides, Cabangbang wrote directly to Huk *Supremo* Luis Taruc on 17 January 1945, identifying himself as "the direct representative of General MacArthur in this area" and offering "to effect a compromise or agreement between the Hukbalahaps and BMA." The next day, he advised headquarters of Huk support when he forwarded a message from Supremo Taruc to President Osmeña: "We pledge our full support for your leadership in the struggle for independence and democracy." Cabangbang posed as a neutral mediator to the Huks by minimizing his friendship with Captain Santos ("It just so happened that I passed by Bulacan and naturally I had to stay here.... I am just unlucky not to be with you"). But he also warned headquarters that the Huks "are and will remain to be communists." Using deception to court their support, he told MacArthur that the Huks "cooperate only on the promise of receiving more arms and we are stalling with excuses that I do not decide on these matters but just inform GHQ [General Headquarters]."[57]

For these impressive accomplishments, Cabangbang was promoted to captain and nominated for the U.S. Distinguished Service Cross. The promotion cited his "splendid services," and his superior's covering letter for the medal called his work "one of the outstanding achievements of the war." Cabangbang responded to these honors by advancing classmates and comrades. After his promotion in December 1944, he radioed MacArthur requesting "that due recognition be given to full cooperation ... which the commanding officer Bulacan [Santos] ... and his men have accorded us." A month later, he used his appointment as MacArthur's direct agent to order battlefield promotions for his classmate Pelayo and several former PMA cadets in his unit.[58]

While Cabangbang orchestrated guerrillas from his mountain redoubt, his classmate Vic Osias commanded a jazz orchestra on Manila's dance floors. Assigned to army flight training at Zablan Field after graduation in March 1940,

Osias, a brilliant athlete who was twice elected class president, topped the pilot's course and was assigned to headquarters at Camp Murphy in Manila.[59] There he joined a group of young officers who played on the army basketball team and attended society parties. Slowly, Osias drifted away from classmates to join this new batch who shared his entrée into Manila society.

The outbreak of war in December 1941 added new dimensions to these boyish friendships. In the opening days of war, as Japanese fighters strafed Nichols Field, Lieutenant Cesar Basa, Osias's basketball buddy, crash landed his fighter and ran from the wreckage. "The Japanese fighter came down and strafed him," Osias recalled. "I ran out to help Basa and he fell down almost into my arms with a bullet in the head and a bullet in the stomach. I had to stand there on the field cradling in my arms a man who in his strength and youth had been a fellow basketball player and I had to watch him die in my arms. A very upsetting, very moving, very saddening experience."[60]

Later, when the pilots joined the Philippine Scouts on Bataan, Osias served with distinction, repulsing a Japanese attack at Aglaloma in January 1942— heroism that won him both the U.S. Silver Star and the Philippine Gold Cross. "Demonstrating leadership of the highest caliber," wrote the historian Uldarico Baclagon of his classmate, "Lt. Victor Osias . . . was able to penetrate into enemy territory and succeeded in wiping out several snipers with his BAR [Browning Automatic Rifle.]"[61]

After Corregidor's surrender, Osias and the rest of the island's garrison joined the Bataan prisoners at Capas. Finding themselves in a disorganized, demoralized prison, their prewar group—Tony Aquino, Charlie Albert, and Osias— swapped bunks with others so they could sleep together in a single row. After several weeks, however, the Japanese command identified officers with parents among the political elite and moved them to a separate building. As sons of prominent politicians, Albert, Aquino, and Osias were held as hostages until they were told that their parents were helping organize the pro-Japanese government.[62]

When he was released two months later, Vic Osias found himself in a difficult situation. His father, legislator Camilo Osias, had become assistant director in the pro-Japanese *Kalibapi* and secretary of education in Jose Laurel's collaborationist cabinet. Educated in the United States, a decorated veteran of the U.S. Army, and the loving son of a leading pro-Japanese collaborator, Osias faced a cacophony of loyalties. He responded by forming a jazz band to dance away the war.

At first Osias took refuge as an instructor in his father's Kalibapi. But when he learned that the Japanese were after him to join the constabulary, Osias started a dance band that became the sensation of wartime Manila—Vic Osias and the All Stars. After recruiting Manila's top musicians, Osias, the band's drummer and arranger, decided to take up the Glenn Miller sound. The All Stars filled daytime theaters and nightclubs with people stomping their feet. His jazzed-up, syncopated version of the Filipino folk song "Planting Rice" ended every concert and never failed to bring down the house. Osias explained his sudden shift from soldier to bandleader, saying, "It was something to do without having to be under those guys. I didn't want to be under those Japanese."[63]

As MacArthur's island-hopping campaign leapt toward the Philippines in early 1944, Osias noticed Japanese forces drilling in the streets with bayonets. Realizing the implications, Osias evacuated his family to Baguio. There Osias made contact with the guerrillas through his classmate and fellow aviator Pedro Baban, a battalion commander in the local resistance. After the U.S. landing on Luzon in January 1945, Baguio City, once a refuge, became dangerous as the Japanese forces began to retreat. Baban, realizing his classmate was at risk, assigned his troops to escort Osias on a six-day hike to a mountain airstrip. From there he was flown to safety behind American lines.[64]

In the final months of war, academy ties thus reached across the battle lines in some surprising circumstances. After their release from Capas, two members of Class '40 were trained by the Japanese *Kempeitai* at Fort Santiago and assigned to the Presidential Guards—a unit charged with protecting top officials of their client regime from guerrilla assassins. These classmates, Tomas Tirona and Licurgo Estrada, were promoted to captain and given command of the two companies guarding Malacañang Palace. Though under close Japanese supervision, the two had regular contact with the resistance through their former classmate, Colonel Benedicto Valenzona, commander of President Quezon's Own Guerrillas II. Periodically, the two smuggled their guerrilla classmate past Japanese sentries to use the palace's powerful radio for communications with Allied forces in Australia. They also supplied the resistance with information about the location of the Japanese, the number of Japanese troops in Malacañang, and details of other enemy garrisons.

As U.S. forces advanced toward Manila in February 1945, these Japanese-trained guards executed a deft, and rather dangerous, change of sides midbattle. Their commander, Colonel Jesus Vargas (PCA '29), made radio contact with the U.S. Army's advance columns and ordered his captains to dig trenches around the palace for a possible Japanese assault. When the U.S. Army columns swept through downtown Manila and approached Malacañang, Tirona and Estrada were surprised to see their former tactical officer at the academy, Captain Napoleon Valeriano (PMA '37), riding atop the lead tank. Instead of an exchange of fire between enemies, there were handshakes among fellow cavaliers. Then, Japanese mortars opened up on the palace grounds and in the heavy shelling another classmate in the Presidential Guards, Epimaco Orias, was wounded by shrapnel.[65]

By 1944, many in the class had joined guerrilla units on islands across the archipelago. According to their *Golden Book* biographies, thirty-two of the seventy surviving classmates joined the resistance. Several commanded battalion-sized forces while still just lowly lieutenants. But most served as staff officers, raising the effectiveness of these disparate guerrilla commands. Indeed, one military history noted the exceptional contributions of Class '40 alumni to guerrilla units from northern Luzon to Panay, Negros, Bohol, and Mindanao.[66] Although many classmates served with distinction, the localized nature of guerrilla warfare removed these activities from the realm of their collective experience.

Through combat, captivity, and resistance, the war strengthened their intertwined values of honor and loyalty. Classmates trusted each other without question whenever chance brought them together. Class ties superseded all other loyalties within the moral ambiguities of enemy occupation. Ignoring the divide of global war, classmates reached across the battle lines to give each other life-sustaining support. Whether black-market merchants, collaborators, or guerrillas, class members were certain that all had somehow remained honorable cavaliers. The war destroyed Manila, smashed the Philippine Army, and traumatized Filipino society. But Class '40 emerged with its solidarity undiminished.

These classmates also came through the war with a strong commitment to the honor code's principles of professional integrity. Combat and captivity somehow affirmed the ideals held by Class '40, deepening their internalization of cadet lessons and forming values that would later withstand the pressures of higher command. The contrast with Class '71 highlights the significance of this second stage in the older class's socialization. Under Marcos's martial-law regime, the young lieutenants of Class '71 would, during a parallel phase in their careers, extract very different lessons as combat officers in a civil war and guardians of a garrison state. Class '40's suffering as prisoners deepened their military socialization, while Class '71's empowerment as jailers and torturers would destroy theirs.

Chapter 4 Career Soldiers

War had strengthened their faith in the academy's values of "integrity, courage, loyalty." But service in a peacetime military would confront Class '40 with difficult political choices, creating a crisis of morale. If war had proved their capacity for courage, then peace would challenge their integrity in ways that they could never have imagined as cadets. Peace, not war, would prove the harshest test of their military socialization. Rising to command in the army of an independent state, they faced constant conflicts between their principles and a pervasive politics. In their last years at the academy, adherence to its standards of honor had, they felt, required organized disobedience to superior orders. At some risk to their careers, they had taken a moral stand that produced a political victory and affirmed their sense of honor.

In the postwar military, however, choices were more complex. Scattered across the armed forces, classmates could still monitor each other, but they could not impose their standards upon the whole military. As cadets, the class had protested with spectacular effect, but as officers they had no opportunity for collective action. Their reunions, held every year after 1947, were large, lavish affairs with wives and dancing that usually not did lend themselves to debates or drafting statements.[1] Faced with a conflict between their ideals and the reality of service in a patronage democracy, classmates would agonize or seek the counsel of close friends. Ultimately, however, they were forced to choose compliance or resignation.

After Japan's surrender, the class returned to an army transformed by war and independence. Thousands of reserve officers with wartime service crowded the ranks, using patrons to win regular commissions. Alumni of the Philippine Military Academy would comprise almost 44 percent of the country's seven thousand regular officers by 1970, but in the immediate postwar years they were a far smaller component. More broadly, the military itself was now subject to a host of new political pressures as it sought to define its role within a newly independent nation. While both colonial and Commonwealth governments had encouraged professionalism before the war, independence now opened the armed forces to political interference.[2]

As they rose to command, Class '40 met growing conflicts between principle and the reality of service in a new nation run by patronage politics. Politicians pressed officers for an exchange relationship outside the chain of command at each turn in their careers. From their PMA training and its American antecedents, the class believed that the military had to be insulated from partisan politics in a democratic society—for the good of both democracy and the military. Such ideals, though essential for a military in a modernizing society, were nonetheless a colonial legacy that faced a rough transition after independence. Since the PMA's traditions were so recently borrowed and confined to this highland campus, they lacked roots or resonance within the country's political culture. However sound they might have been, these principles did not command an ingrained support from the country's political elites trained in the less rigorous college reserves.[3]

Indeed, Class '40's embrace of these ideals did not spring from an uncritical admiration of America. After graduation, they had been angered by the apparent discrimination in pay between Filipino and American soldiers within the same U.S. Army units. In later interviews, moreover, classmates expressed bitterness over their postwar treatment. Although Filipino and American troops had fought in the same army against a common enemy, in 1946 the U.S. Congress passed legislation reducing Filipino veterans' benefits and paying them at a dollar-peso ratio that cut their value by half. "I got wounded during the war," complained classmate Francisco Jimenez, who fought on Bataan and suffered the Death March; "I got only one-half of what the U.S. armed forces get." After the indignity of fighting on Bataan at half the pay and rations of American soldiers, Horacio Farolan was outraged by these postwar cuts, remarking angrily, "No pay, no pay, no back pay, nothing. Nothing, and we were serving in the U.S. Army." His classmates, he explained, had concluded that the Philippines should "not be dependent on the United States. That's the Class 1940 idea—you cannot trust the Americans."[4]

REGULARS VS. RESERVISTS

After its rapid mobilization and sudden collapse into guerrilla warfare, the Philippine military's personnel roster was in chaos. Its seniority system of grade, date, and status—the foundation of any military hierarchy—had been severely disrupted. For seven years after the war, Class '40 struggled to recover their rightful positions as regular officers within the seniority and lineal roster. But events complicated the situation. To prepare for an invasion of Japan in mid-1945, MacArthur's command had mobilized a Philippine Army of 250,000 men, elevating guerrilla officers to major and colonel, well above the lieutenants and captains of Class '40. In the subsequent demobilization, the Philippine military suddenly shrank to a force of 30,000 with only 3,000 officers. Still, many reserve officers remained in the service with temporary promotions that now had to be regularized. Moreover, this adjustment coincided with the transition from a colonial to a national army with smaller budgets and reduced pay. After independence in 1946, a Filipino first lieutenant, for example, found his monthly salary cut by 30 percent at a time of high inflation.[5]

Class '40 had both advantages and disadvantages in this troubled transition. During the war, the Japanese had recruited regular officers for the Bureau of Constabulary and, after liberation, some classmates came under suspicion as enemy collaborators. To clear their names, each appeared before a Loyalty Status Review Board staffed by wartime guerrillas. One classmate who served in this constabulary, Salvador Piccio, faced hostile questioning from these guerrilla reservists. "I think they were interested in . . . driving out all the members of the regular force who were in the constabulary," he explained. But armed with affadavits from wartime guerrillas, Piccio, like most of his class, was cleared without a full hearing. On the plus side, in 1945 the U.S. military sent some forty classmates, over half the class, to America for advanced training at camps such as Fort Benning and Fort Sill. Upon their return, these regular officers were the most qualified for key positions in the different staff sections.[6]

Regularization started in 1947, when President Manuel Roxas approved the readjustment of ranks, a procedure that awarded all members of Class '40 a permanent rank of captain. In the next session of Congress, guerrilla veterans lobbied hard for the integration of reservists into a new roster, denouncing the regulars who resisted as the "old guard" or even as Japanese collaborators. As they pressed their case, the veterans were led by a group of rising politicians—General Macario Peralta, Major Ferdinand Marcos, and Representative Ramon Magsaysay. Indeed, several, notably Magsaysay and Marcos, engaged in aggressive advocacy to build a following among veterans and thus launch their political careers. Barred from public debate, the regular officers lobbied headquarters for legislation that would defend their prerogatives.[7]

Seeking to reconcile the two groups, President Roxas proposed legislation that would set aside the prewar seniority roster to allow limited integration of reservists as regular officers. In the debates at headquarters, seven members of Class '40 joined a group of regulars known as "the Indians" for their opposition to the reserve leaders. Resolution came when the House National Defense Committee under Magsaysay discovered that there were only 718 active regulars among the 2,118 authorized commissions, allowing ample space for the integration of reservists. With passage of a compromise bill, the Armed Forces Officer Personnel Act of 1948 (Republic Act 291), the dispute shifted to the regulations for drawing up the roster. The reservists insisted on date of commission as the basis for seniority, but Class '40 "were fighting like mad" to win credit for their education at the PMA. A year later, new legislation, Republic Act 207, gave the class an acceptable niche on the new roster by minimizing the reservists ahead of them. The process was finally finished in December 1950, when President Quirino established the Armed Forces of the Philippines (AFP) with separate rosters for each of the four major services.[8]

Throughout this debate, the class responded to these threats to their careers with individual resignations and collective action, including the threat of a "mini same-banana action."[9] By acting together within the military hierarchy during an expressly political process, the class successfully pressed headquarters to defend their special status as regulars and advance them, as a group, within the roster.

While the roster restrained reservists, its rigidity fostered internal tensions. Under Act 291, seniority was set for the whole of their thirty-year careers by their academic rank within the class at graduation. Those who had, through heroism or skill, advanced outside this strict seniority were now reduced in rank. Most classmates accepted these demotions in silence, but they still left a legacy of bitterness that later produced tensions, particularly in the navy. By the late 1940s, moreover, strict seniority would encourage resignations by some of the most talented officers.[10]

The class continued to police its own ranks, working to retain fit classmates and ease out those unfit. In 1954, navy classmates became concerned that Pacifico Barrios was drinking heavily. In the spirit of the honor code, two from Class '40 on the selection board voted against their classmate's promotion. After an undistinguished record at the academy, Barrios had had a troubled career that included service in the Japanese constabulary and a trial before the People's Court for the murder of five anti-Japanese guerrillas. After denying him promotion, the class worked successfully for an honorable discharge. Similarly, classmate Manuel Acosta faced mandatory retirement after losing his left arm on Bataan. But the class argued successfully that he was "more qualified with one arm than most men are with three."[11]

For many classmates, resignation represented, in the words of Reynaldo Mendoza, "a silent protest against political interference." A few months after the war, classmate Ricardo Foronda, still a lieutenant, was discouraged by the flood of reservists. "You know," he recalled, "there were so many colonels, lieutenant colonels who were guerrilla officers reporting." During the war, moreover, Foronda had witnessed guerrillas slaughtering civilians in his native Ilocos Sur. "If they could get away with murder . . . and all the atrocities that they committed," he thought, "I don't want to be in the military service." Ashamed at breaking with the class, Foronda resigned, bought a jeepney with his back pay, and disappeared into private life.[12]

A year later, Ramon Olbes, their regimental commander or "class baron," stunned the class when he resigned his commission to head the Bislig Lumber Corporation, tearing at the bonds that held them in the service. Two other classmates also resigned in 1947, including Pedro Yap, who found it impossible to support three dependents on a reduced captain's salary. Not only was he forced to board his family with in-laws in Cebu, but Yap found that the "army was infiltrated . . . by pure civilians," incompetent reservists who degraded the quality of the officer corps. This career path widened in February 1956, when twenty years of service allowed Class '40 optional retirement with benefits. During the next three years, eight classmates took positions in private industry—notably Uldarico Baclagon, who resigned as a "silent protest against the interference of politicians in the assignment of senior army officers."[13]

But the majority served out thirty-year careers that moved them upward through the echelons of unit leadership, staff duties, and command. As they advanced, Class '40, still the largest contingent of regular officers, had an exceptional impact on the armed forces, particularly during the Huk campaign of the 1950s.

OVERTURE TO A COUP

The postwar communist revolt (1947–54) elevated many classmates to command of battalions in a hard-fought civil war. The conflict began in the late 1940s, when five thousand communist-led guerrillas, known as the Huks, launched a peasant revolution in Central Luzon. At first, the government deployed the Philippine Constabulary, a poorly trained paramilitary force. By 1950, the Huks were expanding rapidly, and it was clear that the constabulary's corruption and abuses were alienating the countryside. Moreover, the constabulary was essentially a defensive force with small, company-sized units of sixty to ninety men that were incapable of engaging the heavily armed Huk squadrons. By early 1950, some ten thousand Huk guerrillas had almost succeeded in seizing national power.[14]

Appointed in September 1950, Defense Secretary Ramon Magsaysay, a charismatic leader and wartime guerrilla officer, began a massive reorganization for counterinsurgency. Within a year, he had broken up the military pyramid, designed for conventional defense, into twenty-six autonomous Battalion Combat Teams (BCT)—mobile, self-contained units of eleven hundred men with their own artillery and armored vehicles. Initially staffed by former reservists at the top of the roster, the new BCTs were hampered by what U.S. advisors called "generally incompetent officers." Detached from the regular chain of command, the BCT could not succeed without good leadership. Magsaysay, moving about the countryside tirelessly for impromptu inspections, rewarded exemplary officers with on-the-spot promotions and punished "the inefficient and the corrupt" with summary transfers. With surprising speed, the new secretary, recalled one Filipino officer, "weeded out all the deadwood in the officer corps."[15]

As the secretary reached down the roster for officers who could fight, Class '40 rose to command. "At that time, the assignment of battalion commanders were first by seniority," recalled Colonel Conrado Nano, a classmate who led the Twenty-Fifth BCT. "Now as far as they fail or as far as they don't make good, they are relieved, and relieved, and relieved, until it reaches us, Class '40. We were still very low in the ranking." By the time the Huks were defeated in 1954, seventeen of the twenty-six battalions had members of Class '40 serving as their commanders or executive officers. Under their aggressive leadership, the BCTs fought hundreds of engagements against the Huks between 1950 and 1955, pounding their peasant squadrons and inflicting heavy losses—over six thousand guerrillas killed, four thousand seven hundred captured, and nearly two thousand wounded.[16]

With half the nation's arsenal under their direct control, Class '40 faced the temptation of power. As more of them took command, a classmate in the navy, Ramon Alcaraz, was the first to realize the implications of this extraordinary circumstance. "Don't you know that seventeen of the twenty-five BCTs are commanded by Class '40?" he pointed out at a class social. "So if we can put out a revolt, we can take over the country."[17]

Others soon noticed. When Defense Secretary Magsaysay resigned to run for president in the 1953 elections, many of his partisans were convinced that the incumbent, Elipido Quirino, would resort to systematic fraud. Rumors, persistent and convincing, circulated that Magsaysay would launch a coup if he were cheated out of victory. In October, only a month before the elections, the CIA's chief Manila operative, Edward Lansdale, advised the U.S. ambassador that the president was faced with the "nightmare" of "a Magsaysay-inspired coup d'état." The threat, Lansdale said, was real, and would involve the "cream of combat commanders" of the AFP—an apparent reference to the BCT and their officers. Lansdale advised that in the event of a coup, "most of the AFP men in the initial Quirino force would soon defect to Magsaysay," and warned that "the United States is not in position . . . to stop a revolt." Only days before the elections, the CIA's weekly intelligence digest for the White House reported, as fact, that Magsaysay was plotting a coup.[18] But there was, as it turned out, a short circuit in these communications.

During the presidential campaign, two of Magsaysay's favorite officers, Napoleon Valeriano (PMA '37) and Romeo Honasan ('43), had approached the commander of the Second BCT, Lieutenant Colonel Reynaldo Mendoza. When they raised the idea of a coup d'état, Mendoza, a member of Class '40, replied, "If you start anything like that, I'm going to fight you. . . . That's no way to fix our country."[19]

At headquarters in Manila, there were more considered overtures. The AFP's chief of operations, Colonel Manuel Cabal, who was married to Mrs. Magsaysay's sister, approached his deputy, Colonel Deogracias Caballero, the oldest member of Class '40 and long its moral leader, to discuss a coup. Pointing out that his classmates commanded over half the AFP battalions, Cabal asked Caballero to draw up plans for a coup in the event that Magsaysay lost by fraud. Without contradicting his superior, Caballero thought: "If we do that, if we seize power in a coup, how do we get the army back in the barracks? How do we end the politicization of the armed forces?" In the end, he decided that a coup was wrong and ignored his superior's orders. Fortunately, the elections were a model of probity, and the military, under Colonel Cabal's leadership, mounted a massive operation to prevent fraud or violence.[20] After the Huk Supremo Luis Taruc surrendered in early 1954, the BCTs were quickly demobilized and Class '40's opportunity for power passed.

For those of Class '40 who served out their thirty years, reconciliation of honor with rising political pressure was becoming difficult. Solidarity yielded to ambition in the competition for promotion. As early as 1955, when army officers from Class '40 were slated for promotion to lieutenant colonel, those denied through a shortage of slots turned their anger on Melchor Acosta, the AFP's deputy personnel officer and their own classmate.[21] With each step up the hierarchy, officers found politicians pressing them for compromising favors. Capitulation risked censure by classmates. But there were times, as some began to discover, when the price of propriety was too high.

Every officer promoted to colonel had to pass through a political gauntlet to win a unanimous vote of approval by the Commission on Appointments—a bipartisan, bicameral congressional committee whose decision was final. In 1961, Lieutenant Colonel Ciceron de la Cruz found his paperwork stalled by a single congressman who insisted that every nominee meet him to express personal gratitude. Realizing that this was "one occasion where politics [could] not be avoided," de la Cruz met the congressman at Manila's Santa Ana racetrack to express his thanks—an experience that made him feel somehow soiled. Similarly, that same year eleven members of Class '40 found their appointments as colonel blocked for months by a bitter power struggle between the defense secretary and the AFP chief of staff.[22]

With each step up the narrowing hierarchy, the competition for promotion among classmates intensified, fraying the ties that knitted them together. One well-placed observer, Edmundo Navarro, who had resigned to join the U.S. Veterans Administration in Manila, watched his classmates suffer the jealousies that came with each step. By the early 1960s, he felt that it was "not the same banana anymore."[23] Indeed, one classmate, Joe Mendoza, quit the service in disgust over these unprofessional intrigues. In the first round of promotions to full colonel in 1960, Mendoza found it "quite embarrassing" to be passed over by classmates below him on the roster and even by a few in Class '41. With the promotions list in his hand, Mendoza confronted his superior, General T. G. Fajardo, a West Point graduate who had been his instructor at the academy.

"Okay, general, you know we've been together from the PMA since I was a plebe. You know me and I know you—and you know damn well who these officers are too," said Mendoza with an uncharacteristic bluntness. "Now look, here is my record, you know my 201 file," he added. "I challenge you to prove to me that any one of these guys below me who bypassed me as full colonel has a record better than this."

"Ah, you know, Joe," the general replied with his usual charm, "sometimes there are times . . . you just can't do the things you want. And sometimes these politicians . . ."

"Okay, General," Mendoza replied sharply, "I really believe that this is not the army anymore that you and I were trying to build. If you're beginning to . . . play politics with the careers of regular army officers, that's not the kind of organization I want to belong to. If you don't mind, I'll start working for my retirement."[24]

But Mendoza was determined to retire as full colonel. After failing in his first attempt for want of patronage, he decided to play politics, nakedly and aggressively, when Congress began drawing up the 1961 promotions list. As an officer in the Adjutant General Corps, Mendoza recalled his descent into politics with revealing detail:

> So, I discussed it with my wife. You know, I said: "I got to retire. But before I retire I got to be full colonel, too. . . . If to be a full colonel you have to play politics, hell, I can also play politics."
>
> I just went to Senator [Eulogio] Balao, who used to be my commanding officer. . . . So I told Balao through his aide, who was a friend of mine, [Lieutenant Colonel Vicente] Raval, and said, "Vic, do something for me. My God, can't you do anything so I can also be promoted to full colonel?"
>
> He said: "Sure. I'll try. But first, ask the people in the headquarters to submit your name. We can't do anything unless the secretary of defense puts in your name first, you know."
>
> I had a lot of friends in the secretary of defense office. . . . Anyway, it was no sweat. I said: "Just put in my name there and [I will] handle the rest from Malacañang or Senate." Anyway, my name was submitted there.
>
> But there was a lot of horse trading in that Commission on Appointments. And on the day that the Commission on Appointments was sitting there, here comes Balao and the other members of the commission and suddenly Vic Raval comes to me and says: "My God, Joe, your name is out again."
>
> "What the hell is happening?"
>
> "You know those damn fool politicians, they horse trade."
>
> "You better tell the boss there, Balao, to do something about it."

He said, "Okay, . . . I'll get in there and speak first to Balao."

"Sir," he said [to Balao], "Mendoza's name is out again."

"Oh no, it can't be. Look," he [Senator Balao] told his friends in the commission . . . , "Look, you can't take this guy's name out. He's a good man. . . ."

They say, "Okay, it's so."

So there I was. We were out in the Senate while waiting for things to happen. So, finally the Commission on Appointments comes out and the Senate president reads the names. When I hear my name, bang! That's it!

I felt very bad about the whole thing, you know.

I said, "Bull."

I said, "How the hell can this thing happen to the careers of dedicated so-called regular officers, you know."

I said, "This is outrageous."

So I said, "I'm going to work my way out of this place as soon as I can."

Indeed, three years later, after earning a master's in business administration from the University of the Philippines, Colonel Mendoza resigned.

REACHING FOR A STAR

By 1961, the class members had reached full colonel and entered the second phase of their postwar careers by winning senior staff posts such as regional service commands.[25] As colonels, classmates were now moving toward the apex of the hierarchy, where every act was inherently political in ways that they could not have imagined as cadets or even battalion commanders. Indeed, several classmates became embroiled in a bitter battle over the integrity of the chief of staff, Lieutenant General Manuel Cabal.

After a rapid rise under the patronage of President Magsaysay, his relative by marriage, General Cabal spent his last six months as chief of staff mired in a humiliating investigation of his personal integrity. His nemesis was Colonel Jose Maristela, a hot-tempered member of Class '40 who had been turned back to Class '41 as punishment for fighting with a tactical officer at the PMA. An anonymous *Golden Book* biography describes Maristela as "a person who was not afraid to fight for whatever he thought was right" and praises him as an officer who had once "challenged an AFP Chief of Staff whom he felt had given him unethical orders." Whatever his motivation, in July 1959 Colonel Maristela became a media sensation when he accused General Cabal of corruption, purchasing land with ill-gotten gains, and harassment of subordinates. In defense of this last charge, Maristela complained that he had been, through the general's influence, "yanked out as 6th BCT commander in the thick of the anti-Huk campaign and 'frozen' in the PC School."[26]

When the Defense Department convened an inquiry, the PMA's superintendent, General Manuel Flores, accused Cabal of unprofessional conduct, alleging that he had rejected the nomination of Colonel Maristela as commandant in favor of his own protégé, Lieutenant Colonel Francisco Jimenez, whose performance was supposedly "unsatisfactory." That same day, Cabal retaliated by ordering the AFP inspector general, Colonel Lucendro Galang, to investigate Maristela, an inquiry that soon produced charges of conduct unbecoming an officer.[27] Significantly, the three colonels mired in this sordid affair—Maristela, Jimenez, and Galang—were all members of Class '40.

For the next four months, a presidential committee under retired Supreme Court chief justice Manuel Moran probed Cabal's "unexplained wealth," compiling fourteen hundred pages of transcript—much of it Maristela's relentlessly detailed investigations. Even though Maristela charged that Cabal, when chief of constabulary, had taken payoffs from "vice joints in Rizal," the general refused to testify and clear his name. Even his collector, Lieutenant Josephus Ramas, later notorious as army commander under Marcos, was widely known. By the time the inquiry closed in January 1962, Cabal had resigned in disgrace and Maristela was promoted to chief of the Constabulary Investigative Service.[28]

Such intrigues were far removed from the simple verities of the academy's traditions of loyalty and honor. Whatever friendship these classmate colonels may have felt for each other at the academy was now eclipsed in the media glare of a political prizefight. Galang was forced to court-martial Maristela; and Maristela, in turn, stirred controversy that stained Jimenez. All three somehow survived this bare-knuckle, bureaucratic infighting and continued their advance up the military hierarchy. But they did so as isolated individuals who had to weigh their own moral choices, drawing upon professional ethics learned as cadets and junior officers. Now that they had reached the command echelon, they had, in effect, assumed the burden of state power and were forced to deal daily with elected politicians. Struggling to survive among a brawling political elite, they could no longer rely on their classmates for support or guidance.

In the last year of the Macapagal administration, promotion to general became the next hurdle for the colonels of Class '40. With retirement on the horizon in June 1966 and their stars fast fading, the class became anxious. Hearing reports of their grumbling, Defense Secretary Macario Peralta assembled twenty-nine colonels from Class '40 in his office to explain that he was postponing their promotions to award stars to the colonels of Class '38 a few days after their retirement. As a former U.P. cadet and reserve officer, Peralta was unsympathetic to Class '40's sense of entitlement as PMA-trained regulars. Posing a rhetorical question, he asked: "Are there any reasons why the members of the class ahead of you should not be promoted ahead?" One member of Class '40 stood and blurted out: "Sir, there are 29 reasons right here!" When Peralta asked which among them should be promoted, several classmates, feeling this a stupid question, spoke up: "Why ask us? . . . We are all general material. It's up for you to choose."[29] Promotions did not come until the waning months of the Macapagal administration, when Secretary Peralta finally nominated just seven members of the class.

Finally, in early 1966, only months before mandatory retirement, "the stars fell" on Class '40. In his first year in office, President Marcos promoted seven more classmates to general with commensurate commands—chief of constabulary, flag officer in command of the navy, and armed forces chief of staff.[30] For the Class of 1940, it was an extraordinary achievement to have fourteen of their seventy-nine graduates awarded star rank.

Yet this distinction did not come without cost. At this final stage, Class '40 could no longer escape the imperatives of partisan politics. As their classmate, Colonel Galang, thought as he opted for early retirement, "Getting a star is very much dependent upon your Malacañang connections." Marcos was determined to control the military, and Class '40, now at the apex of the hierarchy, represented an obstacle that he overcame, in his first months, with a strategy of division—promoting allies and punishing enemies with forced retirement. "One night," classmate Ramon Alcaraz recalled, a close Marcos ally, General Ernesto Mata, "sent to our houses applications for retirement for us to sign. And the threat was that if you do not sign, you will be reduced to colonels. . . . All of them signed." At least two generals among the class were thus forced into early retirement.[31] With equal finesse, Marcos cultivated allies among the class by promoting seven to general and offering others civil posts far above their status as retired colonels. When these deft manipulations were done, only potential loyalists among Class '40 remained on active duty, including a group from Marcos's own Ilocos region.

Classmates were forced to compete with each other in this scramble for stars. Many could become colonels, but only one could be the chief of constabulary or chief of staff. As the military hierarchy narrowed at its apex, politics intervened and class solidarity, strong for decades, frayed.

But even in this final phase as political courtiers, Class '40 conducted themselves with an integrity that seems, by later standards, exemplary. The few who played politics to win promotions from Marcos were generally cautious in their compromises, and the class as a whole maintained a posture of professionalism. Indeed, some classmates would come to feel that their forced retirement had removed a major barrier against Marcos's corruption of the military.[32] By contrast, the PMA's first postwar class, 1951, later allied openly with Marcos. Among the ten senior officers who planned the declaration of martial law in 1972, five were members of Class '51. Under the dictatorship, nearly half their active-duty alumni won stars and commensurate posts. Indeed, at their silver reunion in 1976, the fourth year of martial rule, Class '51 celebrated its proximity to power with Marcos himself speaking at their luncheon and presidential cousin General Fidel Ramos, an "adopted" classmate, hosting the picnic.[33] Not even the most ambitious members of Class '40 had courted power so openly.

As senior officers, the men of Class '40 were scattered throughout the military and the final chapters in their collective history are best studied through individual biographies. Instead of selecting subjects at random, I have picked eight classmates, six in this chapter and two in the next, paired from the poles of political choice to throw their decisions into sharper relief. By such comparison, we can gain a sense of the subtle shadings that inclined them toward compromise or capitulation. Though they followed very different professional paths to the apex of the AFP hierarchy, all faced moral choices beyond the simple maxims of the honor code. Almost all struggled to reconcile these principles with the realities of power. This final phase of their thirty-year careers would prove the strongest test of their military socialization.

This biographical approach reveals aspects of military life usually concealed from outsiders. Publicly, Class '40 has always given an impression of solidarity and simple dedication. But these biographies reveal that, at the close of their careers, ambition fostered some bitter internal rivalries. At the peak of the hierarchy, where military and executive power intersect, senior officers also became involved in extralegal operations for their president. These life histories afford us a rare glimpse at this underside of state power in the Philippines.

CLASS "GOAT" GELVEZON

In 1962, Class '40 faced a mingled irony when its "goat," or worst scholar, became its first general. Raised in Iloilo Province, where his father was a newspaper editor and municipal mayor, Ramon Gelvezon grew up in a household permeated with politics. He was a good student at the provincial high school, but his family lacked the finances for his college education. After graduation, he was unemployed for two years, living aimlessly until he passed the PMA entrance exams.

Gelvezon performed well in both academics and leadership in his first two years at the academy. Then, in October 1938, his charm and good looks brought trouble. Assigned to escort a wealthy, well-connected Ateneo University student who was visiting the academy, Gelvezon flirted with his girlfriend, a student at the University of Santo Tomas. As the two visitors were leaving the PMA in a taxi, the Ateneo student, in a fit of jealous rage, stabbed the girl, who was seventeen years old, seventeen times. Gelvezon was deeply disturbed by the incident and was forced to appear as a prosecution witness in the sensational murder trial. His grades fell, he lost his rank as corps sergeant, and he finished at the very bottom of the class, winning the cheers and applause of the corps during graduation as the class goat.[34]

Gelvezon was shamed by this ritual. "When I graduated at the Academy as a goat," he recalled nearly fifty years later, "it was very insulting. They stood up, threw their hats off, and then clapped their hands much more than [for] the one who graduated valedictorian. The reason I don't know, but to me [it was] an insult." With surprising accuracy, however, the class annual, *The Sword of 1940*, predicted that "with winning ways and smooth technique, Ramon will undoubtedly be catapulted into the arms of success some day."[35]

Once commissioned, Gelvezon dedicated himself to rapid promotion as a way of "erasing the stigma" of being the class goat. In his quest for success, Gelvezon, unique among his classmates, carried a peculiar burden that dogged his every step forward—the unrelenting opposition of his family's local enemies, the Tronos. Gelvezon's hometown of Guimbal on Iloilo's southern coast had, since the late 1890s, suffered from bitter factional infighting marked by violence and murder. By the 1930s, the town's politics revolved around an intense rivalry between two factions led by the Gelvezons and the Tronos. When Gelvezon entered the academy, the politician Pedro Trono petitioned for his dismissal on grounds that he had fathered an illegitimate child, forcing the cadet to swear an oath of denial before the superintendent. After Class '40 graduated, Trono filed a formal objection to Gelvezon's commission on the same grounds.[36]

Assigned to Iloilo in the months before World War II, Gelvezon used his PMA training and local language skills to perform well in the mobilization of peasant conscripts, winning accelerated promotion to captain in February 1942. During the Japanese occupation, he rose through guerrilla ranks to major and by the end of the war had won command of a battalion.

During the campaign against Huk communist guerrillas, Gelvezon became constabulary provincial commander in Quezon Province, a communist stronghold, and was awarded a battlefield promotion. In Gelvezon's account, Defense Secretary Ramon Magsaysay took notice of his activities and, "with his penchant for awarding 'spot' promotions to deserving personnel," promoted him to lieutenant colonel, bypassing many senior officers. Privately, some classmates claim that Gelvezon had staged spectacular, albeit spurious, surrenders of massed Huk supporters to catch Magsaysay's eye.[37]

After serving as constabulary commander in several politically volatile provinces, Gelvezon won command of two sensitive investigative agencies, the Constabulary Investigative Service (CIS) and the Constabulary Revenue-Customs (CRC). In late 1958, when his CRC investigators began probing an electrical appliance firm whose owners held influential positions in the government, Malacañang interceded to force his transfer. After the Manila press ran stories of these intrigues, civic and religious groups from his past provincial commands rallied to his defense. Then events overtook this case, sweeping it into a vortex of controversy over a threatened coup d'état. In the midst of this crisis, Gelvezon maneuvered skillfully to mobilize allies—military superiors, classmates, and powerful politicians. By the time the palace had finished a military shake-up that destroyed many careers, Gelvezon emerged unscathed, with his command of the CIS confirmed.[38]

Watching Gelvezon's advance through such acrobatics, members of Class '40 felt a mix of reservation and admiration. Sometime in 1957, for example, classmate Ciceron de la Cruz was trying without success to negotiate a transfer for his brother Ben, a captain and a military surgeon, to the military hospital in their home province of Pampanga. During a social call on Gelvezon, then chief of the CIS, de la Cruz mentioned his frustrations:

> I was surprised when he [Gelvezon] said that was easy. He picked up his telephone and asked me to listen at the extension line as he dialed a number. As the phone rang at the other end of the line, a voice said: "Secretary [of Defense] Balao speaking."
>
> Gelvezon said: "Sir, Gelvy. I had just come from Malacañang. The President [Magsaysay] has a message for you. He wants you to transfer a certain Captain de la Cruz from Eldridge Hospital to Station Hospital No. 3 in San Fernando, Pampanga."
>
> Secretary of National Defense Balao just said "Will do" which ended the short dialogue.
>
> I was apprehensive afterwards because I do not like my brother to be caught in a vise were Secretary Balao to call the President for confirmation.
>
> Gelvezon said: "Don't worry. If the President will deny, I will insist that it was his order. I'll say he just forgot it because he is too busy to remember."

That was some boldness no ordinary individual will take. It takes guts, grit, and gumption to do that. Gelvy got it, being a goat in class notwithstanding.[39]

Not only was Gelvezon bold, but he was also a master of political reciprocity. At each stage in his career, he reached out, almost compulsively, to cultivate allies who might help him in some way in the future. While constabulary commander of Pampanga in 1953 during the Huk revolt, Gelvezon, almost by instinct, courted Representative Diosdado Macapagal, incurring a debt that would assure his later promotion to general. Writing in their 1986 *Golden Book,* Gelvezon told classmates that his promotion "was helped by a little luck when my humble contributions to the Anti-Huk Campaign impressed then Congressman Diosdado Macapagal, with whom I developed a cordial relationship. This later proved a god-send for my future career." In an interview two years later, however, General Gelvezon was more forthcoming.[40]

> At that time, . . . Pampanga [was] under the leadership of Governor Rafael Lazatin. It appears . . . that Macapagal [had] been harassed by these people, by disarming the men of Macapagal in Lubao. So, one day, Congressman Macapagal went over to my office. . . .
>
> He asked me . . . , "Can you help me, Colonel?"
>
> I said, "What's [it] all about?"
>
> He said, "My men are being harassed by Governor Lazatin, and I don't know how I can solve the problem."
>
> I said, "I'll see you tomorrow. I'll drive you tomorrow."
>
> So the next morning, he came around about seven o'clock in the morning and I rode with him to Lubao. . . . I told him . . . "would you please gather all your men, so I can talk to them." So when we arrived there, they were all gathered already in the town plaza, and immediately I lectured to them. . . .
>
> Then I asked Macapagal, I said, "Do you need these men for your protection?"
>
> He said, "Yes, I need them."
>
> "Okay," I said, "I'll make them my agents."
>
> So I gave instructions to my junior officer. . . . I said, "get a list of these people, their firearms, and from now on, you are all my agents. No one can touch you any more, and tell me if anybody touches you." That was the time when, you know, the Central Luzon was under suspension of habeas corpus. . . .
>
> I never saw him again. . . . When he became president, I visited him . . . in the Malacañang [Palace] among those officers who have paid respect to him. And when my turn . . . came around, he introduced me to Mrs. Macapagal, and said, "Mommy, you know Colonel Gelvezon? He helped us a lot. I owe him so much when we are in Pampanga. . . ."
>
> So, [all] of a sudden when . . . the vacancy existed . . . in the Third P.C. Zone, and

> [General Alfonso P.] Palencia was supposed to retire, immediately he [Macapagal] called me up. Asked me to be ready, airplane and all that. To immediately report there and take command without any . . . delay.

The president capped that posting six months later by promoting Gelvezon to general and then pulling strings in Congress to assure his confirmation. Proving his qualifications for the star, Gelvezon had recently reversed his dismal PMA record by graduating number one in his class at the Command and General Staff School. When the president announced the promotion, Gelvezon advised him immediately that there would be "a little hitch because of a congressman in my place, in my hometown" who would oppose his promotion. Gelvezon was referring, of course, to Representative Pedro Trono, who had now risen from town hall to Congress. Fortunately for Gelvezon, Trono was a member of Macapagal's own Liberal Party. The president told the congressman that he had made the promotion based on his evaluation of Gelvezon's capability as an officer and ordered Trono not to interfere with his selection. Despite a lifetime of relentless opposition, Trono was forced to keep silent.[41]

Asked if all these presidential favors had compromised his professionalism, Gelvezon used an anecdote to explain how he had preserved his integrity. During the 1965 elections, Gelvezon adroitly sidestepped pressure from the president's ally Sergio Osmeña to raid the Danao City bailiwick of Cebu's warlord Ramon Durano.

> When Macapagal went to Cebu, he asked me . . . to accompany him to the house of Sergio ["Serging"] Osmeña, then mayor. . . . He told Sergio Osmeña to coordinate with me. I don't know what he meant by "coordinate." The poor fellow, the late Serging, took [it] probably as to take orders from him. . . . So one day before or two days before the election, this was in 1965, he calls me up and gives me orders. "You raid the Danao town, the town of Danao, you raid . . ."
>
> I said, "By what reason should I raid?"
>
> He said, "Didn't you know that you took the order of the President Macapagal that . . . I can give you orders."
>
> I said, "Yes, but if it's right . . . you call up General Olivares, . . . our P.C. chief then, get from him clearance to give me orders to raid, I will raid. But I will not take it on my own to follow your orders. I'll follow my . . . direct commanding officer, okay?"
>
> So, he called up probably General Olivares, in turn Olivares called me up, "Oy, Ramon, what happened?"
>
> "Serging wants me to raid Danao," I said. "You give me the order, I will . . . follow."
>
> "What's the implication?"

I said, "Very much. Imagine raiding a political opponent on the eve of election? . . . I won't do that if I were you, but if you give me the order, I'll follow it."

He [General Olivares] said, "No, I will not give the order . . ."

So that time Serging didn't like me.

Pressed further about the implications of this exchange relationship with President Macapagal, Gelvezon argued with great conviction that he had preserved his professional integrity.

I know Macapagal is a level-headed fellow, who would not utilize you. If he had wanted me to be used, he would have called me confidentially and said, "You make me win in the election." But he didn't. He was too decent to tell that to me, or maybe he was too honest to himself. He never gave me instructions, honestly speaking. In conscience and in God, he never told me and he could have told me.

But would I do it? I will probably play around only because, as I told you, I am a professional soldier. I'm a professional soldier and don't think . . . he will like me or appreciate me or admire me if I will go out of the way for him.

General Gelvezon felt himself untainted by his relationship with Macapagal, but others did not share his view. When President Marcos took office after defeating Macapagal in the 1965 elections, he demanded the resignations of several generals he felt were beholden to his rival. Gelvezon tendered his resignation five months before his retirement date and left the service without the parade or the honors due a retiring general.[42]

Twenty years later, his classmates were still sensitive about his political choices. When the class returned to Baguio for their fiftieth reunion in February 1986, the twilight of the Marcos dictatorship, Gelvezon made eye contact with Chief of Staff Fabian Ver, who had served as his subordinate in Cavite thirty years before. Ver's recognition of Gelvezon angered the class.

On that occasion . . . we were invited to be in the grandstand . . . in a parade . . . where the general [Ver] was supposed to be guest speaker. He noticed me. . . . He saluted me, so I saluted him. To everybody's surprise . . . a civilian [was] noticed and then saluted by a four-star general. So he began his speech saying . . . "I'm happy to see that Class '40 has a sentimental journey headed by my mentor, my mentor General Ramon Gelvezon." And he looked at me. "I owe him a lot of gratitude having been my provincial commander in Cavite and . . . vice chief of CIS. . . . I would not be here right now if it were not for him."

After his speech, my classmates were not happy . . . because at that time there were already anomalies, irregularities in the armed forces charging Ver [with] so much. . . . They were serious. And they all look at me; pointing their fingers to me and say, "Mistah, mistah," because we call each other mistah, "Mistah, you are responsible for what the armed forces are doing now! You are responsible!"

I said, "I've nothing to do with him. He has changed a lot . . . under Marcos and [he was] not that way . . . under me."

But they were serious. They were serious . . . even for that little thing that happened to me. What more [would they do] if you are charged with irregularities and corruption?

So that is one of many restraints among PMA graduates not to do anything wrong while they are in the service. Could you imagine a fellow being ostracized by his own classmates?

Their condemnation was not a simple matter of partisanship. Riding up to Baguio for that reunion on the bus only a week after the 1986 snap elections, the class found itself equally divided between the Marcos loyalists and the "Cory Crusaders."[43] Divided they may have been over politics, but they were united in their opposition to the politicization that Ver represented. Gelvezon's imagined relationship with General Ver violated two central sanctions within the group's definition of honor—corruption and favoritism. Even in retirement from careers with countless compromises, the class still felt strongly, even passionately, that cultivating a political patron was a transgression. The thought that a classmate might somehow be implicated in the rise of General Ver, then the very personification of military corruption, outraged the class—even at this moment of reunion and celebration.

IBALOI GENERAL BABAN

While the members of Class '40 had reservations about Gelvezon's deft maneuvering, they praised Pedro Baban's restraint in his relations with President Macapagal. Born into an impoverished, "ignorant" Ibaloi farming family, Baban spent his childhood working in the family's highland gardens—small plots scattered on the ridges of the Luzon Cordillera amounting to no more than a hectare of land. Although this work made him "strong as a horse," Baban had a mediocre high school education and had to struggle with academics at the academy.[44] Finishing seventy-six among the class's seventy-nine graduates, he was still admitted to Army Flying School and there won his wings. In the last months of war, he led a guerrilla battalion near Baguio and fought the retreating Japanese at Bessang Pass.

After the war, Baban continued to rise in the air force. In mid-1965, he was serving as liaison officer with the U.S. Thirteenth Air Force at Clark Field when President Macapagal nominated him for promotion to general. In his orders to AFP headquarters for nominations, the president had specified one cultural minority from the north and another from the south. The November elections were approaching, and Macapagal was clearly playing politics with the roster to win votes. Advised of his nomination by the chief of staff, Baban expressed reservations about its propriety, saying he did not want to bypass a classmate who was senior to him. In the words of the *Golden Book:* "The Chief of Staff said that in all his life, he had yet to find a person crazy enough to refuse a promotion. He told Baban that it was also the height of impropriety to rebuff the wish of the President who was the Commander-in-Chief." With great reluctance, Baban accepted his star.[45]

Baban still felt compelled to apologize to classmates senior to him on the roster. He called on Reynaldo Mendoza, weeping and apologizing for the promotion. Mendoza urged him to accept. Baban was upset that politics had advanced him above his friend and fellow aviator Horacio Farolan. "Let's go see the president," Baban suggested to Farolan. "I'll tell him that I should not have been promoted ahead of you." So the two classmates marched into the palace, where General Victor Dizon, the senior aide-de-camp and their air force comrade, rang the president. "You fix it up," Macapagal instructed his aide. General Dizon then assured Farolan, "You'll be the next to be promoted." A few weeks later, the palace, as promised, nominated Farolan for general.[46]

In accepting his star, Baban had also accepted, albeit reluctantly, political interference in the lineal roster. Yet the class still celebrated him as a paragon, not because he had avoided politics altogether, but because he had not sought patronage. Indeed, in their scramble for stars on the eve of retirement, Baban would be one of very few who would place solidarity above ambition. Honor, once an absolute at the academy, was becoming relative.

In their struggle to reconcile professional ideals with political reality, classmates not only made different choices, but they did so from different personal positions. When the PMA picked a poor highland farmer like Pedro Baban, family and academy merged to produce an impersonal instrument of the state. For those, like Gelvezon, raised in the households of the political elite, the conflict between politics and professionalism persisted throughout their careers. For the sons of the lower middle class, military professionalism meant simple avoidance of political influence; but for sons of the elite, like Gelvezon or Vic Osias, it meant choice and discretion.

CHIEF OF STAFF OSIAS

The balance between politics and professionalism had been difficult under Macapagal, but it became even more complex when Marcos became president. The first year of his presidency, 1966, also marked thirty years of service for "the batch of 1936," who now faced compulsory retirement. In June, the anniversary of their admission to the academy, the armed forces held an extraordinary passing-out parade to honor the class's many retiring officers. Since military regulations counted their four cadet years toward their thirty years of maximum service, most were still only forty-eight or forty-nine and were, needless to say, upset at quitting in their prime. At least one classmate, Colonel Francisco Jimenez, quietly boycotted the parade, feeling that Marcos had failed to extend their service because the class's prickly idealism was interfering with his plans for the military. "If the Retirement Law was amended to exclude the four years of Cadetship before the retirement of . . . Class '40, '41, '42," he later wrote, "Ex-President Marcos would not have been able to make a 'coup d'état' in the guise of 'martial law.'"[47]

By the end of 1966, only a handful of classmates remained on active duty. For two of these who would later serve as the chief of staff, relations with the executive would test their standards of professional honor. Significantly, both of those elevated to supreme command were, like Marcos himself, native Ilocanos—the northern Luzon ethnic group known for its clannishness.

In January 1967, Marcos appointed General Victor Osias as chief of staff, the highest post in the AFP. Osias had been the outstanding cadet of Class '40—class president, basketball captain, and yearbook editor. As the son of a prominent politician, Senator Camilo Osias, General Osias had tried, following graduation, to be circumspect with politicians. After the Japanese defeat freed him from an irreconcilable conflict of loyalties, Osias quit his wartime jazz band and returned to active duty in 1945. Since he served in the air force, the most technical and least political of the services, Osias advanced through its hierarchy without incident for the next twenty years. But he again faced complex choices after his appointment as chief of staff.

Looking back on these events, Osias felt that the president's motivation for his appointment was political. During the 1965 presidential campaign, Marcos, the unchallenged leader of Ilocos Norte Province, had approached Senator Camilo Osias, the senior politician from Ilocano-speaking La Union, and proposed an alliance to cement the "solid" Ilocano north. Since Senator Osias was "by sentiment something of an oppositionist," he had remained neutral and "did not move close to Marcos." Moreover, General Osias himself had no real affection for the president: "I have always been a pretty good judge of men and their character and I simply didn't trust Marcos." Despite their lack of intimacy, the president picked General Osias as chief of staff, "probably," Osias speculated, "because we are both Ilocanos."[48]

During his first months in command, General Osias observed developments that he didn't like. He noted that Marcos's cousin, Major Fabian Ver, then commander of the Presidential Security Unit, was setting up what Osias called "Marcos's one big private army" inside the armed forces. If an officer faced obstacles on grounds of seniority or merit, it only took a phone call from Ver to overturn an unfavorable decision. Although loyalty had often been a factor in postings, Ver seemed to be building up a faction based on personal loyalties by advancing reservists who were, like himself, alumni of the University of the Philippines and its Vanguard Fraternity. Osias regarded this situation as "a real mess."

A confrontation between the president and his chief of staff was not long in coming. After only six months in command, General Osias received a courtesy call from a fellow pilot, Vietnamese Vice President Nguyen Cao Ky, and escorted him to Veterans Hospital in Quezon City, where Marcos was recuperating from a minor operation. As Ky emerged from the meeting, a Marcos aide approached General Osias, saying: "Sir, could you stay behind, the president would like a word with you." Marcos had suffered a serious setback in the legislature only a few weeks earlier when the Senate elected opposition leader Camilo Osias, the general's father, as its new president.

For the next four hours, General Osias sat alone in the anteroom until the aide emerged, saying: "Sir, the president would like it if you could approach your father and ask him to resign as president of the Senate." Aware of the consequences of refusal, the general replied: "No, it is not possible. My father and I have an understanding: he does not mix in my work and I do not mix in his. . . . I wish you would let me speak to the president and say that to him directly and we can save a lot of time."[49]

When General Osias phoned his father several days later to advise him of Marcos's request, the senator replied, "That's okay. We've all stood on our own, by ourselves, and we can take this too."[50] Father and son would defy the president.

Two months after their hospital encounter, the president punished General Osias. On 13 August 1967, the general's birthday, Marcos announced that he would be retired and replaced by his classmate Segundo Velasco.[51] Sacking Osias on his birthday was revenge, but choosing a classmate as his successor was humiliation. By selecting Velasco to succeed Osias, Marcos signaled that he was not repudiating Class '40 but simply punishing a single member. A week later, on August 21, General Osias stepped down without any ceremony and Velasco took command.

CHIEF OF STAFF VELASCO

Not only was General Segundo Velasco, like his predecessor, an Ilocano, but he was also a native of Marcos's home province of Ilocos Norte. Throughout his career, he rose steadily as a team player, not as a take-charge commander. Velasco was born in Bacarra, Ilocos Norte, in 1918, only six months after Marcos, and the two shared a similar background—middling family in decline, childhood in an impoverished small town, and a vision of law as the path to success. But without money for a college education, Velasco wound up at the PMA. By the end of his second year, he was well liked, though undistinguished in either athletics or academics.[52] Velasco, unlike Osias, gave few hints of exceptional leadership during his cadet days.

After graduation, Velasco washed out of flying school and wound up in the Field Artillery. After fighting on Bataan and suffering nine months of confinement at Capas, he returned home, married, and waited out the war. He rejoined the army in March 1945 and rose steadily, without demerit or distinction: battalion commander against the Huks, 1952; deputy commander of Task Force Jolo, 1955; and chief of constabulary, 1966.[53]

In his rise, Velasco learned the saving grace of patronage. During the Huk campaign in 1953, the army mounted a major operation to capture Huk Supremo Luis Taruc on Mount Arayat. As the cordon squeezed ever tighter up the flanks of this massive volcano, Major Velasco, then commanding the Sixth BCT, became responsible for a key sector. After days of patrolling, his men were exhausted and he failed to post sentries. Late that night, Taruc's security unit attacked, overran Velasco's lines, and escaped. The next morning, nearly fifty of his troops were found dead, most shot in the head while asleep. "One platoon was liquidated," recalled his classmate and fellow BCT commander Francisco del Castillo. "And that should have been a big debacle for him. But he was a *niño bonito* [fair-haired boy] of the generals and so he was always in the good grace of the old generals."[54]

While del Castillo's later rivalry with Velasco raises questions about his objectivity, the incident was still a textbook case of combat incompetence. In his manual for the AFP's Infantry School, military historian Uldarico Baclagon was sharply, albeit anonymously, critical of his own classmate. "Luis Taruc . . . and his well armed men slipped through a weak portion of the cordon held by a unit of the 6th BCT," wrote Baclagon. "Huk leader Luis Taruc should have been killed or captured at this time had it not been for the poor security employed by this unit."[55]

Other classmates were more sympathetic. Right after the war, Ramon Alcaraz had shown his respect for Velasco by inviting him to join an informal *hunta*, a "sort of ombudsman, composed of officers from different branches of the service with great integrity, to monitor officers getting astray." Although Alcaraz admits that this incident cast a shadow on his classmate's career, he disagrees strongly with del Castillo's imputation of favoritism. "When that Taruc incident came to the attention of the hunta, Velasco was replaced by Reynaldo Mendoza not because he was derelict—there was no doubt and later he was given benefit of the doubt. To claim 'niño bonito' in 1953," argues Alcaraz, "is absurd, as our standard of professionalism at that time was at peak with Magsaysay as defense secretary."[56]

In his *Golden Book* autobiography, Velasco glossed over this dereliction. "In December, 1952," he wrote, "I was assigned as Battalion Commander of the 6th Battalion Combat Team, in which post I was promoted to Lt. Colonel in April 1953." Then, he skipped past the disaster on Mount Arayat and moved on to his next success. "In March '54, I became the Intelligence Officer, G-2 of the IMA [Military Area] and, after a period of continuous operations, the surrender of the then Huk *Supremo*, Luis M. Taruc, became inevitable."[57] In Velasco's telling of his own story, Supremo Taruc becomes the vehicle for his ascent, a war trophy whose capture marks him for command. Velasco's contacts in headquarters may well have helped in the recasting of this narrative, turning gross demerit into merit.

The rivalry that inspired this revelation came in the final phase of the careers of Class '40, when competition for command, and pressure for compromise, were most intense. During the November 1965 elections, with mandatory retirement only seven months away, classmates felt compelled to position themselves for promotion. After taking command of the Anti-Fraud Squad to supervise balloting in the southern islands, Colonel del Castillo met with key aides in the Marcos campaign—Alejo Santos, Eulogio Balao, Floro Crisologo, and his own classmate Bartolome Cabangbang. The colonel wanted to cap his career as chief of constabulary, and the Marcos men played upon this ambition. "They promised me I'll be the next chief P.C. if Marcos wins," del Castillo later recalled. "But when Marcos won, I was forgotten." Instead, the president picked the colonel's classmate, Segundo Velasco. "They are from the same province... Ilocos Norte," del Castillo explained. "I am from Ilocos Sur, so I was not taken as chief P.C." Marcos's advisors then assured the colonel that he would become deputy chief, but Velasco exercised his prerogative to pick another, their classmate Salvador Piccio. Sulking, del Castillo thought about quitting the service. Then Cabangbang approached him, saying, "I need you, cavalier. So you join me." Marcos had just named Cabangbang to head an antigraft unit, the PARGO, with cabinet rank, and approved del Castillo as deputy director.[58]

After Velasco served a year as chief of constabulary, Marcos promoted him again, this time to deputy chief of staff. And, in August 1967, when Chief of Staff Osias defied him, Marcos again promoted Velasco. To split this powerful PMA class, Marcos played skillfully upon internal rivalries, winning allies and isolating enemies.

"There is no doubt in my mind about the loyalty of Velasco," insisted his classmate and provincemate Manuel Acosta. "Velasco was selected following Osias because of his being part of the Ilocano group of [ex-senator Eulogio] Balao and the president." Nonetheless, in a 1988 interview Velasco attributed his promotion solely to merit and stated that Marcos had appointed him without any prior consultation. Of course, the general admitted, as "the practical way of looking at it," the post was inherently political. "When you're already being appointed as chief of staff or being groomed..., you must be... the correct kind for a politician, you know. Otherwise, it will be apt for the politician also to name whom they want." Still, he denied any close relationship to Marcos, and insisted that the president did not try to influence him during his term as AFP chief.[59]

Velasco made loyalty to the president the hallmark of his tour. At the parade to honor his assumption of command, the general praised Marcos for deploying the military "in a massive civic action program... aimed at improving the economic posture of the Philippines." He promised that the armed forces would "not fail him in this noble endeavor."[60] Velasco's compliance would stand in stark contrast to Osias's defiance.

The two generals evinced almost diametric attitudes toward the president and his handmaiden General Ver when interviewed in 1988–89—long after his corruption had been exposed. Osias had observed clear signs of Ver's interference in the chain of command by 1967, but Velasco denied seeing any signs of impropriety during his later tour. Osias had detected a "private army" loyal to Marcos inside the AFP, but Velasco, his successor, saw nothing. As one of his classmates said of Velasco: "If he does not like some ideas that Marcos is telling him to do, he just closes his eyes. He does not object because, in the first place, that is the nature of Velasco—he was not an activist."[61] Oblivious to Marcos's manipulations, Velasco saw his career end in scandal when the "Jabidah massacre," the product of an executive covert operation, exploded in banner headlines.

One Sunday morning in March 1968, two Cavite fishermen found a half-dead Muslim soldier floating in Manila Bay not far from Corregidor Island. After treatment for a gunshot wound in his thigh, Private Jibun Arula, age twenty-three, was delivered to Cavite governor Delfin Montano, who called the press. The soldier's story was sensational. A native of Sulu, Private Arula was one of 150 Muslim trainees who had been shipped to a secret commando base on Corregidor in January 1968. After three months of training for sabotage operations inside Sabah, the Muslim recruits, angry over the harsh conditions, threatened mutiny unless they were paid. On March 17, Arula was one of twelve recruits taken to an airstrip where constabulary rangers suddenly opened fire with automatic weapons. As his companions fell to the ground, Arula, hit in the thigh, fled into the jungle and swam into the bay. The Muslim rebel Nur Misuari would later claim that sixty-eight Muslim soldiers had been massacred and that Marcos was planning "to use the trainees as the core of an invasion force to recover Sabah from the Federation of Malaysia."[62]

On March 21, a front page report in the *Manila Times* charged that the military establishment had a secret training camp—"so secret high defense officials don't even know of its existence"—and that this camp was training "armed infiltrators and saboteurs."[63] Marcos was sensitive to the controversy and named a top-echelon investigative panel—Chief of Staff Velasco, his deputy General Manuel Yan, and army commander Romeo Espino. On March 22, only days after the incident broke, the panel delivered a report that appears, in light of later information, to be a cover-up.

> The training going on in Corregidor is part of the AFP training in counterinsurgency. It constitutes a regular part of the course for the special forces, which include *infiltration, sabotage,* escape and evasion, survival, and organization of indigenous forces for defense. Corregidor has always been a training ground for such special forces. . . .
>
> Information from friendly neighboring countries emphasizes the fact that Communist agents are operating in the general area of our southern backdoor and indicates an upsurge in communist activities there. This situation requires the organization of well-trained counterinsurgency forces to meet the threat. The training going on in Corregidor is part of this program.[64]
>
> The shootings and killings were simply mutiny and desertion by trainees.[64]

The palace released the report to the press corps immediately. To lend authority to its soothing conclusions, Marcos invited reporters to photograph him posing at his desk holding the report with the three uniformed generals arrayed before him. At the center of the *Manila Times* photo, Velasco stares directly into the camera lens.[65]

Congressional investigators soon found evidence of a covert apparatus operating under executive control. Major Eduardo Martelino, commander of the operation, testified that he had sent 180 Muslim recruits from Sulu to Corregidor in December 1967 for training in the top secret "Operation Merdeka." Intrigued by this dramatic testimony, the *Free Press* found that the mysterious major, after serving as an undercover political operative for two past presidents, had become head of the Civil Affairs Office under Marcos and, in June 1967, had gone to Sulu, where he converted to Islam, taking the name "Abdul Latif." There he married a young Muslim and, wearing a burnoose like Peter O'Toole in the film *Lawrence of Arabia,* recruited several hundred Tausug trainees for this special mission.[66]

In his testimony, the army's commander, General Romeo Espino, admitted that he had inspected the Corregidor camp in February. However, the *Free Press* charged that Espino had, in fact, first visited Corregidor in December, prior to the arrival of the Muslim trainees, and ordered that the area around the old U.S. Army hospital be turned into a training base code-named "Jabidah" and declared a restricted area. The opposition's leader, Senator Benigno "Ninoy" Aquino, charged that Operation Merdeka was cover for a "secret strike force under the president's personal command, to form the shock troops of his cherished garrison state."[67]

As the controversy intensified, Marcos reacted by abolishing the Civil Affairs Office and ordering a court-martial for the twenty-four soldiers involved in the operation. Although their trial started in April 1968, only weeks after the massacre, it dragged on inconclusively for another three years until the issue faded.[68]

In May 1968, at the height of the media storm, Chief of Staff Velasco became another casualty of the Corregidor controversy. After a close study, Senator Ninoy Aquino charged that "there was a secret army, so secret even the chief of staff had been bypassed in its planning." Indeed, Velasco insisted, more than twenty years later, that he was not aware of any base on Corregidor. In retrospect, he suspects, but is not certain, that it was "an intelligence operation . . . under the control of the commander in chief." While nobody accused him of involvement, it was clear that the general was not in control of his command. On May 25,

Marcos announced that Velasco would retire after only nine months as chief of staff.[69]

Velasco went quietly, praising the president all the way. "I wish to express my most profound gratitude to our commander in chief for reposing his trust in me," he told the troops at the turnover ceremony three days later. "His concern for the development of... the armed forces... has been a source of inspiration to the rank and file. Never have they felt a greater sense of achievement."[70]

Loyalty had its rewards. After Velasco retired, Marcos appointed him to a lucrative post as a director of the Veterans Bank, a position he used to launch a career as head of a half-dozen corporations. Nonetheless, Class '40 felt that their old friend "Gunding" had preserved his honor. "The reason... Gunding served only less than a year was that the president could not cause Gunding to do what's not right—that's Class '40," explained classmate Francisco Jimenez.[71]

In January 1971, three years after the massacre, bombs erupted in Manila with the fingerprints of a military operation. After the headquarters of Esso and Caltex were bombed, *Daily Mirror* journalist Cesario del Rosario, who had investigated the Corregidor killings, reported that "all indications point to the continued existence of this secret military unit." The original Jabidah team, including Major Martelino, had been absorbed into the newly created Home Defense Force under General Espino and his psychological warfare (psy-war) expert Jose Crisol.[72]

Then, on 21 August 1971, there was an explosion at a Liberal Party rally in Plaza Miranda that killed nine people and wounded three opposition senators. After a respite of several months, the bombs began exploding across the city—the Supreme Court, City Hall, and a downtown department store. One general experienced in demolition inspected the blast at the city water main and saw telltale signs of a military operation.[73] Using this chaos as pretext, Marcos, with the support of his new chief of staff, Espino, declared martial law on 22 September 1972.

"DIE-HARD FASCIST" KARINGAL

Martial law clarified, even simplified, political choices for Class '40. At the outset, a blanket repression limited options to quiescence or cooperation. With the resumption of elections in 1978, classmates faced more complex choices, ranging from support to subversion. While Bartolome Cabangbang won national prominence as an active opposition leader, his classmate Tomas Karingal gained notoriety as a tough, even brutal police commander. If Cabangbang, a former pilot, flew across the archipelago to rebuild democracy, so Karingal, an intelligence operative, burrowed into his Manila bailiwick to make it a bastion of repression. Indeed, throughout his career Karingal, in contrast to his former classmates, advanced through espionage and patronage—an exceptional path that others might have chosen had they not been restrained by cadet socialization and class solidarity.

Karingal's personal history seems, up to the point of his dismissal, typical of Class '40. Born in Luzon's impoverished Bikol peninsula, he graduated from his local provincial high school in 1933 and joined the constabulary as an ordinary trooper. He had been admitted to the academy with the original 120 members of the class but then flunked out after a year, thus experiencing only the negative conditioning of hazing. With a conventional career path now blocked, Karingal would win promotion through his flair for covert operations. Up to the moment of his death, when he apparently made a fatal miscalculation, he showed an extraordinary ability to amass local knowledge and translate it into operational intelligence, first in his home province and later in the nation's capital.[74]

After leaving the PMA, Karingal returned to the ranks until headquarters gave him his first clandestine assignment. According to the *Golden Book* biography of another dismissed classmate, Benedicto Valenzona, "he and . . . Tomas Karingal, were assigned an intelligence mission by Gen. [Guillermo] Francisco to conduct surveillance . . . of a known Japanese spy, Madame Panseni." As a reward, headquarters appointed Karingal a reserve third lieutenant in April 1940, only a month after his former classmates had graduated with regular commissions.[75]

After the U.S. Army withdrew to Bataan and the Japanese occupied Manila in January 1942, Karingal remained behind enemy lines and engaged in espionage with Valenzona. Another classmate, Lieutenant Alfredo Filart, separated from his unit in the retreat to Bataan, met the two operatives in Manila and decided to join them. After the U.S. surrender on Bataan, they continued to spy on the Japanese by opening a *sari-sari* store downtown on Rizal Avenue to mask their movements. This underground unit, President Quezon's Own Guerrillas II, conducted espionage and sabotage operations around Manila, working closely with several Class '40 alumni in the Japanese-sponsored Bureau of Constabulary until the city's liberation in early 1945. While Filart was caught by the *Kempeitai* and suffered eight months' harsh imprisonment, Karingal emerged from this dangerous game unscathed. Instead of staying in the capital, where capture was almost certain, he spent much of the war with a guerrilla unit in the hills of his Bikol homeland. At war's end, Karingal, like many classmates, went to the United States for advanced training and spent three months in the Provost Marshal School at Fort Sam Houston.[76]

In the military reorganization of 1947–48, Karingal, along with nine other nongraduating members of Class '40, was commissioned a regular officer. Though he would now rise with his class, he continued specializing in covert operations. During the Huk revolt, Karingal served with the constabulary in his native Bikol for five years, winning two medals and a commendation for aggressive operations that brought, in his words, "the HMB [Huk] situation in southeastern Luzon . . . under complete control." Karingal would later claim credit in press releases for the killing of eleven Huks, including Commander Nabisco, and leadership of an operation that "bagged," or killed, Communist Party leader Mariano Balgos. A year later, he again distinguished himself as intelligence chief for Task Force Jolo, learning the Tausug language to effect the capture of the renowned rebel Hadji Kamlon.[77]

In 1960, President Carlos Garcia, then closely allied with members of Class '40, appointed Karingal police chief of Quezon City, the country's capital and a fast-growing Manila suburb. The post effectively enmeshed Karingal in national politics, adding a second dimension to his career. During the city's mayoral campaign three years later, President Macapagal, apparently feeling that Karingal was tainted by ties to the opposition, ordered his sudden transfer. The Citizens League of Quezon City filed a formal protest with the Commission on Elections demanding Karingal's immediate return. Significantly, the league's mayoral candidate was Carlos "Charlie" Albert, a 1937 Annapolis graduate with close ties to Class '40.[78]

Karingal's surprising reappointment in April 1965 testifies to both his efficiency and political dexterity. After two Quezon City policemen attempted to rape a college professor in her own car, President Macapagal scrambled to contain the political damage in a presidential election year. Blasting the "complacency of the highest officials," opposition candidate Marcos charged that the professor's case "epitomized . . . the perversion of law and authority." Four days after the assault, a front-page photo in the *Manila Times* showed Macapagal shaking hands with his new Quezon City police chief, Tomas Karingal—the very man he had fired only eighteen months earlier.[79]

By the time martial law came in 1972, Karingal had joined the constabulary's Metropolitan Command (Metrocom) and was superintendent of its Northern Police District, a zone that included his old Quezon City bailiwick. When anti-Marcos demonstrations erupted after the assassination of Ninoy Aquino in 1983,

Karingal, then a general in the Integrated National Police, became notorious for his deployment of riot squads with automatic rifles, gas guns, and batons to smash picket lines and protest marches. On 24 May 1984, the general was dining with fellow police officials at the Fisher's Reef Restaurant in Quezon City. Two gunmen entered and shot him fatally.

The next day, the Alex Boncayao Brigade, the "sparrow" assassination unit of the communist New People's Army, issued a press release taking credit for the Karingal killing. Their manifesto called him a "die-hard fascist" and charged him with "blood debts to the Filipino people"—citing his killing of Huk leader Mariano Balgos, his capture of Hadji Kamlon, and his bloody assaults on anti-Marcos demonstrators. After thanking "AFP personnel for assisting . . . in the successful operation against Karingal," the statement concluded: "Let the punishment of Karingal serve as a warning. Revolutionary justice will catch up with the incorrigible enemies of our people."[80]

Marcos honored his loyal servant. The day after the murder, the president promoted Karingal posthumously to major general and, in an emotional ceremony, awarded him three medals for distinguished service.[81]

Although many classmates opposed Marcos, they still honored Karingal at their fiftieth reunion two years later. Their anti-Marcos militant, Ramon Alcaraz, bruited about reports that "the assassination was masterminded by Imelda Marcos because . . . Karingal was supposed to deliver votes . . . and the candidate of Imelda lost very badly." No longer was he the public face of repression, a riot-squad commander who killed strikers and shot student demonstrators. Their *Golden Book* narrative reinvented him as another character in a chronicle of military professionalism—an officer who "served brilliantly" and was the victim of a murder that "remains unsolved."

Through cinema it is their view that has prevailed. Only days before the class's fiftieth reunion in March 1990, Seiko Films released a blockbuster biographical feature, *Ako ang Batas—General Karingal* (I Am the Law—General Karingal), on fifty-one screens across Metro-Manila. Featuring Eddie Garcia in the title role and some erotic simulated sex, the film attracted huge crowds and positive reviews that, like Class '40 itself, ignored Karingal's reputation for repression and celebrated him, in the words of the film, as "a brave soldier, a model father, . . . and a symbol of true service to the nation." As a final step in this process of historical revision, the Philippine National Police have honored his two decades as the capital's top cop by naming its Quezon City headquarters Camp Major General Tomas B. Karingal.[82]

But both views, positive and negative, miss the most significant aspect of

Karingal's career. Karingal was a more important historical figure in death than in life. For had he been Metrocom chief of the Northern District two years later when Enrile and Ramos launched the "people power" revolution, Karingal, with his fierce loyalty and passion for order, may well have followed the president's orders and attacked the crowds around the rebel camps, crushing the revolt on its first day. As it turned out, his death allowed the promotion of General Alfredo Lim, a pliable, acutely political officer who would make a very different decision in those critical hours that changed the nation's course.

CONGRESSMAN CABANGBANG

While most of Class '40 agonized over their relations with politicians, Bartolome Cabangbang resolved the ambiguity by quitting the military and running for Congress. During his thirty years in electoral politics, he established a remarkable record as a three-term congressman, presidential candidate, and leader of a mass movement to win U.S. statehood for the Philippines.

Born into a family of poor tenant farmers on Bohol Island, an impoverished backwater in the Central Visayas, Cabangbang entered the PMA because he could not afford a college education. He placed a low sixtieth in the class academic rankings as a yearling, but he proved himself a leader and an aggressive athlete, a boxer who "dons the gloves with ease and slugs it out toe to toe with another until one of them drops."[83]

Cabangbang went to the United States for flight school in mid-1945 after distinguished war service as a pilot and anti-Japanese guerrilla. After his fellow pilots returned to the Philippines, he stayed behind and joined the U.S. Air Force, rising to the rank of major. In 1947, his former squadron commander, Jesus Villamor, convinced him to return to Manila as his deputy in the Civil Aviation Administration (CAA). After serving as operations manager of Manila's airport, Cabangbang accused his superior, the secretary of public works, of partisanship and resigned, charging that "politics was being injected into the aviation policies of the country."[84]

With characteristic irony, Cabangbang decided to battle politicization by becoming a politician. During the 1949 presidential campaign, he joined Dr. Jose Laurel's team as a stump speaker, using his own exemplary war record to defend his candidate against charges of collaboration with the Japanese. At a party meeting back home in Bohol, Cabangbang suddenly announced his candidacy for Congress when none of the local politicians were willing to run against an entrenched incumbent. With only fifteen hundred pesos, Cabangbang campaigned hard and emerged as the apparent winner on election day. Four days later, however, late ballots from an opposition-controlled bailiwick gave his rival the winning margin.[85]

Defeated by his poverty, Cabangbang decided that integrity demanded he become a smuggler. With an anonymous backer, Cabangbang and his classmate Urbano Caldoza purchased a war-surplus C-47 aircraft and set up Trans-Asiatic Airlines. For a year, the two pilots made repeated trips to Switzerland and Hong Kong, where, according to classmates, they loaded up on high-duty items such as watches and perfumes for the flight back to the Philippines. With his unequaled knowledge of the archipelago's remote airstrips, Cabangbang landed the contraband without inspection.[86]

In a later campaign biography by classmate Baclagon, Cabangbang concealed this blank in his biodata with a fiction that hints at a mystery: "In February 1950 Cabangbang, . . . in utter disgust with the 1949 elections, accepted the job of operations manager of Trans-Asiatic Airlines, Inc., then operating in Burma under contract with the Burmese government." The date, February 1950, is alluring. It was at that time that the U.S. Central Intelligence Agency, after recruiting "a motley band" of aviators with covert-action credentials from World War II, armed nationalist Chinese troops along the Burma-China border for an invasion of China in defiance of Burmese sovereignty. Certainly, Cabangbang's wartime espionage would have impressed the CIA. But it would not have impressed the Burmese military. Within months, the operation proved a disaster, and most of the mercenaries moved on. In another interesting coincidence, Cabangbang's biography says that "after less than a year of self-imposed exile, he returned to the Philippines and served as an airline pilot for . . . Philippine Air Lines, Inc." In effect, he left Manila when the CIA operation started and came back when it was over. Had he been just a smuggler, or was he a smuggler and a CIA operative? Was he really flying watches from Switzerland to Manila, or was he, like some CIA-contract pilots, flying opium from Burma to Bangkok?[87]

Cabangbang built a war chest by smuggling and later won the congressional seat for Bohol's second district in the 1953 elections. During his first term, he finished a law degree at Far Eastern University, flew for Philippine Airlines, and continued smuggling to pay off his campaign debts—an alternative to trading votes for cash. With his usual bravado, he told classmates that he had to be a smuggler to avoid becoming a thief.[88]

In Congress, Cabangbang proved himself, in the words of his biographer, "an incorruptible legislator." He also proved adept at procuring sufficient patronage to win reelection in 1957 and 1961. When Vice President Carlos Garcia, a fellow Boholano, succeeded to the presidency after Magsaysay's death in an airplane crash, Cabangbang became his "virtual hatchet man" in the halls of Congress, a proximity to power that won him a national reputation. In the November 1957 elections, Cabangbang ran unopposed in Bohol and served, simultaneously, as chairman of Garcia's winning presidential campaign. With Garcia's inauguration, Cabangbang exercised extraordinary influence, serving as "presidential mouthpiece" and battling his regional rivals, Senator Mariano J. Cuenco and Sergio "Serging" Osmeña, Jr.[89]

From his vantage point in Congress, Cabangbang became a central figure in the politico-military crisis of 1958. In contrast to his bruiser role in debate, he had established an expertise on military matters as chair of the House National Defense Committee—a position that lent credibility to his explosive exposé of a threatened coup.[90] The real meaning of this incident is still clouded by the intensely partisan controversy of the day. In retrospect, it seems possible that some officers loyal to the deceased president Magsaysay may have considered a coup to recoup their waning influence.

This crisis revolved around three academy classes. At the apex of command, Chief of Staff Alfonso Arellano and Defense Secretary Jesus Vargas had graduated first and second respectively in the Constabulary Academy's Class of 1929, the renowned "class of generals," with four stars among only seven alumni.[91] This older batch was loosely allied with a rising clique from Class '44 that played a catalytic role in the crisis. Opposing their ambitions was a group of middle-ranking officers from Class '40, including Representative Cabangbang.

Entering the academy in mid-1940, Class '44 had been disbanded in the first days of World War II, and many had joined the Hunters ROTC guerrillas, knitting close ties with comrades through risky missions.[92] Several classmates had served with Magsaysay in the Defense Department, and later, along with allies among the Hunters veterans, launched the Magsaysay for President Movement—apparently compensating for their unconventional career path by mixing regular duties with covert operations and politics. After taking office in 1954, President Magsaysay appointed 122 military officers to civilian positions, including a coterie from Class '44 and the Hunters—notably, Labor Secretary Eleuterio Adevoso ('44), Presidential Action Committee head Frisco San Juan ('44), Agriculture Under-Secretary Jaime Ferrer, and Foreign Affairs Under-Secretary Raul Manglapus. When Magsaysay died in 1957, some of his loyalists reportedly met to discuss a military takeover since the vice president, Carlos Garcia, was abroad on a diplomatic mission to Australia. Eventually, they broke publicly with Garcia, sparking rumors of a possible coup d'état.[93]

Significantly, the rumors began in October of 1958 following a succession of unsettling coups in Burma, Thailand, and Pakistan.[94] Tensions in Manila rose on October 25, when former president Jose P. Laurel publicly praised Defense Secretary Vargas in a speech that one paper called "the opening gun in the Vargas-for-President boom."[95] Several days later, liaison officers in the Defense Department loyal to Vargas received reports that President Garcia was meeting with congressmen to consider sacking their boss.

On October 30, one of the defense secretary's liaison officers, Captain Emilio Domingo, a member of Class '44, telephoned a military aide at Malacañang and asked whether or not the president and party leaders had decided to remove Vargas from office. The next day, Captain Domingo went to the palace, where he approached several of the president's aides, seeking details of these discussions and asking where the president had slept.[96] Concerned, the two military aides reported the incident directly to the president.[97] Moreover, the president's senior aide, Emilio Borromeo (PMA '42), also reported that Secretary Vargas had had a meeting with some officers of a suspicious nature.[98]

That same night, Garcia, angered and afraid, summoned Secretary Vargas and Chief of Staff Arellano to the presidential bedroom for a "severe dressing down" that lasted two hours. The palace press corps observed the two emerging "red-faced." In a move that was "almost certainly related," the commander of Military Area I, responsible for the capital's security, stepped down in a rushed ceremony with a valedictory that "scotched rumors of a coup d'état by the armed forces."[99] As coup rumors broke on page one, Secretary Vargas and General Arellano staged a massive "loyalty rite" at Camp Murphy with elite units and the PMA cadets parading before the president. Two days later, the defense secretary banned all military officers from entering the palace without authorization.[100]

With the martial specter looming, Representative Cabangbang, Garcia's closest legislative ally, delivered a secret report, his so-called "white letter," denouncing coup plans by Magsaysay partisans. Plan one was a "special political operation" featuring a massive psy-war campaign to build up Secretary Vargas as the country's "messiah." Plan two, for the "coup d'état," had been preempted by the negative press reaction and was now "subject to future developments." Cabangbang identified the key plotters as Chief of Staff Arellano, Secretary Vargas, and the commanders of several intelligence units. As evidence, Cabangbang claimed that the conspirators were removing neutral officers who could block their coup: Lieutenant Colonel Job Mayo, chief of the Military Intelligence Service; Lieutenant Colonel Deogracias Caballero, chief of the Psychological

Warfare Office; and Lieutenant Colonel Ramon Gelvezon, chief of the constabulary's Criminal Investigative Service. In Cabangbang's sensational account, Gelvezon's impending transfer was "a missing link" in the coup plot.[101] Significantly, all three lieutenant colonels, like Representative Cabangbang himself, were members of Class '40. After releasing his "white letter" to the press with the plotters' names, Cabangbang called for a military shake-up and launched a sustained attack on both Vargas and Arellano.[102]

Once press coverage faded, the president began an unprecedented purge of dozens of senior officers. In mid-December, Navy Captain Carlos Albert, named in Cabangbang's report as a plotter, was relieved as chief of AFP intelligence, as was Chief of Staff Arellano. Finally, in May 1959 Secretary Vargas was forced to resign amid unfavorable press about his arrogance and ambition. The Hunters ROTC contingent, with few exceptions, also resigned en masse.[103]

To assure the military's loyalty, Garcia replaced the "Magsaysay boys" with his own partisans, often favoring fellow Boholanos or other Cebuano speakers from his native Visayas. For example, the new chief of constabulary, General Isagani Campo, had been the First Lady's high school classmate.[104] Now at the peak of his powers, Cabangbang influenced many of the new military postings. To replace Captain Albert as the intelligence chief, headquarters appointed Navy Captain Heracleo J. Alano—the president's provincemate and Cabangbang's classmate. Similarly, Cabangbang backed the appointment of his wartime commander, Alejo Santos, as the new defense secretary and advanced other allies within the AFP.[105] Simultaneously, Cabangbang urged Congress to strip the military of the power that it had gained under Magsaysay, recommending that serving officers be barred from civil posts and intelligence funds be cut to restrain the AFP's covert capacities.[106]

Although caught up in the controversy, Cabangbang's classmates tried to remain aloof from its partisan rewards. The Class '40 colonels who had reportedly blocked this coup won neither promotions nor prime postings. Even Captain Alano, who rose from navy service to a senior staff post, returned to the fleet at the close of Garcia's term and later excised the whole episode from the official narrative of his career.[107] On balance, the active-duty Class '40 alumni acquitted themselves in this murky affair with something akin to honor. Unlike those of Class '44, who were active as partisans and plotters, the cavaliers of Class '40 had worked to block a possible coup and did not seek political reward. Even so, as the class members continued to rise through the hierarchy, their ideals were starting to face pressures far beyond the simple verities of the honor code.

After his patron, President Garcia, lost a reelection bid in 1961, Cabangbang became one of the new president's sharpest critics. Speaking before Congress two years later, he charged that President Macapagal's relatives were engaged in graft on his behalf. Several days later, one of the president's allies, Representative Floro Crisologo, rebutted by implicating Cabangbang in an alleged plot to assassinate former president Magsaysay. The administration then filed formal charges that "Cabangbang, in view of his training as a commando and sabotage expert, authored the Magsaysay plane crash . . . and engineered [an] assassination plot against President Macapagal."[108]

In the 1965 elections, Cabangbang made an unsuccessful bid for his party's vice-presidential nomination, issuing a biography written by his classmate Baclagon that praised him for "remaining true to his convictions . . . and, above all, to the 'Honor Code' of the Philippine Military Academy." Cabangbang's hatred of Macapagal made him a natural ally of opposition candidate Marcos at the Nacionalista Party convention and won him a seat on its senatorial slate. When Marcos took office in 1966, he appointed Cabangbang, narrowly defeated in his Senate bid, to head the executive graft-busting arm, the Presidential Agency on Reforms and Government Operations (PARGO).[109]

As always, Cabangbang shared his success by staffing PARGO's upper echelon with three retired classmates. And, as always, his sidekick and classmate Caldoza followed him into the new administration as assistant commissioner of tourism. Eventually, Cabangbang resigned when he found that the president was blocking prosecution of cronies implicated in PARGO's investigations. Nonetheless, Marcos gave him a slot on the Nacionalista Party's senatorial lineup for the 1969 elections, but Cabangbang was swept away in the opposition landslide.[110]

Breaking with both Marcos and conventional politics, Cabangbang then launched the "Statehood USA Movement." Although the nation's political elite dismissed this campaign as "clearly idiotic," it had strong mass appeal and attracted several million dues-paying members. Martial law eclipsed his crusade in 1972, but Cabangbang reemerged six years later as a leader of the powerful Central Visayan opposition party, *Pusyong Bisaya,* and won a seat in the regime's new parliament, the interim *Batasan Pambansa.*[111]

In keeping with a biography of shadows and voids, Cabangbang's relationship to Marcos from the ranks of the opposition is another mysterious chapter. After six years as the president's trusted aide, it seemed that Cabangbang had cut his ties and gone home to the Visayas, where he emerged as a leader of an independent opposition party. Or had he?

Indeed, the curious progress of the 1978 legislative elections in the Central Visayas' Region VII forces us to question the opposition's autonomy. When the incomplete electoral returns began leaking from the Commission on Elections, or Comelec, on April 15, the opposition *Pusyon* jumped to an early lead and seemed certain to sweep the region's thirteen seats. In these early returns, Cabangbang himself was running fourth among the twenty-six candidates with an impressive tally of 263,565 votes.[112]

But the counting soon turned against the opposition at Comelec, the Marcos-dominated elections board, and four opposition candidates, Cabangbang included, lost what had seemed certain seats. Then, after violent protest demonstrations erupted in Cebu City, Marcos suddenly capitulated with uncharacteristic grace and conceded all thirteen Region VII seats to the opposition. Why did the president give way so readily when his dictatorial powers were still unchecked? Perhaps because he controlled both parties—his own KBL and, covertly, the opposition *Pusyon*. Indeed, the founder and president of *Pusyon*, Casimiro M. Madarang, Jr., was a Marcos appointee to Comelec and had remained on the commission even while organizing his opposition party—raising serious questions about the sincerity of his break with the regime.[113] If *Pusyon* was, in fact, a secret Marcos tool to control the opposition, then the possibility remains that its leaders, Madarang and Cabangbang, were his hirelings.

In this murky world of martial-law intrigue, another of Cabangbang's classmates gained some rare insight into his ambiguous politics. Returning to Manila from America a few months after the April 1978 elections, Ramon Alcaraz was exploring the possibility of an anti-Marcos coup and arranged a secret meeting with Cabangbang and his sidekick Caldoza. Cabangbang claimed that he had won the president's approval for a Philippine-American Friendship Group and was now using it as a facade for arming a special force to initiate a rebellion by a rising number of active-duty officers. Though Alcaraz concluded that Cabangbang was "one of three real oppositionists elected to Parliament," he still had doubts about his classmate's relationship with Marcos. Why, after all, would the president allow him the extraordinary privilege of organizing his friendship group?[114]

When Marcos rescinded martial law and proclaimed his New Republic in 1981, Cabangbang revived his statehood movement and ran for president under the banner of his Federal Party. As he told his classmates, he "admired the Americans, their way of life, and ways of running things," and was not afraid to offend nationalist sensibilities by advocating U.S. statehood. Although no oppositionist could expect to challenge Marcos, Cabangbang did attract a tremendous amount of support, particularly in his own Cebuano-speaking region, and felt that his campaign affirmed "a latent desire of . . . Filipinos to make the Philippines a part of the United States." Despite the omnipresent nationalist rhetoric of the day, Cabangbang, in the words of the newspaper *Malaya*, "astonished many by garnering almost a million votes" on a platform that promised U.S. statehood.[115]

But there is still reason to question the sincerity of Cabangbang's opposition to Marcos in the 1981 elections. Running with all the advantages of an entrenched dictator, Marcos desperately needed a credible opposition to lend legitimacy to his New Republic. On the day before the ruling KBL's nominating convention, the United Democratic Opposition (Unido) embarrassed the president by announcing a "total, active, and vigorous boycott."[116] Since the real opposition would not cooperate, Marcos seems to have created his own. Cabangbang had already launched his candidacy at the head of the Federal Party, but his quixotic "statehood USA" plank denied him credibility. The president then signaled to the leaders of his former party, the now-moribund Nacionalistas, to launch an opposition candidate and they dutifully nominated General Alejo Santos, a close Marcos ally. Although Santos was supposed to be an opposition candidate, his campaign manager was Francisco "Kit" Tatad—recently retired from the regime after eleven years of loyal service as Marcos's minister of information.[117]

The results were predictably lopsided—88 percent for Marcos, 9 percent for Santos, and 3 percent for Cabangbang. Watching these elections from exile in America, opposition leader Ninoy Aquino quipped that Santos had run "for the *fund* of it." But nobody, not even his closest classmates, could offer such a tidy explanation for Cabangbang's candidacy.[118] Still, it is curious, to say the least, that both opposition candidates had been close comrades in the wartime resistance and were former Marcos allies.

This campaign turned out to be his last. Not long after the elections, Cabangbang's classmate and lifelong comrade, Urbano Caldoza, died, depriving his statehood movement of its chief of staff. Although the exiled activist Alcaraz maintained contact in hopes of forging a coup alliance between the Philippine Marines and Cabangbang's commandos, the plot never moved beyond the dinner table. In September 1985, Cabangbang complained of a severe headache and was taken to Loon Hospital in his native Bohol, where he died of a cerebral hemorrhage.[119]

Even death adds to the mystery. The militant opposition daily *Malaya* honored him with a front-page obituary, recalling his 1981 presidential campaign against Marcos. Although Cabangbang was supposed to be an anti-Marcos critic, Class '40 still managed to arrange a state funeral at Camp Aguinaldo with acting Chief of Staff Fidel Ramos, a Marcos cousin, speaking for the armed forces. An impressive delegation of his PMA classmates turned out to honor a man whom many felt was the most distinguished member of the Class of 1940.[120]

In retrospect, these biographies illustrate the importance of socialization for officers when they finally reach heights of command and gain the authority to do great harm. General Karingal's exceptional career, filled with covert missions and extralegal violence, highlights a larger point that should not be lost among these profiles in compromise. Despite their many derelictions, most of Class '40 struggled, with varying success, to conduct themselves within the bounds of the honor code. Without overburdening the limited evidence we have about his life, Karingal, who had quit the PMA before the positive phase in its socialization, offers some indication of what can happen when an officer wins command without such internal restraints.

So intense was their indoctrination at the academy that classmates who finished the four-year course carried, in some form, the constraints of honor throughout their careers. With each step up the hierarchy, political pressures forced ever-starker choices between compromise or resignation. Those who served the full thirty years agonized, rationalized, and even compromised. But they rarely capitulated to the pervasive partisanship and venality that became the hallmark of the later generation of reserve officers under General Ver. Each promotion brought choices that slowly eroded the old absolutes of the honor code. But, for the cavaliers of Class '40, the calculus was usually done with honor as a variable somewhere in the equation.

Within this diverse group of seventy-nine men, each with his own moral compass, most adhered strictly to conventional morality, some stretched the rules, others compromised under pressure, and a few presumed to create their own codes. Osias quit rather than comply, Velasco ignored his commander's crimes, and Cabangbang smuggled in an ironic pursuit of principle. They were not corrupt. They did not become the hirelings of politicians. All tried to live by the honor code, albeit one circumscribed by life under democracy and dictatorship.

Though flamboyant, Cabangbang still tried to fulfill the class vision of honor. Most importantly, he practiced his politics outside the military, not inside like Karingal. Even as a politician, he was never the pliable courtier but always the man of outrageous principle who sought somehow to transcend the contradictions of his society, elevating moral choice to a complex calculus beyond the simple binary of do and don't. And he made these choices in pursuit of principle or the public good, not private gain. While Osias's active integrity contrasts sharply with Velasco's passivity as chief of staff, even Velasco's compromises seem, by later standards, restrained. He was guilty of omission, the passive acceptance of patronage or political pressure, but not of the active venality that became the hallmark of his successors.

Even the handful of classmates who reached the highest echelon, where the price for honor was often prohibitive, conducted themselves in ways that evince an enduring socialization. Indeed, our final pair of classmates, the subjects of extended profiles in the next chapter, showed a moral formation that would weather extraordinary pressures for compromise and corruption.

Ramon Gelvezon, from cadet in 1940 (*above*) to general in 1965. (Author's Collection)

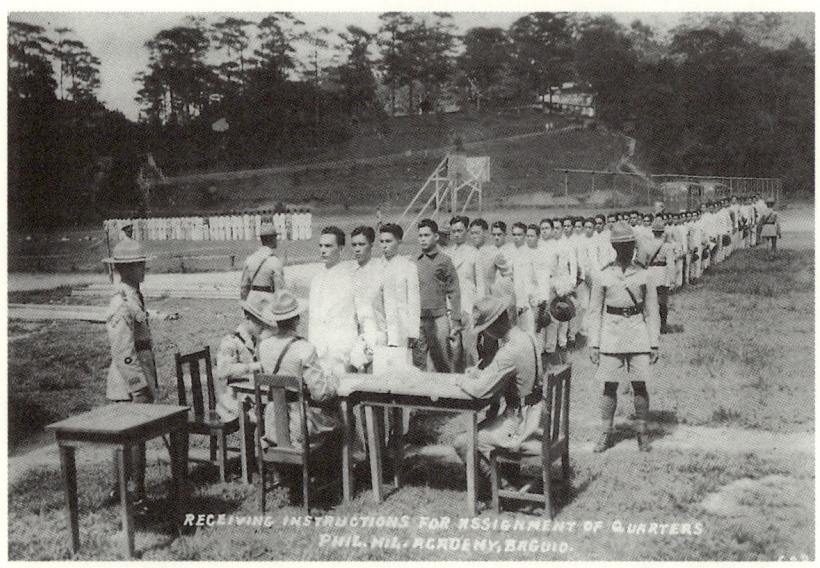

The plebes of Class '40 receiving their room assignments on 15 June 1936, their first day at the Philippine Military Academy (PMA). (Liberato Picar Collection)

Cadet Corps of the PMA at dress parade, Burnham Park, Baguio, 30 December 1936. (Liberato Picar Collection)

The 120 entering plebes of the future Class of 1940, photographed on 15 June 1936, their first day at the PMA. Four years later, only 79 of these would graduate with their class. (Liberato Picar Collection)

The cadet corps ready for review on the parade ground of the PMA on 21 November 1936, only five months after arriving as fresh plebes. (Edmundo Navarro Collection)

Upperclassmen barking commands to incoming plebes as they rush to line up for their "reception," Baguio, ca. April 1938. (Liberato Picar Collection)

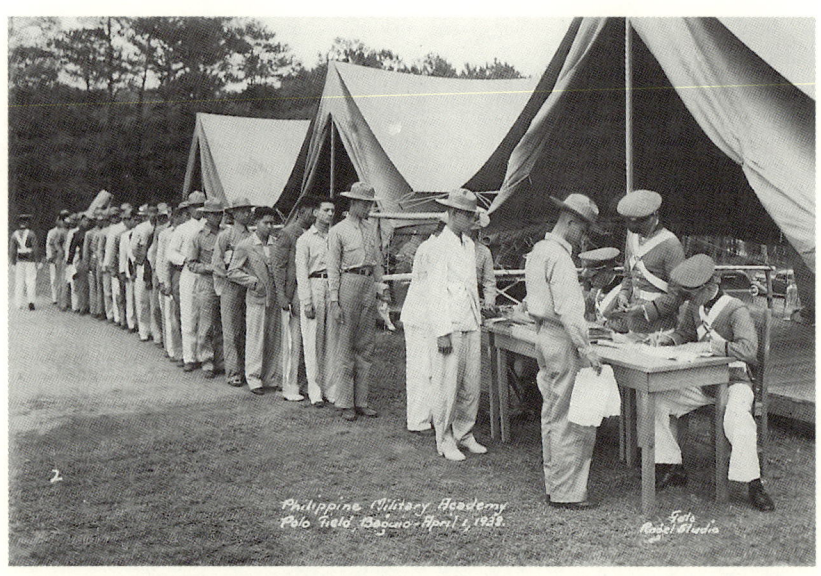

Incoming plebes of Class '42 lining up at the PMA's summer camp for their tent assignments, Baguio, 1 April 1938. (Liberato Picar Collection)

A group of upperclass cadets barking orders at a single plebe as part of the initiation, or fourth-class custom, at the PMA's summer camp, Baguio, ca. April 1939. (Pedro Yap Collection)

Incoming plebes of Class '42 lining up for supplies and quarters at the PMA's summer camp, Baguio City, 1 April 1938. (Edmundo Navarro Collection)

Cadet Corporal Pedro Francisco ('40) hazing Gaudencio Gaddi ('41), a fresh plebe, during the PMA's 1937 summer camp, Polo Grounds, Baguio. (Liberato Picar Collection)

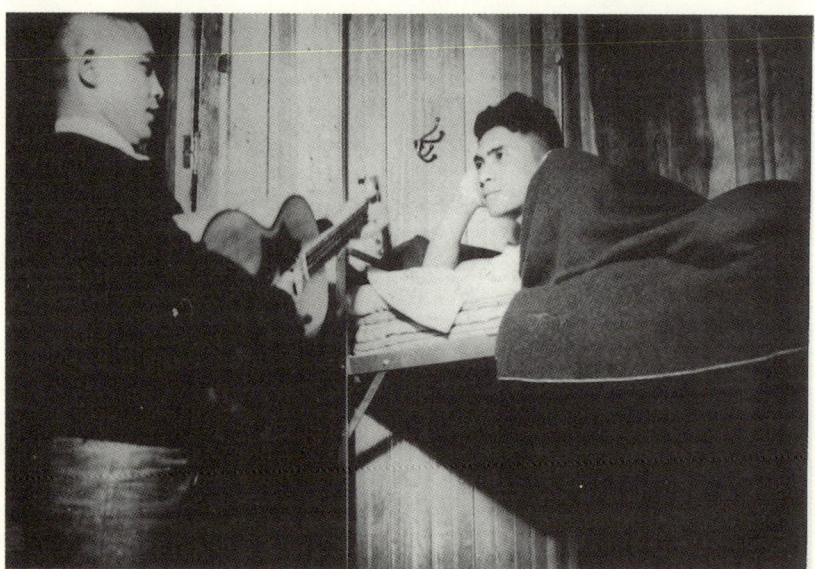

Upperclassman Antonio Evangelista ('41) being serenaded to sleep by an unidentified plebe, a common practice during the PMA's initiation of new cadets during the late 1930s. (Liberato Picar Collection)

An unidentified plebe overcome by the endless cycle of drill and discipline, ca. 1938–39. (Liberato Picar Collection)

Cadet Corporal Gaudencio Gaddi ('41) ordering a plebe, probably Napoleon Mangonon ('42) to brace, a form of discipline during initiation, ca. 1938–39. (Liberato Picar Collection)

Cadet Washington Sagun, a member of Class '40, presenting his solution to an engineering problem. Following the West Point curriculum, every Filipino cadet was required to recite daily in each class. (Liberato Picar Collection)

A dramatized boxing match for a stage play at the PMA in the late 1930s, part of a rounded curriculum designed to mold the cadets into both "officers and gentlemen." (Edmundo Navarro Collection)

Officers of the Battalion Board, a disciplinary committee, interviewing a cadet in the late 1930s. (Liberato Picar Collection)

Class '40 "recognizes" the Class of 1943 at the end of their plebe year as members of the cadet corps, an emotional ceremony often marked by tears and embraces, March 1940. (Liberato Picar Collection)

Cadets cleaning machine guns and artillery, PMA armory, late 1930s. (Edmundo Navarro Collection)

Mess hall, PMA, late 1930s. (Edmundo Navarro Collection)

Meliton Bulan ('41), a graduating cadet, receiving his class ring from an unidentified companion at the ring hop, part of the March Week graduation rituals. (Edmundo Navarro Collection)

President Quezon's eldest daughter, Aurora "Baby" Quezon (*right*), with cadets at the PMA in 1938. Pictured with her are (*left to right*) cadets Aristeo Ferraren ('38), Edmundo Navarro ('40), Pedro Yap ('40), Jose Cardenas ('38), and Renato Barretto ('38). (Pedro Yap Collection)

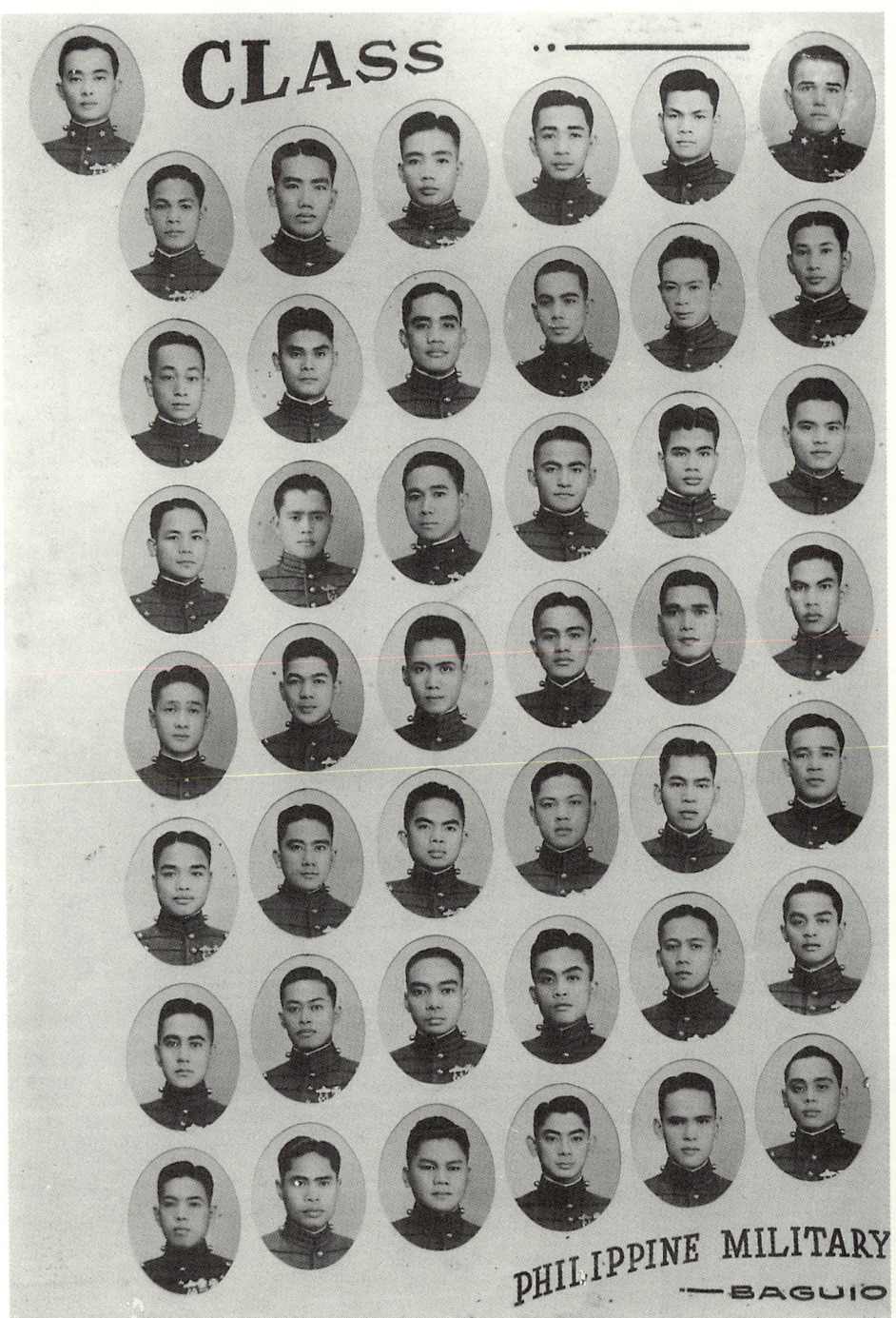

Class of 1940, Philippine Military Academy: (*top row, left to right*) D. Pavon, L. Estrada; (*second row*) A. Acenas, Manuel Acosta, Melchor Acosta, R. Alcaraz, F. Jimenez, R. Olbes, W. Sagun, H. Alano, R. Angeles, F. Apolinario, P. Aragon; (*third row*) A. Aranzaso, D. Argao, P. Baban, U. Baclagon, P. Barrios, P. Bartolome, S. Bayron, P. Bersola, R. Bocalbos, A. Bornales, D. Caballero; (*fourth row*) B. Cabangbang, U. Caldoza, F. del Castillo, F. Causin, C. Corpuz, C. de la Cruz, P. Dulay, L. Ello, N. Escobar, J. Esguerra, Q. Evangelista; (*fifth row*) A. Filart, H. Farolan, F. Fetalvero, R. Foronda, A. Francia, L. Galang, D. Garcia, R.

Gelvezon, B. Genson, D. Iway, J. Javier; (*sixth row*) E. Jamilosa, F. Lumen, C. de Leon, R. Lising, C. Montemayor, J. Mayo, J. Mendoza, R. Mendoza, G. Mercado, C. Monta, P. Francisco; (*seventh row*) E. Navarro, C. Nano, R. Nosce, A. Olayvar, E. Orias, V. Osias, D. Pelayo, A. Perez, L. Picar, S. Piccio, M. Punsalang; (*bottom row*) J. Rodriguez, M. Santos, F. Sebastian, E. Segovia, E. Soliman, T. Tirona, L. Trinidad, H. Tuazon, S. Velasco, F. Vitug, P. Yap. (Edmundo Navarro Collection)

Jose Mendoza.
(*The Sword of 1940*)

Bartolome Cabangbang.
(*The Sword of 1940*)

Uldarico Baclagon.
(*The Sword of 1940*)

Ramon Alcaraz.
(*The Sword of 1940*)

Advertisement for the 1939 Filipino feature film *Punit Na Bandila* (Torn Flag), starring Fernando Poe, Sr., and featuring the PMA cadet corps as extras. Poe's brass buttons and white gloves epitomize the disciplined chivalry of the prewar officers, which contrasts sharply with the lethal masculinity portrayed by his son, Fernando, Jr., some sixty years later. (*Manila Tribune,* 10 February 1939)

Chapter Five

Myth of the Maharlika

When President Ferdinand Marcos began his twenty-year rule in 1966, the alumni of Class '40, then largely retired from the military, soon divided into opponents and supporters of his regime. In one of history's ironies, two classmates, lifelong friends, became antagonists in a high-stakes debate over the central, legitimating myth of the Marcos dictatorship. For two decades, one worked in the palace writing history that celebrated Marcos as the country's greatest war hero. The other spent a decade in exile trying to topple his regime by discrediting those same claims.

These two men were classmates in high school, members of Class '40, and friends for over half a century. They share similar backgrounds, rising from the lower-middle class through the Philippine Military Academy to positions of influence. But they became, in their later lives, a study in contrasts. Commodore Ramon Alcaraz was a flamboyant officer, ambitious for command, who broke publicly with President Marcos in the first weeks of his administration. When the president declared martial law, Alcaraz fled to America, where he became active in the anti-Marcos opposition. By the time the dictator fell from power,

Alcaraz had retired to a wealthy Orange County suburb, where he lived in a large hilltop home with a panorama of the Pacific.

Colonel Uldarico Baclagon was a studious, even scholarly officer, whose only real ambition was to become a military historian. For two decades he worked in the palace as a presidential assistant, propagating the myth of Marcos's wartime heroism. When his president fled into exile, Baclagon retired to an impoverished section of Manila where he rented a dreary walk-up apartment overlooking the city's produce markets.

Despite their differences, the two remained close friends. Even when bitterly opposed over Marcos's war record in the critical, final months of his regime, they sustained their camaraderie across the partisan divide. They may have had strong disagreements at a superficial, political level, but they still shared underlying ideals of personal honor and professional integrity. Alcaraz lived by the academy's honor code in ways that kept his almost consuming ambition within bounds and later inspired his opposition to Marcos. Baclagon, sharing those same principles, sat at the right hand of power but was unwilling, or unable, to reach for any reward. In different ways, these two biographies illustrate the military socialization that restrained the members of Class '40 through their long careers.

ACTIVIST COMMODORE ALCARAZ

Many classmates had reservations about the Marcos regime, but Ramon Alcaraz alone devoted over twenty years to an unrelenting opposition. Calling himself the "first victim of President Marcos," Alcaraz, only days after Marcos's first inauguration, became ensnared in one of the president's complex corruption schemes and was forced to retire from the military under a cloud. Driven by this affront to his honor, Alcaraz moved, by slow stages, from opposition to activism and then terrorism in his struggle against the president.

Alcaraz was born in a Bulacan barrio where his family farmed a modest three hectares and his father drove a *calesa* (horse cart). At the local public schools, he became an enthusiastic scholar in the English-language curriculum of the colonial age and also took up boxing after class, discovering an aptitude for the sport's mix of finesse and aggression. He later worked his way though high school in the provincial capital as a carpenter, graduating as valedictorian and bantamweight boxing champion. In effect, he grew up in two worlds: his parents' Tagalog-speaking village and the English-only classrooms of his American teachers. By the time he left Bulacan, Alcaraz had become bilingual and bicultural, with strong attachments to both Filipino nationalism and American idealism. Watching his boyhood friends quit school to become tenant farmers, Alcaraz was determined to sacrifice for a college education. When the PMA announced its entrance exams in February 1936, he spent weeks in an intensive review and was one of sixty-four admitted in the first round.[1]

Alcaraz prospered at the academy. After years in the ring, the hazing, even the beatings, were tolerable. The curriculum, though demanding, was enjoyable. Showing a flair for leadership, Alcaraz became a cadet lieutenant and, in his third year, led Company B to win the coveted Araneta Trophy for best unit. As a leader of the class's "same banana" strike, Alcaraz, one of only two cadets punished, was demoted to private.[2]

After graduation, Alcaraz joined the Off-Shore Patrol, bassinet of the future Philippine Navy, and quickly developed a strong, even passionate belief in the central importance of his maritime service for the future of an island nation. Throughout his military career he would combine personal ambition, an almost reckless idealism, and deep institutional loyalties.

Alcaraz, unlike most of his classmates, prospered during the war as he had at the academy. In the first weeks of fighting, he won a battlefield promotion for his command of a Q-boat in combat against a Japanese aircraft. He then joined the Japanese-sponsored Bureau of Constabulary and ended the war as commander of a guerrilla regiment that liberated his hometown from the Japanese.[3]

After independence in 1946, Alcaraz, like many of his classmates, began to experience his first real frustrations. As he advanced, Alcaraz found that the idealism that had inspired his military career was now challenged by a pervasive politicization. Protected from invasion by the U.S. Mutual Defense Treaty, the postwar armed forces lost their primary mission of national defense and became a pawn in partisan gambits. While Alcaraz had the social skills to form alliances and cultivate patrons, he was often angered when politicians disagreed with his strongly held views on the nation's defense needs.

When the war ended, Alcaraz spent five years building a new navy for the infant republic. Since all Philippine shipping, commercial and naval, had been destroyed, he worked for two years, from 1946 to 1948, transferring over a hundred ships from the U.S. Navy. His next tours were a succession of successes: selecting officers for the new ships; observing the U.S. Marines in California to plan a Philippine counterpart; and brokering the appointment of his deputy as their first commandant.[4]

Politics was the precondition for these achievements. The navy was able to expand so rapidly because its commander, Jose Andrada, was related to President Manuel Roxas and thus won increased appropriations over opposition from army partisans at headquarters. Indeed, in October 1946 President Roxas transformed the Off-Shore Patrol into the Philippine Naval Patrol with status equal to the army, promoting Andrada to commodore and thus making him the navy's first general officer. Just as Andrada was close to the president, so Alcaraz enjoyed the confidence of the commodore, his prewar commander in the tightly knit Q-boat fleet. Even Alcaraz's success with the Marines sprang from political connections. During the war, Alcaraz had provided guerrilla leader Ramon Magsaysay with a critical radio link to General MacArthur's headquarters, and after the war he gave some of his former "guerrilla boys" naval postings—good deeds that were rewarded when Magsaysay became defense secretary in 1950.[5]

What politics gave, politics could take away. A year after President Roxas died from a sudden heart attack in 1948, the ground-force dominated AFP hierarchy started "ganging up" on Commodore Andrada, stripping him of his command on spurious charges of purchasing an unknown brand of motor oil. After Andrada was forced from command in 1949, AFP headquarters appointed Jose Francisco, an Annapolis graduate whose advance in the air force had been stalled by his activities during the Japanese occupation. As navy chief, Francisco was "a timid leader," and headquarters, eager to avoid any challenge to the army's budget, extended him until his mandatory retirement in 1961. For an extraordinary twelve years, Francisco remained in command and refused all promotions—blocking advancement of navy officers and producing a factionalized service. Commodore Felix Apolinario, a member of Class '40, attributes Francisco's tenacity to his business with the navy under a contract that gave him "the exclusive rights to supply spare parts for the Mercedes-Benz marine engine."[6]

When Francisco finally retired in 1961, mandatory retirement was only five years away for Class '40's navy contingent, making it likely that only one could rise to command. The rivalry between classmates Ramon Alcaraz and Felix Apolinario became intense. Under prewar regulations, Apolinario's graduation at number seven in the class, above Alcaraz, set their ranking within the active-duty roster. In the first weeks of World War II, however, Alcaraz won a battlefield promotion, superseding Apolinario. But when both reverted to permanent rank under a new roster in 1947, Captain Alcaraz was reduced to lieutenant and again became junior to his classmate—an order that was now fixed for the rest of their thirty-year careers.[7]

As Francisco's tenure dragged on, relations between Alcaraz and Apolinario reached the breaking point. Apolinario came to feel that Alcaraz was a "back stabber" who tried to amplify minor oversights into major derelictions that would force his demotion. Indeed, in his memoirs, Alcaraz expressed outrage that Apolinario preserved seniority even after a "derogatory report," arising from a minor mistake, led to his rival's removal from Bangkok as naval attaché.[8]

From his own perspective, Alcaraz believed passionately in the future of the navy in an island nation and was ambitious to realize his vision. As conflicts with Commander Francisco intensified, Alcaraz found himself "deported" to ancillary assignments, first to the U.S. Naval War College and then to Subic Bay as liaison officer with the U.S. Navy.[9]

When Commodore Juan Magluyan took command in 1962, Alcaraz's exile ended and he returned to navy headquarters for a second tour as chief of naval staff under an officer he admired as the last of the "principled pioneers of the Navy." Once Magluyan retired in mid-1964, Alcaraz faced renewed conflicts with his successor, Commodore Santiago Nuval. As attaché in the Tokyo embassy, Nuval had been court-martialed in 1955 for smuggling twenty tons of dutiable goods to Manila on a navy ship labeled as post-exchange or "PX" goods and selling them to businessmen through a partner in crime, Colonel Carmelo Barbero. The court punished the two culprits with a reduction in seniority, a crippling blow to any career. Though Barbero retired early to enter politics in his native Abra, Nuval stayed on and eventually managed to have the "reduction in file" lowered to a "severe reprimand." Ordinarily such a change was impossible, but Nuval, a native of the Ilocos region, had backing from Senate president Ferdinand Marcos, who was then building a following among fellow Ilocanos in the military. Since Alcaraz had helped gather evidence for Nuval's court-martial, the incoming flag officer's first act was to transfer his former adversary out of navy headquarters.[10]

BLUE SEAL SMUGGLING

In his new post as commander of the Naval Operations Force, Captain Alcaraz was responsible for fighting the tide of American cigarettes smuggled from British free ports in North Borneo. He waged a hard-fought campaign, but the enemy proved elusive and surprisingly resilient. As it turned out, Alcaraz was battling not just smuggling, but an invisible international commerce whose scale was almost beyond comprehension.

During the 1930s, Filipino smokers had acquired a taste for mild Virginia tobaccos and consumed over half of all U.S. tobacco exports, starting a romance with American brands like Camel and Lucky Strike. Filipinos called them "blue seals" for their bright U.S. tax stamp. When a bankrupt Republic cut non-essential imports to conserve foreign exchange in 1949 and then banned all foreign tobacco five years later, it created an instant black market for these blue seals. By 1965, the U.S. Department of Commerce would complain that high duty rates and widespread smuggling of Hong Kong-produced cigarettes had eliminated any market in the Philippines for legal American exports.[11]

Almost simultaneously, the new Republic of Indonesia encouraged a parallel smuggling of dried coconut, or copra, from Sulawesi to British North Borneo by overvaluing its currency and thereby punishing exporters. For over a decade, secessionist rebels on sprawling Sulawesi sold their island's copra at world-market prices by smuggling cargoes across the Sulawesi Sea to the port of Sandakan in nearby North Borneo. Similarly, Filipino smugglers carried a large part of the copra production from their southern islands to North Borneo, using profits to buy blue seals from Sandakan's Chinese merchants. By 1957, this tiny port had become the largest importer of American cigarettes in the world.[12] This convergence of smuggling routes at Sandakan—copra in and tobacco out—may have provided a balanced trade and venture capital that fueled the explosive expansion of smuggling of blue seals into the Philippines during the 1960s.

Riding the tides of this vast, illicit commerce, new entrepreneurs sprang up along the arc of islands from Sulawesi to Luzon, seizing profits and power at each strategic point. Blue-seal smuggling grew steadily from the mid-1950s under the protection of corrupt officials in Cavite and Batangas, coastal provinces just south of Manila. By 1963, a poor fisherman from Cavite had been transformed into the millionaire smuggler-king Lino Bocalan, and his tiny village of Capipisa had become the country's largest port for blue seals.[13]

As the illicit imports shot upward throughout 1963, the head of the Naval Operations Force, Alcaraz's classmate Felix Apolinario, made some headway against the smugglers. Instead of dispersing his ships on the Sulu Sea frontier, Captain Apolinario, in his words, laid a "very tight gauntlet" across Manila Bay that slowed the illicit shipments. As the navy blockade cut into the illicit profits, Bocalan "sent his emissary to bribe." When the captain refused to discuss the matter, the emissary, a corrupt constabulary officer, called on Mrs. Apolinario with two Rolex gold watches and a wad of cash, saying: "There's a house waiting there for you. And you go and tell your husband to pick a car. Go to the Mercedes-Benz [dealer] and pick any car he wants." Living in the miserable cluster of Quonset huts that passed for navy quarters, Apolinario was sorely tempted. After days of agonizing and arguing with his wife, he refused. "Remember, in the academy—courage, loyalty, integrity," he explained in a later interview. "Maybe I still have a lot of integrity in me. I couldn't accept it. I couldn't sleep if I had."

Years later, Apolinario ran into Bocalan's paymaster on a Manila golf course. "You know that house, Poling?" the smuggler said, smiling and using Apolinario's nickname; "that house is worth twenty million right now."[14]

When he succeeded his classmate as fleet commander in 1964, Alcaraz, unaware of the enormity of his enemy, plunged into the battle against smuggling with his usual enthusiasm. His first forays into the Sulu Sea yielded little, and he began to suspect sabotage. Defense Secretary Peralta aroused his suspicions by requiring eight copies of his daily ship-position reports, an unwarranted level of duplication. In February 1965, Alcaraz began submitting reports with false data and changed his radio codes daily without notifying headquarters. The results were astonishing, he later said. Instead of seizing one smuggler's ship a day, Alcaraz's barrier patrols now caught ten. During the last half of 1965, his fleet seized P 750,000 worth of cigarettes each month, 95 percent of the total for all government agencies. Significantly, both Flag Officer Nuval and Defense Secretary Peralta repeatedly refused his requests to station a detachment of marines at Barrio Capipisa, the lair of Lino Bocalan.[15]

By late 1965, Alcaraz's intelligence work had produced a portrait of this underground economy. Although the volume of blue seals was vast, the high costs of capital and corruption concentrated the business in the hands of two powerful syndicates—Bocalan of Cavite Province and the Berberabe brothers of Batangas. Operating under the protection of Defense Secretary Peralta and his wife, Pablo and Carlos Berberabe kept a fleet of fifteen *batel* (cargo ships) on the seas between Borneo and Batangas. Protected by the chief of Constabulary Zone II, General Segundo Gasmin, smuggler Bocalan had a flotilla of twenty-five fast ships shuttling between Sandakan and Cavite. Each ship, packed with brands like Lucky Strike and armed with machine guns to fend off pirates, was equipped with powerful engines that could outrun the aging navy patrol craft. With an average of twenty-four hundred cases of cigarettes landing on the beaches of his hometown every year and annual profits estimated at P 600,000, Bocalan was, in Alcaraz's estimation, a formidable antagonist.[16]

In November, armed forces headquarters did a staff intelligence study on the smuggling, discovering a commerce beyond even Alcaraz's boldest estimate. Massive cargoes were shipped through Hong Kong to British Borneo, where they were sold to Filipino smugglers. From 1962 to 1965, these illicit imports totaled some P 156 million ($40 million), equivalent to about 4 percent of the country's "invisible dollar outflow." Indicative of the scale of this traffic, in 1964 the AFP had impounded 169 ships, arrested 2,933 suspected smugglers, and confiscated 1,678,469 cartons of cigarettes. Even so, the military was still seizing only

12 to 15 percent of the total. Since profits were enormous, seizures did little to discourage smugglers. A pack of "C grade" foreign cigarettes purchased in North Borneo for twenty-seven centavos could be retailed in Manila for P 1.20, allowing profits of 80 to 100 percent. An earlier investigation had reported that there were some sixteen hundred suspected smugglers, including fifty constabulary officers and fourteen provincial commanders.[17]

Another classmate assigned to the landward side of the antismuggling campaign also encountered corruption. As head of "Task Force Zapata" in the Cavite-Batangas area in 1964–65, Colonel Francisco Jimenez led constabulary troopers in seizing over a million pesos' worth of smuggled cigarettes in just three months. Through an "intelligence project," Jimenez discovered that the smuggler Bocalan shared half his four-hundred-peso profit on each case with the zone commander General Segundo Gazmin and all the police chiefs between his Cavite beachhead and downtown Manila, making himself "a very popular fellow." When Jimenez gave Defense Secretary Peralta a list of persons involved in smuggling, the secretary ordered his summary transfer to a purgatory known as the Replacement Battalion, pleading "orders of Malacañang." Later, Jimenez learned that Peralta was protecting the zone commander, who "was his provincemate and partner in this business."[18]

Alcaraz ignored the power of these syndicates, plunging onward and doubling his monthly seizures by late 1965. He became something of a public hero and won promotion to commodore, the navy's star rank.[19] Despite these honors, Alcaraz soon met a bitter disappointment.

When Commodore Nuval retired some weeks later, the navy's new flag officer in command was supposed to be chosen from Class '40, a process that produced a final breach between the rival classmates, Alcaraz and Apolinario. Under the rigid seniority system, Apolinario would be appointed unless the Board of Generals found him unfit. According to Apolinario, Alcaraz gossiped among naval staff that his American divorce, not recognized in the Philippines, meant that he was living immorally with his third wife. Suspecting such intrigues, Apolinario was frank when he appeared before the board, admitting his third marriage but insisting that he did not keep a mistress. The strategy worked, and in September 1965 the president promoted him to flag officer in command of the navy.[20]

Apolinario's appointment denied Alcaraz the support needed to weather the political storms ahead. Instead of defending Alcaraz's antismuggling operations, the new flag officer placed navy intelligence officers aboard the ships to monitor his classmate. Apolinario was suspicious of several port calls that the fleet had made in the Sulu Sea, suspecting that smugglers had bribed Alcaraz to secure safe passage through the naval blockade.[21]

Being passed over by a classmate was the bitterest defeat of Alcaraz's career. The disappointment seems to have redoubled his determination to win fame in the antismuggling campaign and succeed his classmate as navy commander. In retrospect, it seems that Alcaraz's ambitions may have blinded him to the pitfalls of the blue-seal business, making him vulnerable to Marcos's manipulations.

During the ongoing 1965 presidential campaign, opposition candidate Ferdinand Marcos accused the incumbent President Macapagal of complicity in the blue-seal racket. In the opening rounds, Marcos charged the administration with failing to stop the illicit import of 3.5 billion cigarettes a year, costing the country P 315 million in taxes. Marcos also accused Defense Secretary Peralta of protecting smugglers, promising "I will name names, which will be so many that President Macapagal will get dizzy," and then releasing a list of forty-eight corrupt officials allegedly protected by the president. Although the palace dismissed these charges, the president replied in a speech simulcast by all of Manila's major radio and television stations.[22] So armed, Marcos launched his candidacy under an anticorruption banner.

Feeling Macapagal somehow responsible for the corruption, Alcaraz was silently sympathetic to Marcos's promises to restore integrity to the military. As Marcos emerged slowly as the winner in the November vote counting, President Macapagal tried to defuse the smuggling issue by making his opponent the nation's new "antismuggling czar."[23] Marcos accordingly summoned Alcaraz to his home for "an informal talk." The commodore was ushered into a private conference room and found himself seated with Alejo Santos, his wartime guerrilla commander, the president-elect, and Ernesto Maceda, a campaign advisor. Marcos got down to business, saying, "Commodore, I want to congratulate you for your antismuggling accomplishment." As Alcaraz later admitted, "of course, my ego was titillated." When Marcos asked for a frank assessment, the commodore answered boldly, denouncing Secretary Peralta and the corrupt constabulary officers. Apparently impressed, Marcos praised him, saying, "I want you to put that in writing. And after that, you go out and catch some more smugglers."

As Alcaraz left the meeting, courtiers in the anteroom gathered round, smiling, slapping him on the back, and saying that his appointment as defense secretary or chief of staff was just a matter of time. Several days later, Marcos asked publicly that President Macapagal relieve both Peralta and the chief of constabulary.[24] With the president-elect as his patron, Alcaraz was confident that he would at last catch up with Apolinario.

Alcaraz returned to his patrol duties feeling "emboldened to take more aggressive action against the smugglers." Instead of towing ships to the nearest ports as regulations required, Alcaraz sank dozens of impounded vessels on the high seas. Despite repeated refusals by navy headquarters, Alcaraz ordered Marines to occupy Bocalan's Cavite base. Cigarette seizures soon climbed to unprecedented levels.[25]

Unknown to Alcaraz, these efforts simply forced the smugglers to take a new partner. Only days after his inauguration in December 1965, Marcos met secretly with smuggler-king Bocalan and agreed to restrain the navy patrols for a share of the profits. By manipulating the commodore, Marcos had deftly redirected the bribes from the base of the political system to its apex. Indeed, years later in Los Angeles, Alcaraz finally came face to face with Bocalan, who, smiling his worldly smile, chided the commodore for his naïveté. All those patrols, all those seizures, Bocalan said, had simply forced him to pay Marcos "millions." Not only had Bocalan survived, but he was later elected governor of Cavite Province with the president's backing.[26]

On 11 January 1966, Alcaraz was, in the words of his biographer, "directing his inspired drive against the smugglers with even greater vigor" when the radio broadcast an announcement from the palace that included his name in a list of notoriously corrupt officers being cashiered. Alcaraz was stunned. Angry at the implications of "hanky panky," he decided to fight.[27]

At the transfer-of-command ceremony, Alcaraz delivered an address dissecting the malaise that afflicted national defense policy, producing a rash of front-page headlines: "AFP Not Equipped to Defend Nation, Alcaraz Claims." Next day, Defense Undersecretary Ernesto Mata, the president's provincemate and close ally, ordered an investigation of Alcaraz for conduct unbecoming an officer—the first step toward a court-martial. In page-one coverage, the *Manila Daily Bulletin* called the case "unprecedented in the annals of the AFP." Other papers thundered their support for Alcaraz.[28]

As the media firestorm intensified, Undersecretary Mata sent an aide to Alcaraz, demanding that he retire and warning that refusal would lead to reduction in rank. "You can reduce me to apprentice seaman," Alcaraz replied. "I don't care. My father was a farmer, I can go back as a farmer." As brother officers rallied quietly, AFP headquarters, not yet under Marcos's full control, subverted the investigation by assigning the case to Inspector General Faustino Sebastian—a member of Class '40 whose close friendship with Alcaraz dated back to their days on the academy's boxing team. Sensing a silent revolt, Marcos suddenly dropped the proceedings.[29]

But Alcaraz felt his honor tarnished and demanded exoneration. On January 25, Representative Rogaciano Mercado, armed with details that only Alcaraz could have known, rose on the floor of Congress to attack the president's antismuggling campaign, producing more damning headlines. Since the controversy threatened to wound his administration in its first month, Marcos decided to settle.[30]

Only three weeks after the dispute began, Representative Carmelo Barbero, Marcos's ally and Alcaraz's friend, convened the House Defense Committee to mediate. The first witness, General Mata, now chief of staff, stated that Alcaraz's relief was not due to any derogatory report and took full responsibility for the decision. In his testimony, Alcaraz proclaimed his exoneration and repeated his indictment of the nation's defense policy. He walked away from the witness table amid applause. With honor satisfied, Alcaraz signed the retirement papers.[31]

Ironically, only four months later, similar pressures forced his classmate Apolinario to step down as the navy's flag officer. The president's brother-in-law, Alfredo Romualdez, a senior officer in the naval reserve, owned a small freighter that regularly unloaded its contraband in downtown Manila under the protection of city police. When naval intelligence reported that the freighter was discharging two thousand cases of cigarettes near Guadalupe Bridge, Commodore Apolinario ordered, "Let them go." Next time, intelligence reported a shipment of ten thousand cases of illegal guns, and the commodore again said, "Let them, let them." In marked contrast to Alcaraz's blind assault upon the powerful, Apolinario ignored this blatant smuggling by the First Lady's brother. But with each shipment, Apolinario felt a rising conflict between his integrity and the imperatives of political survival. When Marcos offered to extend his tour, he refused diplomatically. "Sir," he told Marcos, "I cannot stand some of our congressmen. They will ask me things which I cannot give." With the president's consent, Apolinario retired for the life of an independent businessman.[32]

Meanwhile, after his retirement, Alcaraz invested his pension, a modest twenty thousand pesos for twenty-six years' service, in his wife's pharmacy in Samapaloc and built it into a successful chain, later incorporated as Commodore Drugs. Avoiding partisan politics, he joined the nationalist Lorenzo Tañada and the communist Jose Ma. Sison in the Movement for the Advancement of Nationalism. At first, Marcos had seemed willing to let matters rest. But, over the long term, he proved vindictive. Two years later, when the controversy was forgotten, the National Bureau of Investigation raided Commodore Drugs and arrested Alcaraz on charges of selling stolen pharmaceuticals. When charges failed for want of evidence, two examiners from the Bureau of Internal Revenue arrived to spend months poring over his records before demanding just P 236 (U.S. $76) in back taxes.[33]

As the 1969 presidential campaign began, Alcaraz, concerned by Marcos's corruption of the military and promotion of his Ilocano allies, joined a group of retired officers backing the opposition candidate, Sergio "Serging" Osmeña, Jr. Under the leadership of Eleuterio "Terry" Adevoso (PMA '44), a former secretary of labor, the "Workshop Group" encouraged active-duty officers to remain neutral in the campaign. When Alcaraz lobbied the country's leading military association to speak out for clean elections, Justice Secretary Juan Ponce Enrile retaliated with an investigation of Alcaraz's real-estate transactions.[34]

After Osmeña's defeat in the November balloting, the commodore continued to attend meetings of the Workshop Group, which became "gripe sessions." When Osmeña's court challenges against Marcos turned up a pattern of electoral fraud, their leader, Adevoso, vented his frustration, saying, "the best thing I can think [of] is to assassinate Marcos." More soberly, he suggested an alliance with the radical student demonstrators. But these conversations were "not serious" and were soon forgotten as their anger dissipated. Unknown to his fellows, however, one member of the group reported every word to the Presidential Security Unit. Only weeks after the election, Marcos told his diary: "I have just talked to _____ who is our informer in the Adevoso group planning a coup d'etat." This anonymous agent described the plot as "serious" and urged that the "conspirators be eliminated quietly"—a recommendation he reinforced in the coming months with detailed reports of their ongoing plot to kill the president.[35]

Three years later, when Marcos declared martial law, these activities brought Alcaraz to the attention of the president's security chief, General Fabian Ver. At noon on 15 November 1972, a major arrested Alcaraz at his home and took him to Presidential Security Unit headquarters in Malacañang Park, where he was photographed and fingerprinted eight times. Then a team of intelligence officers began an interrogation that lasted into the night. Alcaraz noted, to his dismay, that two of these inquisitors were recent PMA graduates, members of Class '71. Although they were "very courteous" and addressed him as "commodore" throughout, Alcaraz was disturbed that martial law made these young lieutenants violate the academy's ingrained respect for seniority. "What do we expect from these people," he asked rhetorically in a later interview, "who suddenly tasted [the] power of interrogating a senior officer?"

Their questions focused on the Workshop Group. General Ver's staff had branded the group a "military junta" with plans to assassinate the president and seize power by a coup d'état. While the interpretation was wildly imaginative, the interrogators had a wealth of accurate detail about the group's meetings—who, when, where, what. After hours of interrogation and a warning from the unit's executive officer, Alcaraz was released "temporarily."[36]

To justify his prolongation of marital law, the president would spin these threads into a fabric of threat. In his book-length defense of dictatorship, *Notes on the New Society*, Marcos claimed that the chief danger to the state before martial law had been a plot by "the group of former officers . . . who were conspiring to mount a coup d'état." Embroidering upon this sketch, the president later told an American reporter that "an elite group composed mostly of retired officers . . . conspired to kill the president." He named Terry Adevoso, the Workshop Group's leader, as chief conspirator.[37]

For the next three years, Marcos kept these allegations alive to justify holding the children of exiled opposition leaders as hostages. In July 1975, the regime's mouthpiece, the *Daily Express*, reported that the Workshop Group, funded by Sergio Osmeña III and Eugenio Lopez, Jr., had planned to assault the palace with four thousand troops and assassinate President Marcos. Three of the fourteen military men named as active plotters were members of Class '40—Alcaraz, Jose Maristela, and David Pelayo.[38]

Feeling threatened after his arrest, Alcaraz had his daughter, an American citizen, sponsor his application for a U.S. resident visa. He landed in Los Angeles in early 1973 and launched his third career, as a California realtor, trading properties profitably in the city's long real estate boom. For the next twelve years, the commodore also became an active leader of the main anti-Marcos coalition, the Movement for the Free Philippines (MFP), using the protection of his U.S. citizenship to travel regularly to Manila as liaison to the anti-Marcos opposition.[39]

As the dictatorship strengthened its grip and seemed interminable, Alcaraz became convinced of the need for violence, either by coup or terror, and began working as technical advisor on weaponry to the politicians who led the exile movement, first Raul Manglapus and later Benigno "Ninoy" Aquino, Jr.[40] To this end, Alcaraz collaborated with movement strategist Major Bonifacio Gillego, a retired intelligence officer, to draft a terror campaign code-named "Project Mactan." Arguing that "the Marcos dictatorship will not succumb to popular pressure," the two proposed a modest budget of $180,000 for "small unit strikes against highly selected . . . targets" that would "stir the people to stiffen their resistance to the Marcos dictatorship." Throughout 1979, MFP members set major fires in buildings identified with the regime and its cronies—the Commission on Elections, Philippine Village Hotel, and Sulo Hotel. When the group tried to smuggle plastic explosives, their courier was caught at Manila International Airport, sparking a Christmas Eve crackdown.[41]

While Manglapus had reservations about terror, Ninoy Aquino knew no restraint. Arriving in America from Manila in 1980 for heart surgery, he recovered quickly and assumed leadership of the exile movement. Alcaraz met him in Dallas in July to discuss Project Mactan, and Ninoy revealed that he already had several groups of "concerned citizens" preparing similar operations. Several Vietnam veterans under Ninoy's control began training a dozen five-person demolition teams of Filipino activists in the Arizona desert. For security, the entire operation was, in Ninoy's words, compartmentalized: all members used pseudonyms, and each team was isolated from the others. Once trained, team leaders flew to Orange County for a "sighting" at Alcaraz's office and then, after landing in Manila, appeared at Commodore Drugs to pick up P 50,000 in operational funds. Steve Psinakis, the son-in-law of exiled entrepreneur Eugenio Lopez, Sr., would air-freight detonators and timing devices from San Francisco to cover addresses in Manila.[42]

Bombs began exploding across Manila. "If martial law is not lifted soon," Ninoy Aquino told the Asia Society in New York in August 1980, "some elements have completed plans for massive urban guerrilla warfare." Two weeks later, under the umbrella of the "Light a Fire" group and the "April 6 Liberation Movement," the opposition unleashed a terror offensive. The first casualties came on September 12, when a bomb erupted at Rustan's department store, owned by a Marcos crony, wounding thirty-two and killing the wife of an American businessman. Then came their triumph. On October 19 at the Philippine International Convention Center, Marcos had just finished addressing five thousand members of the American Society of Travel Agents, hailing his regime as "the best example of peace and order." Minutes later, a bomb erupted just fifty feet from the president and the U.S. ambassador, sending shards in all directions and injuring seven. The cavernous hall emptied in minutes as thousands of travel agents raced for the airport to plead for the first flight home.[43]

In the wake of this humiliation, the Marcos regime indicted thirty Filipino exiles as coconspirators—including Ninoy Aquino, Raul Manglapus, and Bonifacio Gillego. Adding weight to the charges, one of Ninoy's team leaders, Victor Lovely, Jr., was caught when a detonator exploded in his room at the Manila YMCA, blowing away a hand, an eye, and an ear. Badly maimed, Lovely named names as the state's star witness but concealed his contacts with Alcaraz, sparing him the FBI harassment that other MFP leaders faced when President Reagan took office in 1981.[44]

As the bombing campaign waned, Alcaraz shifted his efforts to coup plotting—first recruiting conspirators, then building a sympathetic climate in the officer corps. In an early liaison trip to Manila, Alcaraz had identified several promising coup groups and spent the next decade trying to fan these embers. During one of his later visits to Manila, Alcaraz worked with former president Diosdado Macapagal to distribute a "primer on martial law" to all active-duty officers, which argued that loyalty to Marcos "amounts to obeying an unlawful order of a superior that gives rise to criminal liability."[45]

His classmates were deeply divided over martial law. During a 1979 visit home, Alcaraz tried to enlist their support by hosting a class reunion at a Quezon City restaurant. Knowing that many sympathized with Marcos, Alcaraz planned the program with a single speech—his own. He opened with an appeal to that "'same banana' spirit of oneness," recalling their cadet days, when they had "revolted against injustice and won." He urged his classmates to "take part in a democratic revolution with the same ferocity and tenacity of spirit that prompted us to rise as one against a tyrannous [PMA] Academic Board that was out to decimate our Corps of Cadets."[46] The room was silent.

Such caution from classmates reminded Alcaraz that martial law demanded discretion. During his interrogation by the palace guards in 1972, he had been struck by their detailed knowledge of closed-door conversations inside the Workshop Group. He suspected a spy. Years later in Los Angeles, former presidential candidate Serging Osmeña warned him that it had been his classmate, Colonel Jose Maristela—an accusation later "confirmed" by the Workshop Group's chair, Terry Adevoso (PMA '44). Always aggressively independent, Maristela had dropped back to Class '41 after a brawl with a tactical officer and later hounded the AFP chief of staff into retirement on corruption charges. Even so, Alcaraz "could hardly believe" what Osmeña told him. Four years later, however, when the two classmates dined in Los Angeles, Maristela became evasive and broke eye contact when Alcaraz asked about the Workshop Group. Alcaraz was now sure. In this dangerous game, Alcaraz could not trust his class.[47]

As the Marcos regime weakened after 1981, Alcaraz continued his propaganda within the armed forces, focusing on the president's image as the nation's greatest war hero, the source of much of his appeal to the military. From his own wartime experience, Alcaraz knew that Marcos was living a lie vulnerable to exposé. After the war, as Marcos parlayed his legislative power into three medals and then ten more, Class '40 was collectively dubious about his claims. Moreover, their classmate, Colonel Ciceron de la Cruz, who was the AFP adjutant general when Marcos won ten medals in a single ceremony, told them that these were given to ensure that then-senator Marcos would not oppose President Macapagal's reelection.[48]

Alcaraz met with Major Gillego in Los Angeles in 1978, sharing this information and suggesting an interview with Marcos's wartime commander, Colonel Romulo A. Manriquez (PMA '36), now a resident of Washington, D.C. After four years of sporadic research, Gillego finally published his detailed exposé of Marcos's war record in a Philippine-American newspaper during the president's 1982 state visit to Washington. When Manila's opposition paper *We Forum* reprinted the article, Marcos ordered the editor's arrest on capital charges of subversion and then staged a heavily publicized show trial to defend his war record.[49]

Aware of the rising resentments among regular officers, the exile movement targeted active-duty PMA alumni for an anti-Marcos propaganda campaign. In January 1985, Alcaraz published an "open letter to the men of the armed forces," urging all cavaliers to "exert pressure on your superiors . . . to make sure they are guided at all times by professionalism and the honor system as taught in the old Philippine Military Academy." Alcaraz also worked with other Filipino exiles to build opposition among PMA alumni in the United States. In October 1985, he appeared at an MFP press conference at the Dirksen Senate office building to introduce the first defector from the Marcos military, Colonel Alexander I. Bacalla (PMA '62), a member of the general staff who denounced Marcos for "converting the AFP into an instrument of personal rule." When Marcos won the February 1986 elections by fraud, Alcaraz led over a hundred PMA alumni in a call for nonviolent protests.[50]

Only a few weeks later, military rebels in Manila sparked a mass uprising that sent Marcos into exile. In front of the Philippine Consulate in Los Angeles, hundreds of Filipinos paraded with Philippine flags, weeping and cheering. At his Orange County home, Alcaraz hosted a victory celebration with the toast, "Tyranny is over." Landing in Manila a week later, he met with Defense Minister Enrile and Chief of Staff Fidel Ramos to discuss an agenda for military reforms. Alcaraz told the press afterwards that he was "very pleased with the professionalism of the new leadership" and announced that he had accepted a post as military advisor to the Aquino government. Twenty years after an affront to his honor, this was vindication.[51]

PALACE HISTORIAN BACLAGON

When Ferdinand Marcos fled to Hawaii in 1986, Colonel Uldarico Baclagon became one of the few presidential confidantes to fall from power into poverty. For nearly twenty years, Baclagon, the country's leading military historian, had worked in the palace as resident propagandist, using his stature and skills to validate the president's claims to superhuman heroism in World War II. Not only did the colonel insert Marcos's mythic tales into his military histories to please the president, but he did so from conviction. Baclagon truly believed Marcos's incredible stories of sabotage, espionage, and derring-do. Content with his salary as a midlevel Defense Department bureaucrat, he never sought reward for a service that he alone could provide. For Baclagon was the keeper of the Marcos regime's central myth, a historical fiction so important that its ultimate unraveling would threaten the regime's survival.

Baclagon's poverty was not simply a tribute to the PMA's lessons in honor. It was the reward for his lifelong dedication to military history. Its practitioners elsewhere could earn distinction or even wealth. But in the Phillipines military history was not only an obscure field, it was one whose terrain concealed snares and pitfalls that would deny its leading practitioner both income and honor.

Within each of Baclagon's eight books of military history, a biographer can hear their author's faint autobiographical voice. Through them, we can track his military service through World War II and into the postwar campaign against the Huk peasant rebels. We can also trace his career as a historian—from the scholarly distinction won with his first book, *Philippine Campaigns*, a definitive military history, to the discredit that came with his last, *Valor*, a glossy defense of Marcos's medals.

As a child, Baclagon had a disjointed home life and found a refuge in books. His father was a lowly government field auditor, and the family followed him from one provincial capital to the next. By the time Baclagon entered Bulacan High School in 1931, he had lived in seven provinces and grown into a solitary child. Instead of joining the usual groups of male playmates, he read novels, memorized poetry, and practiced his prose. "I first met Uldarico Baclagon in high school," recalled his future academy classmate Alcaraz. "I see him mumbling many times and I thought he was a weirdo, but I later found out that he was mumbling the poems that he was thinking of composing."[52]

As Baclagon was finishing high school, his father retired and soon dissipated his pension in gambling, plunging the family deeper into poverty. To support his mother and siblings, Baclagon put aside college and became a *kargador*, an ordinary stevedore hauling sacks on the Manila waterfront. After a year of sweat labor, he sat the PMA entrance examination and placed in the top ten among the six thousand examinees, surprising his fellow laborers, "who could not believe," that he could pass. For Baclagon, "entering the academy was like being ushered into a strange world." Hardened by brute toil on the docks, he took the demands of plebe year in stride, viewing them as a fair price for fulfillment of what he called "an intense desire . . . to get a free education."[53]

At the academy, Baclagon became a humanist among engineers, a poet among athletes. Instead of the usual cadet pursuits, Baclagon devoted himself to the chapel choir, school plays, and, above all, the monthly magazine, *The Corps*. As its literary editor, he found few contributors and had to fill up the spaces by writing under assumed names, working so hard that he wound up in the bottom half of the class for all subjects except English and military history. Among these many words, what he called his greatest achievement was a composition that became the official "Cadet Prayer." Its stanzas are striking for their humane appeal. "Let the light of Thy divine wisdom," he wrote, "endow our hearts with kindness that we may sympathize with those who sorrow and suffer."[54]

The academy also taught Baclagon his true vocation. He invested himself in the competition for a new school song, but his entry lost. Though his success with poetry and fiction was similarly limited, he won distinction in his final year as the "class marcher" in military history—academy argot for the number one cadet in any subject.[55] At graduation, an anonymous yearbook biographer recalled that Baclagon had been nicknamed "Backy" for his "diminutiveness" and called him "a poet at heart and a songster by temperament" who "sang his way into the hearts of upperclassmen with a lyric tenor."[56]

Despite his unmilitary manner, Baclagon adjusted well to military discipline and found it natural to be "very obedient" to upperclassmen. Not only was he obedient, but he showed a ready respect, even affection, for superiors—a quality that would later facilitate his integration into the chain of command. In a telling example of this trait, Baclagon recalled his abiding affection for the academy's commandant, Captain Jaime Velasquez. When many in the class failed mechanics, he called them together to ask, "What's difficult about the subject?

Give me its problem." When one cadet read out a problem, Baclagon recalled: "He solved it right there. Ten years out of West Point! So we were very much impressed!"[57]

In his half-century of public service, Baclagon would remain an admiring subordinate, advancing by the bounty of his superiors—from his first benefactor, an American general, through his last, President Marcos. Always genuine in his affections, Baclagon cultivated patrons without guile. Yet, he somehow managed to both serve and preserve a certain underlying integrity.

Baclagon also showed a surprising aptitude for combat. After an unexceptional record at the academy, he was passed over for selection into the elite services and went directly into the infantry, where he might easily have disappeared into the swelling ranks of reserve officers. While his poor record in math made him unfit for a technical command, his close study of human behavior through poetry and history would make him an outstanding combat officer.

After graduation, Baclagon went to Manila for infantry school, and there met and married Julieta Basilio, who would remain with him for the next fifty years. The army was continuing to train the citizen reserves, and he wound up on Negros Island drilling raw conscripts from the sugar plantations. When USAFFE began its mobilization in mid-1941, General Guy Fort, commander of the Eighty-first Division, skipped over reserve majors and lieutenant colonels to appoint Baclagon commander of the Seventy-fifth Infantry Regiment. "So I found myself, a twenty-two year old lieutenant, becoming a regimental commander," he recalled. "Well, I succeeded in activating a whole regiment."[58]

While the fighting raged on Bataan in early 1942, the USAFFE forces in the southern islands spent four months preparing for an inevitable Japanese assault. The Negros command divided the island's troops into five subsectors for guerrilla resistance. But when the Japanese invaded in April 1942, the island's commander, an American colonel, was ordered to surrender and only two of his subsector commanders, both Filipinos, stayed in the hills for guerrilla warfare.[59]

As the enemy landed, Baclagon and his wife evacuated into the mountains above Kabankalan with the local sugar planters. When the Japanese began searching for him, he surrendered rather than place civilians at risk and spent months behind barbed wire in a Bacolod City school. One night, Baclagon led fourteen fellow prisoners in an escape into the mountains by "taking advantage of a blinding rain." He spent several weeks moving about the hills above Bacolod before making contact with a resistance leader, Major Ernesto Mata (PMA '37), whom he had known at the academy and would later describe as "young, handsome, and fearless, . . . the most glamorous figure in the Negros force."[60]

Scattered by defeat, the USAFFE remnants on Negros spent the rest of 1942 negotiating a new hierarchy. In December, Mata finally met with the leading guerrilla in the south, Major Salvador Abcede, and the two agreed to merge their forces. Mata yielded leadership to the more senior Abcede and assumed command of the Seventy-second Division, with Baclagon his chief of staff and commander of the Seventy-fourth Infantry Regiment. As Baclagon later wrote, he and Mata "made up an excellent team, both being graduates of the Philippine Military Academy."[61]

In September 1943, the Japanese mounted a major attack on the Negros guerrillas and hit the Seventy-fourth hard, providing Baclagon with his first test under fire. "In spite of being badly outnumbered," as he later wrote, "Maj. Baclagon's force held up the Japanese advances for three days before it was forced to disperse." When five hundred Japanese troops enveloped his headquarters, Baclagon ordered a retreat, "withdrawing under cover of darkness into the jungle fastness that characterized Negros." After walking for three days to find civil resistance headquarters, his group lost its way in the jungle, suffering weeks of hiking and hunger. "With Mrs. Baclagon, who was in the family way, and other women in the party," he wrote, "they hacked their way through the dense jungle growth and clambered over precipitous cliffs and steep promontories. After the third day, they ran out of rations. From then on, they subsisted for fourteen days on whatever food they could extract from the jungle."[62]

By late 1944, the Negros resistance, supplied by U.S. submarines, had expanded to ten thousand trained guerrillas and was ready for an offensive against the sixteen thousand-strong enemy garrison. When the guerrilla command reformed its scattered forces into three combat teams, Baclagon and his classmate Abenir Bornales were picked for command of two. Advancing north toward the capital at Bacolod City, Major Bornales attacked Japanese garrisons in the island's southern sugar districts, assisted by another classmate, Epifanio Segovia, who led the battalion "responsible for the reduction of Isabela."[63]

At the other end of this extended pincers movement, Baclagon led his First Combat Team in an advance along the northern coast that "reduced the enemy garrison at San Carlos after a vigorous attack." With only 4,500 guerrillas against 12,800 Japanese regulars in his sector, Baclagon outfought superior enemy forces to liberate the island's broad northern plains. "So I had to be . . . on my toes, all the time you know," he later remarked with pride, recalling the experience of commanding a regiment in battle at only twenty-six. "But of course, my military training had a lot to do with that, because I had the respect of my battalion commanders. . . . So when I ordered them, they did what they were told."[64]

The battle for San Carlos brought Major Baclagon the temptation of instant wealth. In liberating a local mill, his troops captured sixty-four thousand sacks of sugar that the retreating Japanese had failed to destroy. Rejecting an offer from black market traders "to become a millionaire overnight," Baclagon turned the cache over to members of the local civil resistance government, who to his "disgust," appropriated the sugar for themselves."[65]

The two guerrilla combat teams were already converging on the capital, forcing an enemy retreat into the island's rugged mountain spine, when the U.S. Army's Fortieth Division landed on Negros in March 1945. The American forces assembled north of Bacolod and then turned inland "with three regiments abreast" and the two Filipino combat teams on their flanks, attacking the "formidable mountain defenses" of the Japanese for the next two months until the enemy was "finally annihilated by the combined Filipino-American forces."[66]

The Class of 1940 produced many war heroes on Bataan and in the resistance that followed, but few could equal Baclagon's record. Though still only in his mid-twenties, he led nearly five thousand guerrillas for three years against one of the strongest Japanese garrisons in the archipelago. In recognition of his services, the U.S. Army awarded him the Distinguished Service Cross. Two years later, the Philippine Army decorated him with the Gold Cross medal "for outstanding gallantry in action."[67] In a word, Baclagon was a hero.

PURSUIT OF CLIO

In the midst of mopping-up operations on Negros, the U.S. Army relieved Major Baclagon of command and flew him across the Pacific to the infantry course at Fort Benning, Georgia, where he graduated with distinction. Two years later, he returned to a reactivated PMA as head of its new Department of Social Science, a posting that would launch his career as a military historian.

To build a curriculum in military history, the academy later sent Baclagon to observe at West Point and Annapolis. During this year-long tour, he renewed his relationship with Colonel Jaime Velasquez, the former PMA commandant now assigned to the Philippine embassy in Washington, D.C. The colonel, a brilliant cadet at West Point, urged Baclagon to write a textbook on Philippine military history for the nation's future officers. Joining a succession of well-placed patrons, he eased Baclagon's access to the key archives—the Library of Congress, the National War College, and the U.S. Army's Historical Division. Returning to the PMA with a wealth of notes and insights, Baclagon established its first course on the "History of Military Arts" while writing his definitive military history, *Philippine Campaigns*.[68]

Three years later, Baclagon published this history to supplement the lessons of the great generals, Lee and Napoleon. Both had, as he explained in his preface, fought a "continental type of campaign involving huge armies operating over extensive areas." Instead, his work would teach Filipino cadets about the very different "insular type of campaign which characterized . . . Philippine history."[69] Despite its modest aspirations, Baclagon's work made major contributions to the writing of Philippine national history.

Considering that it was first published in 1952, years before Philippine historical writing revived, *Philippine Campaigns* should rank as a landmark work. In its documentation, analysis, and objectivity it surpassed the writings of established Filipino historians. Through four hundred pages of careful analysis, the volume surveys the country's three major periods of military conflict: anticolonial revolts, national revolution, and World War II. While clearly a nationalist, Baclagon does not let ideology cloud his judgments. Writing of the early revolts against Spain, for example, he celebrates the courage of Filipino rebels but notes their tactical ineptness. Nor he does sanitize the country's colonial past to serve its national present. Instead of excising native troops from the national history, he devotes an entire chapter to Filipinos who served the Spanish army, arguing that their training "contributed much to our military heritage."[70]

Similarly, his account of the revolution was the first full survey of a tangled history—from revolt against Spain in 1896 to defeat by the United States six years later. Baclagon applied his combat experience to analysis of these battles, achieving insights that would elude later academic historians. At the outset of the Philippine-American War in 1899, U.S. forces used their superior firepower to advance relentlessly across the Central Plain, leaving Filipino forces "no other possible line of action except . . . trading space for time." When American troops failed to break the guerrilla resistance, they used "violent and cruel methods" to extract intelligence—tortures that proved counterproductive, Baclagon notes pointedly, because they inspired the guerrillas to fight "with more intensity and fury." Despite its ultimate defeat, Baclagon wrote, the revolution contributed to "the complete unification of our people as one Filipino race."[71]

His survey of World War II analyzes opposing armies with a model objectivity. Though writing when the wounds of war were still raw, he is balanced in his treatment of both enemy and allied commanders, arguing, for example, that General Homma's first attack on Bataan was "untimely and ill-considered" but that his later offensives showed a "keen appreciation of terrain."

It is the book's final section on MacArthur's liberation campaign of 1944–45 that sets it apart. Almost alone within a vast literature, Baclagon overcomes an arbitrary division between conventional and guerrilla warfare. While the American performance in the arduous mountain fighting against General Yamashita was "marvelous and admirable," Filipino guerrillas "effectively increased the combat power of Allied forces" by their "enveloping attacks on enemy flanks and rear." This coordination took an even more devastating toll when General MacArthur launched amphibious assaults on the southern islands. In its invasion of Cebu, for example, the U.S. Army, supported by Filipino guerrillas, lost only 449 dead in fighting that killed 10,221 Japanese. In contrast to a later, largely derivative account by leading Filipino historian Teodoro Agoncillo, Baclagon's sensitivity to terrain and tactics produced an original work of scholarship.[72]

But beyond the academy, Baclagon would remain obscure. During the tumult of the 1960s, students and faculty inside Manila's university belt were engaged in a passionate debate over nationalism and the nation's destiny. Historians and their students focused on the six years of revolution, searching the country's past for lessons to guide its future. Peasants who rose against Spain in 1896 were celebrated; those who rallied to General MacArthur in 1944 had to be forgotten. In an era of revolutionary rhetoric, Baclagon's focus on military issues identified him with the forces of order and denied him an audience.

Baclagon's career as a military historian was cut short even before his book went to press in 1952. In the rapid formation of Battalion Combat Teams (BCTs) to fight the Huk revolt, Defense Secretary Magsaysay was hard pressed for competent officers. When Baclagon's wartime commander, Colonel Abcede, recommended him to Magsaysay, a friend and provincemate, the defense secretary pulled the young historian out of the PMA, saying he was needed more to fight the Huks. Baclagon commanded the First BCT, the army's premier battalion, for over a year, fighting in the communist heartland of Nueva Ecija and earning three medals for "high meritorious service." When this civil war was won, the army command selected him to lead the Philippine battalion at the exercises of the Southeast Asian Treaty Organization (SEATO) in Bangkok, a performance that won him the Legion of Honor.[73]

Between combat assignments, Colonel Baclagon completed his second military history, *The Lessons from the Huk Campaign*—an Infantry School training manual with unique moral and military insights into unconventional warfare. "It must be emphasized," he notes in a pointed criticism of conventionally trained officers, "that the operation is directed towards the destruction of the enemy, not the capture of a piece of ground." Since the ultimate mission of counterinsurgency warfare is "to assert the majesty of the state," he insisted that "under no circumstances, must any information be obtained through threats, intimidation, or third degree method."[74] Significantly, this condemnation of torture in counterinsurgency, characteristic of his classmates, would stand in sharp contrast to the later experiences of Class '71 against Muslim rebels in Mindanao.

As he rose above the middle echelons, Baclagon began to encounter difficulties. His faith in authority had made him an excellent field officer, but it left him ill prepared for the politics of higher command. After he served as a legislative advisor under his classmate, Representative Cabangbang, and an aide to President Garcia, his nomination to head the sensitive Criminal Investigation Service was blocked by "politicians who wanted to protect their private interests." Angered, he decided to quit the service and devote himself to military history—a decision that he later deemed "a mistake."[75]

It was more than just a mistake. In the Philippines, there was no market for books on military history. Outside the PMA, there were no jobs for military historians. Through his blind passion for history, Baclagon became a seller in a market with just one buyer. Unknowingly, Baclagon set himself on a path toward Ferdinand Marcos, a man whose vision of military history was far from the objective social science that Baclagon espoused.

At first, Baclagon tried, without success, to make a living as a historian. He taught military science to ROTC cadets at Far Eastern University, an overcrowded, underfinanced private school in the downtown university belt. After a long day of teaching and chasing jobs, he would try, with little success, to write. After several months, he gave up and moved to Mindanao to work as an assistant manager for a logging company. There he remained, month after month in a dead-end job, until good friend Cabangbang, still a congressman, happened through on a flying visit, saying, "You're wasting your talents here. You'd better join me."

Through Cabangbang's influence, Baclagon was elevated from a bookkeeper in a Mindanao mill to chief of intelligence for the Nacionalista Party's presidential campaign. During the four lean years that followed its defeat in the 1961 presidential campaign, he stayed with the party in minor positions until the next race for the palace in 1965. Then the Nacionalista nominee, Senator Ferdinand Marcos, made him the campaign's intelligence officer, a post that brought them into close contact. In the distribution of patronage after victory, Baclagon became director of the Anti-Smuggling Action Center (ASAC). To establish his image as a reformer, Marcos appointed Class '40 alumni to his action agencies—Cabangbang became head of the Presidential Agency on Reforms and Government Operations (PARGO), del Castillo was appointed his deputy, and Baclagon and Caballero were given top positions in ASAC.[76]

As blue-seal smuggling waned, Baclagon moved on to new duties as deputy presidential assistant on civic action—reporting to Defense Undersecretary Jose Crisol (PMA '42), his former plebe and now the country's top psychological warfare expert. Baclagon kept that same title for the next twenty years and remained a trusted aide in Malacañang Palace. In 1986, only months after Marcos's fall from power, Baclagon offered his classmates a description of his service to the dictator. "Because I worked directly under the President," he wrote in his *Golden Book* autobiography, "there were times when I got orders to do something for him like go to the United States to deliver some war relics and to do research work in line with his instructions that I rectify errors in Philippine military history, which I tried to do to the best of my ability."[77]

MYTH AS HISTORY

Baclagon was being deceptively modest. Starting in his first administration (1965–69), President Marcos wove the slender threads of his war record into a heroic tapestry that would later serve as an ideological backdrop for authoritarian rule. To spread this myth, Marcos mobilized state power to embroider it in mural, film, poetry, monument, and history.

In such a vast project of cultural construction, Baclagon was, in some sense, just one of many craftsmen summoned to the palace. But if we probe for the heart of the Marcos myth, Baclagon, as a military historian, was keeper of the regime's ultimate mystery. Underlying the legal justifications for martial rule was the myth of Marcos as a reincarnation of ancient Malay warriors who fought against colonial conquest. By his heroism in World War II, Marcos, telescoping time, had revealed himself as heir to their heroic mantle of leadership. Just as ancient warriors proved their prowess by taking heads, so Marcos, in a deft elision of historical epochs, showed his leadership by winning medals in World War II. Through heroism in combat, Marcos had become an instrument of history, freed from the restraints of law and answerable only to his own destiny. Baclagon's writing was no mere matter of "relics" or "rectification," but was in fact the warp in the regime's ideological fabric. Indeed, twenty years later, when he failed to defend this myth from partisan attack, the president's power would start to unravel.

Whether by accident or brilliant design, Marcos had sketched the outline for this myth at the end of World War II when he, like hundreds of resistance commanders, petitioned the U.S. Army to recognize his guerrilla unit and reward its members with back pay. In his August 1945 application, Marcos included documentation for his unit, which he called *Ang Mga Maharlika,* Filipino for the precolonial warriors who fought under a Malay chief or *datu.* Most importantly, he attached a twenty-nine-page account of his underground activities, titled "Ang Mga Maharlika—Its History in Brief"—a master text unsurpassed, for the next forty years, in its detail or degree of hyperbole.[78] If this account could be believed, the Maharlika was one of the most remarkable guerrilla armies in the history of warfare.

Read as a statement of historical fact, Marcos's "History in Brief" is sloppy in its chronology and absurd in its claims. To cite the most fanciful of its falsehoods, the idea that Marcos's bootleg toothbrush factory could finance an eight-thousand-man guerrilla army should strain credulity. But, if read at another level, it is a brilliant exercise in historical imagination. While other guerrilla units took names that tied them to America or MacArthur, Marcos alone appealed to nationalism by evoking the heroism of precolonial warriors. With his choice of this single word, Maharlika, Marcos transcended the colonial ambiguity of fighting for America and linked his actions to a Filipino quest for freedom. Through a narrative that fuses an epic structure with the drama of a spy novel, he relates the saga of a young soldier who is tested on the field of battle, executes remarkable feats of espionage, and emerges a great hero.

The account begins at the outset of war, when a young Lieutenant Marcos, a brilliant lawyer and crack rifle shot, showed, in his own words, "outstanding gallantry in action" on Bataan. After the U.S. surrender in April 1942, he suffered "the dragging pain and ignominy of the Death March" and the horrors of a Japanese concentration camp. Out of "such hatred of the enemy as could be quenched with his blood alone" was born the Maharlika guerrilla army. There in "the filth and disease" of that prison camp, Marcos, with his cousin Simeon Valdez and eight officers, "vowed that whomsoever should come out of that festering hole of lingering death should devote himself to vengeance."

After his daring escape, Marcos organized an espionage unit in Manila under the eyes of the enemy. With a toothbrush manufacturing company as cover and a source of funds, the Maharlika grew into the country's most powerful resistance outfit, with a nationwide espionage net and an underground army of eighty-two hundred guerrillas. To rekindle the "flagging hopes" of the Filipino people, Marcos writes, "the Maharlika kept a propaganda machine of its own" that transcribed radio broadcasts of American victories for distribution to thousands of Manila "subscribers" three times daily.

When MacArthur's return seemed imminent, Marcos's Maharlika guerrillas unleashed a devastating sabotage campaign, sinking three Japanese ships on the Manila waterfront and "causing a near riot in districts by the bay." His operatives later sank three Japanese oil tankers, two trawlers, and a supply ship, while damaging a destroyer and a Maru-class troopship.

Marcos fled Manila in September 1944, escaping in a Japanese staff car as Kempeitai troops were about to close in. Since Manuel Roxas, a prewar cabinet official, had asked him to plan an escape route, Marcos agreed to stop off in Pangasinan Province to build a secret runway. There he found his local guerrillas embroiled in bloody territorial battles with American commanders and, despairing of these conditions, Marcos led his Maharlika fighters into the mountains to find an alternative site for Roxas's runway. Several days later, when MacArthur's landings in Lingayen Gulf made return to his unit impossible, Marcos merged his combat forces into another guerrilla unit, the Fourteenth Infantry.

The Maharlika units, in Marcos's account, distinguished themselves in the eight months of fighting that followed. As the U.S. Army raced toward the city on 28 January 1945, the Maharlika broke through enemy lines to deliver "the first detailed map of the enemy defense of Manila," thus assuring the capital's liberation and salvation. As for Marcos, he fought valiantly as a junior officer with the Fourteenth Infantry during the Battle of Bessang Pass in March 1945, winning a second U.S. Silver Star.

The "History in Brief" ends its epic narrative in December 1945, when Marcos, like General Washington, disbanded his men with a farewell address that advised them to avoid "lowly squabbles over rank and remuneration" that have "cheapened the name of the guerrillas" greedy for back pay. "If your name must remain unknown," he counseled his Maharlika, "remember that your greatness lies in this anonymity."

In seeking to overwhelm the U.S. Army with eloquence and evidence, Marcos underestimated the skill of his inquisitors. He made some elemental errors of historical fact in his original application for official recognition that caught the eye of the U.S. Army's Guerrilla Affairs Division. Explaining his decision to abandon his command, Marcos said in his cover letter: "The [U.S. Army] landings at Lingayen [Gulf], Pangasinan, cut off my return to my own organization. I was attached to the 14th Infantry, USAFIP, NL, since 12 December 1944."[79] Unwittingly, he had lost his three-year battle for recognition in its first weeks.

To verify Marcos's claims, U.S. Army investigators questioned Major Harry

McKenzie, an American who had led Filipino guerrillas in Pangasinan. He pointed out the obvious: Marcos's account was "contradictory in itself. . . . Landings a month later could not have influenced his abandoning his outfit and attaching himself to another guerrilla organization."[80] Marcos had blundered badly. General MacArthur's vast armada had, in fact, landed at Lingayen Gulf on 9 January 1945—not on 12 December 1944, as Marcos would have it.

Two years later, after completing a damning report on the Maharlika, the U.S. Army informed Major Marcos that his men were denied recognition or reward. After Marcos appealed, Guerrilla Affairs conducted a merciless review and issued a final, damning report in March 1948 concluding that "no such unit ever existed" and finding his claim "fraudulent" and a "malicious criminal act."[81]

Denial of recognition did not discourage Marcos from using his "History in Brief" as a master text to build an image of superhuman heroism over the next thirty years. In his rise from congressman in 1949 to president in 1965, Marcos, in a brilliant political synergy, used office to win more medals and used medals to win ever higher office—culminating in an extraordinary ceremony in December 1963 when President Macapagal, courting his support for reelection, awarded him ten medals in a single day. Thus, nearly twenty years after the war, Marcos became, in the words of a newspaper report, the "most decorated Filipino soldier."[82]

With twenty-eight medals to attest to his valor, Marcos launched a successful campaign for president in 1965. His campaign biography, *For Every Tear a Victory*, claimed that "without Ferdinand's exploits Bataan would have fallen three months sooner." With a sacred talisman buried in his back and the "intervention of magical forces," Marcos emerged from the bloodiest battles of World War II unharmed: a clear sign of his heroic destiny. A film version of the biography, *Iginuhit ng Tadhana* (Marked for Destiny), was released just weeks before the elections to win him, in the words of his wife's biographer, "at least 300,000 non-intellectual votes."[83]

In his first term, President Marcos made battlefields into national shrines and commemorations into civic rituals. As leader of the country's veterans, he commissioned plans for the *Dambana ng Kagitingan* (Altar of Courage)—a shrine capped by a soaring cross atop Mount Samat at the heart of the Bataan battlefield. In April 1969, the twenty-seventh anniversary of Bataan, the president led a crowd of fifteen thousand to inaugurate a 300-foot, marble-clad cross, which its curator, Colonel Manuel Acosta (PMA '40), described as a "towering structure of steel and reinforced concrete, with an elevator and viewing gallery . . . 555 meters above sea level." Years later, when Pope John Paul II visited

Manila, Marcos would prevail upon him to fly over Mount Samat and bless the monument, an act that resonated deeply within the country's cult of national heroes and its Catholic faith.[84]

As the skeletal steel of the cross began climbing above Mount Samat, Colonel Baclagon made his first contribution to the Marcos myth by publishing a popular war history titled *They Served with Honor*. To accommodate the president's mythic vision, Baclagon abandoned his scientific study of war as a mechanized juggernaut that dwarfs every soldier and reverted instead to a Homeric rendering of battle as epic combat between great warriors. Baclagon inserted the president repeatedly into this epic catalogue, using his medal citations to describe Bataan as "a saga of heroism woven around the exploits of Lt. Marcos . . . unequaled in the annals of the . . . campaign.[85]

Baclagon's twenty years of service to Marcos represented an unlikely pairing of two irreconcilable institutional ideals. At the University of the Philippines, Marcos had imbibed a nationalist vision of history and a view of the military as servant of national salvation. At the PMA, Baclagon had studied military science and internalized an apolitical military professionalism. Marcos's officer was thus an empowered warrior, his history mythic and heroic. Baclagon's officer was a public servant, his history objective and scientific.

Marcos, moving beyond the limitations of print, used the visual arts to depict himself as a man marked by destiny. After his first inauguration, he commissioned national artist Carlos V. Francisco to paint an enormous eighteen-by-twenty-four-foot mural with his six-foot-high head at its center. Nearby, ancient Filipino warriors raise their swords to unleash tongues of blue fire that lead the viewer's eye around the canvas through Marcos's rise from wartime heroics to his presidential inauguration. Just beneath his huge head is a curious doll-like figure with an infinite void where its face should be. The painting seems to say that this form, the spirit of national resistance, has now taken a face. Invested with this spirit, Marcos has become a divinely chosen leader, beyond the court of the present and answerable only to history.[86]

After martial law in 1972, Marcos's wartime heroism became the ideological foundation for his authoritarian regime. Commemorations became a national cult, celebrated every year on the anniversaries of the battles at Bataan and Bessang Pass. State propaganda agencies drummed a constant theme. In leading the Maharlika in "daring daylight raids, death-defying sabotage missions," a military aide wrote in 1977, Marcos had revealed "his uncanny but prophetic vision and foresight . . . [and] his humaneness towards the common man."[87]

Under his dictatorship, the myth of Marcos as the reincarnation of ancient *datu* merged with his larger vision of social reconstruction. Just as datus had ruled independent communities, or *barangays*, before the Spanish conquest, so Marcos would, through constitutional authoritarianism, govern directly through local units now renamed barangays, thereby liberating the nation from its colonial past. In 1976, the peak of his martial-law powers, Marcos articulated these ideals when he led his generals to Mount Samat for the annual wreath laying. "In Bataan, something emerged within the nation which none of us had seen before in our four centuries of colonial subjugation," Marcos said. "Bataan is a symbol of national unity. . . . Yes, not even the revolution of 1896 . . . gave us such a vision of ourselves as a nation."[88]

In the dictatorship's third year, Baclagon published a revised edition of his *Philippine Campaigns* with a new chapter titled the "Saga of Ferdinand E. Marcos," correcting his failure to mention the nation's greatest hero in earlier editions. "Few men, if any, in Philippine history can equal the . . . heroic feats of Ferdinand E. Marcos during World War II," Baclagon begins. "The twenty-six medals . . . are testimonials of his extraordinary combat services that made him the most decorated soldier and the Number One hero of World War II." In contrast to the focus on mechanized warfare in the original chapters, this thirteen-page insert is a hyperbolic celebration of one soldier's heroism. Even so, Baclagon had his limits. His account omits the Maharlika's more absurd exploits and describes it, in a few cursory sentences, as "primarily an espionage organization."[89]

By contrast, on martial law's eighth anniversary in 1980, Marcos's administrative aide, Victor Nituda, invoked the president's medals as intimations of divinity:

> By the grace of God, the spirit of Bataan did not fall
> With the blood that burned in the fire of an April sun . . . ;
> And you were one of them and by the grace of God, you survived . . . ,
> Your iron will becomes the loveliness of us all,
> As testamented by various medals won in the arena of heroism . . . ,
> You have arisen like a new Messiah from browning earth,
> To call unto your bosom all the wayward children of the earth.[90]

That same year, first lady Imelda Marcos celebrated their wedding anniversary by commissioning a lavishly bound, four-hundred-page epic poem by two leading academics—the anthropologist E. Arsenio Manuel and literary critic Florentino Hornedo. Through luminous illustrations of Imelda as the reincarnation of Queen Nefertiti of Egypt and Marcos as the spirit of an ancient Malay datu, the epic juxtaposes the country's mythical past with its political present to reveal a Marcosian conception of historical time as a permeable membrane, joining, rather than separating, past and present.

This sense of temporal transcendence appears in the opening stanzas, with Marcos near death on a Bataan battlefield and Imelda singing to him of history, of warrior datus who fought invaders past.[91]

> Beauty from the heavens who awakened
> The discriminating heart of one who was oblivious
> To the call of love.
> A maiden who possessed all of beauty,
> A girl who is the goddess of all goddesses;
> Attractive, lovable, desirable.
> Singing to him, and energizing him
> Through the core of strength and beauty of our past.
> Courage has returned to the wounded,
> His passion has revived him
> From the blood veins from his courageous ancestors
> Facing up to any attacking conquerors.

The president himself also assembled a secret team of the country's top historians and ordered them to compile a twenty-seven-volume history of the Philippines, in effect creating a past consonant with this epic vision. Titled *Tadhana*, Filipino for "destiny," the first volumes listed Marcos as author and featured his photo on the dust jacket, seated at a desk poised to write beneath an image of Christ. "The search for Filipino identity to solve the centuries of ambivalence in national attitudes," wrote Marcos in the introduction, "gained real purpose and direction in the New Society. . . . All these vicissitudes of polity came to a head in September 1972 with the establishment of 'constitutional authoritarianism' . . . to strengthen barangaic culture."[92]

This unique, at times bizarre, historiography served as the ideological foundation for the president's fourteen years of authoritarian rule. In gestures large and small, sophisticated and crude, the regime sought legitimacy by celebrating Marcos's heroism. He built a two-thousand-kilometer road at a cost of $250 million called the "Maharlika Highway." When the president opened a new parliament in 1978, his ruling KBL Party introduced a bill to change the country's name from Philippines to Maharlika in honor of those who had "fought and died if only to preserve our sacred traditions and ideals." On Armed Forces Day four years later, he reviewed the Manila garrison while a thousand soldiers held up color cards, like those used in American football stadiums, to form a giant portrait of the young Lieutenant Marcos and his Medal for Valor. During his 1982 state visit to Washington, Defense Secretary Caspar Weinberger gave Marcos a case filled with replicas of his U.S. medals and President Reagan hailed his wartime heroism at a White House dinner.[93]

In the midst of this state visit, the *Philippine News*, a Filipino-American weekly, published an exposé by exiled activist Bonifacio Gillego charging that these medals were fake. Gillego, a former intelligence officer and amateur boxer, opened his article with a punishing attack on Baclagon's book *Filipino Heroes*, saying, "The bravest of the brave . . . is none other than Col. Ferdinand E. Marcos," who is credited "with incredible feats of valor and combat prowess." After hammering away at this history sentence by sentence, Gillego backpedals momentarily: "Baclagon should perhaps not be faulted for giving Marcos top billing among the brave men of the 14th Infantry. Baclagon based his narration on official records available in the files of GHQ, Armed Forces of the Philippines." Then, he lunges for the knockout blow: "But should a military historian like himself accept . . . the alleged heroism of Marcos contained in the citations without bothering to subject them to tests of validity . . . ? Are the accounts plausible? Are there supporting documents to substantiate these superhuman feats of intrepidity?"

Unlike Baclagon, Gillego had interviewed the former commander of the Fourteenth Infantry, Colonel Manriquez, who "bristled when told of Baclagon's account of Marcos being recipient of three major awards." Marcos, the old colonel insisted, had been confined to staff duties and "had not even fired a single shot at an enemy."[94]

In October, the Manila opposition magazine *We Forum* reprinted Gillego's charges in full, creating a crisis of legitimacy for Marcos's New Society. In his private diary, the president raged against "the small souls whose vicarious achievement is to insult and offend the mighty," saddened that "their pettiness has besmirched with the foul attention the honorable service of all who have received medals . . . in the last World War." Publicly, the president ordered the arrest of the magazine's entire staff on capital charges of subversion.[95]

Marcos's aging wartime comrades testified at length about his heroism during the show trial that followed. If the U.S. Army had not lost the paperwork, claimed one witness, Marcos would have won the U.S. Congressional Medal of Honor. After the Philippine Supreme Court ruled the arrests invalid on technical grounds, the controversy stalemated into conflicting allegations. But Marcos had made his point. When the *Washington Post* published another exposé of his medals in December 1983, not a single Manila newspaper dared to reprint it.[96]

As the trial proceeded, the president ordered his aides to defend his war record. With Baclagon doing the research and Deputy Defense Minister Crisol adding the adjectives, the two rushed to publish *Valor*—a glossy, large-format book decorated with full-page color photos of all twenty-eight medals. The authors claimed that Marcos had "electrified soldiers of every rank in Bataan" by showing "sheer personal courage that demolished powerful enemy offensives" and thereby slowed the Japanese juggernaut by three full months. Not only did Marcos commission the work, but he helped write it by calling his authors to the palace for war stories. "We talked a lot," Baclagon explained. "He likes to talk about it, you know." Indeed, the book is illustrated with detailed battle sketches and notes in Marcos's own distinctive angular hand. Aside from the president's personal testimony, Baclagon relied upon affidavits in support of his postwar claims for medals, notably a lengthy "narrative account" by Ambassador Nicanor Jimenez, the president's wartime commander on Bataan.[97]

By pairing a full-page photo of each medal with the original citation and a detailed narrative, the book made an exhaustive attempt to prove beyond any doubt that the president's heroism was genuine. Ironically, Baclagon's careful assemblage raises doubts about his larger claim. All the citations and affidavits are dated years after the actual battles. All the documents, maps, and photos are recent reconstructions. In the book's hundred-plus pages, there is not a single scrap of original evidence.

Feeling the book's vulnerability on this score, Baclagon approached the president with a plan. Mrs. Baclagon was dying of cancer and needed costly treatment in the United States far beyond his modest means. If Malacañang could fund his trip, he would search the U.S. National Archives for documents to prove the president's heroism. With the generosity that endeared him to followers, Marcos replied: "I'll have General [Fidel] Ramos give you assistance in going to the States." While his wife was being treated at a Washington hospital, Baclagon worked at the National Archives on Pennsylvania Avenue. Indeed, in January 1985, when this historian signed in at the archives to start a similar search, I was upset to see Baclagon's signature at the top of the page. I imagined him on a relentless search-and-destroy mission for Marcos—searching out incriminating documents and flushing them down the hallway toilet to oblivion. As it turned out, Baclagon had indeed seen some damaging documents, but was incapable, within his conflicted sense of ethics, of doing anything more than ignoring them.

In his travels about America, Baclagon, the presidential aide, stopped in Los Angeles for a reunion with his classmate Ramon Alcaraz, the anti-Marcos activist. The two were bitterly divided over the Marcos dictatorship. But Alcaraz insisted that his old friend stay at his Orange County home, where the welcome was warm. Baclagon recalls, with obvious pleasure, that he "even slept in the room where Cory Aquino stayed." Nonetheless, Alcaraz took his classmate to task for his books about the president's wartime exploits. "Don't you know that, based on what I know about Marcos," Alcaraz said sharply, "what you wrote about Marcos is not true?" Alcaraz admitted giving Gillego leads for his exposé and insisted that Baclagon should "go to Gillego and apologize to him." Baclagon protested that he had relied on documents, statements sworn in support of Marcos's medals and filed at AFP headquarters. Alcaraz dismissed this sort of evidence as "merely affidavits."[98]

Forty years after the end of World War II, the war's history had thus become one of the most sensitive issues in Philippine politics. At the start of the 1986 presidential campaign, speakers in the president's entourage hailed him as the greatest Filipino military hero. The ruling KBL party distributed a Filipino-language comic book titled *Batang Matapang* (Courageous Youth)—its first page showing a young Lieutenant Marcos blazing away with a machine gun, muscular shoulders dripping Rambo-like with ammo belts, and its last frame picturing him at the war's end, with a row of medals on his swelling chest, the "most decorated Filipino soldier."[99]

Two weeks before election day, the *New York Times* published a page-one exposé of Marcos's war record, quoting U.S. army documents, discovered by this historian, that called his Maharlika guerrillas "fraudulent" and his claims of heroism "absurd." Opposition candidate Corazon Aquino seized upon the issue, damning Marcos as a "fake hero" and a "fake president" who could only win "by fake election returns." In his column for *Business Day*, Francisco Tatad, the president's former information minister, called this exposé "the biggest propaganda blow to hit the 68-year-old strongman in this campaign."[100]

Marcos, ailing from regular kidney dialysis, suddenly found himself on the defensive. "I don't know where they got such foolishness," the president told a campaign rally in the Manila district of Tondo. "You who are here in Tondo and fought under me and were part of my guerrilla organization, you will be the ones to answer these people, these crazy individuals, especially the foreign press." Two days before the voting, Marcos announced that Emperor Hirohito of Japan had devoted eleven pages of his memoirs to the exploits of the Maharlika guerrillas—a claim that soon collapsed since the emperor had no memoirs.[101]

At this critical moment, the president turned to his military historians for support. Speaking to the pro-Marcos press three days after the *New York Times* exposé, Deputy Defense Minister Crisol, the military's psy-war expert, charged that "the documents used by . . . Alfred McCoy in portraying President Marcos as a 'fake war hero' are 'fabricated and false.'" To discredit the *New York Times*, he displayed photocopies of real World War II documents "taken by retired Colonel Uldarico S. Baclagon from the National Archives in Washington on April 25 last year," detailing operations by the Maharlika guerrillas.[102]

But Baclagon himself was disarmingly frank in his own front-page interview in the opposition's *Philippine Daily Inquirer*. Baclagon admitted that he had, while researching at the U.S. National Archives, found three letters from American officers refusing recognition for Marcos's guerrilla unit. "But I had no money for reproduction," he explained modestly, "and I had no time to read the voluminous documents." Asked how he found the documents confirming the president's story, the colonel explained: "I selected it [the passages attesting to the Maharlika] because I knew it would make President Marcos happy—I didn't select the ones which will make Marcos unhappy, like the charges of being fraudulent and all that."[103] For a presidential advisor, speaking only a week before election day, this was extraordinary honesty.

Why was Baclagon so restrained? Since he will not answer out of loyalty to Marcos, we can only speculate. Even after twenty years of working for the president, Baclagon may have felt a conflict of loyalties. Bound to his PMA classmates and their rigid honor code, he could only go so far in his defense of dictatorship—even when his president's fate hung in the balance. Honor demanded truth. Loyalty required fidelity to his president. In this conflict between principles, Baclagon struggled to serve both.

Even after Marcos fell from power, Baclagon remained loyal. While Alcaraz or other classmates could have brokered an accommodation with the new regime, the old colonel continued to serve his president. Only six months after the dictator's flight, Baclagon went to Honolulu to offer his services to Marcos. Since the exiled president was slowly dying and his assets were tied up in litigation, he needed doctors and lawyers, not a historian. Besides, the myth that the colonel had long tended had been snuffed out. Later, Baclagon joined the search for the fabulous "Yamashita treasure," returning to Honolulu as an intermediary for a businessman who claimed he could help Marcos recover this mythical horde of gold and gems.[104]

A decade later, when the dictator was dead and the headlines were forgotten, I called on Colonel Baclagon, then seventy-seven, at his apartment in one of the government's low-cost BLISS housing projects, once Imelda's showcase for her "city of man" and now a high-rise slum. His wife had died from cancer the year before, leaving him lonely and depressed. He had suffered a stroke that slurred his speech and made the climb down the building's four flights painful. His social world had narrowed to hospital visits and funerals. He had recently spent the whole day at a crematorium, grieving silently while his classmate Edmundo Navarro was reduced to ashes.[105]

When I asked whether he still believed in Marcos's stories of guerrilla heroism, the colonel's eyes burned intensely as his lips fumbled to form words: "I, I, I don't claim to know everything because . . . he is in the north, I am in the south." From his work as a historian, did he still believe Marcos's claims? "Well, from my point of view, he was telling me the truth. He was telling me everything, you know. He was telling me outright, [the] direct story." Had he been paid for writing that glossy defense of the medals, the book *Valor*? "No, I did not get paid at all." Was he rewarded for his other services to Marcos as a historian? "Well, he, he, he sent me to the States. But that's all." Had there been any bonuses, or other rewards? "I should not be living in a place like this, no?" As he said these last words, our eyes traveled together about the room, taking in the bare bulbs, broken furniture, and dingy walls—conditions that the colonel himself called "squalor."

As these questions continued, Baclagon seemed to feel that I was pressing him to recant and denounce Marcos's war record. He refused, scorning the recent apostasy of Ambassador Nicanor Jimenez, the president's former wartime commander, who had once written an affidavit that won Marcos the Medal of Honor, the country's highest decoration. As president, Marcos, in turn, rewarded Jimenez with a Gold Cross medal and a succession of senior appointments—general manager of the national railways and later ambassador to South Korea. In writing his histories, Baclagon had relied on the ambassador's affidavits since they were "very strong" in stating that Marcos "really did it." But once the president fell from power, Jimenez, in private conversations about Manila, retracted, denouncing the medals as fraudulent. Since he had been Marcos's college contemporary and wartime comrade, Jimenez's apostasy was, within the country's culture of loyalty, unseemly. But flexibility served him well. In the change of regimes, he shifted seamlessly from serving Marcos in Seoul to representing Aquino in Washington, D.C.[106] Baclagon might have had doubts about aspects of Marcos's record, but he was not about to emulate such opportunism.

One could fault Baclagon, as Alcaraz had done, for his naïve faith in Marcos's statements. Baclagon may well have let his sense of loyalty blind him to the absurdity of these superman stories. Indeed, by placing the imprimatur of a historian's authority on these claims, Baclagon strengthened Marcos's grip on power. But it is difficult to fault Baclagon's personal integrity. He did his job as presidential assistant and collected a modest paycheck of eight thousand pesos each month. He did not try to leverage these services into higher positions or special rewards. Once Marcos fell from power, the colonel had the courage to stand by his writing and suffer poverty in silence. It may not have produced great history. But it is, in a certain sense, a performance in keeping with the PMA's demands for "integrity, courage, loyalty."

In different ways, Baclagon and Alcaraz demonstrate how deeply Class '40 had internalized these ideals. Both were keenly aware that their lives had been transformed by their admission to the academy. They engaged cadet life with a passionate intensity and retained, long after graduation, feelings of reverence for their alma mater. Its lessons were tested in World War II, where both, by heroism, affirmed their "courage" and "loyalty." In postwar service, both found their "integrity" tested almost daily. As commander of the navy's antismuggling campaign, Alcaraz had countless opportunities to enrich himself. Similarly, Baclagon had direct access to the president's study during a decade when entrée meant instant wealth.

More broadly, their sense of honor—indeed, the whole of their military socialization—set limits that restrained them, and their classmates, from corruption or capitulation to political pressures. All, of course, compromised. Applying the academy's honor code, with its simple absolutes, to the complexities of military service in the Philippine Republic required moral choice. With wives, children, and responsibilities, they could not simply walk away from thirty-year careers when superiors made unpalatable demands. Thus, Baclagon ignored the implausible and improbable in the president's war record, just as Apolinario failed to stop smuggling by the First Lady's brother. Retreating to a narrowly professional, even myopic, position, Baclagon was willing to write history that pleased Marcos as long as there was some supporting documentation, even affidavits or the president's own statements.

Yet each, in his own way, set limits, things they would not do, lines they could not cross. Apolinario refused a second tour in command of the navy since it meant sanctioning the first family's smuggling. Baclagon, at the hour of crisis when his president desperately needed a defender, proved incapable of lying, or even shading the truth, beyond the bounds of documented fact. Indeed, he insisted on telling the press that he had, in fact, seen documents about Marcos's falsification of his war record. Theirs was not the moral purity of a cadet, or a priest, but it was honor in a way that each found meaningful and their class found acceptable.

One can only imagine all the bargains struck, coups plotted, and damage done had not they, and their classmates, been so restrained. With his driving ambition, Alcaraz, unless bound by principle, might have succumbed to some bargain to win the command that he craved. Apolinario, with his entrepreneurial flair, could have become utterly venal, a shark swimming among the smugglers. Baclagon might have destroyed incriminating documents at the U.S. National Archives or fabricated historical fact to shore up a collapsing regime. Indeed, in the two decades after Class '40 retired, Class '71 would demonstrate, with striking clarity, the high cost to Philippine society when military socialization failed.

Part Two **The Class of 1971**

Chapter Six Torture

On 28 August 1987, Colonel Gringo Honasan stood at the threshold of power. In the confusion that followed Marcos's downfall, people were seeking salvation from their country's endless crisis. Just past midnight, the colonel and his classmates, leading two thousand heavily armed rebel soldiers, breached Manila's defenses and drove for Malacañang Palace. His forces, flying the banner of the Reform the Armed Forces Movement (RAM), attacked the palace gates in darkness. After a seesaw battle, they were finally repulsed by the Presidential Security Guard. As they retreated through the old city's narrow streets, the rebels fired randomly into a crowd of civilians, killing eleven and wounding fifty-four. At that moment, Colonel Honasan's bid to enter history by force of arms had failed.

These colonels and their capricious violence are emblematic of an authoritarian age. From the 1950s to the 1980s, military coups were common across Asia, Africa, and the Americas. By 1985, military juntas held power in over half of the hundred-plus nations that comprise the developing world. With ruthlessness and idealism, these martial regimes often attempted social transformation through systematic violence. The more extreme used terror to wage civil war on their enemies, whether ideological, ethnic, or religious. When their time had passed and these regimes collapsed in the 1980s, the politicized, brutalized armed forces they left behind were often a barrier to democratization—clinging to power, resisting reforms, and struggling to prevent any accounting for the past.

Under the martial-law regime of Ferdinand Marcos (1972–86), the Philippine military spawned several such groups. But none could rival the violent extremes of RAM. Led by a clique of middle-ranking officers, this group plotted a coup d'état against the Marcos dictatorship in 1986 and, failing to take power, launched five more against his successor, Corazon Aquino. Though their attempts all failed, the careers of these colonels still reveal much about the mentality of the military officers who ruled this authoritarian age. Most importantly, they have much to teach us about the impact of state terror upon the armed forces.

Within the Philippines itself, RAM represents, above all, a breakdown in military socialization. Most of its members were regular officers who had graduated from the Philippine Military Academy during the 1970s. There, for four years, they were drilled in obedience and indoctrinated into a belief in civilian supremacy over the military. Unlike earlier generations, these future rebels graduated into a martial-law military and spent their formative years as junior officers fighting a civil war in Mindanao or interrogating dissidents in Manila. As torture and extrajudicial killings rose, they became the instruments of state terror. For them, torture was transformative, freeing them from military socialization and inspiring a will to power.

In their leap from interrogation to coup d'état, these rebel colonels highlight our need to study the torturers instead of focusing solely, as most of the literature does, on the victims. Their history shows how torture can inspire a sense of empowerment in the perpetrator—an unsettling topic largely ignored in the literature on human rights and human psychology. To understand the impact of repression upon the military, we need to examine the ways that the torturers' experience of torture has informed, even inspired, their political visions.

THINKING ABOUT TORTURE

We are only just beginning to understand torture. As military regimes practicing systematic torture proliferated in the mid-twentieth century, the international community reacted with treaties to outlaw the practice and therapy to treat its victims.[1] In 1948, the U.N. declaration on human rights banned torture but provided no mechanism for enforcement. From its founding in 1961, Amnesty International fought torture, but the practice, and the dictatorships that sustained it, spread during the following decade. Thus, in 1972, Amnesty mounted its first Campaign for the Abolition of Torture and, a year later, published a pioneering report.

At the campaign's end, Amnesty, realizing the limitations of its lawyerly practice of documentation and petition, appealed to the medical profession for support. A group of Danish doctors responded with research among victims that discovered a pernicious, often incapacitating form of posttraumatic stress disorder. "When you've been tortured," explained Dr. Inge Genefke, "the private hell stays with you through your life if it's not treated." But the victims did respond surprisingly well to therapy. In 1982, these researchers founded Copenhagen's Rehabilitation and Research Centre for Torture Victims (RCT) and then built a global network of ninety-nine centers that treated forty-eight thousand victims in 1992. Together, Amnesty and the RCT played a catalytic role in turning world opinion against torture—culminating in the 1984 U.N. Convention Against Torture and the 1993 World Conference on Human Rights, where 183 nations condemned torture as "one of the most atrocious violations against human dignity."[2]

Through this global effort, therapists have discovered that torture is a uniquely debilitating experience that leaves lasting psychological harm. While victims often experience fewer physical injuries from torture than, say, a serious auto accident, they nonetheless suffer a deep emotional damage out of any proportion to the physical pain. In 1991, researchers reported that survivors suffer "sleep disturbances with frequent nightmares, affective symptoms (chronic anxiety, depression), cognitive impairment (memory defects, loss of concentration), and changes in identity." Such suffering is persistent, often permanent. In the early 1990s, Polish psychiatrists interviewed victims tortured more than forty years earlier and found that "symptoms of mental disorder were present in almost all."[3]

This research has also explored the role of power in the process of torture. "Essential to torture," Amnesty explained in a 1984 report, "is the sense that the interrogator controls everything, even life itself." A team of Argentinean therapists concluded that victims "experience depersonalization, fear of annihilation, and the destruction of their body image."[4] Similarly, a Philippine therapist found that torture's trauma arises "from the experience of extreme powerlessness" and can be treated by "reempowering the survivor." If torture leaves the victim feeling weak, even crippled, might it not have an opposite impact upon its perpetrators—in effect, inducing an emotion of empowerment?[5]

Scholars have been reluctant to engage this question.[6] Indeed, there has been resistance to the idea that there might be meaning in such inhumanity. During World War II, publication of Bruno Bettelheim's renowned study of Nazi concentration camps was delayed for two years because, as he put it, "the notion of ascribing purpose or intelligent planning to the SS, of taking them seriously, was considered both unwise and unsafe."[7]

Even a half-century later, few scholars have studied the torturers. Even these few have usually asked two rather rudimentary questions: who are these men, and how can they do it?[8] The recent literature offers generally consistent answers. Rejecting the earlier emphasis on the "abnormal" personality, studies since the 1960s have found some common factors: (1) torturers are usually "normal" police or soldiers recruited randomly; (2) many pass through an initiation rite to bind them to their fellow torturers; (3) all are indoctrinated to believe that their victims are "other," that is, deviant and thus deserving of punishment; and (4) the institutional context, with its inherent legitimacy, conditions perpetrators to continue routine torture for months or even years. There is, moreover, an implicit finding: among the passive majority of perpetrators, there are a few master torturers who identify fully with the task.[9]

The Greek dictatorship of the 1970s provides evidence for this view of torture as a problem of compliance rather than deviance. In her study of soldiers who tortured for the junta, psychologist Janice Gibson concluded that all were "mentally normal people" who had been "successfully trained for this horrendous profession." In 1975, at the landmark trial of military torturers, a common soldier claimed that he "was caught up in a machine and became a tool without any will of my own to resist." He added, "I feel the need to tell . . . the Greek people that I am a human being like you, like your neighbor's son."[10]

Research on Philippine torturers also found that torture was "learned behavior." After interviewing active-duty perpetrators in 1995, Dr. June Lopez found military training is "marked by physical and psychological violence" that serves as a gateway to torture. Within these studies of Greece and the Philippines there is, then, an unstated hypothesis: a repressive state needs torturers and trains ordinary soldiers for the task. Torturers are not born; they are made by particular political regimes.[11]

These conclusions find some support in the famous Yale obedience experiments. In the early 1960s, psychologist Stanley Milgram tested forty "ordinary" residents of New Haven, Connecticut, to see if they would, upon command, inflict electric shocks on a helpless "victim." At Yale University's elegant Interaction Laboratory, a uniformed researcher seated subjects before "shock generator type ZLB" and ordered them to activate thirty switches ranging from 15 to 450 volts. At 75 volts, the victim, a trained actor, gave a "little grunt"; at 315 volts a "violent scream"; and after 330 volts he "was not heard from." When the volunteer could not see the victim but just heard his voice, 65 percent flicked the switches to the maximum. If the subject was asked only to assist, compliance was almost 100 percent. But if a nonauthority figure gave commands, obedience was zero. Rejecting the idea that those who delivered maximum voltage were "monsters, the sadistic fringe of society," Milgram found instead that "binding factors" of social convention led normal individuals to accept authority and ignore the victim's pain.[12]

The Stanford prison experiment of the early 1970s refined this view of brutality as situational. After careful screening, psychologist Philip Zimbardo randomly assigned twenty-one "average, healthy American college males" to roles as guards and prisoners inside a simulated campus prison. After only six days, he was forced to terminate his two-week experiment because the guards' brutality was raging out of control. In this prison, power had a transactional quality, with the guards' "sense of mastery and control" matched by the prisoners' "depression and hopelessness." All eleven guards were cruel at least once, but a few "delighted in the newfound power" and demonstrated "great . . . cruelty in the forms of degradation they invented for the prisoners." Everyone is capable of cruelty, but some are more capable than others.[13]

Studies of torture in Latin America have discovered that southern cone dictators used its terror to empower the military and encourage "conformism throughout society." Argentina's junta sanctioned systematic torture marked by "moments of exaltation when the torturer felt as if he were God."[14] After years of treating victims, Argentine psychiatrists also noted a "bipolar . . . omnipotence-impotence logic": torturers inflict "psychological impotence on the victim" and "present themselves to the victim as the absolute owner of . . . their lives."[15] But we are still left with the same unanswered questions. What is the long-term impact of torture upon the torturers? What are the wider political ramifications of their empowerment? We have learned much about the victims and something about the way torturers are trained, but we have still not probed the political logic and legacy of state terror.[16]

Insights from the treatment of Chilean victims tortured under General Pinochet's regime offer a point of entry. Psychotherapist Otto Doerr-Zegers found that victims suffer "a mistrust bordering on paranoia, and a loss of interest that greatly surpasses anything observed in anxiety disorders." The subject "does not only react to torture with a tiredness of days, weeks, or months, but *remains a tired human being*, relatively uninterested and unable to concentrate."[17] These findings led him to a strategic question: "What in torture makes possible a change of such nature that it appears similar to psychotic processes and to disorders of organic origin?" Why, in short, is torture so damaging?

The answer, Doerr-Zegers argued, lies in the "phenomenology of the torture situation," involving (1) an asymmetry of power; (2) the anonymity of the torturer to the victim; (3) the "double bind" of either enduring or betraying others; (4) the systematic "falsehood" of trumped-up charges, artificial lighting, cunning deceptions, and mock executions; (5) confinement in distinctive spaces signifying "displacement, trapping, narrowness and destruction"; and (6) a temporality "characterized by some unpredictability and much circularity, having no end." Thus, much of the pain from torture is psychological, not physical, based upon denying its victims any power over their lives. In sum, the torturer strives "through insult and disqualification, by means of threats . . . to break all the victim's possible existential platforms." Through this asymmetry the torturer eventually achieves "complete power" and reduces the victim to "a condition of total or near total defenselessness."[18]

There are, within torture's phenomenology, elements that seem to evoke theatrical parallels. In a sense, Doerr-Zegers seems to be saying that torture, as done in Chile, was a kind of total theater, a constructed unreality of lies and inversion, with a plot that ended inexorably with the victim's self-betrayal and destruction. In the theater there is, of course, an asymmetry of power between the actors, who know the plot, and the audience, who do not. Working from a well-rehearsed script, the torturers are actors who force the victim to become an audience of one in the drama of his or her own degradation. To make their artifice of false charges, fabricated news, and mock executions convincing, soldiers become inspired thespians. The torture chamber itself often has the theatricality of a set, with special lighting, sound effects, props, and backdrop, all designed with a perverse sort of stagecraft to evoke an aura of fear. Both cell and stage construct their own temporality: the drama collapses and expands time to carry the audience forward toward denouement, while the prison distorts time to disorient and then entrap the victim in a temporal maze that leads inexorably to self-betrayal.

Under the peculiar conditions of torture, victims become the audience. As the torturer manipulates circumstances to "maximize confusion," the victim feels "prior schemas of the self and the world . . . shattered" and becomes receptive to the "torturer's construction of reality." Isolated from others, the victims form emotional ties to their tormentors that make them responsive to a drama in which they are both audience and actor.[19]

Significantly, both torturers and victims use the language of theater to describe their experiences. A Greek victim, referring to his prison and its torturer, said "KESA was a kind of theater of the absurd with Kainich as producer." Two Argentine victims described a sadistically effective interrogator succinctly, saying, "He turned torture into an act of theater." Elaine Scarry notes that "in the torturers' idiom the room in which brutality occurs was called the 'production room' in the Philippines, the 'cinema room' in South Vietnam, and the 'blue lit stage' in Chile." On such a stage, torture became "the production of a fantastic illusion of power, . . . a grotesque piece of compensatory drama."[20]

In Latin America, military torturers used theatrical techniques to attack basic socialization—transforming, for example, women from "Madonna" into "whore." In one such instance, the Chilean military staged a mock drama to torture Elba Vergara, the former secretary to President Salvador Allende, forcing her into multiple betrayals.

> She was taken to a room called the little blue room, a small area lit by only two blue bulbs. Her blindfold was removed. Four hooded executioners broke into the room, shouting, "Bring in the actor!" A covered body was wheeled in on a stretcher. One of them, then turning to Elba, said, "Sit down. Here you are going to see the representation of a bad actor. You have to help him remember his lines."
>
> They uncovered the naked body of a brutally disfigured young man. . . . Elba was asked if she recognized him. She said, "No." She did, in fact, recognize the young man. He had been a chauffeur for a member of the Allende government.
>
> "You are telling us that you don't know him. We shall see whether you know him or not." They took one of his hands and pulled out his fingernails. Elba continued to insist that she could not identify the man. They tore off the one ear he had left at that point. They cut his tongue out. They punctured and emptied one of his eyes. This ordeal lasted three hours. He died then.[21]

There are striking similarities between this technique and Philippine methods, detailed below, that raise some intriguing questions of origin. Under the Marcos regime, methods, in general, were more psychological than physical. Some officers practiced a theatrical torture that might be called a drama of social inversion. Are these techniques, so similar in their psychological refinement, independent improvisations in the torture chambers of disparate continents? Or was there, I asked myself in 1993 when first comparing these Chilean and Filipino torture transcripts, a single set of classified manuals, distributed by U.S. agents or attachés on both sides of the Pacific? Four years later, the *Baltimore Sun* published extracts from the CIA's *Human Resource Exploitation Training Manual—1983*, the latest edition of a thousand-page torture textbook distributed to Latin American armies for twenty years.

For decades, starting in 1950, the CIA funded academic research into "the relative usefulness of drugs, electroshock, violence, and other coercive techniques" to discover new methods of psychological torture, a breakthrough in this cruel science. Instead of a soldier's natural inclination to simple physical brutality, the agency's manual, the apparent fruit of such research, teaches psychological tactics to break down a subject's "capacity to resist": start with a predawn arrest to shock; strip-search to humiliate; and confine in a windowless room to disorient. Through "persistent manipulation of time" by "retarding and advancing clocks, disrupting sleep schedules," the questioner can break a subject's will, driving "him deeper and deeper into himself, until he is no longer able to control his responses in an adult fashion." Significantly, the agency did warn that physical torture weakens the "moral caliber of the [security] organization and corrupts those that rely on it." But the CIA missed an obvious point: any form of torture, physical or psychological, has the same corrosive effect.[22]

These CIA techniques are so similar to Philippine practices that we must ask: did the CIA train these Filipino interrogators? In 1978, a human rights newsletter reported that the Marcos regime's top torturer, Lieutenant Colonel Rolando Abadilla, was studying at the U.S. Command and General Staff College, Fort Leavenworth. A year later, another group claimed that his understudy, Lieutenant Rodolfo Aguinaldo, was going to the United States "for six months to one year for additional training under the Central Intelligence Agency."[23] Were these officers given training in either tactical interrogation or torture?

Definitive answers must await further release of classified documents. At present, we will have to content ourselves with comparison. Though the CIA may have taught torturers on both sides of the Pacific, the methods of the Filipino interrogators, particularly the theatricality of the future RAM officers, seems closer to the spirit of the CIA manual. Whether they read this text or received similar training, Marcos's torturers carried its techniques a step closer to perfection.

Their preference for the psychological and theatrical had important political consequences for the Philippines. By breaking their superiors through psychological manipulation, these officers gained a sense of their society's plasticity, fostering an illusion that they could break and remake the social order at will.

Through their years of torturing priests and professors for Marcos, these officers learned the daring to attack Marcos himself.

The torture cell was a play within a larger play. Inside the safe house, Filipino interrogators acted out their script before the victim, their audience of one. If the plot, through twists and turns, ended with the victim's death, then the interrogators discarded the mangled remains in a public place, a roadside or field, to be seen by passersby. Such displays, called "salvaging" in Filipino-English, became the larger play that made the road or plaza, indeed all of public space, a proscenium of terror.[24] Seeing the marks on the victim's body, or simply hearing of them, Filipinos could read, in an instant, the entire script of the smaller play that had been acted out inside the cell.

"In this liturgy of punishment," says Michel Foucault of eighteenth-century Europe, "it is the prince—or at least those to whom he had delegated his force—who seizes upon the body of the condemned man and displays it marked, beaten, broken. The ceremony of punishment, then, is an exercise of 'terror.'" Similarly, in early modern England, writes historian E. P. Thompson, "the corpse rotting on the gibbet beside the highway" made "the ritual of public execution . . . a necessary concomitant of a system of social discipline where a great deal depended upon theater."[25]

In the modern world, however, torture not only dignifies the prince, it empowers his servant, the perpetrator. "We are God. We are the law," Argentine torturers told their victims. "In this place where you are now," Filipino torturers told a priest, "we are the judge." One Greek torturer asked his victim, a senior navy officer, "Do you know who I am? I am Antonopoulos before whom all Greece trembles." For the torturer, Elaine Scarry concludes, "his blindness, his willed amorality, *is* his power." Such men, who shed civility and embrace cruelty, arouse fear and fascination. They become powerful and perversely charismatic, striking a disturbing chord in our consciousness.[26]

ORIGINS OF TERROR

The rise of RAM with its vision of violence is but one manifestation of Marcos's impact upon the Philippine military. After his inauguration in December 1965, Marcos served as his own defense secretary to place men loyal to him in key positions in the AFP, creating a "parallel command" of loyalists for covert operations.[27] In his second term (1969–72), Marcos reinforced his military authority with extralegal force—both electoral violence and random terror. Between March and August 1972, a climate of fear seized Manila as unidentified terrorists set off some twenty bombs in public buildings, reaching a crescendo in September when six bombs erupted—cutting off the city's water and causing forty-two casualties in a crowded store. The president blamed them all on the communist New People's Army (NPA). In a damaging rebuttal, Senator Benigno "Ninoy" Aquino released a copy of Marcos's secret orders "to sow violence and terror in order to lay the groundwork for the imposition of Martial Law."[28]

Then, on 22 September 1972, gunmen ambushed Defense Secretary Juan Ponce Enrile, raking his sedan with bullets but leaving Enrile himself miraculously unharmed. Marcos would later cite these incidents—the bombings, Aquino's documents, and the assassination attempt—as "the anatomy of a plot against the Government of the Philippines." Thus, at 9:00 P.M. that same day, Marcos, weighing his words with a lawyer's care, issued Proclamation 1081 imposing a state of martial law: "By virtue of the power vested upon me by . . . the Constitution I do hereby command the Armed Forces of the Philippines . . . to enforce obedience to all laws and decrees, orders and regulations promulgated by me personally."[29]

But it was Marcos himself who authored this violence. Fourteen years later, when he rebelled against the president, Enrile confessed to staging his own assassination attempt. A member of Class '40 admitted that his close relative, one of General Ver's deputies in the Presidential Security Unit, had "organized some of the bombings that were done to convince people that there was a crisis and democracy was not working." Only a few days before martial law, Marcos rewarded this officer with an accelerated promotion to general.[30]

The armed forces were no longer the servant of the state under martial law, but the bastion of a particular regime. Backed by his generals, Marcos wiped out warlord armies, closed Congress, and confiscated corporations. The president involved the military in every aspect of authoritarian rule—censorship, repression, and governance. Officers became corporate managers, civil servants, local officials, and judges.[31]

Charged with suppressing dissent, the armed forces were brutal and brutalized. Among the authoritarian regimes of the 1970s, the Marcos government was exceptionally lethal. Films such as *Missing* and *Kiss of the Spider Woman* lend an aura of ruthlessness to Latin American dictatorships that overshadows that of the Philippines. Statistics, however, tell another story. Argentina's junta (1976–83) established an unequaled record of 8,960 documented dead and an estimated 30,000 missing. The Marcos regime's tally of 3,257 killed is far lower, but it still exceeds the 2,115 extrajudicial deaths under General Pinochet in Chile (1973–90) and the 266 dead and missing during the Brazilian junta's most lethal period (1964–79). Under Marcos, moreover, military murder was the apex of a pyramid of terror—3,257 killed, 35,000 tortured, and 70,000 incarcerated.[32]

The administration of repression had a profound impact upon the military. Inserted into every aspect of Philippine life through their new roles, legal and extralegal, military officers became politicized in the most fundamental sense. To sustain these multiple missions, in the first four years of martial law the armed forces budget increased by 500 percent and its strength doubled to 113,000 troops. U.S. military aid also doubled to $45 million annually, providing ample weaponry for an expanded Marcos military.[33]

The armed forces thus strengthened the authoritarian capacities of the president. But, over the longer term, these changes also created an opening for a coup d'état. After a decade of dictatorship, Marcos's politicization of the military split its hierarchy into rival factions and produced rising resentment among middle-ranking regular officers. Sensing their anger, Defense Minister Enrile began courting this echelon for a coup against Marcos. His ambitions and their grievances merged over time to become the Reform the Armed Forces Movement.

By the mid-1980s, RAM would grow into a movement of some three hundred regular officers drawn, with few exceptions, from just eleven PMA classes—1965 through 1975. Within this larger pool of 861 academy alumni, only fifteen officers, largely from just one class, 1971, would exercise overall control. That a small group of junior officers could lead a sustained revolt against the entire military hierarchy reflects deeper changes within the culture of the officer corps.

MILITARY SOCIALIZATION

In the broadest sense, RAM represents a revolt against the essence of military culture—obedience of officers to a chain of command and the loyalty of that hierarchy to civil authority. Any state's creation of a standing army is thus riven with a volatile contradiction. By commissioning regular officers, the state invests a self-selecting elite with near-absolute control over the nation's arsenal. If they rebel, the civil state has no alternative firepower. Its capitulation is almost axiomatic. To assure its survival, the state must socialize its officers into obedience.

The postwar Philippine state protected itself from this threat by indoctrinating cadet officers at the Philippine Military Academy.[34] Once commissioned, young officers on active duty learned subordination to military hierarchy and civil authority—affirming and deepening their academy indoctrination. By swearing their oath, officers surrendered their civil rights and subjected themselves to a military justice whose ultimate sanction, death by firing squad, is reserved for the ultimate crime, insubordination. Killing is allowed, disobedience is not. These external sanctions, though extreme, were never enforced and thus remained far less significant than the internal restraint of socialization.

In the last half-century, the PMA's essential pedagogy has changed little. From a reading of the nine yearbooks found on its library shelves during a 1996 visit, every class seems to pass through an unchanging cycle from degradation to command. Not only is the narrative structure of each class history nearly identical, but even the key words used by these nine class historians are often interchangeable. Juxtaposing quotations across the span of forty years, 1953 to 1991, reveals an underlying continuity in this four-year progression.

All classes found their plebe year a time of testing. "Plebedom caught up with us like the rushing in of a fiery gale," begins Class '53. During the ordeal of beast barracks, '72 felt "our metamorphosis from boneless weaklings into disciplined soldiers was striking." After ten tough months, said '76, "tears of happiness rolled down our cheeks" when upperclassmen offered the handshake of recognition, "an enduring embrace of brotherhood." As the "heavy load of academics" in their second year led to many dismissals, '91 felt the pain of "losing good friends, brothers, who . . . shared our sorrows, joy, pain, and memorable moments." In their third or "cow year," each class prepared for command. At the reception for new plebes, Class '76 was proud when "like thunder we roared to face these young boys." At graduation, wrote '77, "the hep-heps and hurray's ushered the flying caps and shouts of joy." As they left, Class '53 felt "bonded by a staunch tie of brotherhood." Wrote '91: "We would go our separate ways, serving the Filipino nation . . . , but always we would be forever . . . PMA Class of '91."[35]

Through this four-year progression, each cadet is submerged into class, corps, and, ultimately, chain of command. The PMA instills a deep patriotism and esprit de corps in place of ties to family and friends. Although outsiders find them silly, the endless rituals of haircuts, saluting, and recitation are all statements of subordination. Molded by constant repetition, cadets are trained to accept authority without question.

As PMA graduates amassed in the 1950s, their belief in subordination to civil authority became the ideological norm for the armed forces. An American researcher who interviewed senior officers in 1975 found, even under martial law, a strong belief in civil supremacy. One told him, "civilian control is going to continue in our country . . . because it has become engrained in our minds." Similarly, a Filipino scholar reported that retired officers "attributed the absence of a military coup . . . to the rigorous indoctrination of the military in the principle of civilian supremacy."[36]

Such socialization also produced intense loyalties. Educated at the PMA and housed inside military cantonments, Filipino officers became an elite apart from the endemic political pressures of civilian society. Over half the academy alumni interviewed in 1975 listed classmates as their "closest friends." Among classes '70 to '89, some 48 percent had relatives in the military. So strong were these ties that the children of officers killed in combat were often adopted by classmates.[37] Under the postwar Republic, classmates drew upon these ties to enforce an apolitical professionalism upon brother officers. After Marcos imposed martial law in 1972, however, these bonds became the fault lines for factions and coups.

Even under optimum circumstances, socialization at the PMA was apparently a fragile, even flawed process. The Davide Commission found, in its investigation of the causes underlying the many coups of the 1980s, that certain aspects of academy training carried a germ of hostility to the civil state. Its ideological indoctrination emphasized a strong "anticommunist sentiment" rather than a "pro-democracy sentiment," opening graduates to "the appeal of rightist authoritarian rule." More fundamentally, the commission found a certain superficiality in the inculcation of core values. Though the honor code is "drummed into their minds for four years," the denial of free expression among cadets means that "the formation of consciousness and conscience is highly external."[38]

THE CLASS OF 1971

As catalyst in the many coups of the late 1980s, Class '71 is an apt case for study of the breakdown of military socialization. Not only did this class provide the leaders for most of these attempts, but about 15 percent of classmates participated in the massive 1989 coup that nearly seized power.[39] These officers were, in every sense, the nation's military elite. Most had been outstanding cadets—regimental officers, top athletes, and leading scholars. They were natural leaders, men who in the normal course of events would have risen to the highest echelons of command. How can we explain their careers, which are marked instead by torture, terror, and coups?

Like others in the long gray line, Class '71 went through the PMA's usual four-year cycle from "dumbguard" to "immaculate." Its yearbook narrative by class historian Victor Batac, the future strategist of RAM's revolt, is unexceptional, even bland. As plebes they suffered the "aching muscles, ever confused minds" of beast barracks and found, ten months later on recognition day, "each handshake, each tear had a special meaning." As yearlings the class "found the cohesiveness it needed" at summer combat training; as "full-pledged cows" they adjusted to "added responsibilities"; and as "firsties" they took "the reigns of the Cadet Corps."[40] On its surface, this four-year, lockstep progression seems even less exceptional for Class '71 than it had been for Class '40.

Their politicization seems to spring from changes in the academy and the military's role within the state. This class, through the sum of its circumstances, received a problematic socialization. The PMA of their day was caught in the turmoil of change, making it an imperfect mold for the shaping of cadets. But, above all, Class '71 graduated only months before martial law and was commissioned into a military with a mission of repression. Instead of garrison duty or conventional combat, these young officers, in this critical second phase of their socialization, would be thrust into civil war and civil administration. Looking backward from this central experience, we can discern secondary factors at the academy—excessive hazing, curricular changes, and group dynamics—that later allowed torture to act with such a powerfully catalytic effect. In the context of dictatorship, these details of the class's PMA experience compounded the corroding effect of torture and thus gain an added significance.

From its reopening in 1947, the PMA was a troubled institution. A confluence of trends—harsh hazing, social change, and radical nationalism—altered the institutional ethos in small but significant ways. Abusive hazing tarnished the honor code and introduced an element of brutality into cadet life. Prewar cadets, mostly lower-middle class, had arrived at the academy with limited ideological baggage, accepted their indoctrination into civil supremacy without question, and found these lessons affirmed by serving as the training cadre for a citizen's army. By contrast, most postwar cadets entered the PMA already exposed to political ideas by college-educated parents and several years at civilian universities.[41] By the late 1960s, the entering cadets of Class '71 would be influenced by the strong left-nationalist ferment sweeping Manila's university belt. "The literary pieces written during this period possessed a political tone," recalled an alumni history of the PMA. "The clamor of change . . . was also emanating from within the cadets themselves."[42]

We cannot ignore the influence of individuals upon a large, complex organization like the armed forces—particularly in the Philippines, where institutions often seem the sum of personal ties. Indeed, at the PMA several key personalities translated broad ideological currents into institutional change. In their last months at the academy, the socialization of Class '71 was influenced by two exceptional personalities—their new instructor in politics, Lieutenant Victor Corpus, and their class captain, Cadet Gregorio "Gringo" Honasan. Their alliance at the academy had a strong influence upon the class, just as their break fifteen years later would shake the armed forces.

In the mid-1960s, the PMA leavened its traditional engineering program with humanities and social sciences courses taught by "the discussion method." Among the new instructors was Major Dante Simbulan—a member of Class '52, chair of its honor committee, and an acutely political scholar whose doctoral dissertation argued that a Filipino elite of only 169 families monopolized all power. In the heady atmosphere of open discussions, Simbulan developed a close relationship with Cadet Victor Corpus of Class '67, introducing him to the idea of radical social change. As school boxing champion and winner of the Athletic Saber, Corpus had a quiet charisma that marked him for future command.[43] But he was, as it turned out, more than a conventional careerist.

After graduation, Lieutenant Corpus joined the communist youth group *Kabataan Makabayan*, and was strongly influenced by its ideologue, Jose Ma. Sison. In May 1969, Sison went underground to establish a revolutionary force, the New People's Army (NPA). In early 1970, after three years with the constabulary, Lieutenant Corpus returned to the academy as an instructor in political science, deeply disillusioned by the society he had seen. In the first semester, Class '71 took Government 411 with Corpus and found him "a committed and influential teacher who encouraged free thinking." Some recalled that he "junked the curriculum and substituted . . . questions on how to stage a coup d'état."[44]

Outside the classroom, the young lieutenant and his cadets were outraged over what Class '71 called the "despicable attempt by some persons to take advantage of the good name . . . of the Philippine Military Academy by . . . smuggling stereos and other articles" for personal use disguised as donations to the academy. Corpus was angry to find officers grafting from the PMA's mess budget and began doubting the military's ability to live up to its ideals.[45] In a clear breach of regulations, Corpus began meeting outside of class with the leaders of Class '71, including their first captain, Gringo Honasan.

"Lieutenant Victor Corpus was my instructor in political science," Honasan recalled fifteen years later. "He was, in fact, our class advisor. Most of the radical thinking of our class was his influence." The two came from similar backgrounds and formed a close bond. Both were from military families. Both had experienced campus activism in Manila during the late 1960s—Corpus as a young officer and Honasan as a student for two years at the country's most radical college, the University of the Philippines. Their gripe sessions about abuses grew into plans for a mass walkout by the entire cadet corps, which the two believed, incorrectly, would be a first in academy history.[46]

In December 1970, only three months before Class '71 was due to graduate, Corpus and another instructor led their communist comrades in a spectacular raid on the PMA armory. They seized hundreds of infantry weapons and disappeared into the communist underground, where Corpus began training the NPA's fledgling guerrilla army.[47] These defections brought strict security that blocked Honasan's protest and cast a cloud over Class '71.

Despite the furor, the instructor's influence over his cadets remained. "While the bravery shown by Corpus in advancing his principles . . . brought him some admirations from the Cadet Corps," the class wrote in its yearbook, "his actuations against his own Alma Mater somehow united the cadets in denouncing him." Fifteen years later, the leaders of Class '71 would launch the RAM movement with their famous "We Belong" march at the PMA's alumni day—a protest that was strikingly similar to the one that Lieutenant Corpus had planned with Cadet Honasan. Indeed, in his later coup plotting, Honasan seemed to strip his mentor's lessons of ideological content and apply their tactical essentials to a simple seizure of power.[48]

Honasan himself would leave a lasting imprint upon the armed forces. Even as a cadet, he showed a unique capacity for dominance and became the single, preeminent leader of Class '71—a sharp contrast with the diffuse leadership of Class '40. As the son of a career officer, ex-candidate for the Catholic priesthood, and fitness fanatic, Honasan entered the PMA with an appealing mix of idealism and athleticism. Every year for four years, his classmates elected him class president and, in his senior year, the tactical officers selected him as cadet first captain, or "baron." Honasan also exercised moral authority as chairman of the honor committee and twice captured the school title as best debater. Aside from five varsity letters in sports ranging from boxing to basketball, he earned the grade of "sharpshooter" with the M-1 rifle and took a gold medal at the National Wrestling Tournament.

His yearbook biography celebrated these achievements in almost reverential tones: "Gringo!—Comfortably warm and constantly dynamic; enough determination and perseverance in achieving his goals, coupled with characteristic humility and compassion for the less fortunate."[49] That nickname was inspired by director Sergio Leone's 1966 Western *A Fistful of Dollars*, starring Clint Eastwood as a drifter known only as "the gringo"—a hired gun whose cool violence and quiet morality save a metaphoric Holy Family from Mexican villains. Upperclassmen applied the name to plebe Honasan. It stuck. As the film and its three sequels were shown and reshown during their cadet years, Class '71 attached Eastwood's persona to Honasan, who, in turn, cultivated the machismo and martial skills to make himself, in every sense, "Gringo."[50]

The class yearbook offers clues that the future RAM leadership had already formed a tight clique around Honasan. Among the class's six elected officers, three were future leaders of the coup group—class president Honasan, vice president Oscar Legaspi, and class historian Victor Batac. A snapshot in the "Class History" section shows twenty-eight classmates posing informally on the gangplank of a navy ship in a jumble of smiles and sunglasses. Even in this fleeting moment of spontaneity, the class arrays itself to frame three cadets, the future RAM leaders, in a central, dominant position—Eduardo "Red" Kapunan left, Honasan center, Victor Batac right.

HAZING AS THRESHOLD

The germ of campus politics had a long incubation, but the effect of hazing on Class '71 was more immediate. Its violence would serve as a gateway to their use of torture when martial law placed civilians under their control.

Plebes from the prewar PMA classes, 1940 through 1943, went through a rigorous "fourth class custom" of drill and discipline, but were generally spared any harsh, physical hazing. In the decorum of that day, upperclassmen had to get permission before any physical contact, asking "Touch your shoulder, mistah?" Nonetheless, within the intense culture of male bonding, initiation had a constant tendency to excess that could be slowed but not stopped by supervision or reform. Throughout its history, the PMA has suffered an unbreakable cycle of abuse, correction, abuse.

When the PMA reopened in 1947, the administration brought back members of Classes '44 and '45 to play upperclassmen and initiate the seventy-six plebes of Class '51. These returning cadets were brutal. At the start of the war in 1941, these two classes, too young for commissions, had been disbanded. "Class '44 and '45 . . . had only one year and few months service in the academy and they . . . had not tried being upperclassmen for long" explained Deogracias Caballero of Class '40. "All they could remember were the indignities that they had suffered as plebes and they took it out on the plebes." His classmate Reynaldo Mendoza, assistant commandant in 1950–51, concurred: "You see, the academy training is a four-year gradual training course. While you make the hit by an impact or blow in the first two years, you have to regain your sense during third year... and learn more about... leadership... in senior year." Dismissed after plebe year, these acting upperclassmen were still driven by what Mendoza called "anger." Caballero noted that they were cruel in their hazing of Class '51—trying to "electrocute cadets at the tongue," forcing them to take cold showers every hour, and using their bodies as "punching bags."[51]

Over the next three years, the cadets of Class '51 were just as hard on the entering plebes. In their last year, class leaders tried to restrain the rising violence by forming a Fourth Class Customs Board to recommend the complete eradication of hazing. Similarly, the postwar *Rules and Regulations* prohibited hazing, defining it as "any cruelty, indignity, humiliation, hardship, oppression, the deprivation or abridgment of any right."[52]

Nonetheless, Class '51 ignored these prohibitions and played upon cadet solidarity to conceal abuse. In their last beast barracks of 1950, upperclassmen beat plebe Jose Arias to the point that he required medical treatment for rib injuries and internal hemorrhaging. Years later, he recalled regular "socking and slapping," "thrusts in the abdominal area with the butts of rifles," and "starvation treatment." When the superintendent dismissed a guilty upperclassman, Class '51 retaliated against Arias by ordering the "silent treatment." For months, nobody would speak to him, not a word, and he soon left on a medical discharge. Six years later, Arias, sick and jobless, was still suffering periodic nervous breakdowns. Another classmate, Crisostomo Ausejo, resigned after only two months because he could not stand ordeals such as the "human bridge"—being suspended head and heel while upperclassman pummeled his stomach with a steel bar.[53]

These were not isolated incidents. With each initiation, brutality became a cancer metastasizing within the corps. Between 1950 and 1961, the academy dismissed forty-eight cadets for hazing—equivalent to an entire graduating class.[54] "When I became commandant of cadets [in 1956]," recalled Lieutenant Colonel Salvador Piccio ('40), "my first offense was a fellow who was wounded by a bayonet. Well, he landed in the hospital for one year." Although Piccio dismissed ten upperclassmen, including future AFP spokesman Oscar Florendo, the impact of the discipline was blunted when the cashiered cadets were later readmitted, including Florendo, who graduated with Class '62.[55] The next commandant, Lieutenant Colonel Francisco Jimenez ('40), also found that his officers frustrated attempts at reform. "Because, you see," he explained, "if you assign officers there who believe in hazing, they will just turn their heads."[56]

When Class '71 entered the academy in early 1967, the *Manila Times*, echoing the concerns of Superintendent Reynaldo Mendoza ('40), reported that: "Hazing is an ugly act. It is bestial. It is never tolerated by Academy authorities." Indeed, when he took command the year before, Mendoza found that a recent mass dismissal for hazing had produced a "lull" in the practice. Pressured to readmit these cadets, Mendoza refused, sending a signal to staff and students. Thus, the incoming plebes of Class '71 would face what the *Manila Times* called a "stern or serious reception by upperclassmen," but one without physical abuse. When he retired in August 1967, Mendoza gave the corps a valedictory celebrating the fourth class custom of tough plebe initiation as an instance of "true Democratic opportunities" which ensured that "the laborer's brother competes with the senator's heir, the corporal's son gives orders to a general's favorite boy."[57]

The plebes of Class '71 still found their first six weeks a harsh rite of passage. Years later, RAM leader Red Kapunan recalled their first traumatic day at the PMA in words strikingly reminiscent of Class '40's reception over thirty years before:

> We got down from the bus. We lined up in front of the grandstand. No cadets in sight. Then all of a sudden, martial music. From behind us came cadets, all in dress white. . . . We said, "the cadets look beautiful." Then, they form in front of us, as if they are going to eat people. A sudden shout: "Charge!" Suddenly, about face everyone. There were those among us who wanted to run back to the bus or bury their heads in the ground. . . . "Mr. Kapunan, where are thou? Double time!" And we did double time. "Raise those knees! Push up!" We raised our knees and did push-ups. It was real shock treatment! . . . The rest is really hazy for me. I was just following orders. I didn't know what I was doing.[58]

Beast barracks was the first stage in the socialization of Class '71. "The uppies were very 'helpful' too," the class wrote in their yearbook. "They had to shape us into worthy cadets within six weeks." Then, at the end of plebe year came recognition when "each handshake, each tear had a special meaning. What was left was gratitude and understanding for those who shaped us into what we were then . . . a full-blooded yearling."[59]

Thus, Class '71 passed through an initiation that broke down individual identities and built them back up as members of class and corps. In their first weeks, the assault on their individualism was so intense that many, recalled classmate Oscar Legaspi, considered resigning. By the end of plebe year, however, they had accepted the logic of their ritual degradation. "If you are to develop an all-weather officer, you really need to expose him to the pressure cooker," explained Kapunan. "In the field you look at all the men under you looking to you for directions. You cannot crack."[60]

When Class '71 became upperclassmen, hazing revived and the academy tried, unsuccessfully, to suppress it. When new cadets arrived in April 1969, the superintendent arranged a "mild" reception by assigning each plebe to an upperclass "guardian." Even so, after only three days of beast barracks, plebe Miguel Arucan, an athlete who had arrived in good health, was admitted to the hospital, weak and bleeding. Nine days later he died. An initial investigation ruled out hazing and listed the cause of death as "mysterious." But four days after his burial, another hospitalized plebe wrote the *Manila Times* charging that Arucan died of overexhaustion due to hazing and others had been hospitalized with "broken ribs and stomach trouble." Since the death, he added, hazing had become "more sadistic." At the later court-martial, plebes testified about ritual beatings that led to Arucan's death: at dawn they were wakened by blows; they left their rooms through a gauntlet of upperclassmen who pummeled their stomachs and were later ordered to squat on their desks while upperclassmen beat them bloody about the face and torso. After direct blows to the head, Arucan began bleeding profusely from the nose and was taken to the hospital.[61] Five upperclassmen, including a member of Class '71, were later implicated.

To prevent a recurrence, the superintendent ordered that incoming plebes be isolated from any contact with Class '71 during the next year's beast barracks. Nonetheless, during the traditional "reception ceremony" in April 1970, upperclassmen kicked one plebe so badly that he required a medical discharge. A year later, on the eve of the Class '71 graduation, classmate Ruben Cabagnot was finally punished for his role in Arucan's death by a six-month suspension of his commission.[62]

Thus, Class '71 graduated with an experience of brutality that served as an emotional gateway to torture. Through later interviews with military torturers, Dr. June Lopez found that the physical and psychological abuse of hazing gives cadets "some sort of invisible badge of honor" that "instills camaraderie and brotherhood." She concludes with a provocative question: "If they can subject their own selves to abuse, what is there to stop them from inflicting the same, if not greater, violence on a perceived enemy?"[63]

Indeed, an AFP officer charged with torture in the Marcos era replied: "But, sir, I did no more to this NPA suspect than had been done to me as a plebe." Similarly, a student activist detained in 1978 found that his torturers were PMA cadets, suspended for hazing and now using these skills on political prisoners. Their most characteristic torture imitated the PMA's notorious "bridge," with prisoners instead of plebes now being beaten on the back and belly.[64]

YOUNG LIEUTENANTS

Only eighteen months after graduation, Class '71 became the defenders of dictatorship. If they had been molded imperfectly at the academy, then, in this final phase of their socialization, they were fired in the kiln of a dirty war against Muslims in Mindanao and dissidents in Manila. The contrast with earlier generations is striking. The fresh graduates of Class '40 defended the nation against an enemy invasion, experienced the heroism of combat, and were enobled by months of suffering as prisoners of war. The young lieutenants of '71 were forced to treat their own society as enemy, suffered the brutalization of civil war, and were degraded by years of service as society's jailers. The difference between these two classes thus derives, in large measure, from contrasting experiences as prisoners and jailers, or victims and perpetrators. While suffering somehow affirmed the core values of Class '40, torture seemed to weaken military socialization among the leaders of Class '71.

Mindanao bonded and brutalized these young officers of Class '71. After Marcos imposed martial law in 1972, Muslims in western Mindanao launched a secessionist revolt and the military mounted a five-year campaign that threw recent PMA graduates into a vicious conflict. One officer, later a notorious torturer, recalled arriving in Mindanao fresh from infantry training to descend into a hell of heavy casualties, torture, summary executions, and mutilation.[65]

For three years after graduation, Lieutenant Gringo Honasan served in Mindanao, where he earned a reputation as a fearless, sometimes ruthless fighter, winning at least one Gold Cross medal for "heroism."[66] According to one member of a military tribunal, Honasan "participated in the revenge killing of a Muslim leader who allegedly killed his classmate," but was found innocent at an initial hearing. When later evidence established his guilt, Honasan "was able to evade an inquiry . . . because of his links to the defense secretary." Service in this bloody war strengthened bonds among the class. When Honasan was wounded during combat in 1973, Lieutenant Red Kapunan defied his superior and flew his helicopter through enemy fire to rescue his classmate, making the two "bosom friends." Years later Gringo would say of Red: "We have developed a special bond that is only derived from taking risks together . . . and our long association since academy days. We are even closer probably than brothers."[67] After one or more combat tours, many classmates rotated to Manila for staff assignments involving surveillance and interrogation.

These special duties trained the future RAM leaders for political warfare, giving them the skills and experiences that would later inspire their attack on the state. Lieutenant Kapunan spent a year during the early 1970s infiltrating the press corps by posing as a reporter, while Captain Rex Robles (PMA '65) produced black propaganda about the opposition during the 1978 elections. Among the eighty-five graduates of Class '71, at least five practiced torture, six were murderers, and most had combat experience in counterinsurgency. They were the ultimate creatures of martial law.

During the late 1980s, Class '71 became leaders of a military revolt against civil authority. Not only did these coups define them, but the class itself defined the role of the coup d'état in Philippine politics. Apart from providing the leadership for five major coups, these eighty-five graduates supplied fifteen of the seventy-seven officers involved in two or more coups—by far the highest for any single class.[68]

Why did one PMA class play such an influential role in so many coups? Why did so many in this class rebel against their military socialization? In lieu of their own admissions, we must look to events for evidence. Some older alumni were involved in torture and a number of younger classes joined these coups, but only Class '71 exhibits this intriguing coincidence of torture experience and coup leadership. Under martial law, almost all members of Class '71 who later became RAM's core leadership had experiences of torture, interrogation, espionage, or civil war. Breaking the socially prominent so easily in torture may have somehow led these young officers to believe that they could, through violence, impose their will upon society. When RAM tried to capture the state, each of its coup attempts would evince elements that can be traced, through close analysis, to these experiences.

ETHOS OF TORTURE

Under Marcos's martial-law dictatorship, torture became an instrument of state power. In its 1975 report, Amnesty International concluded, in a statement that outraged Marcos, that the Philippines "has been transformed from a country with a remarkable constitutional tradition to a system where star chamber methods have been used on a wide scale to literally torture evidence into existence." In response, the palace announced in November 1976 that some seven hundred military personnel had been disciplined "for maltreating prisoners held under martial law."[69]

Despite this gesture, torture continued. In 1981, Amnesty returned to discover a Philippine gulag of "safe houses" where "members of the Armed Forces . . . had been responsible for acts of unusual brutality." The Philippine Medical Action Group found that 102 of the 120 political prisoners held at the National Penitentiary had been severely tortured—48 subjected to Russian roulette, 19 electrocuted, 17 "fed feces," 5 sexually abused, 4 buried alive. After Marcos fell from power in 1986, an organization of ex-detainees, SELDA, estimated that 35,000 political prisoners "suffered some form of torture."[70]

Why such extraordinary brutality? Even at its peak, the Marcos state, reflecting the underlying poverty of Philippine society, lacked the communication and information systems for a blanket repression. With only 55,000 troops in 1971, the armed forces constituted only 0.1 percent of the country's population—by far the lowest ratio in Asia. Although the military grew to 153,000 troops by 1978, it was still poorly financed and lacked efficient communications.[71] By 1984, the severe fiscal crisis allowed surveillance agents only five liters of gasoline per day and restricted distribution of photographs for wanted dissidents to a single province.

By contrast, the Argentine military, operating in an advanced, urbanized society, tried to eradicate all dissidents. In the words of the governor of Buenos Aires, General Iberico Saint-Jean, "first we kill all the subversives, then we will kill their collaborators, . . . then . . . those who remain indifferent, and finally we will kill the timid." By the late 1970s, the junta was operating a network of 340 secret camps to fight a systematic "dirty war" that left thirty thousand *desparacidos*, those who had disappeared without trace or remains. In a speech to staff at the Navy Mechanics School, the most murderous of these prisons, junta leader Admiral Emilio Massera celebrated this "machine of horror . . . unleashed on the unsuspecting and on the innocent, before the incredulity of some . . . and the stupor of many more."[72]

Instead of a machinery that crushed all resistance, the Marcos regime used the spectacle of violence for civil control, becoming a theater state of terror. In its first three years, the military incarcerated some fifty thousand people. But faced with rising insurgency, the regime soon abandoned this costly enterprise. Moreover, as a lawyer not a career officer, Marcos maintained a facade of legality, speaking with pride of his "constitutional authoritarianism" and responding to President Jimmy Carter's pressure for human rights reform. By 1977, President Marcos had emptied his prisons, leaving only 563 political prisoners.[73]

The regime bridged this growing gap between the fiction of legality and reality of repression by extrajudicial execution. Arrests declined, but "salvagings" climbed. Military murders rose from only 3 in 1975 to 538 in 1984, with most coming after 1981, when Marcos inaugurated his nominally democratic "New

Republic." In striking contrast to Argentina, only 737 Filipinos "disappeared" between 1975 and 1985. But nearly five times that number, some 2,520, equivalent to 77 percent of all victims, were "salvaged"—that is, tortured, executed, and displayed.[74] Indeed, this practice had such a disturbing resonance within the collective consciousness that the Filipino-English dialect coined the neologism "salvaging" to capture the aura of terror.[75]

Marcos's rule thus rested upon a theatrical terror. His officers were not impersonal cogs in a military machine. They were actors who personified the violent capacities of the state. If the president had written a script of terror for his "new society," then these young officers, the future RAM leaders, were his players. When the president ordered the military to enforce authoritarian rule, it was its lowest echelon that became the instrument of his will.[76]

Within the Marcos gulag, there were, by the late 1970s, three archipelagoes of safe houses and prisons. Each rival military faction ran its own apparatus of officers, agents, and interrogation centers. As the president's chief protector, General Fabian Ver controlled the National Intelligence and Security Agency (NISA), Marcos's dreaded spy network. To counter Ver, Defense Minister Enrile organized the National Defense Intelligence Office and also ran a small antisubversion operation through his Security Unit. As chief of constabulary, General Ramos commanded the main civil control units and their notorious torturers—the Fifth Constabulary Security Unit (CSU) and the Metrocom Intelligence and Security Group (MISG).[77]

Officers in these elite units were the embodiment of an otherwise invisible terror. As Marcos's favored instrument, the MISG, in the words of a Manila magazine, "produced some of the most fearsome and brutal cops in memory." Its commander for twelve years, Colonel Rolando Abadilla (PMA '65), "towered over other heavies in that closed, tight-knit, psychotic club of martial-law enforcers—next only to Gen. Ver. . . . in the dictator's trust and confidence." Only his former understudy, Lieutenant Rodolfo Aguinaldo of the Fifth CSU, could rival his psychopathic interrogations. There was an intense competition among these ambitious officers to capture top subversives. For example, Panfilo Lacson ('71) joined MISG right after his PMA graduation and rose through its ranks for the next fifteen years on a fast track to national police power.[78]

As the military mobilized for war on its own society, these elite commands could call upon the intelligence section of every unit, from infantry battalions near Manila to constabulary companies on remote islands.[79] Military units in this state of civil war made arrests without warrants and confined suspects in extralegal "safe houses" for "tactical interrogations" that often involved torture.

Armed with a blanket ASSO (Arrest Search and Seizure Order) or PCO (Presidential Commitment Order), military personnel raided homes on mere suspicion and took suspects to unknown locations for limitless interrogation. In effect, martial law abrogated due process and allowed the military to operate with a de facto immunity to civil prosecution, producing what Amnesty International called "gross and systematic violations of human rights."[80]

The existence of this gulag served as a warning to all who might resist. As officers assumed extraordinary powers, civilians were reluctant to defy them, aware that they were subject to arbitrary, indefinite detention. Even officers who never tortured participated, albeit indirectly, in the aura of power that became invested in everyone in uniform. "Except for . . . Marcos's immediate political cronies," wrote political scientist Felipe Miranda, "the military intimidated civilian politicians and government officials practically everywhere."[81]

Some officers enjoyed their new power over civilians. "We in the military were mud before martial law," an officer told an American researcher in 1975. "Now the people come to the military tribunal seeking justice."[82] We know from the detainees themselves that officers did not torture before 1972. One long-term prisoner, Leoncio Co, noted that the PMA-trained officers who arrested him in 1969 were "very understanding" and even advised him not to incriminate himself. After martial law, however, he found attitudes "very different"—a change epitomized when he witnessed a recent PMA graduate, Lieutenant Rodolfo Aguinaldo, torturing a woman with such severity that she "nearly lost her mind."[83]

Significantly, in the victims' transcripts it is usually lieutenants, not majors or colonels, who appear as the actual torturers. After graduating into a martial-law military, the lieutenants of Class '71 were assigned to arrest, interrogate, and, ultimately, torture civilian dissidents. Later, as they rose through the ranks, they would take command of the regime's safe houses and detention centers.

THE PLOT OF TORTURE

Only the collective weight of torture transcripts can convey the actual work of these juniors officers during this formative phase of their careers. By letting the victims speak to us of their pain, we can gain some sense of its reality. Moreover, these details reveal distinctive patterns in the torturers' methods that seem to have informed their later conduct as coup leaders.

In the victims' statements, the names of the future RAM leaders recur. They did not operate as dispassionate technicians, raising the level of pain to extract information. Thespians all, they assumed the inquisitor's role, using a theatrical torture to heighten the victim's pain and disorientation. Many practiced a pose of latent threat that could, if provoked, unleash unlimited violence. Some appeared as a creator/destroyer whose psychological and physical powers could break any prisoner to their will. A few played the omnipotent by sending detainees off for torture by minions.

Many of these sessions share a similar plot. The torturer begins with a few questions, meets resistance, and then applies coercion, physical and psychological, to elicit cooperation. Within this common script, each inquisitor seems to extemporize around a guiding metaphor that becomes embedded in the victim's recollection. One officer crafted his script around imagery of production; another constructed a near conceit of social inversion.

A rural priest, after torture by one of the future RAM leaders, offered the most acute insight into their methods. Arrested for subversion in October 1972, Father Edgardo Kangleon, social action director of the Western Samar Diocese, was subjected to two months of constant interrogation before breaking down. In his confession, he admitted to being a communist agent and named his fellow clergy as subversives—charges that the regime seized upon to harass the Church.[84] Throughout his long confinement, the priest suffered only limited physical abuse and was instead psychologically terrorized by his chief interrogator, Lieutenant Colonel Hernani Figueroa (PMA '66), later RAM's chairman.

Only a week after his release, Father Kangleon composed a twenty-five-page memoir that describes the aura of terror around Marcos's interrogators. Though he damns himself as a "stinking coward" and "traitor" who had betrayed comrades to become a military vicar, his statement is an insightful, probing analysis of the theatricality of torture. In his account, the cell becomes studio, the inquisitors actors, and the detainees audiences for a psychological drama crueler than physical pain.

After each long day in the military compound with no release in sight, Kangleon found night a time of fear. "For a detainee, the day's ending signals the possibility of facing the nightmare called tactical interrogation," he explained. "Hence, I, fully aware of the secrets the night hides in these portions of the earth, literally sweated it out night after night. Till finally and so unexpectedly, my turn came. And, God, there was nothing more excruciating than to be confronted by that final leap into eternity!"[85] On his fifty-first night of detention, Kangleon was brought before the chief interrogator, Lieutenant Colonel Figueroa, a name synonymous with terror on Samar Island. At the start, the colonel entered like an actor striding to center stage, carrying a prop signifying his role as inquisitor.

> The entry of the dreaded chief intelligence officer, who came in with a thick pile of documents, dashed to the ground the last bit of my hopes to get out of their [sic] "unscathed." His initial declaration: "Father, the general has decided that we start interrogating you tonight" was enough to unleash that fear that was building up inside me for these past two months. I felt cold sweat, sweat broke all over my body, and I thought I was going to faint.
>
> For several hours, predator and prey fenced around verbally, one sizing up the other. Questions were posed and answers of innocence were given. Suddenly the "chairman" changed his approach. He said that since I would not answer his questions without my lawyer's presence, it would suffice if I would just give my biodata. And readily I fall for it. . . . I had fallen into a trap. I was already talking. Hastily, I tried to correct it by sticking to innocent or safe answers.
>
> Sensing that I had caught up to what he was up to and, irritated with the futility of that encounter, Ltc. Figueroa finally said: "Since you refuse to cooperate, Father, we will be forced to use other means. We cannot allow ourselves to be taken for fools."[86]

The colonel projects an image of controlled power. He enters carrying files symbolic of knowledge and power. He starts by probing like a "predator" for an opening to establish dominance. When his priestly "prey" resists, Figueroa announces, with a calm that evokes power, that the priest will be tortured for arousing his "irritation." The colonel seeks not information but subordination to his will.

At the colonel's command, the priest is taken to the nearby offices of the Military Intelligence Group. There, Kangleon discovers a metaphor central to our understanding of the RAM officers:

> Inside I was made to sit on a stool. I felt a small table being placed in front of me. Then, I heard voices—new voices! Three or four of these voices—the more commanding ones—took their places around me. And with [the] actors in their places, the most crucial stage of my detention started to unfold.
>
> "Now, Father, you are going to answer our questions!"
>
> "In court, ha? Now you are detained you are invoking the legal processes of this system. . . . In this place where you are now, we are the judge. And you are going to tell us what we want you to tell us."
>
> "As the NPA utilizes terrorism, we are also willing to use counter-terror. . . . Every time you preached against us, I wished at that time to just shoot you there at the altar."

"What's the name of that sister you used to visit at the Sacred Heart College? She is your girlfriend, *ano*? You are fucking her? How does it feel? . . ."

"For me, he is not a priest. Yes, your kind is not worthy of a respect of a priest."

"OK, take off his shirt. Oh, look at that body. You look sexy. Even the women here think you are macho. You are a homosexual, *ano*?"

"Lets see if you are that macho after one of my punches." A short jab below my ribs.

"Hey, don't lean on the table. Place your arms beside you. That's it." Another jab.

"You, take that stool away from him." I stood up. A blow landed behind my ears. I started to plead that they stop what they are doing to me. I started to cower. More blows . . .

"You better answer our questions or else you will get more of this." With that, a short blow landed in my solar plexus.

I was already quaking with fear. The psychological and physical aspect . . . of my interrogation had finally taken its toll. I finally broke down. "Yes. Go call Ltc. Figueroa. I am now willing to cooperate."[87]

As Kangleon implies by calling the interrogators "actors," his torture is a theater of humiliation far more painful than simple physical brutality. After blindfolding, stripping, and insulting the priest, the soldiers communicate their dominion by beating him, almost playfully, and forcing him to assist in his own degradation. He is, as they tell him, beyond the help of courts and the law.

As their interrogation proceeds, the torture becomes a drama of social inversion. In this Catholic nation, the simple honorific "Father" is a title of great respect, perhaps more prestigious than any military rank except general. After proclaiming themselves "judge," these soldiers strip the priest of his office and his cloth, signifying their superiority. When he resists, they begin with sexual abuse of a celibate clergyman, commenting on his bared body and accusing him of both homosexuality and fornication with a nun. Less courageous but more intelligent than most, the priest sees the inevitable. Inside this cell, he is no longer a priest, honored and protected. Naked and blind, he is theirs. He surrenders to the colonel's power. He calls out the name of his tormentor as his savior.

The colonel required a complete capitulation. After confessing his Communist Party membership, Kangleon was flown to Manila for a press conference where he denied being tortured and denounced his fellow clergy as communists—a sensational exposé that provided the regime with a pretext for repression against the Church. Abandoning his mission to the poor of Samar, Kangleon became a military chaplain and was issued a pistol, symbolically joining the ranks of his torturers. Then, on New Year's Day 1984, the priest was injured in a traffic accident and taken to a Manila hospital where he remained in a coma for three days. There, under mysterious circumstances that prompted several bishops to call for an investigation, Father Kangleon died on January 4.[88]

Among those "communists" in Samar's clergy was Father Pedrito Lucero, arrested in 1983 and subjected to a similarly theatrical torture by Lieutenant Colonel Figueroa's men. The priest's statement notes the props for Figueroa's inquisitorial role—the symbolic chess game, the sound effect of the cocked gun, and the scripted threat.

> 24 May 1983 . . . the military men forcibly carried me to a jeep where Col. Figueroa and Col. Salvador were. We played chess for a while, after which I was blindfolded. At 10:00 P.M., I was brought to Buri Island. Four times I heard a gun being cocked. I could sense that they were under the influence of alcohol. . . . The same question was asked over and over again.
>
> The following day they handcuffed me. After some time, they made me stand. They started to bang at the door. The sound was very damaging to the ears. Not content, they poured water over my body for thirty minutes. The torture lasted till 8:00 P.M. They removed the blindfold and the handcuffs. It was psycho-physical torture. I was not given food nor given any chance to sleep. Then I was brought to Col. Figueroa. I was told: "We don't need any information. What we need is your life." He threatened me. I played for time that I would think it over. So they allowed me to sleep.[89]

In a later deposition, Lucero stated that the three who tortured him were lieutenants—that is, junior officers in the first, formative stage of their military careers.[90]

THE TORTURERS

While Lieutenant Colonel Figueroa delegated violence, Lieutenant Rodolfo Aguinaldo (PMA '72), another future RAM leader, immersed himself in it. A review of human rights under the Marcos regime rated Aguinaldo as its "top torturer," finding him "implicated in the torture of at least twenty-seven detainees." A 1980 report by Task Force Detainees (TFD) described Aguinaldo as a "persistent and systematic torturer" whose "legendary maniacal torture sessions . . . left many detainees permanently injured."[91]

From his victims, Task Force Detainees compiled a portrait of Aguinaldo's transformation. After graduating second in PMA's Class of 1972 just months before martial law, he was assigned to the Fifth Constabulary Security Unit (CSU), where his interrogations were soon distinguished by their relentless determination, physical violence, and psychological cruelty. Beyond inflicting simple physical pain, he had a unique ability to engage his victims emotionally, using his sexuality and aggression to probe for personal weakness. "One of his tactics is seducing wives and sisters of political detainees," noted a TFD report. "He relishes the ensuing misunderstandings and splits between couples and families."[92]

In its first Philippine mission in 1975, Amnesty International identified Aguinaldo as "among the most persistent and systematic torturers at 5 CSU . . . who appeared to have treated prisoners with outrageous cruelty." His methods involved "prolonged beating with fists, kicks and karate blows, . . . the pounding of heads against walls, . . . the burning of genitals and pubic hair with the flame of a cigarette lighter." From interviews with detainees, Amnesty detailed one case of particularly brutal torture:

> Mrs. Jean Cacayorin-Tayag . . .was transferred to 5 CSU where she was kept sleepless for eight days and nights, made to stand several hours naked before a full-blast air conditioner and was slapped hard. . . . Lieutenant R—— A——, her main tormentor, told her that "whether you like it or not," he would take her away from her husband, who was also being interrogated. She said: "He told me he would hurt me where it would hurt most." She was forced to undergo "unwanted caresses." He had threatened to ruin her moral reputation and to spread gossip about an affair. He had threatened her husband and her child.[93]

Many of Aguinaldo's victims over the next nine years reported his obsession with the male organ. Confined at Fort Bonifacio in 1974–75, Father Luis Jalandoni saw men returning from Aguinaldo's session with genitals badly burned. Sometimes the lieutenant inserted a match into the head of the penis and struck. Sometimes he singed the pubic hair. In 1982, Aguinaldo grew frustrated when strangulation, beating, and electrocution failed to extract information from a twenty-five-year-old male prisoner named Marco Palo. "Sonofabitch," the lieutenant shouted at his subordinates, "Make him undress completely and electrocute his balls!"[94] A lay missionary recalled torture sessions by Aguinaldo and Lieutenant Vic Batac with a string of sensory fragments: "electric treatment (penis), water cure, karate kick, strangulation. Russian roulette, asked to masturbate at count of 10 penis did not erect—hit by .45 cal. pistol."[95] Juxtaposed against his usual courtly treatment of female prisoners, Aguinaldo's behavior seems, in a crude pathologic, as if he were trying to destroy the potency of rival men to become his prison's phallic overlord.[96]

Aguinaldo used a mix of physical and psychological means to humiliate social superiors—whether journalists, professors, artists, or activists.[97] "On two occasions . . . between 2:00 to 3:00 A.M.," political science professor Temario Rivera wrote in a later deposition, "I was roused from sleep and interrogated by then Lt. Rodolfo Aguinaldo. During these sessions, he repeatedly punched me in the chest and abdomen. While interrogating me, he would time and again point a .45 caliber pistol in my face and pull the trigger."[98] After his unit took philosophy professor Juan Villegas to a safe house, Aguinaldo administered a dramatized torture. "Villegas was . . . told that he would be taken to a cemetery where he would be killed and buried," read a human rights report. "He was taken by car, presumably to the cemetery, and made to alight. . . . The next morning he was again interrogated by Lt. Aguinaldo who pointed a cocked .45 cal. [pistol] at his head."[99]

Aguinaldo played a leading role in the torture of Satur Ocampo, then business editor of the *Manila Times* and later the NPA's chief negotiator. Arrested in January 1976, Ocampo suffered a two-day torture session at Camp Olivas, outside Manila, with a plot that seems driven by a theme of inversion: junior officers humiliate a media executive, feces become food, genitals carry pain:

> The torturers took Ocampo to the first floor, stripped him naked, poured cola drinks all over his body and forced him to stand up almost throughout the torture period up to the wee hours of the morning. The electrically charged spoon was repeatedly pressed on several parts of his body like the knees, thighs, genitals, groin, abdomen, ears, nipples, shoulders, neck, chin, face and nose. . . .
>
> Besides this punishment on his body, Ocampo was spat on by his torturers. His mouth was forced open and he was fed human excreta. Lighted cigarettes were pressed on his nipples, on the corner of his moth and his toes.
>
> Throughout the torture the jeerers would chorus whenever the electric shock was applied to Ocampo's genitals: "We'll make you impotent for the rest of your life!" . . .
>
> Only when the torturers became physically exhausted did they stop the punishment. Identified among them were Lt. Rodolfo Aguinaldo of the 5th CSU and Lt. Amado Espino Jr. of CSU and Major [Benjamin] Libarnes of the 1st PC Zone.[100]

After this group assault led by Major Libarnes (PMA '67) failed, Ocampo was transferred to Fort Bonifaco in Manila for a one-on-one encounter with Lieutenant Aguinaldo—a revealing psychological duel between a communist leader and Marcos's top torturer. On day one, the lieutenant, unlike others who concealed their identities, removed Ocampo's blindfold and pointed to his own face with forefingers forming pistols, saying, "Look at me. Look at my eyes." Without any questions, Aguinaldo then beat Ocampo and left, promising, "I'll come back tomorrow."

For the next six days, the interrogations followed a pattern. Aguinaldo would try to draw his victim out with tantalizing confessions from a comrade. When Ocampo remained silent, Aguinaldo would mutter something like "*matigas ka masyado*" (you are very hard-headed), and then erupt with dramatically athletic blows crashing down on his victim's immobile body, which was manacled to a steel bed. Recalling his determination to "show defiance," Ocampo said, "I sort of psychologized myself [so] that I don't feel anything. So I kept on staring at him and he kept beating me, so I think it made him unsteady because I wasn't breaking down, I wasn't shouting." Ocampo added, "That enraged him more. . . . He kept cursing me." With a final karate kick to his victim's body, Aguinaldo would leave. On the seventh day of this routine, he disappeared.

After a month's absence, Aguinaldo reappeared, saying "Oh, now let's find out how thick [tough] you are." Unlocking the manacles, he led his victim to the window and, pointing to the Makati skyline, asked "What place is that?" Ocampo, understanding Aguinaldo's need for conquest, answered, "That's Manila Bay." The lieutenant laughed, "Ha, ha ha. You're mad. Sorry you are disoriented." Then, he walked out of the cell and out of his victim's life for the next sixteen years. With insight heightened by his need to survive, Ocampo understood that Aguinaldo's drive for dominance dwarfed any need for actual information. The aim of all torture was, he concluded after a decade in prison, "to terrorize the victims into submission or to break their will."[101]

With his relentless ambition, Aguinaldo scored a major intelligence coup in 1977, capturing the NPA's founder and ideologue, Jose Maria Sison. Following a lead that Sison was hiding on the Ilocos coast, Aguinaldo won funding from his superior, General Ramos, for a six-week surveillance that finally nabbed the country's number one communist on a provincial highway. After the arrest, Aguinaldo was promoted to captain and command of his own antisubversion unit with a motor pool of fast cars and an armory of sophisticated weapons.[102]

Another member of the Fifth CSU, Lieutenant Vic Batac (PMA '71), later known as "the brains of RAM," emerges from his victims' affidavits as a persistent torturer. While Aguinaldo assaulted his victims with a relentless energy, Batac grew too obese for such activity and instead posed as the all-powerful inquisitor, ordering violence and theatrical torture to break his victims.[103] In 1974, only three years after graduation, Batac participated in the torture of a senior officer due the deference of rank. "I was picked up at my residence in Makati at about 9 p.m. on 25 May 1974 by 5th CSU . . . intelligence led by 1st Lt. Batac," recalled Navy Captain Danilo Vizmanos, recently retired after a long tour with the AFP's inspector general. "They had no arrest warrant." After being blindfolded and driven to a safe house, the senior officer "heard the metallic click of an automatic pistol being loaded" and then felt the muzzle pressed to the back of his head "for at least one minute [that] seemed like an eternity." Under "threat of liquidation," he was "ordered to identify certain names." When he refused to cooperate, the captain was locked in "a tomblike cell made of concrete with a solid steel door" for 60 days of solitary confinement. During his 808 days of detention, Vizmanos noted that the "RAM officers who participated in my arrest, detention, & torture were Lt. Batac, Lt. Aguinaldo, Lt. Bibit."[104] For young lieutenants trained to obey superiors without question, these acts—breaking into a senior officer's home, hauling him off without a warrant, subjecting him to death threats and detention—represented a serious rupture in their military socialization.

Among Batac's many victims, it was a journalism student, Maria Elena Ang, who has provided the most detailed description of his methods. In her account, Batac's unit crafted a metaphor of inverted social production: calling their water torture the "NAWASA session" after the national waterworks, electrical shocks the "MERALCO treatment" after the city's power utility, and assaulting her organs of sexual reproduction.

> I am Maria Elena Ang, 23 years old, a senior journalism student and research aide at the University of the Philippines. On the morning of August 5, 1976, I was on my way to Lourdes Church in Quezon City when unidentified military authorities pounced on me . . . and dumped me into the car. . . .
>
> It was about a five-minute trip from my place of arrest to the secret headquarters of ISAFP. . . . Immediately, I was subjected to a most degrading, inhuman and humiliating experience I would never want to relive again. But the memories keep coming back. Up to now in detention, I still have recurrent nightmares.
>
> I remember that while being restrained in a high-backed chair, several men about 10 to 20, swelled the ranks of those already in the room. Immediately, they swamped me with a battery of questions and psywar tactics. They threatened to kill me, get my relatives and friends and torture them in front of me. They kept telling me nobody saw them taking me in.
>
> Failing to answer one of their questions, I immediately received a slap in the face and a blow in the thighs.
>
> By this time, I was able to remove my blindfold and identify two of the officers in the room as Atty. Lazaro Castillo of the National Intelligence Security Authority and Lt. Victor Batac of the 5th CSU or Constabulary Security Unit. . . .
>
> Then, several agents began clamoring that I be given what they called the MERALCO treatment—MERALCO being the supplier of electricity in Manila.

An agent then forcibly removed my blouse and bra and unzipped my fly. Another brought in a hand-cranked electric generator used in military telephone. . . .

Two exposed wires were then tied around the little fingers of my right hand foot. Atty. Castillo, with a sneer on his face, started cranking the generator and fired another barrage of questions. Suddenly, the current shot painfully through my body. I could do nothing but scream and plead and scream but he only turned the crank until I was screaming continuously. . . . The electric shock session lasted for nearly two hours and was repeated in the evening. . . .

After the electric shock session, the military authorities still were not satisfied. . . . This time I was stripped naked and forced to lie on a short table.

At this instance, Major Arsenio Esguerra of the 5th MIG ISAFP entered the room and signaled the start of the water cure, which they laughingly called the NAWASA session—NAWASA being the supplier of water in Manila. . . . This time, besides four men restraining my hands and feet, another formed my hair into a bun and pulled my head down so that it kept hanging on the air until I felt that the water was racing through my brains. I passed out twice but they kept pouring water until I thought I would die.

Beside pouring water, several agents mashed my breasts while another contented himself by inserting his fingers in my vagina after failing to make me masturbate.[105]

Alone among the RAM leaders, Batac has been forced to respond to allegations of torture. While he was speaking at the University of Wisconsin-Madison in October 1986, the local Amnesty International chapter confronted him with the Fifth CSU's record of torture. "Yes," he answered, "we were aware of abuses in the unit." Admitting that comrades "may have been guilty," Batac explained that torture arose from "individual initiatives to get information in a short time." Although he denied any role in Maria Ang's torture, claiming that he was asleep at the time, Batac still insisted on the state's sovereign right to do "anything necessary" to protect itself.[106]

Despite his denials, Maria Elena Ang insists that the lieutenant directed her torture. "Batac ordered me stripped naked and tied to a chair," she stated in a 1989 interview. As soldiers gave her electrical shocks with a crank radio, Batac "sat there, leaning back in his chair with his feet on the table facing me with a smirk on his face. At one point I can recall him saying 'give her the NAWASA treatment.' And they filled me up with water." As she drowned in water and vomit, Batac was "leaning back with arms behind his head with that smirk on his face."[107]

The future RAM leaders may also have engaged in salvaging. After interviewing members of the Human Rights Commission in April 1986, a correspondent for the *Australian* newspaper reported that Colonel Honasan "has been linked with several salvagings in the late 1970s," including one particularly brutal killing in 1983. "Unconfirmed reports strenuously denied by Colonel Honasan," wrote the reporter, "say he played a role in the brutal slaying of a dissident, Dr. Johnny Escandor, whose body was found dumped outside military headquarters in Manila, the brain removed from his skull and his underpants stuffed in the cavity."[108]

The coterie at the heart of RAM also tortured together. During the first nine days after his arrest in 1983, the activist Randall Echanis was blindfolded, handcuffed behind his back, and subjected to round-the-clock interrogation by Honasan, Kapunan, and Aguinaldo. When the prisoner asked Kapunan to unlock his handcuffs, the RAM leader answered coolly, "Sorry about that, we just don't have the key here." The prisoner remained painfully handcuffed for the next nine days. After Echanis proved uncooperative, Kapunan spoke in a tone of threat, saying, "Think about it well because you have just a few days. I won't be able to do anything if the others take over, so you decide." Transferred to RAM's provincial stronghold in Cagayan, Echanis found himself in a cell block with the nine detainees tortured by Aguinaldo, including a couple "whom he had beaten with a baseball bat including the wife."[109]

Group torture encouraged lasting bonds among the perpetrators. From victims' questionnaires collected for litigation, it seems that torturers operated as stable teams inside the regime's elite antisubversion units. At the Fifth CSU, Aguinaldo worked regularly with his PMA classmate Billy Bibit and Vic Batac, forming close ties that would later knit together into RAM. To cite but one example of their daily routine, after a victim was stripped naked and handcuffed, Aguinaldo and Bibit "alternately punched him in the ribs"—a syncopated, rhythmic brutality that may have helped forge an enduring bond between these torturers. Similarly, at the rival MISG, Rolando Abadilla (PMA '65) and two close comrades, Roberto Ortega and Panfilo Lacson ('71), tortured together for over a decade, forming a tightly bonded faction that would fight RAM and then rise together within the police after Marcos's downfall.[110]

TORTURE'S IMPACT ON PERPETRATORS

What is the impact of torture upon the torturer? Since the literature has largely ignored this question, we can only speculate. If officers such as Hernani Figueroa had been involved in only a single incidence of torture, then we could accept it as an aberration with minimal impact upon the officer corps. But when torture becomes military duty and officers spend years in a daily routine of terror, the experience becomes a significant aspect of their socialization. For a young lieutenant to degrade and dominate society's leaders—priests, professors, and senior officers—may well induce a sense of mastery, even omnipotence.

The move from mastery over social superiors to dominion over society seems, within the mind-set of a military torturer, a logical next step. After subjugating strong-willed activists and politicians through limitless violence, the RAM officers may have emerged from the safe houses feeling like supermen capable of seizing state power. Invested with the power of pain and suffering, life and death, the torturer can come to feel himself a protean creator/destroyer. Judging from their later coups, these experiences seemed to foster a theory of social action founded on an inflated belief in the efficacy of violence. In the enclosed arena of the safe house, the future RAM officers played the lead in countless dramas of their own empowerment, rehearsing for a later moment on the national stage.

As lead actors in Marcos's theater of terror, the RAM leaders discarded the military habit of subordination for a flamboyance that seemed to express an expansion of individual will. When I visited Colonel Honasan at the Defense Ministry between coup attempts in July 1986, his security office had taken on the air of Q's laboratory in a James Bond film. In place of standard-issue weapons, officers toyed with crossbows, Israeli assault rifles, and automatic pistols. Rather than the regulation dress uniform for headquarters duty, the RAM boys marched about the air-conditioned corridors in jungle camouflage outfits with quick-draw holsters holding exotic weapons. Instead of short military haircuts, these officers grew flamboyant manes, beards, and mustaches. It was as if they had erupted out of cadet uniforms, with their statement of constrained power, into a costume of lethal masculinity, expressive of a volatile capacity for destruction and an untrammeled will.

Colonel Honasan, RAM's overall leader, cultivated an image of threat. A tank of a half-dozen piranhas greeted visitors to his office. Plastered to his office door was his personal statement: "My Wife Yes, My Dog Maybe, But My Gun Never." He entertained special guests with a jar of dried ears slashed from the corpses of Muslim rebels in Mindanao. He encouraged journalists to write about his exploits: his habit of skydiving with his pet python Tiffany around his neck, his quick-draw shooting practice, his black belt competition in both karate and Filipino *arnis* (cane fencing). His hair was a leonine mane, his uniform custom-tailored jungle fatigues, and his military name patch read "GRINGO."[111]

Instead of rejecting their experiences in the safe houses, RAM invested violence with a romantic power. On day one of the people power uprising, recalled journalist Sheila Coronel, Lieutenant Colonel Kapunan told her the story of "combing the country in search of a hired assassin reportedly out to kill the defense minister. Kapunan found the assassin and later had him killed." At a party for the RAM leaders after Marcos's downfall, several women journalists were stunned when Lieutenant Colonel Tiburcio Fusilero (PMA '71) tried to impress them with a story. "After a few drinks, Fusilero was not that drunk," recalled Coronel. "He . . . said he was given a list of forty people to kill and only two got away." Another journalist present, Jo-Ann Maglipon, wrote that "reformist T. F. admits typing out a two-page hit list of Marcos enemies upon the declaration of martial law in 1972. He noted, with neither pride nor regret, that . . . only one got away; his were professional and clean hit jobs." When Maglipon asked Fusilero if he would kill again, he replied coolly, "Yes, it is my duty to obey."[112]

Through their charm and charisma, the aura of violence around this Class '71 coterie seduced other officers, making this violent machismo the model for RAM's members. Among Colonel Honasan's many followers, few could surpass the devotion of Navy Captain Rex Robles (PMA '65), the group's political theorist. Observing Robles at a party after Marcos's fall, a journalist wrote: "The moment Greg [Honasan] arrived . . . Rex turned into a screaming, hysterical fan. When Greg's celebrated picture clutching an Armalite [rifle] was first published, Rex solemnly called for Col. Honasan who promptly awaited his senior officer's command who only asked that his copy of the picture be autographed." This adulation seems unbalanced in light of the captain's cowardice. During the February 1986 uprising, Robles checked into a luxury hotel, saving himself to become the historical voice for a doomed coup—a made-up mission mocked by his rebel comrades.[113]

Captain Robles reflected on RAM's theory of violence in a July 1986 interview, revealing a romanticized belief in its power and an inflated view of himself as its master. Without prompting, he offered blood and terror in reply to routine questions about chronology and policy. Short, balding, and desk-bound for two decades, he mimed the language and mannerisms of robust RAM leaders like Honasan and Kapunan—combat veterans, torturers, and coup commanders. In an extreme of self-expansion, Robles had broken free from his military socialization to embrace a vision of empowering violence:

> QUESTION: These discussions [of a coup against Marcos] were taking place from about August [1985] onwards?
> ROBLES: One time I remember was in November [1985], when our discussion ran until 2:00 in the morning about the crown of power. It's like when Napoleon said, "there was a crown that was lying in the street and all I had to do was bend over and pick it up with a sword."

I told them, it doesn't end there. You put the crown on your head, and you hold on to this sword, and anybody who even tries or even thinks of getting that crown from your head, you have to be ready to cut his head off....

You have to kill a lot of people to do this. Are we prepared to shed a lot of blood? That's my belief. That is why I am so afraid of this stuff. You have to kill. You have to have the stomach to kill cold bloodedly a lot of people. Because power does not stay on the head of people by itself. It has to be actively maintained—and by blood, especially blood—until people realize that you are serious about it. And they will fall back and say, "Hey, this guy means business. We have to follow him." Unless you do that, I don't think you are going to be very successful. You need to do that.

QUESTION: When did they begin discussing this with you? September, October [1985]?

ROBLES: Yes, around that time.... I could sense that they kept asking me every time we met, about twice a month, they'd bring up the question. I said, "Look anybody in this country who will take over power ... should realize what he is doing. It's taking over power." It's a power grab, and you should know that you want power.... It is not possible for you to say, "We will remove Marcos because it's good for the country." That's good, but once you decide on the method to remove Marcos, then you are making a power grab. And you should love that power enough to be able to hold on to it, otherwise the left will jump on you and where does the country go if you are not resolute enough to kill off the opposition right away?...

History is my bag and I like to read about how ancient kings consolidated power.... You must show people that you want to be their leader. Now, if you want to make a naked grab for power, recognize it for what it is and live up to it. Kill people. Discourage any opposition. Send off your enemies to another country, whatever. But be sure that you want that power. Don't be misled into mixing up idealism with it. I think you will come to grief.

That's when I really, really put the matter to basics. You have to be prepared to shed a lot of blood. When do you stop killing? One of the questions I asked was, "Are we prepared to give ruthless examples? For instance, if a general misbehaves in Region I, we move over and hit him. Execute him in public. So that people will be interested, 'this guy means business....'"

QUESTION: When did you move into high gear and what did you feel your role was? After the press conference in July [1985]?

ROBLES: Well, when I felt that I was being pushed around by Malacañang [Palace], really. My own personal battle with Malacañang was when they tried to come up with all this stuff about my family [background and immorality], and I said, "Wait a minute. I am going to hit back at you guys if it is the last thing I do." That's why I decided to make a good job out of it. I became a very high profile person.

People were surprised. "Rex Robles?" They told themselves, "Rex Robles? I know him. He is not that type of person. Rex Robles, you just leave him alone by one corner, give him a book, and he's okay, he's happy." That's the way I am. But once you get my goat because you've been unfair to me, then I fight you.

I told that to everybody in Malacañang. They said they were out to kill me. I said, "Do it. But do it good, because if I survive, then you have a problem. That's all." I don't mind getting killed. It's all right. Once you realize that you don't mind getting killed, everything else is possible.

Kill me, but then make a good job of it. Once you miss, and I get to escape, then you have problems. I can tell you that. Because I don't start with adults. I start with children. I start with children. I enjoy that. I enjoy chopping the extremities off before finishing people off. That's the side of me that I am afraid of. That's why I never get into discussions of politics and power because I believe in going all the way.[114]

For Captain Robles, power is amoral, a prize to be won by any means and held at all cost. In its pursuit and possession, he and his comrades believed in the transformative capacity of unrestrained violence. Once they had captured the crown, the RAM colonels planned to rule by terror. Mass slaughters and ritual executions, done with a theatrical flair, would, they believed, captivate the masses and strengthen the hold of their junta.

While the bookish Captain Robles theorized, and fantasized, about violence in the abstract, many RAM leaders had lived his fantasy of blood and terror. Through service in the safe houses of the Marcos regime, they were freed from the restraints of military socialization. Torture taught them to embrace violence as both ideology and strategy. It inspired their will to power. Stepping out of the safe houses for their debut on the national stage, they would launch a half-dozen coups with tactics informed by torture. Just as their torture had been theater-writ-small inside cells and safe houses, so their coups would become theater-writ-large with Manila as its stage. But torture also gave them an inflated sense of the political efficacy of violence, culminating in a vision of their last coup as a cleansing cataclysm of mass slaughter. Not only did torture inspire their many coups, but it would preordain every attempt to failure.

Chapter 7 Mutiny

Their coup d'état against President Marcos sprang from a boldness bordering on delusion. Empowered by their mastery over victims in the safe houses of dictatorship, the colonels of the Reform the Armed Forces Movement (RAM) emerged with a will to power and a plot to seize the state. But their coup attempt of February 1986 diverged so sharply from military fundamentals that it had no chance of success. With a strike force of just twenty commandos and a few hundred infantry, the rebel officers planned to capture a palace guarded by seven thousand infantry, tanks, helicopters, and gunboats. As it turned out, both their planning and execution were flawed. Even before its first shot, their coup collapsed into an abortive mutiny by just three hundred soldiers, who were saved when a million Filipinos surrounded them with human barricades for the four days of Manila's "people power" uprising.

In considering RAM's rise and revolt, it is important to recognize an audacity beyond any military rationality. Their extreme self-confidence, particularly after the near disaster of this first coup attempt, indicates that these colonels were not conventional tacticians, soberly weighing their options. Seemingly inspired by a sense of omnipotence, they later launched a half-dozen coups against formidable, often overwhelming, odds. Not only did their sense of personal power spark coup after coup, but it inclined them to haphazard preparation, reckless execution, and near-certain defeat.

In planning for both coups and a later rule by terror, these colonels suffered from an uncritical faith in the transformative power of violence. Unrestrained brutality inflicted upon a torture victim, blindfolded and alone inside a cell, can compel compliance. But in the open political arena, where individuals have allies and alternatives, terror can spark resistance. Violence not only inspired the colonels' will to power, but their illusion of mastery over its praxis eclipsed their tactical judgment. All their coups thus failed from simple military blunders—underestimating enemy strength, equating attack with victory, and failing to plan for tactical retreat. Apparently convinced of their invincibility, the RAM leaders acted as if courage and conviction alone could overwhelm all obstacles. In the end, their embrace of violence blinded them to its limitations.

The colonels not only inflated the potential of their own plans in this strategic calculus, they grossly underestimated the capacities of their enemies. In their initial bid for power, RAM's ultimate nemesis was not Marcos but his chief of staff, General Fabian Ver. Dismissing the general as a military incompetent, the rebel leaders plotted their coup on the misguided assumption that he lacked both the intelligence to detect their plans and the skill to counter them. As it turned out, his abilities were both better and worse than they had imagined.

In retrospect, however, the events of February 1986 did not hinge on the skills of either Colonel Honasan or General Ver. The colonel's boldness and the general's caution, in the end, neutralized each other. Ultimately, it was the refusal of the officer corps to fire on the rebels and their supporters that assured the triumph of people power. A decade later, a senior RAM member, General Jose Almonte, concluded that the "decisive moment comes when the troops of the regime are ordered to fire on its own people." At Manila in 1986, Marcos's military refused to fire on the massive crowds and he fell from power. At Rangoon in 1988 and Beijing a year later, says Almonte, soldiers "did *not* hesitate" and these dictatorships of left and right crushed the demonstrators.[1] From a broader perspective, however, the key factor in these urban uprisings is not simply whether the military shoots, but whether it remains unified. At Manila, the armed forces first split and then failed to shoot, thus losing power. Similarly, at Bangkok in both 1973 and 1992, the Thai army fired on the crowds and then split as shock waves reverberated through the ranks, ultimately losing power.

Clearly, to understand the import of these four days in Manila, we need to ask why some officers rebelled and why most refused to fire. To a surprising degree, the web of loyalties among Philippine Military Academy alumni shaped their response. One military analyst has argued that people power succeeded "mainly because PMA graduates in the Marcos-Ver camp refused to wage war against their fellow 'mistahs' in the opposite camp in defiance of orders."[2] Moreover, hostility between academy-trained regulars and reserve officers split the military into rival factions. Indeed, the cracks within the Marcos military formed not just along these factional fault lines but also along the less visible fissures of class loyalty. At several critical junctures, the outcome of this uprising by a million citizens against the dictator's hundred thousand soldiers turned on the decisions of a few academy classes, each numbering less than a hundred. The internal dynamics of certain strategically placed classes—the solidarity of '71 and a surprising split within '57—bore directly on the unfolding of events. Once the recognized leaders of Class '71 launched the revolt, their classmates, who led key combat units, somehow cooperated: either joining the plot, lending tacit support, or refusing to fire on rebel classmates. Once a few moral leaders of Class '57 broke with the regime, they raised an alternative standard of loyalty, infecting classmates in key commands with doubts that led to their ultimate defection. Viewing people power through the prism of class ties can thus illuminate an obscure, yet significant, political dynamic unseen by the omnipresent television cameras.

The outcome of this mass uprising ramified widely, both in the Philippines and beyond. Not only did the revolt force Marcos into exile, but it also had a powerful "demonstration effect," becoming a rallying cry for pro-democracy demonstrators in Asia, Africa, and Eastern Europe. In the Philippines itself, people power eclipsed RAM's coup, with an immediate impact on the country's future. Had these rebel colonels captured the palace, then the Philippines might have suffered something akin to Haiti's bloody military rule in the five years between the dictator Duvalier and the democrat Aristide.[3] Instead, the millions who massed on the boulevards to stop Marcos's tanks insured that it would be their presidential candidate, Corazon Aquino, and not this violent military faction, who took power after Marcos's flight. Above all, the coup itself signaled a breakdown in military socialization whose troubling ramifications were masked momentarily by RAM's heroic aura as architect of the dictator's downfall.

THE RISE OF GENERAL VER

In his loyalty, venality, and incompetence, General Fabian Ver personified the Marcos military. Beneath a veneer of bureaucratic authoritarianism, the regime was a knot of personal ties to kin and courtiers that the first couple had accumulated through the accidents of their lives—blood relations, school chums, military comrades, and political allies. Within these overlapping networks, several powerful families formed the core of the regime, allying and competing in an incessant contest for wealth and power. In this climate, Marcos's military became politicized and factionalized—with kinship and connections superseding efficiency and experience.

In the first years of martial-law rule, Marcos, then in his prime, was an ironfisted commander in chief who could control his courtiers. As a master of factional intrigue, he divided military authority among trusted subordinates and then played one against the other—his cousin, General Fidel Ramos, chief of constabulary; another cousin, General Ver, commander of the Presidential Security Command (PSC); and a civilian protégé, Defense Minister Juan Ponce Enrile.[4] To restrain their ambitions, the president retained General Romeo Espino, a classmate from college cadet days, as the armed forces' chief of staff from 1971 to 1981, a term of unprecedented length. Marcos invested Espino with corporate wealth seized from the old oligarchy to insure his loyalty. At the next echelon, the president passed over senior PMA graduates and appointed ROTC products to command the army, navy, and air force.[5]

During its early years of vigor and promise, Enrile, Ramos, and Ver cooperated to build the new regime while competing quietly for greater power. As martial-law administrator, Enrile amassed extensive holdings in the politically sensitive lumber and coconut industries. Ramos, the chief of constabulary, took control over municipal police to win a near monopoly over law enforcement. As the president's chief security officer, Ver built a private nexus of palace guards, secret police, and safe houses. From the outset, however, there was a volatile quality to their courtier politics. In 1975, for example, when the president's executive secretary, Alejandro Melchor, Jr., began accumulating power within the palace, Ver and Enrile collaborated to force him from office in a bitter internecine battle.[6]

Ver's proximity to Marcos was not, as some have argued, a mere accident of birth. Born out of wedlock and baptized Fabian Crisologo Ver, the future general saw little of his soldier father, Colonel Juan Crisologo, and grew up in Sarrat, Ilocos Norte, where his mother taught school. Among his many well-connected kin in this clannish region were Floro Crisologo, a powerful provincial warlord, and Marcos, a distant cousin three years his senior. During the 1965 presidential elections, Marcos entrusted then Captain Ver, who had risen through the ranks to become a constabulary investigator, with intelligence work for his campaign. After the inauguration, Ver commanded the Presidential Security Unit, a small detachment at the margins of the military hierarchy, driving his staff hard to collect intelligence and keeping long hours to prove his loyalty. By 1971, he had risen from captain to general and transformed a ceremonial guard into the thousand-strong presidential Security Command, or PSC.[7]

After martial law in 1972, the PSC expanded into a multiservice force of seven thousand men, with tanks, helicopters, and patrol ships. In the regime's fourth year, Ver opened "Camp 1" in his home province to train loyal Ilocano soldiers for the palace. He also awarded his sons key postings in his command as they grew old enough to join the military. Irwin became the PSC's chief of staff, Wyrlo commanded its antiaircraft unit, and Rexor directed the president's close-in security. As concurrent head of the National Intelligence and Security Agency (NISA), Ver transformed it from a small analysis unit into a dreaded secret police that controlled the currency black market, gambling, smuggling, and safe houses for torture-interrogation. Beyond the palace, Ver began influencing postings within the armed forces to such an extent that "discipline began to disintegrate" and the "president lost control over the military."[8]

As Marcos weakened physically and politically, he lost his capacity to balance factions and became more reliant on General Ver. The president received his first hemodialysis for failing kidneys in September 1979, and, in the words of his aide-de-camp, began "dying bit by bit, day by day." While his health slid downward into a spiral of seizure and remission, the first family became obsessed with succession. The president, as his aide-de-camp put it, "began to put his family ahead of his . . . country, and that's when things began to go very wrong." After Marcos fled into exile, it was learned that he had signed a secret decree in 1975 naming his wife Imelda as successor.[9]

To secure this succession, the president moved by degrees to groom General Ver, a reserve officer with no combat experience, as chief of staff of the Armed Forces of the Philippines. Marcos wavered for nearly three years, apparently unwilling to alienate General Ramos, another loyal cousin in line for the post. As General Espino's term at AFP headquarters stretched beyond the usual three years to eight and then nine, the competition between these cousins became intense. Since Ramos had grown up several provinces away from Marcos's birthplace in Ilocos Norte and did not serve him daily in the palace, he felt compelled to prove his loyalty.

As the infighting intensified in 1980–81, General Ramos published a slender volume, *Ferdinand E. Marcos: 77 Days in Eastern Pangasinan*—a testament to loyalty signified, in its curious title, by juxtaposing his family's bailiwick with

Marcos's magical number. Compiled by the general's father, Narciso Ramos, secretary of foreign affairs in the president's first cabinet, the book recalls the closing days of World War II, when he protected Marcos from the Japanese. Beneath its surface, the book seems to say that, just as the president once trusted his survival to the father, so he can now trust his security to the son. The Ramos family could not, however, make an exclusive claim to lifetime loyalty. In the back-pay rosters of Major Marcos's anti-Japanese guerrilla unit, Third Lieutenant Fabian Ver outranked Staff Sergeant Fidel Ramos.[10]

But General Ver did not need a history book to prove his loyalty. When General Espino finally retired in 1981, Marcos passed over Ramos, the next in line, and appointed Ver chief of staff. Pushed from the inner circle, Ramos and Enrile formed an ad hoc alliance. In the ensuing struggle over succession, a factional rift—Enrile and Ramos versus Imelda Marcos and Ver—divided and ultimately destroyed the regime.[11]

General Ver moved quickly to isolate his rivals. Ignoring merit or seniority, he played upon ethnicity, blood, and school ties to pick favorites for key commands. As an alumnus of the University of the Philippines reserve program, Ver became the head of its Vanguard Fraternity and his *brods* now won key posts—notably, the new army commander, General Josephus Ramas. In the decade before his promotion, Ver had also cultivated a clientele among subordinates "sidelined" in intelligence units, particularly the NISA and the Intelligence Service of the AFP (ISAFP). Now he catapulted these loyalists over seniority hurdles to top postings—Admiral Brillante Ochoco, flag-officer-in-command of the navy; General Roland Pattugalan, commander of the Second Infantry Division; Colonel Pedro Balbanero, chief of the military police; and General Artemio Tadiar, Marine commandant.[12] As Ver's men took command, garish mansions, monuments to money and bad taste, began rising from Corinthian Gardens subdivision in the shadow of Camp Aguinaldo, the military's main headquarters.[13]

Ver forged a centralized command in the name of presidential security, creating a powerful praetorian guard and crippling counterinsurgency operations. By 1985, with Marcos isolated in the palace, his health failing, about a third of the AFP's combat forces were allocated to presidential security while growing numbers of communist guerrillas roamed the countryside unopposed.

When Marcos went through two kidney transplants in 1983 and 1984, the first a failure and the next a success, competition among courtiers intensified. This "smoldering hostility" erupted in July 1983, when the First Lady stood before hundreds of delegates at a KBL party caucus in the palace's Ceremonial Hall to attack Enrile's defense policy "long and harshly," leaving him "beet-red and bristling with pent-up anger." A month later, Enrile and Ramos met with the president to protest Ver's maladministration of the military, threatening to resign unless their authority was restored. Marcos rejected their advice but, with his usual deft hand, retained their services, thus heightening the tensions that the U.S. Embassy wryly styled "the Ilocano intramurals." Within weeks, tensions worsened when Ver's soldiers shot opposition leader Benigno "Ninoy" Aquino, Jr., as he deplaned at Manila International Airport.[14]

With two of its largest military bases in the Philippines, Washington became increasingly concerned about political instability and the rising "communist threat." The Reagan administration promised increased military aid in early 1985 "to allow the AFP to move, to communicate and to shoot" while pressing Marcos to retire his "overstaying" generals. But the president refused to sack Ver. By the end of the year, the United States was openly opposed to Ver and quietly supportive of his enemies, Enrile and Ramos.[15]

General Ramos, a respected professional and loyal Marcos cousin, presented a difficult obstacle for Ver's ambitions. As chief of staff, Ver created Regional Unified Commands (RUC) to incorporate, and thus supersede, the constabulary's regional commands, leaving only one Ramos loyalist among the twelve new area commanders. Then, in August 1983, Marcos transferred Ramos's Integrated National Police to Ver's headquarters, stripping Ramos of his last operational units. Moreover, Ver centralized the armed forces' budget, denying all services, including Ramos's constabulary, control over their own finances.[16]

In the regime's last months, however, Ramos struggled to recover some of his lost authority. When Ver was indicted for Ninoy's murder in late 1984 and went on leave, Ramos, now acting chief of staff, tried to restore "moral discipline." But Ramos soon found that his orders "were reversed," arousing an antipathy so deep that he would later remark, "I don't know how I managed to exist next to somebody like General Ver." Though supposedly on leave, Ver still pulled the levers of command from inside the palace security compound. Throughout this long trial, moreover, the president restrained Ramos's ambitions with reminders of his responsibility for a militia massacre of demonstrators in the southern town of Escalante. "He kept saying that I could not be assigned as chief of staff," Ramos later recalled, "because I had this case against me."

When the courts found Ver not guilty in December 1985, Marcos immediately reinstated him as chief of staff, sparking a sharply negative reaction from Washington. Two months later, the president announced that Ver would retire and Ramos would replace him. But the next day, Marcos reversed himself. "It became very clear that there was no intention of removing General Ver from his office," Ramos recalled bitterly. When Ramos's father, Narciso, a member of the president's first cabinet, died on 3 February 1986, only weeks before the impending revolt, the tensions in the funeral cortege were obvious. Marcos and Imelda walked alongside Fidel Ramos, but his sister Leticia kept her distance, angry that the president had refused to appoint her U.N. ambassador. By mid-February 1986, General Ramos, a lifelong Marcos loyalist, was ready to break with his cousin.[17]

As a graduate of West Point, Ramos did not have a network of PMA classmates, making his the smallest among the three military factions. Instead, during his fourteen years in command of the constabulary, he cultivated a following among its senior officers, notably his deputy, General Renato de Villa (PMA '57).[18] Ramos also built ties to other services through frequent field inspections. "I know pretty well General Ramos," explained Marine Colonel Braulio Balbas, who later played a key role in people power. "He had been very kind to me when I was brigade commander. He visited me very often in Palawan and in the Bicol area." While Ramos often recommended promotions to the president, Ver usually blocked them, making "the prospective officers grateful for Ramos's concern and hate Ver." Air Force Colonel Antonio Sotelo, another key figure in the uprising, was outraged when he had to pay a courtesy call on Ver in mid-1985 to win a wing command. "The mere idea of going to him," Sotelo recalled, "as if to get his blessing for my assignment—a small unit for which one knew one was well qualified—was very demeaning."[19]

While Ramos remained quietly hostile, his ally Enrile was openly antagonistic to General Ver. As the architect of martial law, Defense Minister Enrile had been, at the outset, the second most powerful man in the Philippines. After 1978, the palace, now viewing him as a threat to Imelda's succession, removed Enrile from the chain of command, making him the first defense secretary in thirty years unable to order troop movements. Three years later, after Ver's promotion to chief of staff, Enrile began building his own faction within the AFP officer corps. Though an Ilocano like Marcos and Ver, he realized that they monopolized these loyalties and instead staffed his security unit with Ilongo officers from the Western Visayas, a region that the dictator rightly considered hostile. Ver promoted former reservists, called "integrees," so Enrile recruited academy-trained regulars, many from Class '71, who had close ties to classmates in combat commands.[20]

Although Ver's power seemed absolute, his system ignored certain fundamentals: the near monopoly of PMA graduates on field commands, and the intense loyalties among these classmates. Although academy alumni comprised only 44 percent of regular officers in 1970, they still dominated command echelons and relegated integrees to support functions.[21] By retaining his appointees after their mandatory retirement, Ver, in the words of Marcos's aide-de-camp, Colonel Arturo Aruiza (PMA '67), "froze the upward movement of younger men who were capable, ambitious, and professional." Compounding their grievances, Ver played favorites with the few promotions still available. In 1981, for example, he elevated his son Irwin (PMA '70) to full colonel over the heads of classes '67 to '69—a promotion that Marcos's aide called "disgusting to the officer corps."[22] One colonel told an American researcher in 1975 that "the big faction within the military is the regular versus the reserves," while another added, "I call the PMA thing our own silent war."[23]

By making patronage the basis of promotion, Ver's system alienated a group with formidable networks for action. In effect, he had reinforced the vertical lines up the chain of command to AFP headquarters, ignoring the strong lateral ties among PMA classmates. As disgruntled regular officers moved into opposition and then revolt, their activities would coalesce around class loyalties and a shared identity as academy alumni.[24] The outcome of the people power uprising in February 1986 would, at several key points, turn on these ties to class and corps.

THE ORIGINS OF RAM

As the regime split over the succession, the Reform the Armed Forces Movement emerged as an alliance between Defense Minister Enrile and a clique of regular officers. With their careers becalmed in the middle echelons, PMA alumni of Classes '65 through '75 harbored rising resentments against Ver and his reserve-officer faction who blocked their advancement. Sensing this anger among regular officers, Enrile used them to mobilize for a coup against Marcos.[25]

By the mid-1980s, these overlapping strands knitted into RAM, giving it all the attributes of a textbook-perfect coup group—ranks between captain and colonel, ambitions beyond rank, and the leadership of a few key activists. A handful of regular officers assigned to the Defense Ministry, most members of Class '71, became the group's driving force: Gringo Honasan, his deputy Red Kapunan, and their friend Vic Batac, who worked nearby at Camp Crame. As Honasan himself said of RAM, "most of the twenty participants in the planning stage were, like myself, members of Philippine Military Academy Class '71." Under Enrile's protection and patronage, their reach extended beyond its natural network. "Alone, Honasan's influence would have been limited to his classmates and juniors at the academy," wrote Marcos's aide, Colonel Aruiza, himself a member of Class '67, "but since he was Enrile's man, even those senior to him by many years were attracted. Rex Robles and Felix Turingan of Class '65, Tirso Gador and Hernani Figueroa of Class '66, Hector Tarrazona and Ruben Ciron of Class '68."[26]

Despite their rhetoric of reform, RAM's leaders prospered from their alliance with Marcos's defense minister. Enrile used his authority as martial-law "sequestrator" of the assets confiscated from the old oligarchy to reward RAM followers with corporate sinecures: Gringo Honasan, president of Beatriz Marketing; Colonel Hector Tarrazona, general manager of Rajah Broadcasting; and Lieutenant Colonel Red Kapunan, senior investigator for the Philippine Coconut Authority.[27] The RAM leaders spent a decade in Manila under privileged circumstances while their peers rotated through remote combat posts.

These origins were the source of an unresolved contradiction. Though advocates of military professionalism, the RAM leaders seemed, at times, to embody martial rule. Almost all were staff officers assigned to political operations—interrogation, intimidation, torture, disinformation, and penetration. As covert operatives close to the seat of power, RAM leaders were rewarded with special privileges such as graduate study, foreign training, and corporate salaries. Though they became the instruments of Enrile's bid for power, they denied his involvement in RAM to brother officers—a duplicity alien to the PMA honor code. Despite a public commitment to democracy, the RAM boys, shaped by the martial-law experience, fought to strengthen military control through a coup d'état.

The country's deepening political crisis transformed this small coterie into the larger RAM movement. In late 1981, believing that General Ver had formed a "hit team," Enrile moved to build a strong security force under Colonel Honasan. After a controversial combat record against Muslim rebels, Honasan had transferred to the Defense Ministry in 1975 as a senior aide-de-camp, forging a close bond that would make him Enrile's surrogate son. As threats to his safety worsened, Honasan took command of the minister's three-hundred-man security unit, recruited classmates and comrades, and, with British Special Air Service (SAS) instructors and Israeli weapons, began counterterrorist training. At first, the unit's middle-ranking officers began meeting for informal "gripe sessions" about the military until, in 1982, these discussions coalesced into a small, secret group called the Reform the Armed Forces Movement.[28]

With the help of protégés in customs, Enrile smuggled over a thousand advanced weapons into Manila, mainly Uzis and Galils. As relations with Ver grew tense, soldiers in Enrile's northern Luzon bailiwick, a force known as the "Cagayan 100," began to prepare for offensive operations.[29] Sometime in 1983, General Ramos formed a similar unit, the P.C. Special Action Force, and Honasan shared his weaponry, building ties to officers at constabulary headquarters, just across the EDSA highway, through his classmate, Colonel Vic Batac. By December, the two units were engaging in joint training exercises.[30]

In mid-1983, General Ver's agents spotted a RAM squad under Kapunan scaling the sides of the Philippine Coconut Authority's building and learned that Honasan was involved in "irregular arms acquisitions."[31] To counter this threat, Ver hired Israeli instructors for his own special security unit. Tensions rose throughout 1984 to the staccato of distant gunfire as rival factions fired off two hundred practice rounds a day.

Fully organized in early 1985, RAM was, at best, a loose organization, with an eleven-man steering committee and a general membership that was open and informal. Although it never had more than three hundred members, RAM soon won the support of a majority of AFP officers, particularly the PMA regulars. In September 1984, the group, as a trial balloon, circulated a magazine under the signature "Armed Defenders of Democracy," citing abuses and criticizing the leadership of the AFP. Then, in early February 1985, Batac convened three meetings, with up to twenty-three officers present, to draft a "Preliminary Statement of Aspirations" distributed that month at a PMA alumni meeting. Still anonymous, they described themselves as "a group of young PMA graduates" driven by their "youth and idealism . . . to seek reforms in the AFP." They attacked "the prevailing military culture . . . which reward[s] boot-licking incompetents," and called upon other cavaliers to join them in working for reforms.[32]

General Ver's partisans mocked their anonymity, calling them cowards. Stung, RAM's leaders decided to make a public stand at the PMA's March Week graduation. As the alumni gathered in Baguio a few weeks later, the reformers distributed leaflets calling for an end to promotions from "favoritism or *bata-bata, padrino* system and other personal considerations."[33]

This appeal had a stunning effect. During the annual alumni parade, when Class '71 reached the reviewing stand, its members, wearing T-shirts stamped "We Belong," stepped sideways out of the column and unfurled a banner proclaiming "Unity Through Reforms." Then, in a remarkable show of solidarity, some three hundred alumni from Classes '72 to '84 lined up behind them. "It was like being jolted from a deep slumber," said one lieutenant colonel. The next day, in his address to the cadets, General Ramos seemed to echo RAM by proposing changes to revive the armed forces for the fight against the communists.[34]

Suspicious, Marcos summoned the RAM leaders to the palace and launched into a rambling, forty-five-minute dressing-down. When their turn came, the reformers complained that General Ver's son Wyrlo had been given accelerated promotion to major. "These things passed through the promotions board," Marcos snapped. Retreating from further criticism, the RAM spokesman, Captain Rex Robles, spoke up: "Sir, we lack copies of your book . . . which we use as guide and source of our inspiration." The president was, in Robles's view, charmed by the praise for his book, *Towards a Filipino Ideology*, and autographed a copy warmly, "To our Reformists in the Armed Forces." To General Ver's dismay, RAM emerged from the meeting with Marcos's unintentional imprimatur. After a similar nine-hour session, General Ramos sent a confidential report to Marcos saying he had been assured that the group's actions would all be "overt."[35]

With such tacit support from their superiors, the RAM colonels went public in mid-1985, posing as modest reformers and producing a surge of favorable publicity. At a press conference in May 1985, the RAM "chairman," Lieutenant Colonel Hernani Figueroa, announced, "the military is rotten and we want to do something about it." The prominence of Figueroa, a notorious torturer, led one human rights group to question "the integrity of the AFP reform movement." The opposition weekly *Mr. & Ms.* commented that "it was during Col. Figueroa's tenure in Samar that the unsolved murder of Dr. Bobby de la Paz happened. Another unexplained case is Fr. Edgardo Kangleon's 'vehicular accident death.'" Surprised by these revelations, RAM's leadership recoiled and stripped Figueroa of his title, saying: "We were misunderstood. We have no chairman."[36]

To contain the damage, RAM issued a statement on human rights in the July edition of its newsletter, *We Belong*. "For more than a decade, the AFP and INP [police] have been the focus of a popular sentiment that they have violated human rights," the document began. RAM, claiming that the problem was largely one of public misperception, proclaimed its "fundamental view that the dignity of the human person defines the meaning of his existence" and urged the government to affirm "the upholding of human rights."[37] Despite an apparent condemnation of torture, the statement is, at closer reading, a mix of denial and evasion.

RAM thus concealed both its origins and aspirations. Most of the media ignored their dismal human rights record and continued to build their image as reformers with uncritical coverage. At the National Press Club on July 6, for example, Robles denied any coup plans, promising that RAM "will never, never resort to violence in pursuit of its aims."[38] But these same leaders were already meeting secretly to plot a coup d'état.

THE PLOT

Boldness both inspired and doomed the colonels' coup against Marcos. Dominion over countless victims in the regime's safe houses had nurtured their will to power. But years of such unconventional service denied them the tactical skills needed to command a successful coup.

By mid-1985, RAM had become a Janus-faced organization. While their spokesmen met the press to urge reforms, a clandestine core group began plotting a coup. Concealed within the larger reform movement, this group of fifteen officers served as a hidden inner circle and a blind that allowed Enrile to exercise ultimate control. Its members were, with few exceptions, academy alumni assigned to the Defense Ministry—security officer Honasan, his deputy Kapunan, Captain Felix Turingan ('65), psy-war expert Rex Robles ('65), and Colonel Tirso Gador ('66).[39]

Encouraged by the sudden burst of support for their March Week demonstration, RAM's colonels shifted from open agitation to coup planning. In August, Honasan and Kapunan suggested a coup to Defense Minister Enrile. "He asked us to give him about a week to think things over," Kapunan recalled, "then finally he called us up and said, 'I fully agree with our plans.' He volunteered to make the political sensing because we did not have enough political astuteness." Meeting with the RAM leaders in September, Enrile considered all the variables and set the coup date, tentatively, for Christmas or New Year's Day. While the RAM leaders planned operational details, Enrile formed a "political committee" to direct the overall strategy.[40]

In November, Marcos issued his unexpected call for "snap elections" in just three months, forcing the conspirators to reconsider their Christmas coup. In a tense meeting, they tried, as one put it, to "match the military option with the political climate." Elections, they felt, would be the last chance for a peaceful change of leadership. And to move before the balloting, explained Colonel Honasan, "would have been perceived as a military coup."[41]

When the presidential campaign started, RAM launched a crusade for free elections, called *Kamalayan '86* (Consciousness '86), using it as a cover to make contact with military officers, politicians, and businessmen. On December 12, the first day of her campaign, the RAM colonels met secretly with opposition candidate Corazon Aquino to offer security for her public appearances—an arrangement that would foster their ties to civilian organizations. The elections also gave the rebels access to military camps across the country, allowing them to test the loyalties of brother officers and build their image as an arbiter of public probity. While Kamalayan '86 campaigned openly for "honest, clean, fair and free elections," for RAM's leaders it was, as Kapunan explained, "a mobilization exercise."[42]

In the final weeks of the presidential campaign, Enrile and his colonels realized Marcos would win by fraud and resumed their plotting. The defense minister streamlined operations, entrusting military operations to RAM's core group and creating a parallel political committee to control overall planning. Meeting frequently at Enrile's Makati mansion, the political committee—Enrile, his aide Silvestre Afable, Robles, Turingan, Kapunan, and Honasan—developed a scenario for a post-coup government. Once the palace was captured, Enrile would appear on television to announce a seven-member National Reconciliation Commission that would include Cory Aquino, Salvador Laurel, and Cardinal Jaime Sin. In reality, the commission would serve as a facade for a military junta that would consolidate its control through spectacular violence.[43] The commission would supervise elections and cede power to a new government after a transition period of two to five years.

Simultaneously, Colonel Honasan's "core group" refined their tactics for the projected coup. Trusting only their fellow cavaliers, the colonels spent countless hours poring over maps of Manila and a model of Malacañang Palace. Their strategy, said Kapunan, was simple: "Capture Malacañang." Honasan concurs, saying: "We realized that we had to take control of the symbolic and real seat of power, which was Malacañang."[44]

RAM's plan was stunningly audacious, reducing Manila's sprawling maze of streets to its strategic essentials. The key objective was, of course, the palace, the symbolic seat of government and the home of the Marcos family. The palace grounds straddled the Pasig River at the heart of the old colonial city—the first family's residence and office buildings on the north bank, and a fortified compound for General Ver's Presidential Security Command on the south. Arcing about the city like an eight-lane rampart was Epifanio de Los Santos Avenue, or EDSA, Manila's main artery. Villamor Air Base, with its helicopter squadrons, and Fort Bonifacio, the home of marine and army headquarters, anchored the avenue's southern end near Manila Bay. Pushing north out of the city, the highway passed between the main military bases: Camp Crame, home to the constabulary and its riot police, and Camp Aguinaldo, headquarters for both AFP and the Defense Ministry. Beyond the city was an outer ring of defenses: fighter squadrons at Basa Air Base, sixty kilometers north; the army's Fifth Division, nearby at Camp Aquino; the powerful Second Infantry Division at Camp Capinpin, thirty kilometers due east; and the main navy base at Sangley Point, just ten kilometers to the southwest.

This phalanx was stitched together by sinews of kin and classmates. Inside the formal AFP hierarchy, there was an informal chain of command—a skein of personal ties to the president and his wife. Weakened by kidney dialysis, Marcos delegated command authority to his cousin General Ver. So empowered, Ver's security offices inside Malacañang Park became the AFP's de facto general headquarters, where his three sons held exceptional authority as ranking senior Presidential Security Command officers. In the event of riot or coup, their first line of defense was the constabulary's Metrocom, or Metropolitan Command, headed by another presidential cousin, General Ramos.

If the situation taxed the Metrocom's riot squads, then General Ver could call upon his strategic reserves—the military police, the First Scout Rangers, the Marines, and the navy gunboats in Manila Bay. Within this second tier of defense, the main strike force, the army's Second Division, just an hour away at Camp Capinpin, would mobilize under the command of General Pattugalan, a loyalist married to the president's first cousin. In this familial hierarchy, Pattugalan cemented ties to his younger battalion commanders by treating them like "his own sons."[45] As a cavalier of Class '57, General Pattugalan could call upon four strategically placed PMA classmates if reinforcements were needed: General Pedro Balbanero, chief of military police; General Felix Brawner, commander of the First Scout Rangers; General Fidel Singson, head of intelligence (ISAFP); and Commodore Tagumpay Jardiniano, commander of the Naval Patrol Force.

In this game of jacks, the queen was wild. Intimacy had introduced an element of volatility into the AFP hierarchy. As a former reserve officer, General Ver's military authority rested on his relationship with the president and First Lady Imelda. With only tenuous ties to the PMA regulars, Ver cultivated his fellow reservists, notably army commander Josephus Ramas, his "brod" in the U.P. Vanguard fraternity. Serving Ramas as staff officers at army headquarters were other reservists, Ramon Cannu and Cirilo Oropesa—a tight clique bonded by their service in the Scout Rangers during the 1950s and covert operations for General Ver before martial law. Through his access to Ver and Imelda, General Ramas had, in turn, built a special countercoup force of six hundred troops and light tanks, holding it close inside his headquarters at Fort Bonifacio. But Imelda could alienate as well as captivate. During people power, when Ver ordered the Fifth Division to reinforce the palace, its commander, General Antonio Palafox, would refuse because, "I was a victim of hate by the then-first lady Imelda."[46] Moreover, the president's overreliance on a few PMA classes placed his regime's fate in the hands of networks beyond his control. During the uprising, the Marcos phalanx would suffer a critical blow when the moral leaders of Class '57 sided with the RAM rebels and broke the solidarity of this large, powerful class.

In his judo-like strategy, Honasan, a martial arts champion, planned to defeat Marcos with his own military might. RAM's leaders would work through classmates and comrades in the weeks before the coup to win allies inside the Manila garrison. After subverting just three or four army battalions, RAM's coup force would then rise up within these circles of steel for a sudden strike at four key objectives: the palace, to take the president hostage; a television station, to win civilian support; a military camp, for logistics and communication; and Villamor Air Base, for air power and troop movement.

Timed to the minute, the plan was, RAM's leaders felt, a bold combination of lightning commando raid and mass military revolt. But from their own description, this plan required that they capture the palace, neutralize seven thousand soldiers, and secure a metropolis of eight million, all in just ninety minutes. Its success demanded a superhuman precision, without a moment's margin for error or enemy resistance.

01:30—At "H-hour minus thirty," Colonel Honasan and twenty commandos cross the Pasig River in rubber rafts and, guided by allies in the palace guard, break into the sleeping quarters to arrest the President and First Lady.

02:00—At "H-hour," as commandos secure the palace, Lieutenant Colonel Kapunan's hundred-man strike team penetrates the security compound on the south bank, hurling smoke grenades to sow confusion and detonating bombs to kill General Ver.

02:20—With these explosions as signals, two motorized rebel columns, backed by ten light tanks, break through the gates of the security compound.

02:20—Posing as pro-Marcos reinforcements to enter the main palace gate, Major Saulito Aromin's Forty-ninth Infantry Battalion reinforces Honasan's commandos and secures the palace.

02:30—The operations officer of the Presidential Security Command, a rebel sympathizer, begins transmitting false orders to the eight pro-Marcos battalions in Manila, immobilizing them during these critical hours.

02:30—Simultaneously, Colonel Tito Legazpi captures Villamor Air Base and radios rebel units in outlying provinces to commandeer aircraft and depart immediately for Manila.

03:00—With the palace secured, Enrile issues "Proclamation Number One" establishing a revolutionary government.[47]

Although RAM's colonels were convinced of their plan's perfection, more experienced officers, such as General Rafael Ileto, felt that it had the makings of disaster. Their strategy fused two irreconcilable military operations—a commando raid requiring perfect secrecy and a mass military uprising needing widespread knowledge. The opening assault could be compromised by even the smallest leak. Yet the second phase required the participation of over two thousand rebel troops—a number so large that secrecy was humanly impossible.[48] As it turned out, Honasan's plan would have led his commandos to certain death.

Not only was their plan flawed, but its execution was less than perfect. After a decade in air-conditioned offices, RAM's leaders lacked the command experience for such a complex operation. Convinced of the absolute loyalty of his comrades, Honasan and his core group suffered leaks almost from the start. In the week before the coup, its details would be known to almost every espionage agency in Manila—the CIA, the U.S. Defense Intelligence Agency, Australia's ASIO, and General Ver's PSC. Indeed, on February 17, five full days before H-hour, the secretariat of the ruling KBL party warned Marcos about "coup efforts to be perpetrated by rabid rebels AFP (i.e., Minister Enrile, General Ramos, RAM)."[49]

The leaks sprang from Honasan's overconfidence. In January 1986, RAM's core group, desperate for wider military support, abandoned its strict security to hold open planning sessions. With up to fifteen officers in the room and the entire plan on the board, Honasan and Kapunan directed these meetings like a barrack's bull session. In attendance was Colonel Jose Almonte (PMA '56), a brilliant, rather mysterious intelligence operative close to Alejandro Melchor, Jr., himself an Annapolis graduate with strong ties to Marcos and top-level access to the U.S. intelligence community. Moreover, Rex Robles and Boy Turingan suspected that Almonte had his own CIA ties dating back to Vietnam, where he once infiltrated the Viet Cong on a special mission. Fearing that their plans would leak from Almonte to the CIA to Ver, these senior RAM plotters eventually forced Honasan to compartmentalize information on a need-to-know basis.[50] But by then, the damage had been done.

After Marcos won the February 7 elections with fraud and violence that left ninety-three dead, the rebel leaders accelerated their coup planning. By this time, Honasan, working through PMA contacts, had already won the loyalty of three officers with troops inside the capital—Colonel Teodoro "Jake" Malajacan (PMA '71) of the Fifteenth Infantry stationed at the palace gates; Major Saulito Aromin ('74) of the Forty-ninth Infantry; and Major Ricardo Brillantes ('72) with a new Ranger battalion. Pilots from the Fifteenth Strike Wing at Villamor Air Base were also committed. But the plotters still lacked key weapons and personnel.

In a frantic search for the missing elements, RAM further compromised the security of their coup. Seeking light antitank weapons, known as LAW missiles, to knock out Marcos's tanks, Lieutenant Colonel Aguinaldo approached his CIA contact. The American agent, expressing some interest in the coup, extracted details of the plot from Aguinaldo. In the words of one RAM colonel: "You spill a few beans, you get some antitank weapons. It just so happens no antitank weapons came, but we spilled some of the beans." By the time the interview was over, the CIA had the names of fourteen key conspirators.[51]

Unknown to the colonels, the leak to the CIA station was only one of several betrayals. According to General Artemio Tadiar, then commandant of Marcos's Marines, General Ver was already receiving details of RAM's plans through a double agent inside the Presidential Security Command. That agent, according to Tadiar, was Major Edgardo Doromal, a key member of the palace security detail.[52]

Honasan, with characteristic overconfidence, mismanaged this recruitment in ways that compromised the coup. Once planning had revived in January, Honasan made a careful approach through "friends, fellow officers, relatives" to Captain Ricardo Morales (PMA '77), chief security officer for the First Lady, who agreed to guide the commandos into Imelda's bedroom.[53] After surveying other possibilities among the palace guard, Honasan selected Major Doromal, a former first captain at the PMA ('74), now chief of Malacañang's perimeter security. Unlike earlier recruits, Doromal held a key post. He had to say yes.

When Doromal refused the first overture, Honasan applied the ultimate pressure and arranged for him to meet with the many PMA first captains, or "barons," involved in the coup—Captain Boy Turingan ('65), Colonel Tirso Gador ('69), Honasan himself ('71), Major Arsenio Santos ('72), and Major Noe Wong ('75). "We told him, how can seven class captains be wrong about a moral issue?" Honasan recalled.

"Sir, it's hard, it's not time," Major Doromal replied.

"You have to make a choice," insisted Honasan.[54]

Under such exceptional pressure from his fellow cavaliers, Doromal agreed to provide intelligence about palace security. Honasan had, however, made a fatal error of judgment. Torn between loyalty to comrades and commander, Doromal resolved the conflict by serving both, thus becoming a double agent. Returning to duty at the palace, Doromal, in Honasan's words, "started becoming a nervous wreck." Only days after meeting the class captains in December 1985, Doromal broke down and confessed to his commander, Colonel Irwin Ver (PMA '70), chief of staff of the Presidential Security Command.[55] "We knew all the plans from him," said Tadiar. "So that every move that they made, General Ver knew." Not trusting Doromal, Colonel Ver assigned two aides, both fellow cavaliers, to keep him under constant surveillance. They returned with "alarming" reports of Doromal's continuing contacts with the RAM leaders at soccer games.[56]

Reacting to such intelligence, Marcos called his service commanders to the palace on February 14—General Ver, army chief Josephus Ramas, Metrocom chief Prospero Olivas, and General Ramos. They canvassed a range of possible responses, including mass arrests of opposition leaders, reprisals against RAM, and elimination of Enrile. In the following days, Ver's aides drew up detailed plans for a massive crackdown on the opposition that constituted a reimposition of martial law.

With tensions mounting, Enrile and his colonels met on the weekend of February 15 to set the coup date. Sensing that public anger over the electoral fraud was at the boiling point, they set their attack for the following week. Only a few days later, however, Kapunan and Honasan received an urgent call from the assistant military attaché at the U.S. Embassy, Major Victor Raphael—a friend of the RAM officers and, they believed, the resident agent of the U.S. Defense Intelligence Agency (DIA). According to the colonels, Raphael had a message "from the highest level of the U.S. government." Intrigued, Kapunan and two others went to his house. According to Kapunan, the American major said: "One, the U.S. government will not recognize nor look kindly on any unconstitutional move by RAM. Two, in very strong terms, the U.S. government has told President Marcos not to persecute RAM. Three, in case you have to make moves along the line of enlightened self-defense, the U.S. will understand it."[57]

To lend force to his message, major Raphael had a list of fourteen Filipino officers involved in the coup. Realizing, in a general way, that this information was important, Honasan and Kapunan informed Enrile about the meeting. But refusing to admit a critical security breach, the RAM leaders decided, in an act of foolhardy audacity, to continue with their coup plans unaltered, interpreting the message simply as, "the U.S. is betting on all the horses."[58]

Indeed, U.S. Embassy operatives seemed to be engaged in such balancing, tilting first to one side, then the other. In the week before the coup, the CIA's station chief, Norbert Garrett, met with General Ver and told him to "lay off the RAM guys." Then, only a few days later, Garrett again contacted Ver with a strong warning for him to strengthen Malacañang's defenses. Within hours, Ver placed the palace guard on red alert.

On Wednesday, February 19, Enrile's political committee met to set the attack. All agreed: the launch time, called "H-hour," would be 2:00 A.M. the following Sunday, February 23. According to plan, RAM's core group sent coded messages with the time to all operational units. Several reformists, including Colonel Almonte, briefed General Ramos informally on these plans. When RAM's operators phoned Major Doromal at the palace, the message was intercepted and reported to General Ver. Two days before the surprise attack, all the major players in Manila knew the coup's exact time and date.

The next day, Thursday, at 4:00 P.M., General Ver gave an emergency briefing on the details of RAM's plot to eight of his senior officers. On instructions from Marcos, Ver announced he would not interfere but was setting a trap. A navy demolition team was weaving a cat's cradle of detonation wires along the palace riverfront to a cluster of five-hundred-pound bombs and mines. Instead of arresting the plotters, Ver would let Honasan's commandos launch their rubber rafts and paddle toward the palace. The rebels would be blinded as they approached by the sudden glare of powerful spotlights. Then Marcos's son, Ferdinand, Jr., would stand up with a loud hailer to give them one last chance to surrender. If they refused, the river would rise from its banks in a thunderclap, vaporizing the rebels. The next day, at a planning session of this same inner circle, Marine Commandant Tadiar confronted Ver. "If you let this attack go through, the effect on our country as far as the outside world is concerned would be disastrous," he warned. "[Malacañang] is the symbol of the presidency," Tadiar added. "This is the seat of power!" Ver silenced the Marine sharply and the loyalist generals dutifully accepted their assignments.[59]

At midnight Friday, unaware of Ver's plans, Enrile's political committee met at his Makati mansion to put the final touches on a television speech that he planned to read thirty-six hours later announcing Marcos's overthrow. At 3:00 A.M., as the meeting was breaking up, Colonel Tirso Gador, commander of the Cagayan 100, announced, "I have your Uzis in my car." The would-be rebels stared blankly. "We told him we didn't know how to use an Uzi," recalled

Captain Felix Turingan, adding, "I'd been planning to use my Armalite." Nonplussed, Gador held an impromptu weapons course under the streetlights outside the defense minister's home. It was a prophetic moment.[60]

TELEPHONE WAR

That same Friday night, less than twenty-four hours before "H-hour," Honasan and Kapunan were on a final patrol of strategic points around the palace. As they cruised the maze of narrow streets near the river, they were shocked by the number of troops massing around Malacañang. They moved on to the suburb of Pasig for a meeting with Lieutenant Colonel Jake Malajacan, a rebel battalion commander, who told them of his certainty, from things overheard around Second Division headquarters, that the plan was "blown."[61]

Returning to the palace area, Honasan saw a Marine battalion camped near the spot where his rubber rafts were planning to land. Indeed, in the past thirty-six hours two full Marine battalions with over a thousand troops had taken positions near the palace. Other combat units—the Fourteenth Infantry Battalion from nearby Nueva Ecija and the Fifth Marine Landing Team from far-off Zamboanga—were already moving into the capital. By noon, there would be six combat battalions with some thirty-six hundred troops ringing Malacañang and more on the way. "Every inch of the palace was occupied, literally," recalled General Tadiar. "You didn't have room to maneuver." Clearly, the coup had been betrayed. "By Saturday morning we were convinced that the whole plan was compromised," Kapunan recalled. "Something was terribly wrong."[62]

Honasan, driven by a sense of invincibility that seemed to cloud his tactical judgment, was still inclined to plunge onward in defiance of this overwhelming evidence. Instead of issuing an immediate "freeze order," the two colonels wasted precious hours searching desperately for a scenario that would allow the attack to proceed. Then at 9:00 A.M., in the midst of these frantic deliberations, Colonel Rolando Abadilla (PMA '65), the dreaded Metrocom torturer known for his proximity to the president, dropped by with a message from General Ver to back off. "He reminded me that Ver and my father were friends, that [Ver's son] Irwin and I were friends," Gringo recalled. Shaken by the visit, Honasan finally admitted defeat and decided to abort. But they suddenly realized they had no plan of retreat, a violation, as Honasan later admitted, of basic tactical procedure.[63]

After losing twelve hours by indecision, Kapunan and Honasan began making frantic calls at 10:00 A.M. to freeze the attack on the palace. Since retreat ran counter to the psychology of their reckless assault, these communications were poorly prepared and their coded messages, often left with maids and children, went astray, leading to the capture of key plotters. Commanders of the main attack forces—Aromin, Brillantes, and Malajacan—had no sign of trouble until that evening, when their commander, General Pattugalan, ordered their arrest. Such bungling also led to the capture of rebel recruits inside the Presidential Security Command. At the palace that afternoon, Imelda's chief security officer, Captain Ricardo Morales, was caught taking firearms from the palace armory for the attack. His aide was out getting laundry and had not relayed the freeze message. Brought before Colonel Irwin Ver for interrogation, Morales almost died when a loyal presidential aide, Colonel Aruiza, became "so incensed" that he drew his gun and "nearly shot him."[64]

At noon, Marcos received U.S. Ambassador Stephen Bosworth. The ambassador was escorting special envoy Philip Habib, who was about to leave for Washington, D.C., to report to President Reagan on the Philippine crisis. Marcos told them of "an impending coup d'état" but their faces remained "impassive, betraying nothing."[65]

At that hour, Honasan and Kapunan were speeding to Enrile's house in Makati with news of disaster. When informed that their plot was "blown," the defense minister decided that their only chance was to proclaim an open revolt against Marcos. In effect, RAM's three hundred troops and two tanks would now challenge General Ver's powerful palace guard, with seven thousand soldiers, fifty tanks, fighter wings, helicopter squadrons, and naval gunboats.[66] The situation was grim.

Enrile phoned General Ramos, the one man who could inspire defections from the Marcos military. Over the past six months, Ramos had rebuffed every overture from the RAM colonels, reacting angrily whenever they suggested a coup. But several of his aides were involved in the plot and promised that the general would join them "at the last minute" even if, as Captain Turingan put it, "we have to physically drag him." Indeed, Ramos's own security chief, Major Avelino Razon, Jr. (PMA '74), was a member of RAM's core group and had briefed him on their plans.

Foolishly, just that week, Marcos had pushed Ramos to the breaking point—first announcing his appointment as chief of staff, then wavering, and finally withdrawing it during a brusque palace interview. Ramos left that showdown, as he later put it, "doubting I had a future." So, that Saturday afternoon, when Enrile asked Ramos to join his mutiny, the general, to nearly everyone's surprise, answered, "I am with you all the way."[67] Not only did Ramos hold Camp Crame with its 450 headquarters troops, but, most importantly, he commanded the loyalty of the constabulary forces that were the capital's first line of defense.

After changing into jeans and getting his Uzi out of the closet, Enrile, accompanied by the colonels, flew by helicopter to the Defense Ministry compound inside Camp Aguinaldo. There rebel forces broke open their cache of arms and prepared to defend the building. There too, at 6:30 P.M., Enrile and Ramos met the press. "I cannot in conscience recognize Marcos as commander in chief of the armed forces," Enrile told a room packed with reporters, recorders, and cameras. Speaking to the military, General Ramos added: "I would like to appeal . . . to the dedicated and people-oriented Armed Forces of the Philippines . . . to join us in this crusade for better government."

When one journalist asked if Enrile would accept the authority of Corazon Aquino as president, he replied evasively, "I am not making a conclusion." Realizing the implications of his reply, a reporter asked, "Is a coup d'état a part of your options?" Enrile, answering a question about the future in the past tense, replied, "We never had any plan to stage a coup d'état. . . . What I have done is an act of contrition to atone for my participation in the declaration of martial law."[68]

With this slip of the tongue, Enrile signaled the central tenet of RAM's political position during the four days that followed: there was not, nor had there ever been, any plans for a coup. Their action was, they said repeatedly, a spontaneous uprising against Marcos's military abuses and electoral fraud. With Enrile's words, the coup died and "people power" was born. To counter the rebel propaganda, Marcos would struggle desperately and unsuccessfully for the next four days to prove that RAM's principled mutiny was, in fact, nothing more than an aborted coup.

General Ver learned of the mutiny at a wedding inside Villamor Air Base. He was surprised. Confident of his double agent inside RAM, Ver expected their coup at two o'clock the next morning. Every detail of his defense plan was in place. Suddenly forced to attack rather than defend, he wavered, seeming confused by the mutiny and Ramos's unexpected defection. Ver rushed back to the palace for a hasty meeting with Marcos, and from there placed a call to Enrile, finally reaching him around 9:30 P.M. It was then that the palace made a fatal error. Obsessed with the president's security and no longer certain of the rebel strength, Ver, speaking in their shared Ilocano language, communicated his confusion, assuring Enrile, "We never had any intention to harm your group." The rebel defense minister, sensing an opening, replied, "Why don't we talk in the morning . . . and maybe we can find a solution." He added, "Just don't allow your men to approach our area." Ver agreed, almost pleading, "You must also tell your men not to attack the Palace." The revolution began with a cease-fire.[69]

The rebels spent the next twelve hours working the telephones, trying to neutralize Marcos's crushing military superiority. In these critical hours, Ramos was the key to their survival. "The division of powers between Minister Enrile and myself was clear," Ramos later explained. "He would take care of the political ... I would command the military operation."[70] After the press conference, the general crossed the street to Camp Crame and began phoning subordinates in the constabulary's chain of command. "We called first, of course," Ramos explained, "those who were most likely to join us." Knowing the Manila riot police, the Metrocom, were Marcos's first line of defense, Ramos dialed their chief, General Prospero Olivas (PMA '53), saying simply, "We are counting on your support." Olivas replied, "Yes sir"—a verbal salute. "It occurred to me that in our previous conferences on the security of Metro Manila," Olivas wrote in his after-action report to Ramos, "the PC METROCOM ... shall be primarily responsible, and the military elements shall reinforce us only upon my request."[71] In short, if Metrocom did not attack, then the rest of the military, under Marcos's orders, would not react. Then, Ramos called Commodore Tagumpay Jardiniano ('57), commander of the Naval Patrol Force, whose gunships were anchored in Manila Bay. Knowing the Scout Rangers were a likely attack force, the general spoke to several of their commanders—men who, he later recalled, had trained with him as junior officers.[72]

In these critical hours, the rebels pulled on academy ties, tugging at the sinews of Marcos's defenses. At the Defense Ministry, RAM's leaders were dialing desperately to contact classmates and comrades. Across the highway in Camp Crame, Ramos's chief of operations ordered his officers to "contact their classmates," pleading for support or neutrality. "I was calling the commanders," Ramos said, "my young lieutenants and captains were calling their friends, their wives were calling the wives ... on the other side. ... [There were] classmate to classmate calls." In the first twenty-four hours, this incessant phone campaign gradually persuaded "major components of the AFP." Most importantly, calls from Ramos and his deputies won over several key supporters in Class '57— Commodore Jardiniano, General de Villa, and General Balbanero—who would, over the next three days, weaken the resolve of classmates strategically placed inside Manila's defenses. These supporters had been class leaders at the PMA—Jardiniano chaired the honor committee and commanded a company, while de Villa was a top regimental officer and cadet captain, regarded by classmates as "a commanding personality ... manly and far sighted."[73]

At 9:00 P.M., Cardinal Jaime Sin addressed Manila's faithful over the Church-owned Radio Veritas. Enrile's wife Cristina had placed a call to the cardinal earlier in the day, pleading for his support, and he had promised to help. "I call the people to come out from their houses and to protect our friends, the soldiers," Sin announced.[74] As phones rang across the city, political networks, primed by the recent elections, reacted quickly. A crowd of some fifty thousand people gathered within hours at the gates of Camp Aguinaldo.

Over the next six hours, Marcos made five personal calls to General Olivas, demanding that the Metrocom troopers cordon off Camp Aguinaldo and disperse the crowds. With instincts honed by twenty years in power, Marcos read the situation correctly and knew that crowds would complicate any attack. Feigning compliance, Olivas, his blood pressure rising dangerously, replied "yes, sir" each time, but then countermanded any orders while a human barricade formed for his mentor, General Ramos. Frustrated, Marcos then issued the same orders to General Alfredo Lim, the Metrocom district commander. But he too was loyal to Ramos and failed to act.[75]

At 10:00 P.M., Enrile finally reached Cory Aquino in Cebu City, 550 kilometers south of Manila, where she had appeared at an anti-Marcos rally. He asked for her help and she offered her prayers. Aware of the coup plans but suspicious of Enrile's motives, Aquino's supporters were planning to establish a provisional government in Davao City. She was weighing options, considering this refuge. There, in this remote city on Mindanao's southern coast, local businessmen had convinced the Marine commander, Colonel Rodolfo Biazon (PMA '61), to protect her with his brigade. Thirteen hours later, with a night's sleep and chance to reflect, Cory would announce her support for Enrile's revolt, calling upon the people to support the military rebels.[76]

At 11:00 P.M., the president called his own press conference live from Malacañang Palace to expose Enrile's mutiny as a failed coup. With rows of uniformed generals in their gold epaulets at his back, Marcos spoke calmly, pointing authoritatively at the four captured plotters seated submissively in military fatigues to his left. "Their actions . . . do indicate now that they were part of an aborted coup d'état and assassination against the president and the First Lady, which was supposed to take place tonight," Marcos told his television audience. At his command, Captain Ricardo Morales, Imelda's security officer, stood at attention and read a statement detailing, quite accurately, RAM's aborted attack plan. "I would like to appeal to our people to remain calm," Marcos concluded. "We are in control of the situation."[77] Many viewers dismissed Marcos's story as incredible, another of his media confections. But watching that night, rebel

Colonel Red Kapunan said to himself, "For the first time in his life, Marcos was telling the truth."[78]

The mass uprising that followed was a drama in four acts. Its stage was Epifanio de los Santos Avenue, EDSA, the eight-lane highway that rings Manila and runs between Camp Crame, Ramos's constabulary headquarters, and Camp Aguinaldo, the site of Enrile's Defense Ministry.

Day one, Saturday, had ended in a stalemate. By proclaiming their revolt at a press conference, the rebels had sabotaged Marcos's countercoup plans, allowing them twelve critical hours to scramble for the mass and military support they needed to survive. Unsure of the rebel strength and the loyalty of his own troops, Marcos had hesitated.

PEOPLE POWER

Day two, Sunday, brought "people power." In the darkness after midnight, the propaganda war continued. At 1:00 A.M., Marcos, determined to prove himself the victim of a coup plot, again appeared on television, this time exhibiting Major Saulito Aromin, another of the captured RAM leaders. An hour later, Enrile lied and told reporters that the president's charges of a coup were "a bunch of bull."[79]

These hours before dawn were decisive. With its crushing superiority, Marcos's military could still mount a devastating attack. But in the stillness of a city bracing for civil war, nothing happened. Within the rigid AFP hierarchy, all hinged on the president's command and Ver's execution. Fortunately for the rebels, Marcos was a shell of his former self. His aide-de-camp, Colonel Aruiza, described him as "so ill he was beyond grasping fully the significance of the situation," so dulled by medication that he was "unfocused much of the time." Not only was the illness devastating, but it was also publicly denied, barring any alternative chain of command. With Marcos incapacitated, command responsibility devolved to General Ver, an officer, in Aruiza's words, whose "years of obeying had robbed him of the ability to think" and whose only combat experience was the "tanks in Hollywood movies." Paralyzed without orders from Marcos, Ver relied on his son Irwin, commander of the Presidential Guard, who urged a passive, defensive strategy of reinforcing the palace.[80]

At 3:02 A.M., General Ver finally assembled some twenty general officers and their aides at the navy officers' club in Fort Bonifacio to plan an attack. With over a hundred officers of all ranks present, he gave a textbook lesson in military incompetence.[81] As the meeting lurched aimlessly from coup briefings, to defense preparations, to attack plans, Marine Commandant Tadiar said to himself, "We're lost." There were now over eight thousand men packed so tightly in the narrow streets around the palace that they had no room to maneuver, and reinforcements were still arriving.

"You don't have to fire a single shot, if you secure the perimeter," General Tadiar suggested to Ver. "Cut off all their communications, put off the water and light." He was ignored. Instead, Ver appointed the army chief, General Josephus Ramas, an ambitious protégé with limited combat experience, to lead the assault on the rebel-held Camp Aguinaldo.[82]

At 5:30 A.M., only minutes after Ver's meeting broke up, General Ramas convened a staff conference at nearby army headquarters to plan the attack. Although he had three army battalions standing by, Ramas announced that General Tadiar's Marines would take the lead—a decision that defied all military logic and outraged Tadiar. His Marines were tied down defending the palace. It would take hours to extract them from their current positions and reassemble them for an advance.

As dawn broke over Manila, the crowds protecting the rebel camps had dwindled down to just a few hundred. From intelligence, Tadiar knew they could crush the revolt if Marcos's forces moved quickly. But he was forced to wait three hours for permission from the Presidential Security Command to withdraw his Marines from the palace perimeter. At one point, he shouted at General Ramas, "This is insane! I am still moving [troops], yet you are ready to move out!" Tadiar suspected, and other observers later concurred, that Ramas was holding his own army troops in reserve, waiting for an opportune moment in the confusion to seize the palace himself. After the release orders finally came at 8:30 A.M., Tadiar spent the next five hours shuttling his Marines from Malacañang to their base inside Fort Bonifacio.[83]

At 2:15 P.M., the Marines, finally loaded in an armored convoy at the southern end of EDSA, were now ready to move out. As the tanks crossed Guadalupe Bridge, a car with tinted glass passed carrying Cory Aquino, who had arrived only minutes before by aircraft from Cebu. Rolling north up the wide, deserted highway, the powerful cavalcade of seven tanks, ten armored transports, and three thousand troops would be stopped in its steel tracks when it reached Camp Aguinaldo.[84]

In the eight wasted hours since dawn, the crowds outside the rebel camps had grown from some five hundred to over five hundred thousand. News of the Marine advance had prodded Enrile into a dramatic move. Realizing that his force of three hundred troops could not defend the sprawling grounds of Camp

Aguinaldo, he decided to cross the highway and join General Ramos in Camp Crame. At 2:30 P.M., Enrile, with an Uzi slung over one shoulder and the RAM colonels at this side, marched across the highway, the crowd parting as he advanced waving and smiling. It was the first time the colonels had come face to face with the sheer mass of their supporters. Looking at the endless humanity that spilled down the sloping highway, Honasan, walking at Enrile's side, said to himself, "There's no way we can lose."[85]

Just south of the rebel camps, General Tadiar and his battalions were stalled by that same mass of humanity. The Marine commandant radioed General Ramas for instructions. "Ram through," Ramas ordered. "Ram through the crowds, regardless of casualties." But in front of Tadiar were thousands kneeling in the path of his tanks, nuns in white habits reciting the rosary, children in his firing line. His uncle's voice pleaded with him over the radio to turn back. His bishop's voice came next, saying, "We're all Filipinos." His former PMA superintendent, General Manuel Flores, urged him not to kill "classmates and fellow alumni." Three military wives pushed their way through the Marines and lunged forward, gripping Tadiar's arm. "Temy," said Aida Ciron, the wife of Enrile's aide Ruben Ciron (PMA '68), "you also have a wife and children, please don't do it." Another, Vangie Durian, cried out, "Temy, you know me—we were neighbors in Navy Village." Referring to her husband, Commander Jesus Durian (PMA '60), Tadiar asked, "Is Jess there?" Yes, she replied, her husband was inside the camp with the rebels.

Led by nuns, rows of demonstrators kneeled before the tanks, praying and reciting the rosary. Suddenly, without warning, their engines roared, spewing a black cloud, and the steel treads lunged forward one meter. The crowd did not budge. The tanks stopped. The crowd cheered. As night fell, the Marine troopers, hungry and dispirited, became tense, cocking their guns. By now the sea of humanity had swelled to over a million. Fearing that anything might trigger a burst of fire, General Tadiar decided to withdraw. It would take his Marines nearly seven hours, crawling through the crowds, to travel the twelve kilometers back to their base at Fort Bonifacio. People power had stopped the tanks.[86]

This was the turning point. Although neither realized it, both Marcos and Enrile had lost. Meeting throughout that long afternoon with Enrile inside Camp Crame, opposition leaders won his grudging support for a provisional government headed by Cory Aquino. Enrile's coup plotters had failed to capture the palace, but Cory's crusaders had taken the streets. The people had placed the crown on her head and Enrile would soon be forced to bow.

The first cracks were already appearing within the Marcos phalanx. As the tide turned, his commanders, picked for loyalty, began bickering like the courtiers they were. Sensing that their authority was ebbing, General Ver's "favorite son" Irwin argued that "it was necessary for the officers and men to be convinced of the plot to attack the palace." Showing the captured plotters to the troops would, he said, "inspire the men to do their job." That evening, General Ver, senior commanders at his side and three rebel officers in tow, hosted propaganda rallies at Fort Bonifacio and Villamor Air Base. Even at this moment of solidarity, jealousy took command. With heavy sarcasm, the navy chief, Admiral Brillante Ochoco (PMA '55), mockingly introduced General Pattugalan, commander of the captured rebels on display, as the "brilliant General Pattugalan."[87]

SHOOTING WAR

Monday, day three, was the battle. At dawn, six air force helicopters under the command of Colonel Antonio Sotelo took off from Villamor Air Base and landed at Camp Crame. "This was a major turning point in the revolution," recalled General Ramos. "Suddenly, we had air power."[88]

Offshore in Manila Bay, Commodore Tagumpay Jardiniano (PMA '57), chief of the Naval Defense Force, called a dawn conference of fifty officers in the wardroom of a patrol ship. The graying commodore apologized to his men: "As early as Saturday, I committed my unit in support of the minister [Enrile] and General Ramos for what I believe was a cause worth fighting for." After a momentary silence, the officers jumped to their feet and applauded. At 9:00 A.M., a rebel frigate dropped anchor at the mouth of the Pasig River and trained its guns on the palace, only three kilometers downstream. Marcos had lost his navy.[89]

The president finally declared war on the rebels in an early morning broadcast. Again accusing Enrile and Ramos of organizing a coup, Marcos pronounced them "guilty of rebellion and inciting to rebellion," adding that he was "duty-bound to execute the law and the Constitution."[90]

At first light, a full brigade of Marcos's Marines, with riot troops and tear gas clearing a path through the "still sleepy people," broke into the rear of Camp Aguinaldo riding in a column of six armored vehicles and twenty-eight trucks. By 8:30 A.M., the Marines had positioned their howitzers and mortars to shell and slaughter. The rebel forces, just across the highway inside Camp Crame, prepared to die. They said prayers and sang their alma mater's song, "PMA, Oh! Hail to Thee," while, as one RAM leader recalled, "our eyes flowed with tears, our voices broke, and our lips quivered."[91]

At 9:00 A.M., General Josephus Ramas gave the "kill order" to the commander of the Fourth Marine brigade, Colonel Braulio Balbas (PMA '60). The colonel hesitated. Looking down from the high ground of Camp Aguinaldo across EDSA's eight lanes, the Marines had massive firepower "bore sighted" on the rebels inside Camp Crame only two hundred meters away—three 105-mm howitzers, six 90-mm recoilless rifles, eight 81-mm mortars, twenty 60-mm mortars, six heavy rocket launchers, sixty .50-caliber machine guns, and nearly a thousand M-16 rifles. Colonel Balbas, a veteran combat officer, was known among brother officers as "the cool-headed type." If he gave the order, he knew that his Marines, battle hardened by years of jungle warfare, would fire without hesitation. The howitzers would level the camps' buildings and the mortars would cover its grounds with a hail of shrapnel. In such a barrage, thousands of demonstrators, packed shoulder to shoulder on the pavement between the camps, would be slaughtered. Uncertain that Marcos would ultimately back him for the killing of these innocent civilians, Balbas hesitated, telling General Josephus Ramas, "We are still positioning the cannons." The general barked back, "The president is on the other line waiting for compliance!"

After Balbas put down the field phone, Colonel Jerry Albano (PMA '71) approached the Marine position at the head of an armed headquarters company, saluting and calling out, "How are you, sir?" Suspicious, the Marine said, "*Prankhahan tayo* [Let's be frank]. You lay your cards, I lay my cards. What side are you on?" Colonel Albano replied, "Sir, you know pretty well that I belong to Class '71 and Class '71 belongs to the reformist movement. So, I am on the side of [Camp] Crame." Knowing that these combat Marines could slaughter his lightly armed guard unit, Albano suggested, "Okay, coexistence *na lang tayo* [Okay, let's just coexist]." The Marine nodded agreement and Albano withdrew.

Thirty minutes later, the Marine commandant, General Tadiar, entered the palace and met General Ver, who confirmed the order to fire. Tadiar picked up a phone and told Colonel Balbas, "I think the order of Ramas is cleared. So you may fire." The colonel relied, "Sir, if I may, the people have been let inside [Camp] Crame already, and we will be hurting a lot of civilians." Tadiar paused, "Then hold your fire and use your discretion." Ten minutes later, General Ramas called and, for the fourth time, gave Balbas a direct order to fire. The colonel again stalled, this time saying, "We were looking for maps and positioning the cannons and mortars."[92]

Only minutes later, panic erupted in the palace. Rockets crashed into the grounds, exploding near Imelda's bedroom. Flying low over Manila Bay, one of the rebel helicopters had jumped over the post office and skimmed along the

Pasig River to fire a barrage of six rockets. The elite antiaircraft artillery units ringing the palace under Major Wyrlo Ver had failed to fire a single round. The attack, in the words of Marcos's aide, "broke our men's will to fight."

Colonel Balbas got a frantic call within minutes from Colonel Irwin Ver, chief of staff for the palace guards, ordering a "full attack." Balbas replied that he "was still positioning the cannons and the mortars." By now he had decided that "only an insane military commander" would kill "hapless civilians." He was not going to slaughter his "superior officers, classmates, and other friends" inside Camp Crame. The colonel was certain that his PMA classmates, men like Oscar Florendo and Pedro Sistoza, were in there with Ramos. An hour later, sensing, as Balbas later said, that "I would turn over the troops to Camp Crame if I were pushed further," General Tadiar ordered him to disengage and return to base. Marcos had lost his Marines.[93]

In a rage over the rocket attack, General Ver himself radioed the wing commander of the F-5 jet fighters then in the air over Manila. "This is General Ver," he said. "Bomb Camp Crame immediately!" The squadron leader, Francisco Baula (PMA '73), a RAM member, replied wryly, "Yes, sir, roger. Proceeding now to strafe Malacañang." Marcos had lost his air force.[94]

At 9:15 A.M., RAM's Task Force Delta, led by Major Rodolfo Aguinaldo (PMA '72) and three members of Class '71, attacked the government's television studios at Channel 4, killing two soldiers in the assault and capturing the station. The fighting would have been far bloodier had not one of the rebels, Lieutenant Colonel Teodorico Viduya (PMA '71), recognized the officer commanding the loyalist defenders. "Classmate," the rebel officer called out to Lieutenant Colonel Arthur Balmaceda ('71), "don't shoot, the people are with us!" When the Marcos troops continued to fire, Viduya shouted, "Classmate! What's the matter?" The cavaliers negotiated and the loyalist troops withdrew.[95]

The RAM strike force reached the control panel inside Channel 4 at 9:56 A.M. At that moment, Marcos was hosting a live press conference in Malacañang Palace. As he began speaking into the camera to assure the nation that he was in full control, his face fluttered and then faded. The screen went black. While the government still held other stations, the revolution had captured a powerful television transmitter. Marcos had lost control of television.

At noon, three rebel helicopter gunships took off from Camp Crame for the short flight south to Villamor Air Base, still under Ver's control. As he circled the base, the rebel commander radioed the flight crew on the tarmac, "I ask you to vacate the area . . . because . . . I have orders to destroy the helicopters." The ground crew replied, "Come and get it." As the rebel gunships raked the runway with fire, one helicopter exploded and all five were crippled.[96]

At 2:00 P.M., Cory Aquino emerged from hiding and visited the barricades outside the rebel camps, where she led the crowds in singing "Our Father." That morning her advisors had warned that Enrile was no longer mentioning her name, and the appearance was a bid to claim the revolution's moral leadership. As she moved through the masses, her campaign chant of "Cory! Cory! Cory!" erupted like the sound of distant thunder, heard clearly inside rebel headquarters in Camp Crame. "Up to now, I think, Johnny Ponce Enrile disputes that I was there," she would recall bitterly years later.[97]

Inside the palace, General Ver flailed about for a counterattack. In these final hours, seeing Marcos "grope from hallway to bedroom," his aide-de-camp realized that "the momentum of events was now too swift for the president's reflexes."[98] At 1:00 P.M., Ver ordered air force chief Vicente Piccio to strafe Camp Crame. "But, sir," Piccio replied, "we have no more gunships. They have just been destroyed." At a 3:00 P.M. staff meeting of loyalist commanders inside the palace, Admiral Ochoco ordered General Felix Brawner (PMA '57) to launch an immediate attack on rebels inside Camp Crame and Channel 4 with his First Scout Rangers and the Marines. Emerging from the meeting, Brawner took a call from his wife, who, he later recalled, "informed me that Commodore Jardiniano and many other classmates were already on the other side." Indeed, apart from their moral leader, Jardiniano, his classmates in the constabulary, notably Renato de Villa in Bicol and Romeo Reciña in Davao, were now leading the revolt for Ramos. Another call came from his deputy saying that the men were no longer accepting "orders from higher headquarters." Under a direct order from General Ver to report to his unit and persuade them to strike, Brawner took the pulse of his troops and found only twenty willing to attack. "Then and there," he said, "I told them that it was my decision that we would not carry out the mission." Marcos's phalanx had cracked.[99]

At about the same time, General Tadiar was meeting with his officers at Marine headquarters about his orders to join General Brawner's attack on Camp Crame. After thirty minutes of discussion, it was agreed that they should "no longer participate in subsequent military operations." The Marines strung a line of machine guns across their base to repel all forces, government or rebel. That evening, General Ramos calculated that 60 percent of the troops in the field had either declared their support or promised to refuse orders. Marcos had lost his military.[100]

It was time for politics to take command. That afternoon, Enrile began negotiating with Cory Aquino's advisors over the division of power. Abandoning his plans for a junta, Enrile now recognized her as the revolution's "moral leader." While Aquino's emissary, mining executive Jaime Ongpin, shuttled back and forth between her residence and rebel headquarters, Enrile gradually made concessions about the future government. But one sticking point, the location of her inauguration, remained. "The whole night," Ongpin recalled, "both Johnny Ponce [Enrile] and Eddie Ramos were almost insistent on having the swearing-in ceremonies at Camp Crame."[101]

CROWN OF POWER

Tuesday, day four, was denouement. At 7:00 A.M., Ongpin arrived at Aquino's residence to explain the compromise hammered out in the all-night meetings. "I have taken a risk throughout the whole campaign," she told Ongpin. "No way am I going to be installed in a military camp for my proclamation because that would just give everyone the wrong impression." She refused, Ongpin later explained, to be "sponsored into office by a military coup." And, as Aquino herself later explained, she objected to Camp Crame's identification with martial law: "Today it may be a place of heroism, but unfortunately a lot of tortures, executions, and summary detentions took place there in the past." Instead, she insisted on taking her oath at the Club Filipino, the historic home of party politics, to symbolize the popular mandate she had won in the February elections.[102]

At 10:45 A.M., President Corazon Aquino took her oath of office at Club Filipino and then, in her first official act, appointed Ramos chief of staff and Enrile minister of defense. When asked why she appointed Enrile, she replied that "there was no one in the opposition who would have been accepted and respected by the military. Also I wanted to show my gratitude." Once the man who would command a junta, Enrile now settled for a single cabinet seat. Several hours later, Enrile introduced Gringo Honasan to an ecstatic crowd, and the colonel reiterated RAM's line. "We did not plan any coup d'état or assassination," he insisted. "Our action was purely for the purpose of survival."[103]

The palace was the scene of an eerie pathos. At 5:00 A.M., Marcos had his last contact with the Reagan White House when Senator Paul Laxalt called to advise, "Cut and cut cleanly." At noon, Marcos was sworn in as president by the chief justice inside the Ceremonial Hall, establishing a legal basis for a later claim to power. Appearing on the balcony, the first couple sang and waved to the thousands of loyalists below who chanted, "Martial law! Arrest Enrile! Hang Ramos!"

After the ceremony, Marcos retreated into his bedroom, where he was revived by his medical entourage to prepare for flight. Secretaries shredded papers and aides stuffed bundles of cash into duffel bags. Marcos, unable to recall the combination to his bedroom safe, left it loaded with financial documents. The crowd became a mob in the streets around the palace, beating anyone who emerged. "See that the barricades stay intact," Marcos ordered General Pattugalan. "Prevent the people from coming in at all costs." Although the presidential helicopter was fueled for flight, the pilots, handpicked by General Ver, had disappeared. The navy patrol craft assigned to the palace had defected. At 6:00 P.M., Marcos's son-in-law advised the president that the chief U.S. military attaché, General Theodore Allen, had offered U.S. aircraft and ships to evacuate the palace. An hour later, the president emerged from his bedroom and "shuffled forward, slowly and interminably," his first steps into exile.

At 9:05 P.M., four U.S. helicopters lifted off from the palace grounds carrying Marcos and his entourage to the U.S. Air Force base at Clark Field, just north of Manila. As news of the flight spread, a crowd of some forty thousand charged the palace gates, chanting "Cory! Cory! Cory!"

Just before dawn the next day, the Marcos entourage boarded U.S. military transports for the flight to Hawaii. Among the ninety-two passengers, the entire Ver clan, numbering twenty-six, was by far the largest contingent. Half a day later, as the president's party deplaned in Hawaii's pleasant morning sun, two officers in the entourage requested an immediate return flight to Manila. Both were members of Class '71. Marcos had lost power.[104]

CONTRITION

A string of fortuitous events had transformed a suicidal mutiny into a mass uprising. Locked into a bitter struggle for power, both the rebel colonels and General Ver ignored the complex variables in this fast-changing equation. Mesmerized by the Marcos factor, the game of capturing the castle and its king, both Ver and RAM lost sight of the people's determination to shape their own history.

With the audacity that any coup demands, RAM's leaders convinced themselves that Ver was a military blockhead who could never divine their plans. In fact, he knew most of the details. Ver was equally confident that his agents would give him ample warning of RAM's moves. When they changed plans, he had no intelligence on the extent of their military support. Once he learned of the timing of the attack, Marcos, with a confidence that exceeded even Ver's, ordered that the coup proceed undisturbed to provide a pretext for repression.

In the end, it was not the feint and thrust of rival military factions that decided the outcome. Each side underestimated the other and made countless miscalculations. On day one, Ver and Enrile, by their respective bungles, stalemated this game of generals. Driven by the divine or the dialectical, history, in the form of Cardinal Sin, intervened to call out the masses to protect the rebels. By responding in such vast numbers, the people transformed an aborted coup into a mass uprising. Enrile's plans for a junta became irrelevant once the crowds assembled. No longer commander of a rebel army, he had become the agent of a people's revolution and thus had to submit to its moral leader, Corazon Aquino.

Though admitting minor errors, Colonel Honasan came to feel that he had foreseen every possibility. Indeed, in our interview five months later, he portrayed himself as the master of events, the architect of history. He refused to admit that his bungled recruitment of Major Doromal had produced a damaging leak, blaming it instead on Lieutenant Colonel Aguinaldo's inept meeting with the CIA. He even dismissed the significance of "people power," insisting that RAM had prepared for just that sort of mass support. For him, there was no mystery to the way that a million people stopped a dictator's army—his planning had predicted every eventuality, his actions had shaped every event.

Even if Marcos's tanks had attacked and crushed the people beneath their treads, Honasan was confident that he had effective countermeasures in place. Perched on nearby skyscrapers, his snipers were, he insisted, ready to pick off the drivers' heads since they were not trained to drive "buttoned up." If the tanks got past the snipers, his air force technicians were ready.

> QUESTION: What was your estimation of the capability of the [Marcos] tanks without infantry support?
>
> HONASAN: Well, it was very open to our initiative also. We had small units outside, composed of police, civilian volunteers, and other pockets of military groups. We could communicate with them to conduct an attack. We even had provisions for cutting the power lines and using these tanks as conductors for fifteen thousand volts.
>
> QUESTION: Really?
>
> HONASAN: Yes. It's a relatively simple matter. You cut the wires, or lay them on the ground, open wires. We sort of speed up the process by putting water on the road, which makes sure that power comes in contact with the metal parts.
>
> QUESTION: And what would the effect be?

HONASAN: It would have been something like going through a microwave oven. . . . [Laughter, head thrown back showing teeth.][105]

In light of RAM's actual performance in the February revolt, Honasan's claims to omniscience seemed close to delusion. Throughout the revolt, the RAM leaders underestimated both Marcos and Ver. Honasan thought his penetration of the palace perfect, but Ver turned RAM's agent into his double agent. Honasan planned a daring commando raid, but Marcos turned it into a death trap. Honasan then gambled on a mutiny, but Marcos moved adroitly to crush the rebels—ordering riot troops to surround their camps and mobilizing his tanks for an assault.

The rebels were saved by luck and circumstance, not by their own tactical acumen. After months of refusing RAM's overtures, General Ramos, after his confrontation with Marcos, unexpectedly joined the revolt in its first hours and used his patronage to prompt defections among key commanders. When Ramos's constabulary clients refused to crush the mutiny and disperse the crowds, over a million civilians surrounded the rebel camps and made an assault impossible.

The historical factors that transformed a doomed mutiny by three hundred soldiers into a mass uprising by a million civilians are so elusive, so mysterious, that many Filipino Catholics have explained the event as a miracle. Some felt the presence of the Holy Spirit. Others, of a more literal faith, reported sightings of an angelic "blue lady" who hovered protectively over the massed humanity on EDSA. In an interview six months later, Manila's primate Cardinal Sin attributed Marcos's downfall to the hand of God: "In the Book of Chronicles it says, if God wants to punish a country, he sends a bad leader. The moment the people return to God and make penance . . . then that bad leader is removed and a new one is given. This is really what happened." The cardinal concluded, "God was the scriptwriter. And all of us played our own roles."[106]

On other occasions, the cardinal was more fundamentalist: "The troops were stopped, the tanks were halted, not by any force of weapons. . . . The outward miracle was only the manifestation of what was taking place within, of deeper forces which were the ones truly at play."[107] Similarly, a Filipino priest, Father Antonio B. Lambino, S.J., felt that human agency alone could not explain these events: "There were so many so-called miracles all along the way. Miracle in the sense of an event not explicable completely by human causes."[108]

Swept along in this tide of spirituality, former Marcos officials joined Enrile in describing their defection from the dictator as religious conversion. Only a few weeks after the uprising, for example, the former labor minister Blas Ople announced that he was resigning from Marcos's KBL and founding a new political party as "an act of contrition."[109]

So rapid were these conversions that some of the religious had reservations about the quality of contrition. On Sunday, when people power stopped the tanks, some Franciscan sisters were disturbed by the tide sweeping the military toward heroism and absolution. When Enrile and his RAM guardians crossed EDSA, the crowd reached out to touch "the military heroes." But one sister, Cres Lucero, found that she could not even raise her hand. Some twelve hours later, these sisters were awakened from their sleep on the pavement to make way for military defectors, surrounded by a protective cordon of priests and nuns. "Here they are now," said Sister Cres to herself as the soldiers passed, "and they are protected by the sisters and by the priests and by the same cross with which we have committed ourselves? Is this the Church coming to protect the torturers of our people?" In defense of her reaction, Sister Cres later explained that these were the same soldiers who had oppressed the people. "I do believe in conversion," she said. "But . . . how could they be converted overnight? Or are they just taking advantage of the situation, and saving their neck?"[110]

Ignoring these mysteries, Colonel Honasan and his RAM rebels extracted evidence from these awe-inspiring events to confirm their own sense of empowerment. Where Cardinal Sin saw the hand of God guiding human affairs, the RAM officers saw only their own mastery over men and history. They had not bungled, they had triumphed through their own brilliance and boldness. Where Sister Cres saw a need for the military to pause for penance, RAM would plunge onward in its pursuit of power. Not only did they part with the Church, but these rebel colonels, in their bid to seize the state, had repudiated the academy's indoctrination into subordination and civil supremacy. The events around EDSA, even people power, thus confirmed their self-image as omnipotent creator-destroyers—destroyers of the Marcos regime and, through a future coup, creators of a new political order.

Members of the Class '71 active in the Reform the Armed Forces Movement (RAM) posing on top of a tank to commemorate their catalytic role in the 1986 people power revolution against President Ferdinand Marcos. In the second row (*left to right*) are RAM leaders "Tito" Legaspi, "Jake" Malajacan, and "Red" Kapunan; standing (*left*) in the third row is "Gringo" Honasan; at the rear, face shadowed, is RAM strategist "Vic" Batac. (Pete Reyes, *Veritas*)

Oscar Legaspi.
(*The Sword of 1971*)

Gregorio Honasan.
(*The Sword of 1971*)

Victor Batac.
(*The Sword of 1971*)

Eduardo Kapunan, Jr.
(*The Sword of 1971*)

General Fidel Ramos (with microphone), military leader of the people power revolution, speaking to crowds massed outside his Camp Crame headquarters on Epifanio de los Santos Avenue in February 1986. (Noli Yamsuan, *Veritas*)

RAM officers presenting their agenda for military reforms at a Manila press conference, Jade Vine Restaurant, Manila, 7 January 1986. Pictured (*left to right*) are Lieutenant Washington Javier, Captain Rex Robles, moderator Joey Rufino, Captain Felix Turingan, Colonel Hector Tarrazona, and Lieutenant Diosdado Valeroso. (Luis Liwanag, *Veritas*)

President Marcos presenting RAM officers captured while plotting a coup against his government, Malacañang Palace, 23 February 1986. Standing (*left to right*) are Lieutenant Colonel Jake Malajacan, Major Saulito Aromin, and Major Ricardo Brillantes. Behind Marcos, holding a paper, is Armed Forces Chief of Staff General Fabian Ver. (Romeo Gacad, *Agence France-Presse*)

Defense Minister Juan Ponce Enrile (with microphone) proclaiming victory for the people power uprising against Marcos outside rebel headquarters at Camp Crame, Quezon City, 24 February 1986. To his left is his military ally, General Fidel Ramos; standing between these two is Enrile's security officer, Lieutenant Colonel "Gringo" Honasan. (Sonny Q. Camarillo)

Rebel soldiers, led by Major Rodolfo Aguinaldo, preparing to attack the Marcos regime's television station on day three of the people power uprising, Quezon City, 24 February 1986. (Albert Garcia, *Bulletin-Tempo*)

People power blocking the advance of President Marcos's Marine tanks on Epifanio de los Santos Avenue as they move toward the rebel military headquarters at Camp Crame, 23 February 1986. (Erik de Castro, Reuters)

General Fidel Ramos (*front left*, wearing glasses) ordering Manila's police and constabulary commanders to secure Malacañang Palace and government installations from looters, Camp Crame, Quezon City, 7:00 P.M., 25 February 1986. President Marcos's flight into exile is just hours away. To the left of Ramos, in white shirt and windbreaker, is General Alfredo Lim, the Metrocom commander whose refusal to attack the rebels on day one of people power saved their revolt from certain defeat. (Sonny Q. Camarillo)

Soldiers loyal to President Marcos holding back crowds outside Malacañang Palace only hours before his flight into exile on 25 February 1986. (Bullit Marquez, Associated Press)

Heroic portraits of the leaders of RAM, taken just after their mutiny sparked the people power revolution that drove Marcos into exile: Colonel Gregorio "Gringo" Honasan (*above*), Lieutenant Colonel Eduardo "Red" Kapunan (*below*), and Lieutenant Colonel Oscar "Tito" Legaspi (*facing page, top*). (The Honorable Jaime Zobel de Ayala)

Captain Abraham Purugganan (*right*), leader of the RAM rebel forces that occupied Manila's Makati financial district during the 1989 coup attempt against President Corazon Aquino, Davao del Norte, June 1985. (Joe Galvez, Jr.)

President Corazon Aquino and Chief of Staff Fidel Ramos presenting awards at the PMA graduation, 22 March 1986, only weeks after she took power. Behind the president is Cristina Enrile, wife of Defense Minister Juan Ponce Enrile. (Erik de Castro, Reuters)

Defense Secretary Rafael Ileto (USMA '43, *center*) supervising an oath of allegiance by the Armed Forces of the Philippines (AFP) after several abortive coup attempts against the government of President Corazon Aquino, Camp Aguinaldo, Quezon City, late 1986. (Jun Aniceta)

Journalists covering RAM's press conference at the Intercontinental Hotel boarding a V-150 light tank in the rebel-occupied Makati financial district, 5 December 1989. (Boy Cabrido, *Philippine Daily Inquirer*)

Government troops taking cover from RAM rebel forces on the Airport Road in Parañaque, Manila, during the coup attempt, 1 December 1989. (Joseph Bernabe, *Philippine Daily Inquirer*)

Rebel soldiers marching out of the Makati financial district back to Fort Bonifacio after the defeat of their coup attempt, 7 December 1989. (Roger Caprio, *Philippine Daily Inquirer*)

A rebel soldier greeting a government soldier as the defeated RAM forces march back to their barracks in Fort Bonifacio, 7 December 1989. (Boy Cabrido, *Philippine Daily Inquirer*)

Governor Rodolfo Aguinaldo (arm raised) declaring war on President Corazon Aquino before thousands of followers outside the Hotel Delfino in Tuguegarao, capital of Cagayan Province, 4 March 1990. Aguinaldo took AFP spokesperson General Oscar Florendo hostage, sparking a firefight that would leave at least fourteen dead. (Val Handumon)

General Rodolfo Biazon, defender of Manila against RAM's December 1989 coup attempt, being lionized by autograph seekers after a military march down Ayala Avenue in the Makati financial district, early 1990. (Ernie Sarmiento, *Philippine Daily Inquirer*)

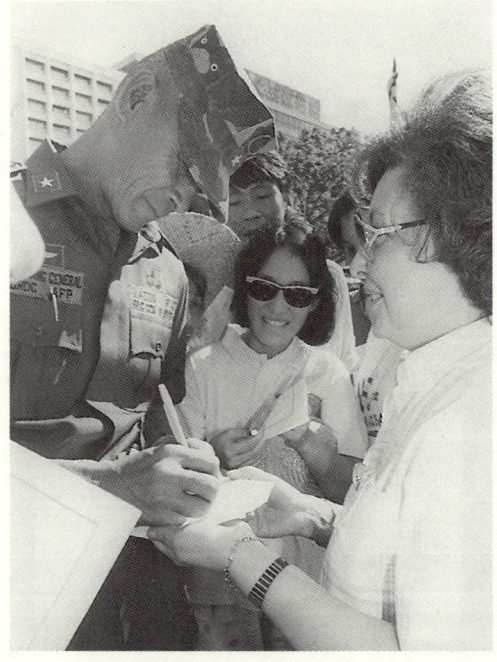

Colonel Alexander Noble (*forefront*) and his rebel troops just before their surrender at the end of a three-day coup attempt that failed to establish a RAM enclave on the island of Mindanao, Camp Evangelista, Cagayan de Oro, 5 October 1990. (Author's Collection)

Billboard on South Superhighway, Manila, March 1995, advertising the latest Filipino action film starring Fernando Poe, Jr., whose violent masculinity replaced the disciplined chivalry of his father's era. (Jose Duran)

Chapter 8 Coup d'Etat

For five years after Marcos's fall, the Philippine state was besieged by its own military. Coups and the threat of coups dominated and distorted its politics, blocking social reform and slowing economic recovery. Among the nine coups d'état, more than in any other nation during this decade, two were serious and came within a rifle shot of the palace. All these attempts involved, in some way, the Reform the Armed Forces Movement, which mobilized hundreds of disaffected officers for a sustained revolt against civil authority.

How can we account for this rebellion by an officer corps once known for its restraint? The answer seems to lie in its experience of martial rule under Marcos. The RAM colonels, liberated from their military socialization, were single-minded in their reach for power, exploiting their position inside the military hierarchy to mobilize coup after coup. Their attempts drew support, passive and active, from an officer corps clinging to privilege won under martial rule and trying, above all, to avoid prosecution for human rights violations done in service to dictatorship. Viewed in a global perspective, this turmoil is symptomatic of a near-universal process known as "impunity." As authoritarian regimes around the world collapsed in the 1980s, the old order, particularly its military, struggled to win amnesty for past excesses—whether torture, murder, or mass incarceration.

The outcomes of these struggles have covered the full spectrum of possibilities. Only in rare cases, South Africa and South Korea, have new democracies persisted with inquiries seeking to purge society of past violence and heal the collective trauma. In some nations, such as Chile and Guatemala, where the transition to democracy was negotiated, the military won formal immunity for past crimes. Where the transition was less tidy and civil-military tensions remain unresolved, as in Argentina or the Philippines, officers have threatened turmoil to extract amnesty from civilian rulers.

Among the many nations making this transition, Argentina offers the most revealing parallels for the Philippines. After eight years of military rule, President Raul Alfonsin took office in 1983 determined to punish those responsible for torture and mass murder in the dictatorship's "dirty war" against dissidents. When the federal courts indicted just one of the two thousand soldiers charged with serious human rights crimes, Lieutenant Colonel Aldo Rico, a hero of the Falklands War, occupied the Buenos Aires infantry school to demand a "political solution." After an agonizing debate, Alfonsin signed an amnesty bill in June 1987, exonerating over a thousand soldiers and launching a process of impunity that concluded two years later with full pardons for over two hundred indicted soldiers and the junta's former leaders.[1]

The Philippines suffered similar tensions in its return to democracy after fourteen years of dictatorship. Swept into office by the euphoria of "people power," Corazon Aquino's government was determined to punish the armed forces for past human rights violations. Six years and nine coup attempts later, a shaken civil state would abandon any attempt at prosecution and instead try to placate a rebellious officer corps. As a weak transitional figure like Alfonsin, Corazon Aquino was forced to balance the ideal of justice with the reality of her dependence upon the same military that had served Ferdinand Marcos.[2]

In this transition, the Philippine military, like its Argentine counterpart, seemed wracked by lingering tensions from a long immersion in torture and terror. The Filipino officer corps refused to accept punishment for its crimes in service to dictatorship. By either joining coups or failing to move against them, the military signaled its demands for full amnesty, increased appropriations, and aggressive counterinsurgency. With such support, the RAM rebels were able to launch coup after coup against Aquino.

Not only did these colonels show exceptional audacity in their plotting, but their execution suffered from an incompetence equaled only by Argentina's generals, who, through their slaughter of unarmed civilians in the "dirty war," gained an "illusion of victory, of glory, and of omnipotence" that crashed spectacularly into the British army in the Falkland Islands.[3] In like manner, RAM's leaders emerged from the torture of manacled civilians with a superman sense of mastery—an illusion that collided with military reality when each of their ill-planned coups met tanks and troops on the streets of Manila.

A REACH FOR POWER

In February 1986, President Aquino took office determined to restrain a military grown powerful under Marcos. With seven leading human rights lawyers in her administration, she moved aggressively on this issue in her first months—establishing the Committee on Human Rights, abolishing Marcos's martial-law decrees, and signing the United Nations Convention Against Torture. Under the leadership of rights activist Jose W. Diokno, the committee began documenting thousands of past violations and filed hundreds of cases, largely against serving military officers. Once the new constitution was approved in early 1987, the president appointed an even more powerful Commission on Human Rights to investigate "on its own or on complaint by any party, all forms of human rights violations."[4]

"They were out to get our blood before, they'll be out to get our blood now," complained one RAM leader, a member of Class '71, about these human rights activists only weeks after Aquino took power. "What do the people want us to do? Flog ourselves?" demanded RAM's Eduardo "Red" Kapunan in reply to a question about investigations into his record. "When we talk about reconciliation," he added, "then we are talking about *not* spilling blood. But if the people want blood, so be it." Indeed, one RAM analyst argued that the coups against Aquino were caused, ultimately, by a Human Rights Commission that "only concerned itself with . . . violations by soldiers."[5]

More broadly, the Aquino government, focused on economic growth, ignored the military's requests for increased pay. As the real value of soldiers' salaries declined steadily, the officer corps, after decades of favored treatment under Marcos, perceived such indifference as hostility. By mid-1987, a survey of regular officers found 72 percent felt "many high government officials . . . do not think highly of the military."[6]

Not only had senior military lost their access to the president, but they now answered to a commander in chief who challenged their perception of the military as masculine. "The ascension of a woman Commander-in-Chief... must have required a major adjustment in the ... psychological world of the military," commented the Davide Commission, an official inquiry into the causes of these coups. "But apart from being a woman, the military only knew her as the wife of Ninoy Aquino, a political prisoner who was later assassinated while under military custody."[7]

Aquino aggravated the military's alienation by offering solace to their enemies. Instead of fighting the communists, she released some five hundred political prisoners, most of them leftists, and opened negotiations with the New People's Army (NPA). "You waged war against Mr. Marcos because he was the embodiment of the worst injustice, greed, and cruelty. I fought Marcos for the same reasons," she said in an appeal to the communists at the PMA graduation in March 1986. "Now that the evil has fled from the land, I shall soon call on you to come out and rejoin your people in rebuilding our country."[8]

Aquino's liberal agenda provided an opening for her enemies. Appointed defense minister in her first cabinet, Juan Ponce Enrile moved against her on two fronts—organizing his own political party and plotting a coup with RAM's support. Now openly allied with the colonels, Enrile used them to build a following in the officer corps and to strengthen his northern Luzon political base. After the people power uprising, he assigned RAM's most notorious interrogators to constabulary commands in his bailiwick—Lieutenant Colonel Aguinaldo in Cagayan Province and Lieutenant Colonel Hernani Figueroa in Isabela. In his continuing coup attempts, Enrile relied on the RAM leaders, still in the Defense Ministry and still his main assets in the quest for power.

As tensions developed within the cabinet, the RAM leaders used their media contacts to conduct a subtle propaganda, questioning Aquino's mandate from the February 1986 elections. "Let's make something clear," Lieutenant Colonel Gringo Honasan told *Veritas* magazine in September. "The military will never be able to determine, now or in the future, if our commander-in-chief... won the election." Indeed, a later survey of PMA-trained officers found that only 29 percent thought Aquino the winner versus 26 percent who thought it Marcos. Under such circumstances, RAM felt that Enrile's leadership in the people power uprising should earn him "an equal sharing of power with President Aquino." Indeed, a long-time Enrile aide, Colonel Ruben Ciron, charged that Aquino herself caused the coups when she decided to "set aside 'pacts' formally and informally agreed upon ... with the dominant forces that brought her to power."[9]

To claim their rightful place in this revolutionary regime, where power sprang from history not ballots, the colonels began proclaiming themselves the real heroes of people power. In this propaganda campaign, RAM's most powerful weapon was Colonel Honasan, who attained superhero stature in the post-Marcos media. Yet the celebration of the Gringo myth cannot be understood within the narrow confines of a political debate over power sharing. Something less literal was transpiring, something in the realm of political mythology.

During his twenty years in the palace, Marcos had used state propaganda to portray himself as the country's greatest war hero and the reincarnation of ancient Malay warriors, effectively playing upon the nation's need for heroism to assuage centuries of colonial subjugation. With the full resources of state power, he created a political space for a myth of the leader as warrior hero. When Marcos's face faded from television screens during people power, his aura dissolved but the mythic frame remained. Indeed, his dramatic defeat inspired new, competing visions, secular and religious. Viewed in the frame of Marcos's own mythology, he, the ancient warrior, had fallen to Honasan, the modern commando, in epic combat on EDSA. Seeking, like Marcos had once done, to replace democracy with martial rule, Honasan would use such mythology to reorient reality in ways that would allow him to suspend, even transcend, the legal foundations of legitimacy. While the Marcos aura drew upon folklore and nationalist history of the 1930s, Honasan's would use Filipino and foreign film, the metatexts of his own age.

To win these laurels for their hero, the RAM colonels were sedulous in their courtship of the media. After people power, they were omnipresent in public forums and private gatherings, cultivating editors and reporters for laudatory copy. Significantly, the Davide Commission found that they benefited from "media adulation after the EDSA Revolt, painting Honasan and his group as larger-than-life heroes, thereby muting the unseemly aspects of their careers."[10]

Some coverage strained the limits of the English language for hyperbole. "Colonel 'Gringo' Honasan was the most publicized and popular 'plotter' of the three-day people's revolt which toppled the seemingly invincible Ferdinand Marcos," read a feature in the popular weekly *Mr. & Ms.* Some copy crackled with a suffused sexuality. "When he opened an art exhibit recently at the University of the Philippines," wrote reporter Eric Carruncho in a "Gringo Mania" cover story in *Mr. & Ms.*, "he was literally mobbed by young coeds asking for autographs and dying for a closer look at an authentic hero." In a coloful profile, one reporter explained the origins of Honasan's nickname by quoting his sister: "He is like Clint Eastwood in that film *The Good, the Bad and the Ugly*. He just sits there and bang!"[11]

Foreign correspondents joined in this mythmaking. In a book on people power, *Globe and Mail* reporter Bryan Johnson likened Honasan's Defense Ministry office to "the props department of a Rambo movie," and described the colonel himself like this: "Quick-witted and armed with a mischievous smile, he has a reputation as a one-man death squad . . . and his airborne exploits are legendary: [Like] the time he leapt from an airplane with a pet python named Tiffany double-wrapped around his neck."[12] In an interview with Pulitzer Prize winner Lewis Simons of the *San Jose Mercury News,* Gringo projected a more malevolent image. "The leaders were divided over how to treat Marcos," Simons reported. "Honasan, the most swaggering of the RAM leaders, was inclined to kill the president, in one way or another, perhaps sadistically. 'What we wanted to do was separate him from his dialysis machine and watch him go,' he said."[13]

This coverage was calculated. In constructing their "Gringo" collage, RAM's psy-war experts seemed to splice together fragments from cinematic images, indigenous and imported, that filled Manila's movie screens—the Filipino "action genre" of gunman-heroes who take up arms against injustice; the early Clint Eastwood persona as cool Western killer of quiet moral authority; and Hollywood's Rambo rebel as high-tech death-delivery system. During people power, Honasan wore a flak jacket covered with a brace of automatic pistols, ammo belts, and an Uzi submachine gun—a human weapon encrusted in armament. Afterward, he displayed a menagerie signifying command over nature—python about the neck, tank of pet piranhas outside the office. All these fragments merged into a media bricolage: karate black belt, quick-draw champion, academy baron, Mindanao war hero, coup commander.[14] So constructed, Gringo radiated an aura of power as destroyer of a corrupt regime, progenitor of a new social order.

While much of the "Gringo" persona was propaganda, the colonel's leadership abilities were real. Over the next six years, his mystique would resuscitate a dying rebellion time and again—persuading unlikely units to rebel and convincing brother officers to sacrifice careers for a hopeless cause. The saga of Captain Felix "Boy" Turingan, one of Honasan's more unlikely conquests, illustrates the quality of his charisma. As a career computer expert in the navy, Turingan worked with Honasan at the Defense Ministry and, like others there, joined both RAM and its revolt against Marcos. Then, in a decision that mystified brother officers, he followed Honasan as his deputy through several doomed coups and into the rebel underground, sacrificing career and pension.[15] Honasan's seniority could not have inspired such subordination since, as a member of PMA's Class of 1965, Turingan stood six years above him in the hierarchy.

Nor was Turingan a born follower. Like Honasan, he had been the "baron," or first captain, in his academy class. How can we explain such uncommon devotion to a junior officer?

In an interview six months after the people power uprising, Turingan replied to bland questions with matter-of-fact answers, which, when closely examined, reveal something of Honasan's extraordinary capacity for dominance.

> TURINGAN: In fact, during the [people power] revolution, we heard that [Lieutenant Colonel Jake] Malajacan and [Captain Rick] Morales were already killed by [General] Ver.
> QUESTION: What was your reaction to that?
> TURINGAN: We were of course mad. But we decided not to tell Honasan about it because he might react differently to Red [Kapunan]. He might turn violent. Greg [Honasan] is capable of becoming utterly violent if he senses something like that. He could be insanely violent. . . .
> QUESTION: Greg is wired.
> TURINGAN: Oh yes. Definitely. . . . He is like that. Enormous energy. You know he lost twenty pounds in the four days [of the uprising].
> QUESTION: Twenty pounds! He is very fit anyway.
> TURINGAN: Yes, he is very fit. . . . You know the one he put on?
> QUESTION: The flak jacket?
> TURINGAN: Yeah, the flak jacket with all the arms and everything? He was carrying about seventy pounds. He had an M-14 [points to shoulder], an Uzi [circular motion about neck], a 9 mm [indicates pistol holster on chest left], eight magazines for the Uzi [sweeping motion with hand across torso], eight magazines for the M-14 [indicates other side of waist], a bolo knife [chopping motion to back behind right shoulder], two grenades [touches chest left], and so many other things, radio [points to right shoulder again]. He was carrying seventy pounds.[16]

A comparison with a photograph of Honasan shows that Turingan's memory was exceptionally accurate: pistol at chest left, radio at chest right, two ammo belts on waist, M-14 rifle over shoulder, and Uzi around neck. Instead of eight Uzi magazines in the ammo pouch, however, I count only six—a small error considering that it had been twenty-nine weeks since the event. It was as if Honasan's "Gringo" persona had imprinted itself into Turingan's consciousness.

Periodically during our interview, Turingan, a graying computer analyst in his early forties, would reach for an automatic pistol among his data printouts and Protestant theology texts. Assuming a marksman's stance, legs well spread and knees bent slightly, he would dry-squeeze several imaginary shots at a target pinned to the wall, the silhouette of a human torso.

But the Gringo image also carried embedded within it the seed of its own subversion: the issue of human rights. Sensitive to allegations about torture, RAM tried to preempt the issue in the months following Marcos's flight. "We were aware of human rights violations, but we were not the ones committing them," said Honasan in an unprompted denial in his interview for the *Mr. & Ms.* "Gringo Mania" feature. "We tried to influence the situation by developing close contact with other operating units. Some we were able to influence, but others . . . " Similarly, in an interview with Bryan Johnson in July 1986, Honasan projected threat while protecting his flank. Wrote Johnson: "Together they [RAM] had killed dozens of men, 'but honorably, professionally, on the field of battle, in line with a job description which requires us to both kill, and die, on command,' as the defiant Gringo Honasan later put it."[17]

In this season of celebration, reporter Fe Zamora was one of the few who raised the human rights issue. Significantly, the journalist asked about torture in general, but the colonel, in a revealing slip, answered with the names of his victims.

> ZAMORA: There have been some accusations against you of human rights violation, some picturing you as torturer and hitman.
>
> HONASAN: We, Red Kapunan and me, we have been in counterinsurgency ever since. We are used to black propaganda like this. It's a risk on the job. As part of the MND security, we get special tasks. Take the Randall Echanis case. Remember this guy was operating in Region 2, the Minister's [Enrile's] region. We captured him and he was detained. There are witnesses against him for murder. I got my meritorious promotion as colonel in 1984 for drug busting, not for arresting Echanis or detaining him. I don't know this case about Dr. Johnny Escandor.[18]

Though soon superseded by RAM's coup activity, the human rights issue would hound the rebels for the next decade, becoming the driving force behind many of their political maneuvers.

MANILA HOTEL COUP

The first sign of RAM's continuing bid for power came in July 1986, when Marcos loyalists seized the Manila Hotel, the symbolic center of party politics, to proclaim a new government. Though this bizarre, one-day affair collapsed without a shot fired, its absurd aspects mask its significance as an event that shattered Aquino's fragile legitimacy and encouraged a five-year succession of coups.

Unnoticed at the time, Defense Minister Enrile's top aides had prior knowledge of these plans and yet made no effort to interfere. At 10:40 A.M. on July 6, six hours before the coup, Colonel Red Kapunan, the minister's intelligence chief, arrived late at his home in Camp Aguinaldo for our scheduled interview. He excused his tardiness, remarking that his office was on "red alert" because the "loyalists were going to take over the Manila Hotel."[19] Just after 4:00 P.M., the colonel's casual remark became the first sign of RAM's break with Aquino. As I stood on the steps of the Manila Hotel while thousands of Marcos loyalists surged past doormen into the lobby, a question came to mind: If the defense minister's intelligence officer knew about this coup hours in advance, where were the soldiers to prevent it?

As it turned out, RAM was a party to the plot. A month before, Marcos's trusted agent, the notorious Colonel Rolando Abadilla (PMA '65), had met with Honasan to recruit him for this loyalist coup. Honasan was noncommittal but maintained contact via Abadilla's aide, Panfilo Lacson ('71). On the day before the attempt, Abadilla had again visited Honasan seeking his support. Still noncommittal, the RAM leader had readied his troops to ride the tide.[20]

At first the coup seemed promising. At 4:00 P.M., some five thousand Marcos loyalists, led by a hundred armed soldiers, marched from their weekly rally at the Quirino Grandstand and occupied the lobby of the nearby hotel without any resistance. Marcos's former vice president, Arturo Tolentino, proclaimed a provisional government within an hour and rebel troops began to arrive. After taking his oath as interim president in the driveway, Tolentino proclaimed his cabinet in the lobby—including, in absentia, Enrile as defense minister. Marcos called from Hawaii to congratulate.[21]

Most of the officers who rallied were diehard Marcos loyalists like Colonel Abadilla and General Jose Ma. Zumel (PMA '59). But RAM's Lieutenant Colonel Rodolfo Aguinaldo also made an appearance—the first clear sign of a rapprochement between these rival military factions. Moreover, many of the 490 rebel soldiers who rallied that night were members of the Guardian Brotherhood, a mutual-aid society close to Colonel Honasan.[22]

In the next few hours, a parade of loyalist sympathizers trooped to Tolentino's suite through a lobby filled with reporters watching rebels, spies watching reporters, and waiters serving drinks to all. By midnight, however, the tide was ebbing. The arrivals slowed, then stopped. The bar closed, the air conditioning shut down, and the camera crews packed up. After dawn, when government troops surrounded the hotel and pro-Marcos forces failed to rally, the rebels surrendered without firing a shot. By afternoon, they emerged, bleary-eyed and unshaven, to be trucked off to a gym at Fort Bonifacio. The entire fiasco had lasted thirty-six hours.

There was more to this coup than met the eye. Throughout, Honasan updated Enrile regularly, and the minister proclaimed his loyalty to President Aquino only when the outcome was clear. Then, in the coup's waning hours,

Honasan and Kapunan approached the rebels with Enrile's promises of amnesty and extricated Marcos's key plotter, Colonel Abadilla. Most importantly, Defense Minister Enrile preempted the president's authority by announcing a blanket amnesty. Speaking to the rebels at the Fort Bonifacio gym on July 8, he said: "No punishment will come your way. . . . Let us forget as though nothing had happened." As punishment for the capital crime of mutiny, Enrile ordered the troops to hit the deck for thirty push-ups.[23] Nonetheless, neither Enrile nor the RAM colonels were investigated. In retrospect, the Manila Hotel incident seems to have convinced them that they could plot against Aquino with impunity.

But the historical opening for revolt was closing quickly. As a revolutionary leader, President Aquino had taken power without any constitutional or electoral mandate. To repair this instability, her administration devoted itself to a step-by-step restoration of legitimacy—a Constitutional Commission in June 1986, plebiscite in February 1987, and legislative elections in May. Popular support for any coup d'état would, as RAM's leaders knew well, wane with each step. As this eight-month window began closing, the rebels would become more decisive, and desperate, in their coup plotting.

GOD SAVE THE QUEEN

As the new constitution took final form in November 1986, RAM officers mounted their first real coup, the twisted "God Save the Queen" plot aimed at reducing Aquino to a figurehead in their military regime. In its study of six major coups, the Davide Commission called this plot "the most dangerous threat to the Aquino government, if it had been executed as planned—a chain-of-command takeover."[24]

The rebels began courting the commanders of the army, air force, and Marines in the weeks before the coup. Instead of captains or colonels, RAM would rely upon generals at the apex of command. Many senior officers were nervous about Aquino's pursuit of human rights violations, and many more were suspicious of the leftists in her cabinet.[25]

Few troops actually moved during these weeks of tension. Rather than a conventional assault, the RAM leaders, inspired by their experience in torture and black operations, crafted a coup that fused psychological warfare, terror, and feint. As veterans of such unconventional combat under Marcos, RAM's psywar tacticians, Captain Rex Robles and Colonel Vic Batac, were subtle in their manipulations of both myth and media to prepare the public for the president's overthrow. Assigned to Minister Enrile's Special Study Group in the mid-1970s, Robles, for example, had produced black propaganda against Benigno "Ninoy" Aquino for the 1978 elections.[26]

At meetings in the defense ministry during September 1986, the RAM colonels had planned a preparatory phase of terror to shake public confidence and propaganda to discredit President Aquino. Two years later, the National Bureau of Investigation (NBI) would report that these officers had plotted "surgical operations by death squads, namely the liquidation of selected targets to destabilize the Aquino government."[27] A member of the rebel planning group, Captain Ricardo Morales, later revealed that, "The first phase was assassination. *Assassination*, not capture, of high ranking government officials in cabinet level. Some were identified with the left, some of them were there simply because RAM did not like their faces." To prepare for these hits, RAM agents commenced surveillance of selected targets, notably labor leader Rolando Olalia.[28]

The RAM colonels juxtaposed past and present in their psy-war campaign, manipulating collective memory of the 1971 Plaza Miranda bombing and the 1986 people power uprising—the omega and alpha of Philippine democracy. In contrast to the visually eclectic Gringo persona, this propaganda was textual and specifically historic. In effect, the RAM colonels, using psy-war skills learned under Marcos, were fixing a historiographic frame for political change.

As an opening tactic, RAM's leaders began claiming credit for Marcos's overthrow, a clear signal, within the rhetoric of the day, for the president to concede some power. As head of a revolutionary regime, Aquino's authority rested on her symbolic leadership of "people power" and the primacy of this mass uprising in the dictator's downfall. To attack this fragile legitimacy, RAM leaders jettisoned past denials of their coup plot against Marcos and now, in countless media interviews, proclaimed their coup's catalytic role in his downfall. As one veteran observer explained, the group "decided to revise EDSA history by claiming they handed victory over to Mrs. Aquino at EDSA, and therefore deserved to be partners in a ruling coalition." On October 21, Enrile met privately with Aquino to demand a share of power and a cabinet shake-up. One Manila newspaper noted that "some of the ministers who RAM wants removed included former human rights campaigners." She refused.[29]

In response, RAM escalated its campaign with an eerily evocative terror. In the days that followed Enrile's demands, three bombs exploded around Manila, a grenade was shot into Aquino's campaign headquarters, and gunmen fired into a crowded Wendy's restaurant. "We seem to have been transported in a time capsule to the early 1970s," the *Manila Chronicle* editorialized, "when President Marcos was inciting the people against the Communists and when bombs were exploding at public buildings."[30]

As these blasts reverberated, RAM revived the memory of the Plaza Miranda bombing, seeking to implicate President Aquino's martyred husband, Ninoy, in an act of terror that still resonated within the collective consciousness. Fifteen years before, in August 1971, grenades had exploded during an election-eve rally at Plaza Miranda, the symbolic center of popular democracy, killing nine and wounding eighty-five. With the exception of Senator Ninoy Aquino, who was still en route, all opposition Senate candidates were injured, several seriously.[31] In the emotionally charged aftermath, rumor blamed President Marcos and he, in turn, denounced the communists publicly and accused Senator Aquino privately, giving his tardiness a sinister cast. If the bombing was not, as most Filipinos thought, done by Marcos to prepare for martial law, might it not have been a plot by Ninoy Aquino and the communists? In one blow, RAM's propaganda could both destroy Ninoy's martyrdom, the ideological foundation of his widow's government, and justify an anticommunist junta.

RAM's leaders circulated rumors of Ninoy Aquino's involvement in the bombing for several months before their coup. At a social gathering of journalists in July 1986, for example, Colonel Kapunan poured himself a large scotch, turned the conversation to Plaza Miranda, and took the table into his confidence about secrets gleaned as an intelligence officer in 1971. "He interviewed Senator Aquino's driver," I wrote in my notebook later that night, "who stated that Ninoy had given him instructions to delay his approach to Plaza Miranda and that Ninoy seemed determined to arrive late. RK [Red Kapunan] interprets this fact to mean that Ninoy and the CPP [Communist Party of the Philippines] planned the bombing."[32]

As their November coup approached, RAM's chief propagandist, Captain Robles, planned to use an incriminating document about Plaza Miranda as the detonator in their destabilization campaign. Through connections and coincidence, Robles had gained possession of a sensational letter from Lieutenant Colonel Victor Corpus blaming the Communist Party for the bombing.

The letter began its journey into the captain's hands one morning in August 1986 when the colonel turned up unannounced at the home of Jose "Pete" Lacaba, a writer working on a feature film titled *The Victor Corpus Story*. A true-to-life account, the script recounted how a young Lieutenant Corpus had defected to the communist guerrillas in 1970 and fought with them for several years until his capture. Now Corpus pleaded for a major rewrite, saying the film should end with his surrender, not his capture. To press his case, Corpus handed over a letter addressed "Dear Pete," explaining why he had surrendered. "I was present when some leaders of the [Communist] Party headed by Joema [Jose Ma. Sison] plotted the bombing of the LP [Liberal Party] rally at Plaza Miranda," he wrote. "Why did the Party leadership order the bombing . . . where so many innocent civilians were killed and wounded?" Angry at his communist comrades over the bombing, Corpus had "contacted a PMA classmate" and surrendered himself to the Military Security Unit.[33]

Despite the officer's characteristic sincerity, scriptwriter Lacaba was unconvinced. "When Vic told me that story in 1986," he recalled, "I was dismissive because I thought that was just some black propaganda from the military." Feeling the issue dangerously volatile, Lacaba handed back the letter.[34] Several days later, Corpus mentioned the incident to his superior, Captain Robles, who warned him that Lacaba was an active communist and the letter would surely reach the party. Fearing that his life was at risk, Corpus made multiple copies and deposited them with friends, the captain included.

On November 3, as RAM's coup plans firmed, Captain Robles called Corpus to his office and tried to draw him into the coup plot by revealing its details. Corpus asked, "What about people's power?" Robles answered, "We'll fire a few bursts of automatic weapons, and when bodies start falling, people will flee." If the defense minister refused to support the coup, then Robles had a pistol "set aside just for Enrile." The attempt to recruit this former communist for a right-wing coup was remarkably misguided. Without hesitation, Corpus reported this meeting to his superior, General Renato de Villa, the constabulary chief loyal to General Ramos.[35]

The following day, November 4, Enrile convened a meeting of senior officers sympathetic to the plot at the home of the Marine commandant, General Brigido Paredes. They agreed to use "a commando team to raid Malacañang, capture President Aquino, and pressure her to yield the powers of the presidency." Enrile set the coup date for a week hence, on the eve of the president's departure for Japan on a state visit.[36]

November 7 was a turning point in this game of generals. Showing the power of propaganda in this volatile period, the administration's first countermeasure was a press leak. After days of speculation about General Ramos's loyalty, the morning newspapers headlined his warning that "military adventurists" should not attempt their "surgical operation"—a clear reference to the impending coup. At a constabulary press conference later that day, Victor Corpus blamed the Communist Party for Plaza Miranda, seeking to blunt the impact of the

"Dear Pete letter" that RAM was already leaking to the media. Afterward, General Luis Villareal, chief of military intelligence, began releasing a second, "Dear Rex letter" that Corpus had just written to Captain Robles. "That night of Nov. 3 when we met in your office was perhaps the most frightful day of my life," Corpus wrote. "When you asked me to join your coup attempt (my part being to expose the Plaza Miranda Bombing incident . . . to trigger a crisis and start the ball rolling for the coup), I was shocked to say the least." Later that day, General Ramos played this letter as a trump card in a meeting with Enrile that seemed to give the defense minister pause.[37]

Only hours before her departure for Tokyo several days later, the president denounced the plotters and called for people power to fight any coup. Angered by her attack, Enrile ordered his RAM followers to raise the inverted flag of revolt over the Defense Ministry. A tense standoff ensued for the next two days as Honasan massed eight hundred troops and ten armored vehicles inside Camp Aguinaldo. Finally, General Ramos and his four service commanders called personally on Enrile to break the deadlock. When army chief Rodolfo Canieso (PMA '56) declared that "the whole AFP would take measures in favor of the government," Enrile backed down, promising that he would not attack.[38]

As their plot stalled, RAM used terror to regain momentum. Responding to the president's call for people power, labor leader Rolando Olalia had declared that his million members would fight any coup. Two days later, November 13, his body was found in a Manila suburb. One journalist noted the signs of salvaging: "the mutilated face, the empty sockets from which the eyes had been gouged, the mouth set in a scream of pain, the bound hands, the absence of trousers." His driver's body was discovered some distance away near a roadside. An autopsy found six bullets in Olalia, four in the driver.[39]

In the following months, a police task force found evidence implicating RAM. The first break came on the day of the killing, when police found Olalia's car, a white Mitsubishi Lancer, abandoned in Quezon City.[40] After months of work, an investigator located a missing car seat at a repair shop in Nueva Vizcaya Province stained with type B blood, the same as the victim's. A mechanic claimed that a Defense Ministry agent named Gilberto Galicia had left the seat for safekeeping. In February 1988, the NBI finally arrested Galicia, discovering that he had worked as Enrile's aide and maintained a provincial safe house for Honasan and Kapunan. Under interrogation, the suspect implicated two RAM officers—Lieutenant Colonel Oscar "Tito" Legaspi (PMA '71), commander of Enrile's security, and Commander Elpidio S. Layson (USNA '75), a navy officer assigned to Isabela Province. On February 19, NBI director J. Antonio Carpio presented Galicia to a packed press conference, announcing that RAM was responsible for the Olalia murder. Prosecutors filed murder charges against the rebel leadership—Legapsi, Honasan, Kapunan, and Robles.[41]

The investigation also uncovered evidence linking the group to another destabilization exercise. On November 15, only two days after Olalia's death, Mitsui executive Nobuyuki Wakaoji was kidnapped and held for 137 days until his company paid a $3 million ransom. The NBI's Manila director, Salvador Ranin, later charged that "the Wakaoji kidnapping was done by the RAM to embarrass (then President Corazon) Aquino who was scheduled to go to Japan for a state visit." One business analyst later commented that the kidnapping "effectively shut off the country from the massive migration of Japanese investments into Asia."[42]

If RAM was indeed guilty of murder and kidnapping, what were its motives? After a prolonged inquiry, the Davide Commission reported that these were "simulated events . . . to effect the tense and unstable atmosphere necessary for a coup."[43]

TWO DAYS IN NOVEMBER

Only nine days after Olalia's death, military intelligence discovered another coup, now aimed at a full seizure of power. On November 21, a member of Class '71 who had infiltrated RAM for General Ramos reported a clandestine alliance between Enrile and his erstwhile enemies, the Marcos loyalists. After a round of bombings and assassinations, RAM troops would, this double agent claimed, seize Manila while Marcos's KBL party occupied the new parliament building to declare the February 1986 elections invalid, thereby dismissing President Aquino.[44]

The intelligence was accurate and RAM was already engaged in last-minute preparations. Most importantly, their leader, Boy Turingan, approached General Ramos to join their coup. Instead, Ramos publicly affirmed his "unswerving loyalty" to the president. Confusing his diffidence with indecisiveness, the rebels had failed to gauge the depth of his "commitment to civilian supremacy."[45]

The plot began to unfold on November 22. Just after midnight, RAM rebels signaled the coup's start by killing Ulbert Ulama Tugung, a Muslim leader pledged to the president. Over the next few hours, General Ramos's headquarters received reports of rebel troops converging on Manila: two battalions from Bicol under Colonel Vic Batac, and constabulary troops from the Cagayan Valley under Lieutenant Colonel Aguinaldo and Colonel Hernani Figueroa. At the

Defense Ministry, Honasan had assembled a strike force of two hundred heavily armed troops, backed by ten Scorpion tanks and several V-150 armored vehicles.[46]

Throughout that long day, General Ramos countered their every move. When rebel troops mobilized in the morning, the chief of staff moved his forces into blocking positions to deny Honasan access to tanks and aircraft under Defense Ministry control. With the rebels immobilized, the coup became a telephone battle for the loyalty of uncommitted troops, still the bulk of the armed forces. By evening, Ramos was ready for a counterattack and ordered all troops in the capital to don full battle gear.[47]

At Fort Bonifacio, army chief Canieso, long courted by Honasan, sided firmly with the president. Instead of giving the go signal for a coup, he confronted the three battalion commanders, all RAM activists, readying troops for the rebel attack. "If you move," said General Canieso, "I will fight you." The attack stalled.[48]

The coup collapsed shortly after midnight, only twenty-seven hours after it began. Realizing that they "were completely boxed in," the rebel troops began returning to barracks by 3:00 A.M. By dawn on Sunday, Ramos had won.[49]

After a sleepless night, the president convened her cabinet, minus Enrile, at 8:00 A.M. She called him to the palace six hours later and demanded his resignation. Her face drawn with fatigue and tension, she told a national television audience that she had fired Enrile and in his place appointed his deputy, General Rafael Ileto. In coming days, to placate the armed forces, the president balanced these moves by dismissing four leftist cabinet ministers and downgrading her commitment to human rights.[50]

Within hours, RAM troops rallied at the Defense Ministry in full battle gear and posted snipers on the roof. Through his friendship with Gringo Honasan, the son of his PMA classmate, Defense Minister Ileto convinced the rebels to withdraw without a fight. Then, Ileto began to dismantle the power and personnel that Enrile had amassed in his seventeen years in office, quietly removing nearly five hundred officers and the ten tanks housed at the ministry for some future coup. "When I took over here, this building was a snake pit full of his people," Ileto later recalled. "If I had done it any other way except transferring his people out slowly, they would have killed me."[51]

Although Aquino had fired Enrile, she was surprisingly lenient toward his RAM subalterns and reassigned them to new posts—in retrospect, an unwise decision. Honasan became a commander in the special forces school at Fort Magsaysay, and Kapunan opted to become an instructor at the PMA. Both posts gave the RAM leaders control over young, idealistic trainees, the ideal recruits for later coup attempts.[52]

After a plebiscite approved her constitution in February 1987, President Aquino finally won the legitimacy to weather future coups. Her rival Enrile campaigned hard for rejection and predicted an overwhelming "no" vote—thus making this plebiscite a de facto presidential referendum. But even this victory held seeds of trouble. While 76 percent of the electorate voted to approve "Cory's Constitution," 58 percent of the military rejected it. A survey of active-duty PMA alumni found that 51 percent voted no and only 35 percent yes.[53] Clearly, there was still sufficient discontent within the ranks to fuel future coups.

AN AUGUST COUP

In August 1987, as the Aquino administration consolidated, Colonel Honasan unleashed a major assault on Manila. As in February 1986, the coup's chief aim was, in the words of Ramos's report, to "establish a provisional government under a military junta."[54] Opinion surveys had detected growing support for a coup within the military. In May 1987, 34 percent of officers surveyed had agreed that "an incompetent civilian leader could justly be ousted by military men," versus only 33 percent who disagreed.[55]

As they had once used safe houses for dramas of empowerment, so RAM's leaders now made Manila a proscenium for terror. With only two thousand soldiers to occupy a metropolis of eight million, their coup had to rely upon psychological shock rather than the sheer force of an overwhelming assault. A new round of terror bombings had rocked Manila in the weeks before the attempt.[56] The coup opened with an explosive overture, rose in intensity as cavalcades of rebel troops advanced in a crescendo of gunfire, and climaxed with a noon television broadcast by a chorus of young lieutenants. When the citizenry were not subdued by the spectacle and jeered its actors, rebel officers dropped their masks and turned their weapons on the audience.

In the aftermath of his November 1986 debacle, Honasan's political base had shifted from headquarters, with its senior officers and security forces, to regular combat units. Assigned to Fort Magsaysay as a training officer, his "charisma cast a hard spell," and his students, the First Scout Rangers, were soon drilling in shirts stamped with the motto "Kill With No Mercy." In only a few months, Honasan forged close ties to commandos whose tactical mobility made them an ideal coup force. Within their ranks, he found a new protégé in Major Abraham Purugganan (PMA '78), a Ranger officer deeply politicized by his reading of captured communist documents. "It was a meeting of minds," said Purugganan of the RAM leader. "I was simply inspired, *kasi* [because] here's somebody *na talagang* [truly] handling the cudgel of the poor."[57]

On the evening of 27 August 1987, Honasan packed his trainee troops aboard commandeered buses and drove south down the expressway toward Manila. Left behind in the classrooms were blackboard maps with Malacañang Palace and Camp Aguinaldo marked as targets. From nearby camps, troops joined the cavalcade at midnight near the Santa Rita tollgate on the city's outskirts.[58] As the rebel columns rolled through the suburbs at 1:00 A.M., detonations resounded in the quiet city and Manila woke to become an audience.

With over a thousand troops from disparate commands, Honasan led a night attack on Malacañang Palace while two deputies assaulted lesser objectives—Villamor Air Base and several television stations. Across the archipelago, over a hundred military detachments led by RAM members raised the inverted flag of revolt.

The assault on the palace was a disaster. Although Honasan had superior force and surprise, rebel troops wasted both advantages in three uncoordinated attacks that spectators dubbed "acoustic warfare" for the deafening bursts rebels fired into the air as they advanced. At 1:45 A.M., he ordered two hundred men, backed by several V-150 armored vehicles, to advance down the narrow curve of J. P. Laurel Street toward the palace. Then, fifteen minutes later, Lieutenant Colonel Reynaldo Ochosa (PMA '72) led his Sixty-second Infantry Battalion in an uncoordinated attack on nearby Nagtahan Bridge and was driven off by government fire. At 2:30 A.M., the rebel Fourteenth Infantry Battalion under Lieutenant Colonel Melchor Acosta, Jr. ('71) made another uncoordinated advance into the Malacañang area and occupied the same bridge that Ochosa had just vacated. Hearing a burst of fire from the palace, Acosta's men shot a few flares into the air and withdrew to Camp Aguinaldo.[59]

Though outnumbering the defenders, Honasan was unable to concentrate his troops and was pushed back from the palace gates.[60] Supported by a reserve of Marine units, the Presidential Security Group unleashed a fire that forced the RAM forces to retreat down the narrow roads fronting the palace. In the frustration of this street fighting, rebels ambushed a car carrying Benigno "Noynoy" Aquino III, wounding the president's son and killing three passengers.

At 4:00 A.M., after failing to take his major objective, Honasan led his troops in a retreat toward Camp Aguinaldo in nearby Quezon City, apparently hoping to regroup and gather defectors. Onlookers began jeering as the soldiers moved along the tight streets around the palace toward Nagtahan Bridge. The rebels fired into the crowds of unarmed civilians with sustained bursts from their automatic weapons, killing eleven and wounding fifty-four.[61]

After bluffing their way past the guards, the rebels fanned out to occupy Camp Aguinaldo. Instead of concentrating forces for a coherent defense, Honasan posted troops on a few gates, issued press statements, and sat by his home phone waiting for calls of support that never came. Pausing for an interview mid-coup, Honasan told a reporter, "We feel betrayed by our superiors." He added, "I have no ambition or thirst for power. After a new chain of command is installed, perhaps I can go back to Fort Magsaysay."[62]

At noon, a group of handsome young officers in pressed jungle fatigues broadcast an appeal to the people of Manila from the studios of Channel 13. In his scripted statement, Navy Lieutenant Robert Lee (PMA '81), a former member of Honasan's security group, claimed that the rebels had captured Camp Aguinaldo, held the countryside, and would soon control the government. They were not, he said, fighting for any faction or individual. He attacked the "overindulgence in politics which now pervades in society," and promised that a military regime would "initiate the struggle for justice, equality and freedom"—something, he said, "our senior officers had failed to do."[63] But, as news of the Nagtahan massacre spread, RAM's appeal quickly faded.

During the coup's first twelve hours, General Ramos scrambled desperately for reliable units. As the attack unfolded, the AFP command found that only 35 percent of the troops in the Manila area had remained loyal. At one point, the Second GHQ Battalion under Colonel Luisito Sanchez (PMA '67) refused to fire on rebel forces led by fellow PMA cavaliers, saying: "The Cory government is not worth dying for." At noon, an infantry advance into Camp Aguinaldo collapsed when the ad hoc array of progovernment troops, failing to make radio contact, fired on one another by mistake. As Ramos worked the phones to form another assault, many commanders expressed doubts about the loyalty of their troops. Like General Ver before him, Ramos turned to the Marines.[64]

Headquarters had readied its counterattack by late afternoon. Reinforced by airlifts from Mindanao only hours before, some three hundred Marines drove up EDSA with two tanks in the lead and halted before the high walls of Camp Aguinaldo. "We are Marines, we stick together," their commander, Major Emmanuel Teodosio (PMA '72), told the troops who circled before him. "We were honorable at EDSA because we stuck together. Today we are not fighting for one personality or faction. We are following orders as professional soldiers. Let us show that we marines have unity." While T-28 aircraft dive-bombed Camp Aguinaldo, the Marines broke through RAM's flimsy barricades and secured the area in just four hours. Escaping in a helicopter, Honasan and several aides landed at nearby Villamor Air Base and disappeared into the sprawl of Metro-Manila. General Ramos called the rebel leader "a coward" for fleeing the field of battle.[65] Once again, Honasan had failed to form a plan for retreat.

By the time the fighting died down later that night, 53 were dead and 358 wounded. Property damage totaled P 450 million (U.S. $22.5 million), and the cost to the business climate was incalculable. "They would rather reduce the country to rubble now," said Rex Robles of his fellow RAM allies, "and see it rise again than watch its current slow death."[66]

The coup was a disaster for RAM. The massacre of civilians at Nagtahan, in the words of the Davide Commission, did "irreparable damage to the rebels' reputation and credibility."[67] Though apparently spontaneous, the massacre nonetheless reflected RAM's long-held belief in the utility of violence and was strikingly reminiscent of their plan, framed in January 1986, for firing squads to quell any opposition to a postcoup junta.

Those minutes of terror, whether planned or not, cost the rebels popular support. Only weeks after the coup, a public opinion poll showed 46 percent expressing strong disapproval of Honasan versus an approval rating of 43 percent for General Ramos. The RAM leader maneuvered to repair the damage in the following weeks. On September 5, he told a national radio audience that "the reported death of civilian bystanders was regrettable and inexcusable, but was never condoned by our group." He condemned the Commission on Human Rights, saying its "blanket authority to conduct a witch hunt within the military" was a prime cause of the coup. Several days later, he gave an interview to journalist Leonie Pagulayan, denying any role in what he called "that atrocity," moderating his Gringo image to show humility.

> QUESTION: You and your group have been accused of firing on civilians in the Malacañang area, and causing casualties.
>
> HONASAN: No matter what information has been given the Filipino people, please believe me when I say that we did not commit that atrocity, that firing on civilians in the Nagtahan area.... The truth is that the first ones to open fire were some armed men riding in a car whom we believe were companions of those [government soldiers] inside Malacañang, with snipers and other armed groups in buildings near Nagtahan Bridge. They opened fire on us, and we soldiers naturally took cover and dropped to the ground. Civilians cannot do this without training.[68]

In retrospect, the coup's collapse exposed Honasan's limitations as a commander. The key to a successful coup d'état in Southeast Asia is a regular unit, such as the Thai First Division, with integrated armor that can concentrate force quickly to seize the capital, site of population and power. Ignoring this imperative, the colonel drew two thousand troops from some twenty-five units and then tried to form an ad hoc concentration on the day of the coup. By fragmenting his forces for assaults on secondary targets, he denied himself the strength to take the palace, the operation's main objective. Overestimating the capacities of infantry, the colonel failed to recruit any armor, essential in breaking through well-defended positions such as the palace gates. While Honasan fumbled for an order of battle on the march, General Ramos was able to mobilize the Marines, a coherent force, for a counterattack that smashed the disorganized rebel columns.[69]

The composition of his forces confirmed that RAM's political base was narrowing. Instead of the senior officers of past coups, Honasan could only mobilize classmates or former underclassmen. Of the 158 officers involved, 65 percent were captains and lieutenants. Above all, this August coup was a Class '71 outing. Classmates led assaults on the main objectives—palace, headquarters, television, and air base. In the critical two-hour battle for the palace, Honasan had relied on three lieutenant colonels, all alumni of classes '71 and '72. Outside the capital, it was, with few exceptions, his classmates who rebelled.[70]

In his after-action report on the coup to the president, General Ramos framed his explanation of its causes around this single factor. "The rebels' mentality is a mystery in itself," he wrote. "In the words of a PMA classmate, Honasan's . . . thinking reflects a mind in a state of madness, especially when he asserted on one occasion that 'only their class can save the country now.'"[71]

While the government had reason to be encouraged by the coup's outcome in Manila, events at the PMA showed strong underlying support for rebel ideals among the next generation of officers. Indeed, these years of ferment at the academy, 1986 to 1988, would produce a new cohort deeply committed to RAM's radical vision of social change.

During the people power uprising against Marcos, the entire cadet corps, in the words of the Class '87 yearbook, had "pledged . . . support" to the rebels and the news of Marcos's flight made them "jump with joy." Within weeks, RAM leaders Honasan and Kapunan appeared on campus to brief the corps "on the new . . . policies of the New Armed Forces of the Philippines," in effect, legitimating their rebellion and celebrating its heroism. Even after their dismal God Save the Queen coup, RAM's leaders still had the admiration of these idealistic cadets. "We looked upon Honasan and Kapunan as the heroes of EDSA," explained Ruben Basiao, baron of Class '88. "We didn't know EDSA was a coup."[72]

In the months before the 1987 coup, several members of Class '71 active in RAM joined the academy's staff and won the cadets' respect. In March, five months before the coup, the academy's isolation from Manila's turmoil was shattered when a bomb erupted during graduation rehearsal, killing four and wounding thirty-seven. When the rebels attacked Manila on August 28, the cadets alarmed the academy authorities by mounting a silent demonstration of support. "In behalf of the Cadet Corps, Armed Forces of the Philippines," said Cadet Allen Paredes ('88) at noon on Radio Bombo Baguio, "we hereby affirm our stand by the principle of anticommunism the forces of Colonel Honasan are fighting for." After voting by company to support the coup, the entire corps, all 863 cadets, donned combat gear and, in a statement read over Radio DZWT at 4:00 P.M., declared their "open support for the rebel cause."[73]

Inspired by a midnight briefing from two members of Class '71, Lieutenant Colonel Kapunan and Assistant Commandant Nelson Eslao, the cadets boarded trucks and headed down the driveway to occupy the city. Guards at the main gate advised them that the coup in Manila had collapsed. The cadets executed a quick about-face back to barracks.[74]

After a two-day suspension of classes, the corps was disciplined—ninety days confinement for all cadets and ninety punishment tours for first classmen. Such sanctions did not dim their rebel sympathies. In their yearbook published three years later, Class '90 insisted their participation had ended "only after we had our dialogue with Vice President Laurel . . . and other government officials with our demands being fulfilled." Their account closed with a defiant invocation of RAM's battle cry: "One thing is sure, 'our dreams shall never die.'"[75]

SHIFTING ALLIANCES

In the coup's aftermath, Honasan's fortunes changed with surprising suddenness. To rebuild his shattered movement, the colonel expanded his base beyond the original RAM membership, reaching out to both revolutionary left and right. In October 1987, the Manila press reported a formal alliance among rightist groups—Honasan's RAM and a coalition of Marcos loyalists called Soldiers of the Filipino People—under the abbreviation "RAM-SFP," which would be spray-painted on tanks in their next and most dangerous coup. After his alliance with the lees of the old dictatorship, Honasan's public opinion rating plunged to 8 percent favorable and 74 percent unfavorable.[76]

Honasan tried to repair the damage in his frequent encounters with the press. In an interview "from the underground" with reporter Criselda Yabes, he defied the government to capture him in tones evocative of his "Gringo" image. Referring to the Clint Eastwood police character, Honasan confessed to his "Dirty Harry" fantasy of capturing General Ramos after a victorious coup d'état. "What will I do? What will I say first? General Ramos, make my day," he said laughing faintly. "I also know where General Ramos eats Japanese food. He does not go there anymore," he added roaring. "He knows I know," he concluded to a chorus of laughter from followers.[77]

As regular officers used to steady pay and logistics support, the rebels did not adjust well to life in the underground. In December 1987, Honasan was captured, hiding under a maid's bed in the same Manila villa where he had met reporter Yabes.[78] To prevent his escape, Honasan and the other key RAM members were jailed on a navy ship anchored in Manila Bay. With his arrest, only eighteen rebel officers remained at large under a deputy, Captain Boy Turingan, who lacked his charisma. The rebel movement seemed to be expiring. Then, only four months later, Honasan won over his jailers, members of the elite Navy commandos. "I told them," he recalled with characteristic bravado, "to shoot me when I attempted to escape . . . or to join me." Together, they sped off in a rubber boat into the urban underground, where Gringo began organizing for a third coup attempt.[79]

In his effort, Honasan would have the support of a new rebel faction, the Young Officers Union (YOU), organized in August 1988. After RAM's dismal August 1987 coup, the movement's junior officers had grown critical of Honasan's leadership. Realizing the "political necessity" of a new group, Lieutenant Colonel Vic Batac encouraged his "mimeograph boys" in constabulary intelligence to form the YOU. Its aggressive radicalism appealed to younger academy alumni, classes '78 to '88, and soon won active support from over 60 percent of this latter class. Since its leader, Lieutenant Diosdado Valeroso ('82), was a Batac protégé and a former RAM member, the two organizations would remain closely allied.[80]

While Ram courted allies among the powerful, YOU sought a dialogue with the revolutionary left. These younger rebels had imbibed a mix of nationalism and socialism from Nilo Tayag, the Communist Party's former secretary general, who had served out his parole from prison in Vic Batac's office during the dictatorship. Abandoning Marxism, Tayag had become an advocate of Marcos's "Filipino ideology" and stayed close to Batac after the regime's downfall. When YOU formed, Tayag was enthusiastic, believing that with 250,000 potential recruits from the armed forces it would soon become a major revolutionary force.[81]

While these younger officers embraced socialism, Honasan's faction, in the months before their bloody 1989 coup, came to a violent praxis that defied such simple classification. Inspired by the Indonesian army's slaughter of a million civilians in 1965, the colonel now envisioned his next coup as a cleansing maelstrom of mass violence that would give birth to a new, disciplined nation. "Honasan said the problem of the EDSA and subsequent revolutions was that they had been achieved . . . with few casualties," a U.S. Embassy attaché reported after dining with Honasan in February 1989. "The 'price' was too cheap. He said the Filipino people needed a national catharsis of great proportions, so they would wake up and be disciplined." A revolution with "half a million casualties" could, Honasan felt, make the people "frightened" enough for fundamental change.[82]

Honasan held a press conference at a Manila safe house to explain his theory of transformative violence only ten days before his 1989 coup. "We have learned our lessons and we have learned them well," he said. "Looking back to August 28 [1987], we were all willing to die, but we were not willing to kill." Next time, he warned, there would be a "bloody" rebellion "if this is required to generate the necessary cathartic effect, so that we can once and for all go to zero and begin (again)."[83]

In the years between RAM's two major coups in 1987 and 1989, the Aquino government tried to preempt the threat by courting the armed forces. In the aftermath of the 1987 coup, Congress raised military appropriations to double the salaries of middle-echelon officers. Instead of cutting the bloated Marcos rosters, Aquino raised troop strength by 8 percent to a historic high of 171,000. Abandoning her soft policy of conciliation, she ordered the AFP to wage "total war" on the communists.[84]

Military abuses revived as the president muted her advocacy of human rights. Between 1987 and 1990, her Human Rights Commission produced only eleven convictions from 7,994 complaints. By the end of her term, Amnesty International concluded that the military was "beyond the reach of law." Despite these concessions, she failed to win support from a majority of officers who believed that her top officials were "leftists or communist sympathizers."[85]

With few allies in the all-male hierarchy, Aquino, as a civilian and woman, faced formidable difficulties in building support. Instead of reaching out widely to a factionalized military, she had, from the outset, narrowed her potential base by allying with the faction under her chief of staff, General Ramos. Though by no means an ideal solution, their alliance, over the long term, assured both her survival and his succession to the presidency. In January 1988, after defending her from six coup attempts, the general called in his debts, pressing Aquino to appoint him defense secretary and promote his protégé, his "carbon copy," General Renato de Villa (PMA '57), to succeed him as chief of staff.[86]

Only one man blocked Ramos's two-step progress through the defense department into the palace—the incumbent defense secretary, Rafael Ileto (USMA '43), a fellow West Pointer who, in Ileto's own words, had treated Ramos "as a father would a son." Despite a record of integrity, Ileto sat astride a generational juxtaposition of Classes '43 and '71 that made him vulnerable to accusations of disloyalty. As a member of PMA '43 before entering West Point, the defense secretary was classmate to both Romeo Honasan, father of Gringo ('71), and Gonzalo Batac, father of RAM strategist Vic ('71). "She wanted me to attack Gringo," said Ileto of the president. "I like Gringo and am close to him. His father and I were in the same PMA class and fought against the Huks in Bicol. We were promoted to lieutenant colonel by Secretary Magsaysay on the battlefield in Bicol on the same day. So I picked Gringo's father to be the godfather of one of my daughters."

In January 1988, the president summoned Secretary Ileto to ask for his resignation and offer the salve of several lucrative appointments. Ileto was, in his words, "rather rude and refused her on the spot." But his deepest resentment was reserved for Ramos. "I never would have thought that a fellow West Pointer like Eddie Ramos would have done this to me," he remarked bitterly on the day of his dismissal. "All those months . . . he was maneuvering to get my job." More broadly, Ileto was concerned that senior promotions from a single faction "could pose a threat to stability" and warned, with prescience, that "another coup is possible at any time."[87]

By inaction and indecision, Aquino fostered conditions for another coup. Her administration failed to reform the armed forces to prevent future coups, regarding every one as the last. By 1988, her diffident administration had plunged Manila, the site of any potential coup, into a miasma of crippling power failures, rising fuel prices, and severe unemployment that, economists warned, "may paralyze the people's resolve to support the present government."[88] Apart from these long-term trends, the president refused, for two months before the 1989 coup, to allow Marcos's body home from Hawaii for burial. For a president who had come to power as a grieving widow, such hardness seemed vindictive, producing what one commentator called a "growing isolation from the masses." Her approval rating slid downward from its 1986 peak of 78 percent to just 38 before the 1989 coup.[89]

When this coup finally came in late 1989, the public would waver, denying both sides a clear mandate in its first, critical hours. Then, on day two of the fighting, as people grasped the seriousness of the situation, thousands would rally to block rebel advances on Manila, and, at critical junctures, individuals would take risks to slow the coup. But the RAM rebels, blinded by their inflated sense of personal power, had misread Aquino's declining popularity as their rising support. Indeed, a survey taken in the coup's aftermath would find that an overwhelming 82 percent disapproved of the rebels.[90]

CHRISTMAS COUP

In December 1989, Colonel Honasan launched a coup that, for the first time in five attempts, came surprisingly close to capturing the palace. Although their essential strategy did not change, RAM's leaders had profited from past failures to improve both planning and execution.

The Davide Commission found a remarkable continuity in coup leadership. As before, Honasan was in charge of the overall planning, while "operational, tactical, and implementation responsibilities" were the work of his old RAM core group—Vic Batac, Red Kapunan, Boy Turingan, Tito Legaspi, Billy Bibit, and Abe Purugganan. The RAM plotters greatly expanded their base of military support by an alliance with the pro-Marcos officers of the SFP. Through the loyalist General Zumel, the coup group recruited key Marcos supporters still on active duty, notably Lieutenant Colonel Arsenio Tecson and Lieutenant Colonel Romelino Gojo—both former aides to General Ver and classmates of his son Irwin ('70). Most importantly, the Marcos loyalists provided funding. Throughout the preparations, Honasan met regularly with Enrique Cojuangco, the brother of Eduardo "Danding" Cojuangco, the powerful Marcos crony who returned from exile in California only six days before the revolt. Moreover, Marcos liaisons within the loyalist movement, such as Luis Tabuena, delivered up to five million pesos in cash for each rebel general and promises from Imelda of ten million more if the coup were successful.[91]

Honasan's boldness during the planning stage again compromised security. Acting as "deep penetration agent" for the AFP chief of staff, General Alejandro Galido (PMA '58), head of the Southern Luzon Command, had asked an aide to make contact with the RAM underground after Honasan's 1988 escape. These overtures bore fruit and, in January 1989, Enrique Cojuangco drove General Galido to a meeting with Honasan at a Makati parking structure. As they sped down the South Expressway, Honasan, insensitive to General Galido's ulterior motive, revealed his plans for a coup and invited him to join. On the eve of the coup, when the plotters phoned Galido to advise him that "H-hour" was set for midnight, he briefed AFP headquarters on the rebel plans instead of joining them.[92] Honasan's hubris had cost him the most critical element of success in any coup—surprise.

Despite these blunders, the rebels still mounted a carefully planned assault that, if properly executed, could have captured the palace. Most importantly, instead of fighting the Marines, Honasan made them his main strike force. As a self-contained heavy infantry with eight thousand troops, armor, and artillery, the Marines were the ideal coup or countercoup force. In both his 1986 and 1987 attempts, Honasan had found himself inside Camp Aguinaldo confronting disciplined Marine battalions that were—under both Marcos and Aquino—the palace's praetorian guard. After his 1987 defeat at their hands, Honasan had gained a sense that his third cousin, Lieutenant Colonel Gojo (PMA '70), the Marines' operations chief, was open to persuasion. For fourteen years, Gojo had served in the Presidential Security Command, where he was known as General Ver's sidekick. Now Honasan, with the support of the Marcos loyalists, convinced Gojo that the new coup's chances of success were high and played upon his contacts to win other Marine commanders.[93]

Honasan had won converts among the Scout Rangers from his days with this elite unit at Fort Magsaysay. Again, the loyalist alliance was useful in winning a key officer, Captain Danilo Lim, loyalist General Zumel's former aide-de-camp at the PMA. In effect, Honasan had made the AFP's most effective countercoup units, the Rangers and Marines, his own coup force.[94]

For the first time, moreover, RAM finally succeeded in mobilizing a broad spectrum of officers. In each of their past attempts, the rebels won the support of a single echelon: majors and colonels in February 1986; generals in November; and lieutenant colonels in August 1987. Now Honasan would field 523 officers from all echelons—214 lieutenants, 155 captains, 58 colonels, and 7 generals. While Class '71 had provided 20 percent of all regular officers in the 1987 attempt, this Christmas coup had officers from more than thirty PMA classes. Beyond the RAM core of thirty-five alumni from classes '71 to '75, some eighty-two junior officers from '81 to '85, many of them members of the Young Officers' Union, also joined. Significantly, the academy supplied only 37 percent of officers for the 1989 coup, indicating Honasan's first inroads into the ranks of the reserve-officer integrees.[95]

Despite this expanded support, the original RAM core group was still in command. As in past coups, Class '71 would lead many of the key rebel units. On day three, AFP headquarters released a list of "leading putsch leaders" showing

Coup Leaders Identified by the AFP on 3 December 1989

Name	PMA Class	Faction	Service	Role in Coup
Col. Gregorio Honasan	'71 (baron)	RAM	Army	Commander
Gen. Jose Ma. Zumel	'59	Marcos	Air	Organizer
Com. Domingo Calajate	'60 (baron)	RAM	Navy	Unit leader
Gen. Jose Comendador	'59	RAM	Air	Cebu leader
Gen. Edgardo Abenina	'58	RAM	Constabulary	Cebu leader
Gen. Marcelo Blando	'60	RAM	Army	Unit leader
Capt. Felix Turingan	'65 (baron)	RAM	Navy	Unit leader
Lt. Col. Victor Batac	'71	RAM	Constabulary	Tactician
Lt. Col. Reynaldo Ochosa	'72	RAM	Army	Plotter
Lt. Col. Eduardo Kapunan	'71	RAM	Air	Unit leader
Lt. Col. Oscar Legaspi	'71	RAM	Air	Unit leader
Lt. Col. Franklin Brawner	'71	RAM	Army	Plotter
Lt. Col. Rafael Galvez	'71	RAM	Rangers	Unit leader
Lt. Col. Luisito Sanchez	'67	RAM	Rangers	Unit leader
Lt. Col. Tiburcio Fusilero	'71	RAM	Constabulary	Cebu leader
Lt. Col. Romelino Gojo	'70	RAM	Marines	Unit leader
Lt. Col. Arsenio Tecson	'69	Marcos	Army	Unit leader
Capt. Proceso Maligalig	'69	RAM	Navy	Plotter
Maj. Abraham Purugganan	'78	RAM	Rangers	Unit leader
Lt. Col. Alexander Noble	'69	RAM	Army	Unit leader
Maj. Domingo Oliveros	none	RAM	Army	Unit leader

Source: *Philippine Daily Inquirer*, 4 December 1989.

seven of the twenty-one were from Class '71. Significantly, all but one were PMA alumni, indicating that these ties were still the foundation of the rebel movement.[96]

In contrast to the poor planning of 1987, the rebels had secured all the elements for victory in their Christmas coup: a coherent strike force, superior forces, and control of the air. That this attack, with such overwhelming advantages, would stall at the palace gates is remarkable, and attributable, in retrospect, to the distinctive mentality of its leadership. Although Honasan had sufficient forces to capture the capital, he would divert them from key military objectives to symbolic actions. As in August 1987, RAM's theatricality became its fatal flaw. Honasan and his classmates still suffered from that superman sense of boldness that, at critical points, clouded sober tactical judgment.

Honasan unleashed his attack at midnight on 1 December 1989 with superior force and near-precision. Since Manila had been stripped of troops to clear the capital of potential rebels after the 1987 coup, AFP headquarters held only a small defense force of constabulary and Marines.

As planned, the rebels struck decisively at their key objectives—Villamor Air Base, Sangley Point Naval Station, and Fort Bonifacio. Slipping into the city from camps across Luzon, the rebel Rangers were formed up and battle ready by 8:00 P.M. under the command of Honasan's close ally, Major Purugganan. Massed at Fort Bonifacio on Manila's southern edge, the Rangers and Marines were by far the strongest force in the capital. At midnight, an armored column of rebels drove out of Fort Bonifacio, crossed over the South Expressway, and attacked nearby Villamor Air Base. Within the hour, a force of seven hundred Marines under Lieutenant Colonel Gojo assaulted air force headquarters, while the Rangers captured the nearby runway, which had sixteen helicopters lined up, combat ready. But it was an illusory victory. As it turned out, their pilots would refuse to fly and the air force commander defended his headquarters compound, denying the rebels effective use of the base.[97]

In this first wave of attacks, Lieutenant Colonel Kapunan and Lieutenant Colonel Batac led eighty Marines and three armored vehicles in seizing another key objective, Channel 4 in Quezon City. Kapunan set up a command post at 4:00 A.M. in the studio penthouse to broadcast twelve videocassettes with messages from Honasan. But the station manager, before slipping out, had disconnected a critical mechanism, blocking any transmission. Instead of abandoning the studio, RAM's leaders wasted precious hours trying, without success, to broadcast Gringo's image.

Again, as in 1986, television transmitters became a bloody battleground. When government troops approached Channel 4 around 7:00 A.M., two rebel T-28 bombers strafed the convoy, killing a number of civilians and soldiers. Two hours later, rebel officers moved a full Marine battalion out of Villamor Air Base to Channel 4, where they remained, tied down defending a marginal objective, for the balance of day one. For a full twenty-four hours, the most critical period in the coup, the rebel command wasted valuable time and resources—aircraft, armored vehicles, and an infantry battalion—to hold a television studio. Had these forces been used to attack Malacañang Palace, the decisive objective, the coup's outcome may well have been different.[98]

Lieutenant Colonel Rafael Galvez ('71) and his Scout Ranger battalion occupied much of Fort Bonifacio's sprawling grounds in these first hours. Although they controlled most of the area, army artillery and armor units refused to join the revolt, denying the rebels the firepower they would need for their planned assault on Camp Aguinaldo.[99]

South of the city at Sangley Point, Lieutenant Colonel Tito Legaspi ('71) and Captain Boy Turingan ('65) led a column of two hundred mutineers from the Sixteenth Infantry Battalion in a sudden seizure of the naval air station. After overwhelming the guards on the main gate just after midnight, the rebels drove straight to the flight line, capturing twelve T-28 fighters and several helicopter gunships. With key points occupied, Navy Captain Danilo Pizarro ('63), formerly captain of Marcos's presidential yacht, marched his rebel forces into navy headquarters to demand, without success, the cooperation of Fleet Commander Proceso Fernandez ('59). At 11:40 A.M., five hundred rebel soldiers under Lieutenant Colonel Arsenio Tecson ('70), a former aide to General Ver, docked at Sangley's pier after crossing Manila Bay on a chartered fishing boat, giving the mutineers absolute control over the base.[100]

The rebels had, in the coup's first four hours, captured Manila's air defenses by occupying Villamor Air Base and Sangley Point. With most of the capital's air, naval, and ground forces now under their control, their tactical superiority was absolute. Surveying the coup chessboard from the vantage of historical hindsight, Aquino's position that morning was far more desperate than anyone outside the palace could have imagined. The city was largely undefended. Camp Aguinaldo, the seat of military command, was guarded by only a hundred soldiers. Yet, through what seems a mix of arrogance and incompetence, RAM's leaders would fritter away the hours that could have given them victory. "The failure of the rebels to launch a massive and sustained attack against . . . Camp Aguinaldo during the early hours of the coup enabled reinforcements . . . to get themselves organized," the Davide Commission reported.[101]

As the sun rose on day one, the rebels began wasting their resources in a series of flamboyant, symbolic moves. Instead of using their superior ground forces against key objectives, the rebels launched aimless air attacks. At 6:45 A.M., just as an emergency cabinet meeting was breaking up, rebel Colonel Tito Legaspi led three T-28 bombers and a helicopter in a series of rocket attacks on the palace grounds, sending officials scrambling for cover. The president herself had taken refuge in the basement near the racks of Imelda's shoes, and rebel pilots fired rockets directly at her hiding place. These bombings, the Davide Commission concluded, "were apparently intended to kill President Aquino."

Simultaneously, a hundred rebel soldiers, backed by civilian mercenaries armed with antitank weapons, set up roadblocks outside the palace, seeking to isolate it from reinforcements. The apparent leader of this feeble assault was Colonel Alexander Noble ('69), the former deputy chief of the palace Security

Group, who later admitted that he was willing to kill the president "if necessary" to prevent her from taking refuge in the U.S. Embassy.[102]

At 9:15 A.M., rebel aircraft, in acts of pointless violence, bombed the constabulary headquarters in Camp Crame and Chief of Staff de Villa's home in Camp Aguinaldo, sending up spectacular clouds of flame and smoke.[103] While the RAM leaders engaged in aerial arson, their infantry stood idle, failing to attack the only objectives that could bring them victory: Malacañang Palace, the seat of government, and Camp Aguinaldo, the headquarters of the armed forces.

Throughout the day, the rebel position weakened. For hours, a potent strike force of 350 Scout Rangers just waited. "As early as 4:00 A.M. Bonifacio was ours. *Walang nakagalaw* [No movement]," recalled Major Purugganan, leader of the rebel Rangers. Without any contingency plan, they were waiting for General Marcelo Blando ('60) to arrive from Fort Magsaysay, north of Manila, and "take his seat" as overall commander of the rebel army. "None of us had the 'personality' to do that. Blando was needed there," explained the major.[104]

After committing himself to Colonel Honasan before the coup, General Blando now hesitated, leaving the Rangers leaderless. At 1:30 P.M., nearly fourteen hours after the coup had started, the general was forced to engage in deception to get his troops moving, telling them, "It will be useless for us to go to Manila if we end up joining the rebels." Then, while driving down the highway at the head of two full battalions, Blando found that they still "refused . . . to fight the government forces." Unable to act for either side, his powerful combat force took a neutral position by driving into the Greenhills shopping center and occupying the parking lot.[105]

By midday, the tide began to turn. At noon, government forces captured Channel 4, denying the rebels their media outlet. An hour later, a lone F-5 fighter from Basa Air Base made a series of perfect bombing runs over rebel aircraft on the runway at Sangley Point, damaging the tail sections but leaving their engines intact. Angered by rebel bombing of Manila, the pilot, Major Danilo Atienza, was determined to disable their aircraft. On his last pass, the sudden eruption of a nearby fuel dump engulfed his F-5 and he crashed fatally. Thinking that rebel ground fire had downed their comrade, two more fighters from the major's squadron blasted the runway repeatedly with bombs and cannon fire, destroying all aircraft and sparking a sudden panic among rebel forces.[106]

By the time two U.S. Air Force Phantom fighters made a low pass over Manila at 2:04 P.M., the RAM leaders had already lost control of the air. At dawn, when rebel aircraft had first bombed the palace, U.S. Ambassador Nicholas Platt phoned to offer support. Concerned at the loss of air superiority, Defense Secretary Ramos, acting on the president's orders, made a series of calls to American officials requesting "persuasion flights by U.S. aircraft in order to deter . . . hostile (rebel) aircraft." After the embassy forwarded his request to Washington, the U.S. National Security Council endorsed it to President Bush, who approved while midair on Air Force One bound for the Malta summit. Two Phantom fighters took off from the U.S. air base at Clark Field at 1:51 P.M. and made a short flight over Manila thirteen minutes later, without firing any weapons or engaging the rebels in any way.[107]

As it turned out, these flights had a limited impact on the coup. Even though RAM had already lost its control of the air, the U.S. decision to play deus ex machina probably demoralized the rebels, who seemed to both fear and resent this unscripted American intrusion. Whatever actual impact they might have had, these flights later provided the RAM leaders with an excuse for defeat. "We were about to take over the government," rebel General Edgardo Abenina later complained. "Then the U.S. warplanes appeared. We simply cannot hope to win against the strong power of the United States Air Force."[108]

During the day, moreover, government checkpoints ringing the city began to hold and, reinforced by thousands of civilians, barred rebel reinforcements from entering the capital. After declaring himself for the revolt on Radio DZRH in Cagayan, Governor Rodolfo Aguinaldo led a convoy of five hundred militia toward Manila but was stopped at a checkpoint in Nueva Ecija.[109]

Without effective leadership, the rebel coalition cracked along factional fault lines and officers began refusing commands. Around 2:00 P.M., RAM's Captain Turingan, shipboard in the middle of Manila Bay, radioed the SFP's Colonel Tecson, then holding Sangley Point with five hundred troops, and ordered him to attack the Western District police headquarters, which was responsible for Malacañang's defense. Tecson refused, insisting that he would take orders only from General Blando, who was still moving down the highway toward the city. When government troops drove through Sangley's main gate an hour later, his five hundred rebel troops were already scattering and surrendered without resistance.[110]

The Presidential Security Group, backed by seven armored vehicles, counterattacked and cleared the precincts around the palace of rebel forces by early evening. Tied down all day at Villamor Air Base, a secondary objective, Lieutenant Colonel Gojo decided on his own that it was time to attack the seat of military power at Camp Aguinaldo. At 8:00 P.M., he ordered his Marines to smash the helicopters on the runway and move out. After hiking for hours northward through Manila's side streets, some six hundred Marines reached the White Plains suburb outside Camp Aguinaldo well after midnight. "They could have made us attack [Camp] Aguinaldo," Gojo later complained of the RAM command. "We should have attacked Aguinaldo (that day)."[111]

On day two of the coup, the rebels maneuvered slowly, allowing the government time to mobilize reinforcements. Without a clear plan, RAM commanders improvised tactics on the march. After waiting pointlessly inside Fort Bonifacio for thirty-six hours, Major Purugganan led his rebel Rangers out of the camp at 1:00 P.M. and occupied the towers of the Makati financial district—another detour on the road to victory. The gung-ho major, still convinced of his tactical skill, told a journalist a month later: "Now, if you take note, there are 40 embassies there and . . . 109 multinational centers in Makati. And . . . the place is like a fortress." By 2:00 P.M., some five hundred Rangers had occupied the nation's financial district and posted snipers on the rooftops.[112]

At Enrile's instigation, the opposition Nacionalista Party held a press conference within an hour at the Intercontinental Hotel, the epicenter of the new Ranger redoubt. The Nacionalista leaders denounced the government for "giving cause to the military rebellion" and charged, falsely, that U.S. fighters had "shot rockets that killed innocent Filipino civilians." Blas Ople, Marcos's former labor minister, then called on the president to resign, while Enrile expressed optimism that the coup would triumph. From Hong Kong, Vice President Salvador Laurel condemned President Aquino's use of U.S. air power as "a desecration of the . . . constitution." In a noon television address, Aquino answered her critics, telling the rebels "to surrender or die."[113]

On day three, it was the coup that died. In the darkness after midnight, the rebel Marines finally attacked Camp Aguinaldo. As they advanced from the surrounding subdivisions where they had been sitting idle for a full day, the battle for Camp Aguinaldo raged in earnest until dawn. After an intense barrage of artillery and mortar fire from positions in the residential suburbs to the east, a rebel Marine tank smashed through the camp gates at 1:45 A.M. Then, a thousand rebel soldiers fought their way across the camp grounds to just thirty feet from AFP General Headquarters. The opposing forces blasted each other in close combat for nearly two hours in the dark, "shaking the ground and lighting up the sky." Gradually, the two thousand defenders pushed the rebels back with point-blank fire from machine guns, recoilless rifles, mortars, and three 105 mm howitzers. In this heavy bombardment, a rebel V-150 personnel carrier was destroyed and an LVT tank took a direct hit from a 90 mm recoilless rifle, igniting flames that killed all of the eighteen-man crew. As the rebels fell back into the White Plains subdivision at the edge of the camp, sunrise lit the government's counterattack. Helicopter gunships and F-5 jets bombed and strafed the Marines in the early light, while government ground forces swept the rebel troops away from the approaches to the camp.[114]

Pinned down by heavy fire inside the camp, some two hundred of the rebel Marines held out for four hours, refusing to raise a white flag. To force their surrender, General Rodolfo Biazon ('61), commander of the capital's defenses, warned that was he ready to "wipe them out." This was a poignant confrontation. As a former Marine brigade commander, General Biazon, as he put it, knew "a lot of these guys . . . by their nicknames." And they knew him well enough to take his threat seriously. The Marines surrendered at 11:00 A.M.[115]

The coup's commanders were quickly rounded up: General Blando surrendered; Lieutenant Colonel Gojo was arrested at a checkpoint; and Lieutenant Colonel Tito Legaspi was caught trying to escape. Surveying the smoking battlefield at Camp Aguinaldo later that day, General Ramos announced: "It is now 1400 hours, the third day of December 1989. We can declare that the attempted coup by rebel soldiers has been crushed. What remains to be done is the mopping up."[116]

But it would take another three days to cleanse the Makati financial district. After placing snipers atop twenty-two high-rise buildings and positioning machine guns to sweep Makati's wide boulevards, the rebel soldiers waited for an attack. On the morning of December 3, a formation of government tanks advanced down EDSA, but rebels rained mortar and machine-gun fire from their posts atop the Intercontinental Hotel and nearby buildings. As the tanks approached the Manila Garden Hotel, a rebel soldier darted out and fired a 90 mm recoilless rifle at a Scorpion tank, disabling it with a direct hit, while rebel snipers maintained a heavy covering fire. Chastened by the force of this firepower, government troops pulled back to a secure perimeter.[117]

At the Intercontinental Hotel on December 5, the triumvirate commanding the rebel Rangers—Lieutenant Colonel Galvez, Major Purugganan, and Captain Lim—demanded Aquino's resignation. "This government is going to get a big spanking," warned Captain Lim (USMA '78). Outside, as government forces continued their cautious envelopment of Makati, sniping and firefights crackled, causing casualties on both sides.[118]

Negotiations began that night. After twenty-four hours of round-the-clock talks, Lieutenant Colonel Galvez finally agreed, over the militant opposition of Major Purugganan, that the Rangers would return to barracks.[119] The next day, December 7, hundreds of Scout Rangers, bodies swathed in ammo belts, paraded down the glass-curtain canyon of Ayala Avenue, the heart of the financial district. Waving, smiling, and shaking hands with spectators, they crossed EDSA to enter their barracks at Fort Bonifacio, looking more like victors than vanquished.

While Manila returned to normal, rebel forces continued to hold Mactan Air Base, closing Cebu International Airport, the country's second largest air terminal. The Cebu coup had begun on day one, when some three to four hundred rebels landed from Mindanao on a commercial interisland ferry, the *M.V. Sweet Pearl*. After evading a blocking force, they occupied Mactan Air Base. On day two, the leaders revealed themselves at a press conference: General Edgardo Abenina, a former regional commander cashiered after the August 1987 coup; General Jose Comendador, Second Air Wing Division commander; and Lieutenant Colonel Tiburcio Fusilero ('71), the mercurial RAM conspirator.[120]

The Cebu rebels refused to surrender, convinced that their revolt would spread spontaneously across the country. For nine days, government and rebel forces engaged in a tense face-off from their barricades at opposite ends of the bridge that arcs high over Cebu's harbor, linking the airport to the city. Although General Comendador raved on local radio about being "ready to die and ready to kill," not a single shot was fired. But the general held superior government forces at bay by threatening to blow up the thirty-three military and commercial aircraft grounded at Mactan—including two Boeing 747s, four F-5 jet fighters, and eleven combat helicopters.[121]

On December 9, after days of phone negotiations, the coup leaders, led by Lieutenant Colonel Fusilero, finally agreed to return to their barracks in Mindanao. That evening, as the rebel troops drove through the city in a cavalcade of cargo trucks, Fusilero, in sunglasses and a red bandanna, rode alongside on his 750 cc motorcycle, with General Comendador clinging to the tandem seat and seven other motorcyclists from his club, "The Cycluns," following in his exhaust. People cheered and the rebel soldiers waved their rifles while the column passed in a cacophony of high-pitched revs.[122]

President Aquino convened a mass celebration that same day beneath the steel statue of Our Lady of Peace that looms above EDSA at the symbolic center of the 1986 revolt against Marcos. With snipers on the rooftops and a combat helicopter overhead, Aquino addressed the throng of half a million chanting "Cory, Cory, Cory." The president attacked the coup's ultimate patrons: Vice President Laurel, Senator Enrile, and Marcos crony Eduardo Cojuangco. Calling the vice president by his nickname, she asked mockingly, "Should I even bother to mention that 'Doy' Laurel? Or should I just flick him away like a fly?"

According to one press account, "the crowd roared."[123] Despite reservations about her administration, popular antagonism to military rule evidently ran deep.

Some of the more reflective RAM leaders later admitted their lack of popular support. "When we launched that coup," an anonymous RAM officer conceded in a later interview, "we thought the people would follow. Everything pointed to it. Times were hard, and the people were bitter.... In 1989, we thought we were the spark. Nothing lit."[124] In a survey of five hundred PMA-trained officers for his dissertation, Colonel Ruben Ciron ('68), a RAM sympathizer, found 46 percent felt that the main reason for the coup's failure was lack of popular support. Significantly, support for an anticommunist coup among regular officers had slipped sharply from 62 percent before the December debacle to just 24 percent after.[125]

With a toll of 99 dead and 570 wounded, the coup was over. The economic damage was almost incalculable. The Central Bank released $230 million to cover panic withdrawals from the country's banks. Economic growth skidded downward from six to zero percent, crippling the country's economic recovery. Observers estimated property damage at a billion pesos and total economic losses at thirty billion, including two hundred thousand jobs that would not be created. The president proclaimed a state of emergency.[126]

Colonel Honasan again disappeared into his urban underground in the coup's waning hours. "We can see the dawn of our approaching victory," he said in a press release several days later.[127] Several weeks after that, thirty RAM leaders held a clandestine postmortem. In a political volte-face, the group cut ties to the right-wing loyalists of the SFP and allied with the left-leaning Young Officers' Union (YOU). Agreeing that they had failed to mobilize popular support, RAM renamed itself the *Rebolusyonaryong Alyansang Makabayan* (Revolutionary National Alliance) and issued a call for land reform, an end to U.S. bases, and legalization of the Communist Party. With these changes, RAM transformed itself from a military faction into a movement for social revolution.[128]

The Manila press was sharply critical of Honasan's pretensions. "Most of us know about his tank of piranhas, his pet python . . . , and the jar of dried Muslim rebels' ears that he used to show off to friends," editorialized the *Philippine Daily Inquirer*. On a similar note, presidential press secretary Tomas Gomez remarked pointedly that "these are the very guys who were guilty of the grossest human rights violations."[129] Manila television stations began broadcasting photos of the RAM leaders with prices on their heads.

In July and August of 1990, RAM launched a terror campaign in Manila, setting off six plastic charges, the first of more than fifty bombings linked to the rebels.[130] Barred from the military camps, the group no longer enjoyed easy recruitment through the chain of command and began seeking a new path to power.

THE MINDANAO ENCLAVE

At this low ebb, RAM's chief planner, Colonel Vic Batac, developed a new strategy called the "enclave concept." Under this novel doctrine, RAM would spark rebellion in isolated pockets about the countryside, setting off a revolts like a chain of detonations. Unable to hold outlying provinces, the government would, in theory, collapse and capitulate. By forming secure rebel zones, this strategy would overcome what Batac called "a tendency for forces committed to join to postpone participation until such time that a clearer picture has developed."[131]

Meeting secretly in Manila, the RAM executive—Honasan, Kapunan, Batac, Turingan, and Zumel—approved the strategy and decided to apply it in provinces where their local leaders were strong. Writing to a supporter in August, Batac showed what seems a naive faith in the fragility of the Philippine state. "We here believe that if you succeed in taking over Mindanao and establish a provisional revolutionary government and a separate armed force," he explained, "the rest of the country will just follow suit and we will have a bloodless coup."[132]

As it turned out, there would be just one enclave—the vast southern island of Mindanao. Unknown to the coup plotters, government agents soon captured RAM's documents and were ready to counter their final coup. In December 1989, the head of the National Intelligence Coordinating Agency (NICA), General Rodolfo Canieso, had told a Senate committee that rebel Colonel Alexander Noble (PMA '69) would soon launch "Phase III" of RAM's coup plan to divide the country by forming a junta in the Visayas and Mindanao. Significantly, the NICA chief calculated that only 30 percent of troops were divided between government and the rebels while the great majority of troops, some 70 percent, were fence sitters "being courted by the warring forces."[133]

The RAM coup followed both plan and prediction ten months later. While the rebels distracted the government with terror bombings in Manila, Colonel Noble, who had spent the past two years training tribal militia, moved constantly about Mindanao to prepare the RAM "enclave." In July, when the colonel fled into the forests of Agusan del Sur, he was soon joined by two hundred "fiercely loyal" Higaonon tribal militiamen, who fought off six army and Marine battalions for several months.[134]

In October 1990, Colonel Noble launched the coup by leading two hundred rebel troops in the capture of the 402d Infantry Brigade headquarters, twenty-six kilometers from Butuan City. The next day during lunch hour, the rebel chief, with flaring beard and magical amulets dripping down his neck, led a cavalcade of six armored vehicles and three trucks loaded with rebel soldiers into Cagayan de Oro City while crowds lining the streets cheered.[135]

But there were signs of weakness beneath the bravado. The crowds soon evaporated. Noble's bid for support from the Manobo and Higaonon was reportedly rejected by tribal leaders. Instead of sparking a mass uprising-cum-mutiny, Noble only attracted three hundred soldiers and six hundred civilians. One of the rebel leaders, Lieutenant Vicente Batac (PMA '86), the younger brother of RAM's tactician, later described the coup as "premature and haphazardly done." Noble surrendered two days after the revolt had started, telling reporters, "I gave myself up without any conditions."[136]

Chastened by defeat, Noble wrote his comrades with a dismal assessment of RAM's future. Their terror campaign was, he said, "double-edged" and "could alienate us from the people." Moreover, he said, "organizational bungles have negated the effectiveness of our forces nationwide," while the "classic coup . . . could be a thing of the past." Surrender was their only option. "Let's tone down our aggressiveness," he suggested, and "make a formal offer for talks," then "go for a general amnesty during bargaining." To dress up this demand, RAM should "harp on issue[s] of reconciliation, unity and peace."[137]

In the coup's aftermath, a survey of academy alumni found only 8 percent were satisfied with the civilian leadership. But the rebel colonels had failed, in five coup attempts, to devise tactics that could translate this dissatisfaction into action.[138] Their only option was, as Colonel Noble advised, to negotiate a "general amnesty" for their past crimes of torture, murder, and rebellion. Finding the Aquino administration unwilling to negotiate, Colonel Honasan would spend two more years in hiding before changing political circumstance would allow negotiations.

While RAM's leaders retreated into an urban underground of condominiums and take-out meals, the junior militants of YOU captured headlines. In April, Lieutenant Valeroso had led thirty YOU rebels in a midnight raid on the Manila city jail to release Lieutenant Colonel Billy Bibit, a RAM activist captured during the Christmas coup. Two days later, YOU published a full-page manifesto in Manila's newspapers, attacking the Aquino government as a "monster rapist" and promising to save the nation from the "wounds and scars left by the past" of colonial subjugation. When asked for a solution to corruption at an underground press conference, one of these radicals suggested: "We could impose the mandarin agenda—kill off one to intimidate the rest."[139]

But the young rebels failed to match rhetoric with action, producing a series of botched operations that indicated the slow, inexorable closing of the revolutionary option. As they grew more isolated within the military, YOU and RAM opened a dialogue with the communist left. Over the next three years, these conversations continued, in the underground and in prison yards, providing better arms for the terrorist left and a populist rhetoric for the military right.[140]

Such contradictory alliances signaled an unraveling marked by capture, surrender, and dwindling support. As arrests accelerated in 1990–91, revolutionaries of left and right found themselves in the same prisons, united by a common failure and endless free time. Inside Fort Bonifacio, a handful of military rebels led by Boy Turingan broke a hole in a dividing wall for months of basketball and conversation with captured communists—NPA negotiator Satur Ocampo, his wife Bobby Malay, and Rick Reyes. Reflecting RAM's nagging fears of prosecution, the group's ideologue, Alan Paje, proposed that "we agree to forget all about human rights violations." Others, notably Lieutenant Colonel Jake Malajacan (PMA '71), chimed in, blaming Marcos and saying that the soldiers were "merely implementors." As a torture victim, Ocampo objected, saying, "that's very difficult for the group, because we do not only speak for ourselves . . . but we also speak for the people . . . who have been victimized so many times."

As conversation turned to future coups and the enemies of radical reform, Major Purugganan, ever the impulsive radical, suggested, "Maybe the best solution is to line them up on the wall and shoot them down." But Ocampo, referring to his communist comrades, replied: "You know, we don't even plan to do that. We'll have to put them on trial and specify the crimes." The major, chastened and reflective, admitted, "We still have a lot to learn."[141]

In their five-year revolt, RAM's rebels had demonstrated the significance of military socialization, particularly its second, sensitive phase, when cadets become junior officers. Once internal restraints broke down, these dissident officers, through their privileged access to the nation's arsenal, could mobilize major coups that plunged the country into instability and recession. As these coups made clear, the rebel factions never grew beyond a minority within a few PMA classes. The majority of regular officers, still committed to civil supremacy, put aside their personal reservations about the regime to crush these coups. Even so, RAM still came remarkably close to taking power. Although they were defeated, the costs from this rupture in socialization were high—hundreds of dead and wounded, over a billion dollars in lost growth, and stagnation in the midst of Asia's economic boom. Over the next five years, the political price of ending their revolt, though less direct and dramatic, would prove just as high.

Chapter 9 Impunity

After six coup attempts in five years, the colonels were further from power than ever before. During the late 1980s, the Reform the Armed Forces Movement had launched more coups with less success than any army anywhere. They had exhausted every option within the tactical canon of the coup d'état—revolt of the generals, attack on the palace, assault on the capital, urban terrorism, and rural revolt. All had failed.

By the end of 1990, surrender remained their only option. Facing charges for crimes of murder, rebellion, and torture, the RAM colonels, like their peers elsewhere in the developing world, were determined to lay down arms in ways that would guarantee immunity against future prosecution. In the end, through a mixture of bluff and violence, they not only won an absolute amnesty but had also placed their leader in the Senate, the country's most powerful legislative body. Like its counterparts in Argentina and Chile, the Philippine military would win "impunity" for crimes and coups.[1]

As a recent phenomenon, impunity is a little-understood process with far-reaching ramifications. After military dictatorships collapsed around the globe in the 1980s, euphoria was soon tempered when uniformed criminals began manipulating weak states to escape punishment. At the Sixth International Symposium on Torture at Buenos Aires in 1993, delegates defined *impunity* as "the fact that, even in countries where dictatorship has given way to democratic rule, many torturers and other violators of human rights go unpunished." In some nations, the military wins impunity by direct negotiations and in others by forcing a political stalemate.[2] Expanding upon this narrow definition, torture therapists have reported that impunity prevents the full rehabilitation of their patients and contaminates the whole of society.[3] At the Seventh International Symposium on Torture at Capetown in 1995, delegates noted the "enormous emotional power" of impunity as societies are torn between "a long-term vision of reconstruction" and "a natural desire for justice."[4]

At a fundamental level, such a struggle often entails a debate over remembering and forgetting. Atavistic elements, usually within the military, urge forgetting in the guise of national reconciliation, while their victims demand an accounting. "The same powers that resorted to State terrorism are actively promoting collective oblivion," wrote two Argentine psychiatrists in 1996, "the usual way that the winners tell the story."[5] Since the "social body has been deeply wounded" by the years of repression, collective trauma, unless confronted through remembrance and redress, will, they argue, be "transmitted from generation to generation."[6]

Most visibly, this debate is carried on in tribunals, local and global, that seek justice for the victims of authoritarian rule. While the surface discourse revolves around points of law and fact, there is, embedded in these transcripts, a deeper debate over memory. To cite one example, South Africa's Truth and Reconciliation Commission seeks not retribution but confession to correct the historical record and thus shape collective memory.[7] Similarly, in Chile a democratic government published a report on the killings under General Pinochet; in Greece, the prosecution of military torturers "functioned as a catharsis"; and in Uruguay, where a plebiscite to rescind amnesty for torturers failed, public debate "allowed an inscription of all that history into the collective memory." By contrast, a society like the Philippines that tries forgetting can find itself trapped in the traumatic past.[8]

In this struggle between justice and impunity, the sites of conflict are many—the courts, the ballot box, commemorations, and cultural productions. By threatening the stability of a fragile democracy, military officers can win amnesty for grave crimes. To support these maneuvers, advocates of forgetting contest public representation of the past to alter the transcript of collective memory. Victims usually use public space, protests, and the arts for acts of remembrance. Collective memory is thus the site for a wide-ranging process of cultural construction and competition, a struggle to reconcile past and present in ways that complement rival visions for the future. So pervasive are these debates that they give new meaning to Ernest Renan's dictum that all nations are "constructed on the basis of great rememberings and great forgettings."[9]

The Philippines provides an example of extreme impunity. In the decade after Marcos's downfall in 1986, evasion of the traumatic past persisted, becoming a malaise that subtly shaped the political process. During her first months in office, President Corazon Aquino (1986–92) appointed four human-rights lawyers to her cabinet and seemed strongly committed to the issue. But battered by repeated coup attempts, she abandoned any attempt to prosecute the military for past crimes of torture and murder. Her successor, General Fidel Ramos (1992–98), transformed this impunity from a de facto to de jure status, bestowing the imprimatur of legality upon what had been, under Aquino, an ad hoc compromise. As chief of constabulary for fourteen years under the Marcos dictatorship, Ramos had commanded the elite units responsible for repression and torture. He dismissed the human rights issue as defense secretary in 1989, insisting that "93 percent of all cases against the military are baseless."[10] As president, he effected the integration of former torturers into society and their elevation to positions of power. More broadly, Ramos gave the military unprecedented civil authority, appointing five retired officers to his cabinet and a hundred more to senior positions in his government. Many had been implicated in Marcos-era repression, and their influence, over time, reversed Aquino's commitment to human rights and redress for the victims. After Ramos's vice president, Joseph Estrada, a movie star and Marcos loyalist, won the presidency in the May 1998 elections, his new government perfected the process of impunity by offering the dictator's surviving cronies both legal and symbolic absolution for their crimes.

In the country's rightward drift, commemorations and cultural productions recast the past, transforming torturers into heroes. Gradually, remembering became stigmatized as subversive, and consensus came to favor forgiving, and forgetting, in the name of national unity. Denied justice at home, some ten thousand Filipino torture victims filed suit against the exiled dictator in the U.S. courts. Between the poles of local impunity and global justice, Philippine politics emerged from this contested decade with the lingering paralysis of collective trauma.

The reintegration of rebel officers into society, unpunished, complicated the rebuilding of democracy. Just as these many coups dominated the Aquino administration, so the negotiations over their end would, less visibly, distort the political process under her successor, Fidel Ramos. Moreover, resolution of this revolt blocked military reforms, leaving the legacy of state terror untouched. In both the cost of their revolt and the complexity of its resolution, the RAM colonels showed the importance of military socialization for any society with a standing army.

IN PURSUIT OF RAM

After the failure of their last coup in 1990, the RAM colonels maintained their image of threat while maneuvering for a chance to negotiate. President Aquino, the target of their many coups, insisted on unconditional surrender and ordered her generals to effect their capture. The rebels were thus forced to survive in the urban underground until the 1992 presidential elections brought a more sympathetic administration.

The armed forces formed a special unit, the Counter Intelligence Command (CIC), with seventeen hundred troops, to track down the RAM fugitives. Led by officers screened for any ties to the rebels, the CIC hammered the rebel underground. During a massive operation in February 1991, the task force captured two top rebel leaders, Victor Batac and Abraham Purugganan, living comfortably in Manila. Not only was Batac RAM's so-called "brains," but Purugganan was an explosives expert held responsible for the fifty bombings that terrorized Manila in mid-1990. Reduced to just thirty-four officers and ninety-three men, RAM felt the losses and retaliated by reviving its bombing campaign, producing another round of terror.[11]

As the rebel organization collapsed, surrender negotiations accelerated. In September 1991, the AFP chief of staff, General Lisandro Abadia, met RAM leaders Eduardo "Red" Kapunan and Gregorio "Gringo" Honasan to reach what he called a "gentleman's agreement" that there would be no coups while contacts continued. Within six weeks, Kapunan staged a formal surrender in full uniform armed with a .45 caliber pistol, saying, "Our actions have been propelled by our wholehearted desire to be of service to the Filipino people." Within days, General Abadia had restored Kapunan's rank, saying that he wanted to be remembered as the "AFP chief of staff who united the military."[12]

Left and right joined in a chorus of protest. One brigadier general complained that the deal was "unfair to [those] who fought against them." Several days later, the Crispin Tagamolila Command of the communist New People's Army (NPA) issued a statement denouncing the RAM leaders as torturers. "Specifically," they charged, "Gringo Honasan, Red Kapunan, Boy Turingan, and Figueroa . . . plotted the deaths of Fr. Kangleon, Dr. Juan Escandor, KMU Rolando Olalia, Dr. Bobby de la Paz . . . and scores of others during and after Martial Law."[13]

In the May 1992 presidential elections, General Ramos, running with President Aquino's support, scored an upset victory, introducing a new variable into RAM's calculus. As a combat veteran, Ramos could assuage the martial masculinity that his predecessor had so clearly offended. With a career officer in Malacañang for the first time in the country's history, the idea of launching a coup to establish a military junta now seemed redundant, even to the most militant of rebel officers.

After his inauguration in July, President Ramos established the National Unification Commission (NUC) to negotiate the surrender of all rebel forces—Muslim, communist, and military. RAM was at the edge of extinction, with only twelve members in its underground, when the group's leaders began meeting secretly with negotiators in October. Nonetheless, their representative, Commodore Domingo Calajate (PMA '60), boldly issued "talking points" demanding major reforms such as "nationalist economic development."[14]

Through the good offices of the National Unification Commission, the rebels were home for Christmas. On 18 December 1992, after six years underground, General Jose Ma. Zumel, leader of the pro-Marcos loyalists, surrendered at the Manila Polo Club, insisting on his loyalty to the constitution and claiming that it was only Corazon Aquino's "vindictiveness and vengefulness" that forced him to rebel. Observing PMA protocol, Chief of Staff Abadia ('62) saluted Zumel ('59) and addressed him as "sir."[15]

When Honasan and his remaining followers finally surfaced five days later, President Ramos hailed them as "prodigal sons who had returned to the fold." In a noon ceremony at the University of the Philippines, Defense Secretary Renato de Villa led a delegation in full-dress uniforms to meet rebels in maroon polo shirts proclaiming the RAM slogan, "Our Dreams Shall Never Die." Both signed a preliminary accord stipulating a "complete cessation of hostilities" and the suspension of court-martial proceedings. On Christmas, the government released forty-three rebel officers, and RAM reciprocated by surrendering nine of its stolen light antitank weapons.[16]

Honasan used the media to recast his image during this season of reconciliation. In an interview with the *Chronicle*, the colonel, with a studied humility, admitted receiving movie offers but insisted that he would sacrifice profit to make the film "less cinematic and more substantial." Speaking with what the reporter called his "beatific smile," Honasan said that his only ambition now was "to grow old with my wife."[17] Six months later, however, the *Philippine Daily Inquirer* revived the human rights issue by publishing a list of the "top twenty" torturers under the Marcos regime, noting that many had recently been promoted. Among these twenty, five were RAM leaders, including Hernani Figueroa and Victor Batac. Indeed, number one on the list was Rodolfo Aguinaldo, who was, the paper said, "implicated in the torture of at least 27 detainees."[18]

Despite such revelations, RAM's leaders maneuvered the peace talks deftly toward impunity. In March 1993, both negotiating panels agreed on a qualified amnesty for "all acts committed in furtherance of rebellion" such as "murder, homicides, kidnappings." But after Honasan made veiled threats of further coups, Unification Commissioner Haydee Yorac recommended, at the end of the first round in July, an "absolute and unconditional amnesty to all rebels." Even so, RAM refused any discussion of their light antitank weapons as "premature," though they still had enough of these missiles to destroy the AFP's entire tank force.[19]

As the talks crawled through a first year into a second, RAM's flurry of press releases and protests indicated that they were seizing any pretext to prolong negotiations. Under the terms of surrender, they had a semisovereign status as belligerents, allowing them government protection and financial support. Instead of signing an agreement and fading into obscurity, RAM's leaders were using this public presence to launch their political careers. Through Rotary luncheons, television appearances, and feature films, they courted both public support and powerful allies. Gradually, the rebels moved steadily toward amnesty through a mix of public charm, closed-door recalcitrance, and periodic threat.

GOVERNOR AGUINALDO

In RAM's quest for impunity, Lieutenant Colonel Rodolfo Aguinaldo was the first to test the tolerances of the new democratic order and discover its capacity for political absolution. Only months after Marcos's fall, he began winning new notoriety as the warlord of Cagayan, an impoverished province at Luzon's remote northern tip.

Cagayan was best known as the birthplace and bailiwick of Marcos's former defense minister Juan Ponce Enrile. Indeed, Aguinaldo was first posted there as a junior constabulary officer in 1984 through Enrile's influence. During the people power uprising, Aguinaldo had led the famous "Cagayan 100" in RAM's assault on Marcos's television stations, and was rewarded with promotion to constabulary commander for Cagayan. As the communist presence in nearby mountains grew in 1986–87, Aguinaldo used defense funds to form a private army of twelve hundred Negrito tribesmen and surrendered NPA guerrillas. With an arms cache from the anti-Marcos coup, he had all the requisites for a provincial warlord. He soon built an independent economic base of logging, smuggling, and gambling, making him, in the words of the Davide Commission, the RAM leader who "earned the most from illegal activities."[20]

In this remote northern province, Aguinaldo bonded nightly with his militia in their rugged camp, drinking until dawn before blazing bonfires and firing off rifle bursts into the darkness. By early 1987, Cagayan was a redoubt for right-wing rebellion. During the short-lived August coup attempt, Aguinaldo seized the regional military command and broadcast his support for his RAM comrades—an act of treason that led to his removal as provincial commander.[21]

After resigning his commission in December, Aguinaldo broke with his patron Enrile to run as an independent for governor against the leader of a powerful political dynasty, Teresa Dupaya. Aguinaldo showed an unexpected flair for fiery populism, advocating land reform and public probity. "Throughout my adult life," he proclaimed on the stump, "I have always been fighting a bloody and senseless war. Filipinos killing brother Filipinos. What for? . . . The time has come to bring peace to our poor and weary people." Speaking before massive crowds chanting "Agui! Agui!" he delivered violent rhetoric that threatened all who stood in his path, even inspectors from the Commission on Elections (Comelec). "If a Comelec man does something foolish," he told journalists on election eve, "he will be the first to leave the earth. We have 300 firearms right here in Tuguegarao. At a given signal, we chop off the heads of anybody who is foolish."[22]

Aguinaldo won by a landslide in the January 1988 balloting. Opponents charged him with terror tactics, and Comelec consequently delayed his oath-taking. "I'm in no hurry," he told an Australian reporter with his signature smile. "But if they cheat me out of victory—I'll kill them."[23] Eventually exonerated and sworn in, the new governor soon became a powerful provincial leader.

During the December 1989 coup, Aguinaldo again declared for the RAM rebels on Cagayan radio and advanced on Manila with hundreds of militia until roadblocks stopped him. A month later, the secretary of local government,

Luis Santos, ordered the governor's "preventive suspension" for giving "aid and comfort" to the rebel forces. But Aguinaldo was defiant, ignoring the directive and threatening to retreat into the mountains for guerrilla resistance. Determined to assert her authority, President Aquino sent troops to Cagayan in January 1990. For eleven tense days, hundreds of Aguinaldo's supporters kept the soldiers at bay by forming human barricades around the provincial capital.[24] To break the standoff and arrest Aguinaldo, the government dispatched ten thousand troops and four helicopter gunships to Cagayan's capital, Tuguegarao, with the popular General Oscar Florendo as mediator.

On March 4, after only a day of negotiations, the situation spun out of control. While thousands of supporters rallied in the streets, rebel troops led by a defector from the Scout Rangers, Captain Feliciano Sabite, marched into the Hotel Delfino and took General Florendo hostage. Some 150 government soldiers stormed the lobby, producing a crossfire that left Captain Sabite wounded and General Florendo dead. As troops raked the capital with gunfire that killed fourteen and wounded twenty, Aguinaldo returned to the mountains with a militia three hundred strong. With a million-peso reward on his head and an AFP task force on his heels, Aguinaldo spent the next hundred days on the move, finally surrendering at the provincial capital before a crowd of supporters chanting "Agui! Agui!"[25]

Although Aguinaldo had joined two coups and killed a popular general, the Aquino administration seemed powerless to punish him. Released on bail, he ran for reelection as Cagayan's governor in the May 1992 elections. Comelec cited pending charges to bar Aguinaldo's candidacy at the behest of Enrile, now congressman from Cagayan and a bitter rival. But Aguinaldo defied the order and won reelection by a landslide, scoring 170,382 votes against only 54,412 for Enrile's favorite. In light of this popular mandate, the Supreme Court ruled Comelec's ban on his candidacy "a grave abuse of discretion" and confirmed his election. Encouraged, Aguinaldo successfully petitioned a lower court for dismissal of rebellion charges from the 1989 coup—thus becoming the first of RAM's leaders to win impunity.[26]

In the midst of this campaign for exoneration, some of Aguinaldo's former torture victims, now emerging from prison, represented a threat to his position. The governor encouraged their silence with charm and an intimidating directness. In early 1992, when the former NPA negotiator Satur Ocampo seemed due for release, Aguinaldo started a two-year campaign to still his recollections of their torture session sixteen years before.

The first contact came at Fort Bonifacio, when RAM leader Billy Bibit, Aguinaldo's PMA classmate and torture partner from the 1970s, called on Ocampo, saying, "Rudy wants to see you. He wants to discuss about the past. He is now governor of Cagayan and he would like to go on with his life and would like to patch things up with you." Ocampo replied carefully, filtering his words through the memory of Aguinaldo's pathological violence. "I'm willing to forget, but I cannot speak for the hundreds of others he had tortured. He has to be accountable to them." After his release some months later, Ocampo was finishing a television interview about torture when a studio operator handed him a message from the governor: "I would just like to let you know that was part of the past and would like to propose to Mr. Ocampo if we can forget the past."

Two years later, in December 1993, Ocampo ventured into Cagayan Province on a speaking tour and found himself the involuntary guest of Governor Aguinaldo. As Ocampo breakfasted with local human rights leaders, Aguinaldo unexpectedly joined the activists at their table. "This guy is my boss," he said, using the honorific he once reserved for his military superiors. "I learned from him. I became a revolutionary because of him." Then, addressing his former victim, Aguinaldo said earnestly, "You know, boss, I am now implementing your program here in the province."

That same day, Aguinaldo invited himself to a human rights luncheon for Ocampo, announcing, to the astonishment of the activists, "I owe what I am now to my boss." That night, he arranged a dinner for the speaker and the human rights activists. Throughout this long day, Aguinaldo did not mention the torture he had inflicted on Ocampo. Instead of "the taunting beast he was in the torture chamber," said Ocampo, Aguinaldo was now "courteous," even humble. Still, the effect was chilling. A local radio station had planned to interview Ocampo about his torture experience, but changed topics when it learned that the governor was monitoring it.

What can we make of Aguinaldo's actions: simple intimidation or a plea for forgiveness? Whatever the motivation, the outcome was clear. In the 1992 elections, Aguinaldo, with ample funds and an armed entourage, had been reelected governor of Cagayan by a wide margin. As his second term drew to a close in 1996, the Ramos administration charged him with graft, an apparent ploy to ensnare him in litigation until he switched his affiliation to the ruling Lakas-NUCD party. By capitulating to this pressure, he won a de facto exoneration from these charges and direct access to the president.[27] Under two successive administrations, Aguinaldo had manipulated the democratic process through charisma and cunning to win impunity for crimes of torture, murder, and rebellion.

EYE OF A CAMERA

In RAM's political makeover from violent rebels to social reformers, the Filipino cinema played a central role. These officers, with their mix of idealism and violence, were ready-made for roles in the Tagalog action genre and its romance with the gunman as agent of social justice. In the years after their surrender, the industry selected just two among RAM's many flamboyant leaders for film-length biographies—Rodolfo Aguinaldo and Billy Bibit, academy classmates and close comrades since their days as torture interrogators under Marcos. In this season of national reconciliation, Aguinaldo, through the medium of cinema, recast the collective memory of the Marcos era, transforming his violent career into the stuff of heroism.

In August 1993, only months after his reelection, Regal Films released "Philippine cinema's biggest action picture ever," a P 25 million epic titled *Aguinaldo: The True-to-Life Story of Gov. Rodolfo Aguinaldo of Cagayan*. "The exploits of Aguinaldo are interesting," explained action star Lito Lapid, who played the title role. "He's got guts. He stands up for himself. His life is very cinematic."[28] The film opened on fifty-seven screens in Metro-Manila, launching the action epic as one of the year's major box office successes.

Within the acrobatic hyperbole of the action genre, the events of Aguinaldo's life were reinvented, inverting, even subverting, his notoriety as the nation's top torturer. Indeed, torture serves as the film's central dramatic device, its defining image of evil. Through his struggle against a brutal torturer, the cinematic "Agui" becomes a moral center between the NPA's revolutionary violence and the oppression of the country's ruling elite.

Blurring the boundaries between person and persona, the film opens with footage of the actual Aguinaldo, shot in slanting sunlight, seated behind a broad politician's desk, saying: "My beloved countrymen, I can assure you that this is the true story of my life." Then, after segueing through a long sequence of blazing shootouts and flaming car crashes, we pause to witness Agui's capture of Communist Party founder Jose Ma. Sison—the actuality of the event signified by lettering typed onto the screen, "Pagdalagan Norte, San Fernando, La Union, November 10, 1977." In reality, the historic Aguinaldo arrested Sison on a provincial road while the communist was riding a small motorcycle. In the film, this traffic stop becomes a major battle with a helicopter, forty professional divers, and dozens of heavily armed extras. While the actual Jose Ma. Sison was then a dynamic revolutionary in his thirties, the film shows a timid man on the downslope of sixty.

Cut to the Luzon coastline, where Agui swings into the NPA's hilltop hideout from a helicopter, blasts dozens of guerrillas skyward, and captures "Chairman Sison" hiding in a hole, shuffling his papers. "Mr. Chairman," says Agui with quiet authority, "you are under arrest." Wearing the white dress shirt of a petty clerk, Sison remains silent and stooped as Agui's muscular soldiers grip his arm and march him away—not just to prison, but into historical oblivion. With his strength and charisma, Agui is challenging the NPA guerrillas for command of their own social revolution.

Cut to constabulary headquarters, where General Fidel Ramos promotes Agui to captain. Cut back to Cagayan, where Agui confronts Congressman Tito Andaya, a powerful warlord who controls illegal logging, sugar milling, and local government. Angry over the seizure of his logging truck, the politician assigns his chief goon, a knife-wielding sadist, to even the score by kidnapping Agui's soldiers.

Fade to a vast cave, where Aeta tribal scouts with blazing torches guide Agui through the cathedral-like cavern to skeletons hanging from a rock wall, identifiable by their dog tags as the kidnapped soldiers. Plunging deeper, Agui finds another of his missing men mangled by torture, his face a swarm of knife wounds, his arm hacked away to a bleeding stub. The soldier pleads for release. As the camera zooms to a single, swollen tear slipping down Agui's cheek, he embraces the man and squeezes off a gently fatal pistol shot. Then an abrupt segue to grainy television footage of the 1986 people power uprising showing the real General Ramos, RAM rebels, and the massive crowds of people power.

Why does the film take Agui and his men into a cavern, and do so just before people power? As every Filipino schoolchild knows, on the eve of the 1896 revolution the nationalist hero Andres Bonifacio led his secret society, the *Katipunan,* into the Pamintinan cave near Manila, where they scrawled "Viva La Independencia Filipinas" on the wall—an emotional experience that inspired their later revolt against Spain. Placing Agui in a cave on the eve of people power seems to mark him as a latter-day Bonifacio, the leader of a modern social revolution.

After footage of people power on EDSA fades, Agui returns to Cagayan as constabulary commander, ready to confront the twin forces of oppression—the communist NPA and Congressman Andaya. Segue to an NPA safe house, where constabulary agents capture the lovely communist amazon Ka Melissa and bring her to Aguinaldo for questioning. Instead of his trademark sexual torture, there is an electric ideological debate in which her Marxism withers before his democracy.

In a final, epic combat, Agui battles Congressman Andaya's thug, a psychopathic torturer who has become the film's personification of evil. The hero and his men enter the congressman's sugar mill in darkest night, scaling the walls commando-style with ropes. After blazing gunfire cuts down their minions, Agui and the goon come face to face like epic heroes for the final battle. The goon raises his jagged knife for a death blow, but Agui quick-draws for the fatal shot with his .45 pistol, the symbol of just violence in the hands of an action hero. The camera zooms in as the knife, icon of torture, fills the screen and then slips from the goon's dying grasp.

In the film's final reinvention of the recent past, the death of General Florendo is blamed on President Aquino. When the Marines arrive to arrest Agui for his role in RAM's 1989 coup, the provincial capital erupts with the masses chanting "Agui, Agui," echoing the cadences of "Cory, Cory" during people power. As Cagayan 1990 resonates with EDSA 1986, Agui glares into a television camera and speaks directly to the president, demanding to know why she is sending her soldiers to attack him instead of the communists and challenging her claim to the spirit of people power.

Rejecting Agui's attempts at resolution, the Marines, machine guns blazing, riddle General Florendo's body with bullets and slaughter the crowds. Cut to the mountains, where an eagle screams. "Aguinaldo spent one hundred days among the people of Cagayan," a voice-over intones, "and the government offered a 1.2 million-peso reward for his death, capture, or surrender. Agui came to know the masses better. The Supreme Court later ruled that . . . the warrant of arrest should be voided." The camera pulls back to show Agui leading his band of humanity, militia and peasants, riding water buffalo through the rolling grasslands toward the horizon, into the future.

"Whatever he did," commented star Lito Lapid, "you can't deny the fact . . . that he is beloved of the poor and the people and that's what makes Aguinaldo an interesting and colorful figure."[29] But there was much more than love and color at stake. Through its juxtaposition of historical references, the film projects Aguinaldo as the nation's new hero—the embodiment of Bonifacio's nationalist revolution, Sison's communist revolt, and Cory's populist uprising. With the collapse of the Marcos myth through defeat and exile, Gringo Honasan had been contending for his crown within the popular imagination. Suddenly, Agui intrudes into this telemythic realm to reach for the mantle of the warrior-hero.

A year later, as the 1995 Senate campaign was getting under way, Gringo Honasan's long-awaited movie biography finally appeared in a disarmingly modest package. After months of well-publicized negotiations with the studios, Honasan in the end agreed only to endorse the story of Billy Bibit, a secondary RAM leader best known for a bungled aircraft hijacking during the people power revolution. Released under the title *Colonel Billy Bibit—RAM!* the film featured leading man Robin Padilla as Gringo and his lesser-known brother Romel as Billy. With the megastar taking second billing, the film became, in effect, Gringo's story.

In September 1994, as the colonel was making his transition from coup plotter to Senate candidate, the film opened on sixty-seven screens across Metro-Manila. At the premiere, Honasan dominated the press coverage, posing for photographs and proclaiming star Robin Padilla his favorite actor. Explaining the film's political import, Honasan told reporters, "We have been the victims of attacks from various sectors. We have no chance to defend ourselves."[30]

While lacking the rich symbolism of the Aguinaldo story, this film's recasting of history was even bolder. No longer torturers and coup plotters, the RAM characters became idealists fighting for social justice against a corrupt Aquino government. In a marked departure from the dictates of the action genre, the film sacrificed violence for Gringo's long, populist speeches.

After the opening credits scroll through the 1986 hijacking, the film flashes back to a Mindanao hilltop in the mid-1970s where Billy Bibit's detachment is under attack by Muslim rebels and all seems lost. Suddenly comes a helicopter with commandos sliding down ropes into the battle. One officer, identified only by his telltale Gringo mustache, leads his men in acrobatic combat that ends with the slaughter of the encircling Muslims. As the smoke rises from the body-strewn battlefield, Gringo salutes Billy in a dramatic bonding gesture.

Newsreel footage flits past, signaling Marcos's downfall and RAM's failed 1987 coup against Aquino. Then the scene shifts to a Manila mansion late at night. Seated around a table in uniform, the rebels listen carefully as Gringo intones, "Comrades, the fake freedom will not bring the end of the poverty of the nation. We won't stop until we attain victory." After dismissing the others, Gringo pulls Billy aside to share his revulsion at the Aquino government, saying: "History will say that the Filipino soldiers did not abandon the nation but saved it from oppressive and repressive government by people who are greedy and hungry for power."

After RAM's abortive 1989 coup, Billy is confined in a Manila prison until rebel commandos free him in a daring midnight raid. As a helicopter carries him off into a glowing dawn, the text of an epilogue rolls, an advertisement for the rebels' amnesty demand. "After his escape," the announcer intones, "Colonel Billy Bibit went back to the folds of his comrades in the RAM. . . . With the present administration's sincerity, the RAM sees the break of a new dawn for a true peace and national reconciliation."[31]

Though surrender had stripped him of men and weapons, Honasan could slip through the camera lens into a cinematic imaginary to regain an aura of power undiminished by defeat. In this wide-screen universe with all its narrative power, film recast recent history, transforming the RAM leaders from coup plotters into crusaders for social justice. By the time they slipped back through the lens onto the national stage, the colonels were dressed for their new roles as political candidates.

RAM POPULISM

In the long run-up to the 1995 elections, RAM passed slowly from rebellion, to protest, and, finally, candidacy. A dozen military leaders campaigned in these local and legislative elections, and several would score major upsets, winning a mandate for a concession the government had long refused—unconditional amnesty. Through this synergy of media, cinema, and politics the colonels would win impunity.

In the two years between surrender and elections, the RAM leaders, swinging from hard right to far left, repackaged themselves as militant advocates for social justice. A few months after their surrender in late 1992, public opinion polls showed that Honasan and his comrades had hit rock-bottom, with most respondents expressing strong disapproval of their right-wing violence.[32] A year later, in February 1994, they launched their new strategy by joining left-wing protests against President Ramos's imposition of a one-peso tax on gasoline sales. As the opposition gained momentum, RAM leaders appeared on radio and television, speaking eloquently for the masses.

Interviewed on Radio DZRH, Honasan and his fellow rebel, Captain Proceso "Boy" Maligalig (PMA '69), were at once modest, sympathetic, and empowered. "Many of us are already suffering or are in extreme poverty," said Honasan in reply to a question about the gas levy's impact. "What is their plan to lessen the hardships of the people?" Captain Maligalig added: "Why is it that the masses of our people are the ones asked to sacrifice?"

After the show turned to callers, one Jun Hernandez of Quezon City asked if the colonel would run for the Senate in the next elections. Instead of denying ambition, Honasan weighed his chances, asking: "Will the contest be fair . . . for you to have an equal chance . . . of winning? Or do you still need so much money, a complicated political machinery, and in some cases an armed force?" In closing remarks, Honasan urged reconciliation with the moderate tone that would later serve him well in his Senate campaign: "Let us unite. Let us help one another in a sober manner before we think of other ways of addressing these problems."[33]

When dozens of "cause-oriented groups" mobilized twenty thousand students and workers for a rally to protest the tax at the EDSA monument, RAM was there in T-shirts reading "Our Dreams Shall Never Die" amid the left-wing crowds with banners calling for: "Rollback! Rollback!" "Trash the tax!" "Destroy the World Bank!" In the end, the president retreated, canceling the oil levy and validating RAM's new populist image.[34]

During the annual celebration of the EDSA uprising a few weeks later, the rebel colonels continued their careful repositioning on the political spectrum. With each step in this elaborate, four-day ritual, RAM and the Ramos administration sparred subtly to redefine the meaning of people power and claim its mantle of heroism. On February 22, the first day in the commemorative cycle, the president used a wreath-laying ceremony at the *Libingan ng mga Bayani* (Heroes Cemetery) to blame RAM's many coups for the slow pace of economic growth. In an interview to set the historical record straight, presidential advisor General Jose Almonte (PMA '56), himself a former RAM member, claimed that it was Ramos who had turned the military against Marcos. "Without his instruction, perhaps only a few would follow us," Almonte said.[35]

Simultaneously, RAM leaders led a prayer rally behind Camp Aguinaldo to honor comrades killed in their 1989 coup against Aquino.[36] Again, Honasan struck a softer note, saying that EDSA "was not a real revolution because it has not filtered down to the grass roots level." He appealed for unity, noting sadly that today it was "each to his own group, each to his own monument."[37]

Nobody mentioned RAM's record of torture during these rival rituals. Only one commentator, Antonio Nieva of the *Philippine Daily Inquirer*, recalled the rebel massacre of civilians during their 1987 coup:

> That INQUIRER photo showing Gregorio "Gringo" Honasan and other RAM leaders piously laying a wreath Feb. 22 at White Plains for their coup dead provoked . . . memories of a night of terror six years past. . . .
>
> In the pre-dawn darkness, people had collected on the side-walks, . . . mothers with their children, fathers squatting on the curb. . . . The [RAM] convoy had stopped, and the soldiers were getting off, and firing at moving shadows. Others were

already fanning out into alleys and side streets, firing.... I looked, and saw a soldier aiming his M-16 at a fleeing group....

At least 17 were killed... that early dawn of Friday, Aug. 28, 1987.... There are no wreaths for the innocents, the civilian onlookers, that the RAM mutineers had vent their ire on after failing to take Malacañang.[38]

Only a month later, President Ramos, in a deft move, forced RAM leaders out of their comfotably ambiguous position as state-sanctioned belligerents by issuing an amnesty for all insurgents—communist, Muslim, and military. Under the terms of Proclamations 347 and 348, rebels had six months to make individual applications to a National Amnesty Commission for all offenses except "acts of torture, arson, massacre, rape, or other crimes against chastity."[39]

The decrees sparked an intense debate. Secretary of Justice Franklin Drilon argued that civilian dissidents could win amnesty for human rights violations but military rebels should not. "The reason is that, technically, only agents of the state can commit violations of human rights," he explained. An officer of the Free Legal Assistance Group (FLAG), Ma. Socorro Diokno, charged that the proclamation would exonerate officers guilty of brutal crimes such as salvaging.[40]

When the decrees moved to Congress for approval, the executive, responding to criticism, added an amendment barring amnesty for soldiers guilty of "serious human rights violations." In the upper-house discussion, Senator Rodolfo Biazon, a former AFP chief of staff who had assisted in the drafting, explained that crimes "like 'torture' and 'massacre' are generic terms that are not found in the Revised Penal Code, the Articles of War, or special laws." Indeed, in the subsequent debate Senator Wigberto Tañada announced he would oppose amnesty because it failed to offer any "definition of what serious human rights violations are."[41] Despite these objections, Congress approved and the new commission soon issued a call for amnesty applications.

RAM, with good reason, was outraged. While the original surrender agreement had given the group authority to negotiate amnesty for its membership, the new decree, as Senator Tañada pointed out, downgraded its status, "rendering the negotiations senseless." RAM's spokesman, Lieutenant Colonel Jake Malajacan, announced, "We will not apply for amnesty. Amnesty is not even on the agenda." But faced with a choice of amnesty or prosecution, more than fifty RAM members soon broke ranks to file applications, including Governor Aguinaldo. By November, fifteen members had won amnesty, notably the Mindanao coup leader Colonel Alexander Noble.[42]

As their organization began to unravel, the rebel leadership responded by moving into the political arena. In May 1994, the first of RAM's top leaders took the plunge into politics when Red Kapunan resigned his commission to run for Congress against a powerful incumbent. At his press launch in May the rebel revealed that the national security advisor, General Jose Almonte, was backing his candidacy, assuring him nomination by the ruling party. Known as the president's "unofficial grand vizier," General Almonte, one of RAM's founders, had once been charged with diverting government funds for the group's coups against Aquino and was now emerging as their chief patron. "The rebels came here last month and we had discussions for one whole day," Almonte admitted indiscreetly to a Singapore newspaper. "I said, '. . . if you stay as idealistic as you are and you don't compromise your idealism, then we will have a group who can threaten these people who are resisting reforms; otherwise it will be very expensive for the government to organize another group like you to threaten them!'"[43]

When the campaign opened in late 1994, RAM pressed for reforms that would even their electoral odds against entrenched local dynasties. To build public support, Honasan gave a wide-ranging interview to the *Philippine Daily Inquirer*, sidestepping a direct question about his Senate candidacy but showing sharp insight into the obstacles he might face. "What political party or alliance will give me access to the machinery that will ensure my victory in the archipelago?" he asked. "Do I have the goons, the armed people who will ensure my victory in remote areas?" Almost thinking aloud, he added, "If you are popular or notorious, the first problem is how to translate your popularity into votes. Second, how to guard the votes that were cast in your favor. Last how to count the votes." He closed with a surprising statement about the future of the rebel factions. "With regard to the RAM-SFP-YOU, the sooner we dissolve it the better."[44]

With RAM now committed to an electoral strategy, Honasan's Senate candidacy seemed an inevitability. In February 1995, Honasan finally agreed to a "tactical alliance" with opposition dynamo Miriam Defensor Santiago and became a "guest candidate" of her People's Reform Party.[45]

Eventually, RAM would field ten candidates in the May 1995 elections, including: Red Kapunan, Billy Bibit, and Zosimo Paredes for Congress; Alexander Noble for provincial governor; and five more for local offices.[46] Although no longer counted among the RAM candidates, the group's original patron, Juan Ponce Enrile, was running for senator with the ruling coalition and Governor Aguinaldo was up for reelection. Moreover, the two actors who had played RAM officers in their recent films, action star Lito Lapid and movie "bad boy" Robin

Padilla, campaigned for governor and vice governor of major provinces. In sum, RAM, its allies, and cinema shadows were a substantial presence in the 1995 elections.

The Manila media was cautious, even critical, of the group's political debut. At the launch of his campaign in early February, Honasan felt compelled to apologize to "the families of civilians killed during the coup attempts." When Kapunan announced his formal candidacy, reporters asked about his "reported cases of human rights violations." But he replied bluntly, "The issue of human rights is not a very important issue for me."[47]

Although running as an independent on a weak opposition slate, Honasan brought some important assets to a contest that was uniquely suited to his talents. Since the country's twenty-four senators were elected at large on a national ballot, it was the only race where the country's iron laws of local politics did not apply. Unlike the lower house, where legislators swap patronage, the Senate is a national forum for statesmen and often serves as a showcase for future presidents. In such a campaign, Honasan's name recognition was a powerful asset.

The "Gringo" persona had no equal in this personality contest for the twelve open Senate seats. With his physical grace and distinctive mestizo features, Honasan was a stunningly handsome presence on the stump. He sparked a near frenzy among crowds on the campaign trail, striking a particularly responsive chord among younger women with his lethal aura as coup leader now amplified by its wide-screen portrayal. Long known for a high, thin voice that seemed to mock his robust physique, Honasan, in his first appearances, now spoke with what one columnist called "that new husky timber in the voice," hinting at "a voice coach hidden somewhere."[48]

Among his major political assets, Honasan could count on a strong vote from his native Bicol region, an isolated peninsula inclined to vote only for Bicolanos. Although running on an opposition ticket, Honasan had the tacit support of the Ramos administration. Not only was he close to the president himself, but he had, as a career officer and army brat, close personal ties to the many PMA alumni who held top posts in the administration, including General Almonte, now the president's chief campaign strategist.[49]

In mid-February, Honasan's made-over media persona made a captivating appearance on the EDSA anniversary edition of *Public Forum,* a popular television talk show. Seated round a glass table, the panel spanned the political spectrum—host Randy David, sociologist and socialist; Senator Aquilino "Nene" Pimentel, grass-roots leader of the anti-Marcos opposition; Gringo Honasan, former coup leader and populist; and Representative Juan Ponce Enrile, Marcos's former defense minister and leader of the far right.

Though still a political neophyte, Honasan, by word and gesture, overshadowed his veteran rivals. Wearing a candy-red windbreaker that highlighted his slumping shoulders and dyed hair, Enrile spoke with faint gesture and wrinkling brow, quoting the Bible like an elderly priest. Zipped into a sky-blue bush jacket stamped with a dove at sunrise, Pimentel, slouching in his chair, belly protruding, spoke with staccato hand gestures and hoarse voice, making him look the aging political warhorse. Honasan, with denim sleeves rolled above massive forearms, was braced in his chair, muscular body ready to spring, hair coiffed in an immaculate, leonine mane, one sensual lock turned downward to graze his forehead. Brow wrinkles, crow's feet, and neck sinews fluttering with the cadences of his speech, Honasan seemed to invest his whole being in every word.

Host David noted, in his introduction on the meaning of people power, that "the basic legacy of EDSA has slowly faded away." Asked to comment, Enrile, mindful of his weak following in the Ilocos region, where voters were still loyal to the dictator's memory, insisted that "President Marcos also did some good things," like roads and rural electrification, and added, "We had no intentions of hurting the family of former President Marcos."

This reinvention of the past was too much for Senator Pimentel, who reminded everyone that under Marcos "the human rights situation was really bad." Seizing upon that opening, host David spun and lunged like a fencer, asking Enrile if he were not, as Marcos's "martial law administrator," responsible for these "human rights violations." Deftly, the former defense minister shifted the blame to Marcos's chief of staff, the exiled General Ver. Since they were now running on the same ticket, Pimentel granted Enrile and his former comrade President Ramos a general absolution, saying that they had "exculpated themselves" through the "cleansing process" of the EDSA revolution.

In the midst of this sparring, Honasan, bowing his head deferentially to his elder Senator Pimentel, appealed for an end to fault finding: "If we really want to bring out the spirit of EDSA, the spirit then was unity. . . . If we continue to blame each other, eh, then all of us should take the blame for what has become of our country."

After a commercial break, host Randy David focused on his star guest, demanding to know if Honasan was "accepting responsibility for the deaths during the coup attempts." Very carefully, the ex-colonel both admitted and evaded responsibility. What about deforestation and flooding that kills thousands of people? What about wasted public money needed to feed dying children? Is that not, he asked, violence? "So, if we count the losses . . . during the elections in past years," he argued, "these will amount to more than the combined losses in the coup attempts." Then, suddenly humble, Honasan apologized, saying, "We accept responsibility. . . . We didn't want anybody to die." But in the next breath, he waved away any culpability, suggesting that "this is a part of change."[50] As the show closed and the credits rolled, Honasan sprang from his chair into a wrestler's spread-leg stance, smiling full-faced as his arm swept wide for a triumphal handshake with David that showed the sinews of his forearm flexing with the grip.

Honasan's campaign literature displayed his integrity, heroism, and physique. In the poster competition that papered every city street, Honasan's was an eye-catcher. Wearing a traditional white *barong* buttoned neatly at the neck, he smiled faintly, eyes crinkling, that lock of hair curving sensually to touch his brow. "The Filipino people are my party," read his motto.

His campaign handbill featured a photo of the vice president, aging mestizo action star Joseph "Erap" Estrada, raising Honasan's hand on the stump with a bold red caption, "Erap Says It's Gringo for the Senate." On the flip side, above the slogan "Gringo Honasan Independent. No Godfather," voters read of his sincerity. "Since the formation of . . . RAM in 1985, Gringo has kept close watch over the state of the Filipino nation." Concerned about the thousands "bowed down by hardship" and "the destruction of our forests," Honasan has decided to run for the Senate as an independent candidate, "Without any party. Without any patron to be free to work for reforms."[51]

Honasan's first rally was an extraordinary event, even in this nation with a century-long history of flamboyant campaigns. In early March, Miriam Defensor Santiago led her People's Reform Party senatorial slate in barnstorming across her native Panay Island at a grueling pace of seventy-one rallies in just six days. When her road show reached Bacolod City, Honasan made his appearance, arousing the crowd of thirty thousand to near frenzy. After Miriam's fiery rhetoric, Honasan's speech was lackluster. But, as he finished and moved toward his car, the crowd erupted in excitement captured in this dispatch by the *Manila Chronicle*'s local correspondent:

> Honasan was swamped with "love notes" or short personal messages from residents here, mostly young women. . . . His own speech was interrupted several times by loud screams of "I Love You" from women and, of course, from the third sex. . . . Even Honasan's battle-tested security personnel, all members of the Rebolusyonaryong

Alyansang Makabayan (RAM) which Honasan led, were helpless before the hordes of people that flocked the candidate.

It took the leader of seven failed coup attempts a full 20 minutes to reach his waiting vehicle from the stage because of the enormity of the crowd that wanted to kiss or just hold him.... "They just keep on coming from everywhere. This is a security nightmare for us," an unidentified Honasan aide said.[52]

How can we explain the intensity of this response? It seems more complex than a matter of mere physical appeal. Though fit and handsome, Honasan was approaching fifty and lacked the youthful glow that commands box office in the Filipino film industry. Perhaps RAM's reputation for torture, murder, kidnapping, and coups added an erotic element. With looks and manner eliding with power and violence, Honasan seemed to project a seductive aura of subtle threat. Whatever the appeal, thousands became frenzied in his presence. Though large and enthusiastic, these campaign crowds were not, however, representative of women voters, who, on election day, rejected him in overwhelming numbers, producing a marked gender gap of 10 percent.[53]

OPERATION DROP-ADD

With the balloting a month away, ex-president Corazon Aquino emerged from political retirement to stop Honasan. Mobilizing her constituency, she launched the "Never Again Movement" to fight the senatorial candidacies of Honasan and Ferdinand "Bongbong" Marcos, Jr., the former dictator's only son. Two weeks before election day, Aquino, reviving the symbols of the anti-Marcos protests, led some five thousand demonstrators in yellow T-shirts through the Makati financial district, ending at a rally before the bronze statue of her martyred husband, Ninoy Aquino, at the moment of his death, a dove taking flight from his shoulder. There, she urged voters to punish the two candidates "for the evil things they did in the past." In a press release, Honasan dismissed the rally as elitist and, in an ironic twist, damned Aquino for the "worst human rights violations in Philippine history."[54]

While the protest had limited impact, Honasan's campaign still entered its last days in deep trouble. Only two weeks before the balloting, a preelection survey found him finishing fourteenth—well outside the twelve-seat winner's circle. His campaign, supported by an army of citizen volunteers, intensified its pace. Working through the radical officers of the Young Officers' Union, Honasan built a strong constituency on the campuses and forged an alliance with several ex-NPA guerrillas with mass support in the provinces. By election day, his spokesman, Captain Boy Maligalig, claimed two million campaign workers led by "a nationwide alliance of women's organizations."[55]

On election day, May 7, Honasan flew by private helicopter to his hometown of Bulan, Sorsogon, in a symbolic appeal to his fellow Bicolanos.[56] Then, like the rest of the nation, he sat back to watch the twelve-hour carnival of thirty million voters, 60,311 candidates, and twenty-seven election-related murders. In the minute-by-minute media updates, the early trends looked grim for Honasan and his fellow RAM candidates. Only hours after the polls closed, the ABS-CBN network's "quick-count" of Senate returns showed Honasan and his patron Enrile at fourteen and seventeen respectively, well outside the winner's circle.[57]

In early local and congressional returns, it was clear that most RAM candidates had suffered fates ranging from defeat to humiliation. With masterful understatement, veteran observer Amando Doronilla wrote that RAM's "challenge to the traditional elite has not been completely successful." A younger member, Leovic Dioneda (PMA '78), was reelected mayor of his hometown, and Governor Aguinaldo, no longer affiliated with the group, also won reelection.[58]

The rest of the RAM candidates polled badly. In the Mindanao province of Agusan del Sur, Alexander Noble, leader of RAM's 1990 coup, lost by a wide margin in the governor's race to the entrenched Plaza family dynasty. On election eve, his campaign suffered a double blow when the Comelec chair stripped his goons of their guns and a widow charged that Noble's men had gunned down her husband, a town mayor and Higaonon tribal leader, at church fifteen years before. All three RAM candidates for Congress lost. Despite strong support from the president, Red Kapunan suffered a humiliating defeat in Capiz's second district, winning only seventeen thousand votes to the incumbent's forty-eight thousand. His classmate Zosimo Paredes lost by a crushing margin in Baguio City's lone congressional district, and was later charged with inciting to riot when five hundred supporters attacked election officials. In Makati's second district, Billy Bibit lost despite high name recognition from the release of his Tagalog action film.[59]

Honasan remained RAM's only hope for victory. For the next four weeks, as Comelec tallied the returns province by province, the Senate rankings shifted almost daily, prompting a flurry of protests by angry candidates. By May 16, the ninth day of counting, Comelec began to show a troublesome pattern—the top six candidates were clear winners, but the next nine were clustered together with only tiny margins separating those inside and outside the "magic twelve." While Enrile at number eight had well over three million votes, he was only forty thousand ahead of Bongbong Marcos at number nine, who, in turn, was only twelve thousand ahead of number ten.[60]

The slow count and small margins separating the candidates created circumstances ideal for systematic fraud. On May 15, Senator Nene Pimentel charged that Enrile, a member of his own coalition's slate, was involved in "Operation *Dagdag-Bawas*" (Operation Drop-Add)—bribing Comelec canvassers in the provinces to shave votes from one candidate and add them to another. On May 16, coalition candidate Rodolfo Biazon, a former AFP chief of staff, filed a formal complaint with Comelec that 10,000 of his votes from two provinces "were added to the names of Juan Ponce Enrile and Gregorio Honasan in the certificates of canvas."[61]

By May 24, with 99 percent of the vote tallied, Honasan had won a Senate seat with a lock on ninth place. Enrile, with over eight million votes, had barely slipped into the magic twelve—just 35,000 votes ahead of Nene Pimental at fifteen. Insisting that Operation Dagdag-Bawas was still running, Pimentel filed a formal protest, claiming that Enrile had been "erroneously" awarded over 130,000 of his votes.[62] Despite signs of fraud, on June 6 Comelec proclaimed the twelve winning senators, including Honasan and Enrile.[63]

Even though cheating may have given him a final push into the winners' circle, Honasan's victory was still impressive. In less than two years, he had moved from hunted rebel to one of twenty-four members of the country's highest legislative body. Pressed to explain his unprecedented win, observers cited two factors—personal charisma and ethnic bloc voting. Using exit poll data, political scientist Temario Rivera noted that Honasan, as expected, had scored votes from 38 percent of his fellow Bicolanos, but also did surprising well in the Ilocos region with 57 percent. When final results were in for Ilocos, Honasan had scored 135,000 votes, second only to favorite son Bongbong Marcos. One analyst attributed Honasan's win there to the Ilocanos' "macho culture and love for warriors."[64]

Only days after Comelec proclaimed the winners, the defeated Pimentel filed for a recount of dubious returns and formed the Foundation for Clean Elections to finance the high cost of these proceedings. Similarly, in mid-June, the defeated coalition candidate Rodolfo Biazon charged that the president and his advisor General Almonte were behind "a concerted drive to kick us out of the winning circle." He claimed that 100,000 votes had been subtracted from his total, with many of them given to Honasan. When reporters asked if General Almonte had sufficient resources for such a complex maneuver, Biazon, himself a former chief of staff, answered: "If you are national security advisor, you have all the facilities." Biazon explained that Almonte wanted "to give Honasan a free hand" in the Senate: "They know that if I am around, I would be a force to contend with against Honasan."[65]

After following the elections closely, political scientist Rivera argued that the palace had backed opposition candidates Enrile and Honasan because their personal ties to the president made them more reliable legislative allies than his own party's candidates. A year later, these suspicions were corroborated when a memo was leaked from the palace press office stating that General Almonte and Executive Secretary Ruben Torres had rigged the results "to favor certain [pro-Ramos] coalition bets by granting them more votes than they actually got."[66]

But these intrigues, particularly the Honasan-Biazon dispute, had a deeper significance overlooked by media commentators. The military was now so immersed in electoral politics that the key actors in this controversy were all career officers—Ramos (USMA '50), Almonte (PMA '56), Biazon ('61), and Honasan ('71). Despite partisan bickering, they shared an underlying faith in the military. As one leftist commentator explained, "RAM-YOU in fact now occupied the same political space as the core faction of the Ramos regime, belief in the military as a base for economic and political reform."[67] Twenty years before, PMA graduates had tried to remain aloof from partisan politics; now they immersed themselves in it. Once they tried to serve the civil state; now they fought for its highest offices.

IMPUNITY

Honasan's triumph was soon eclipsed by sensational revelations in an almost-forgotten murder case. In August 1995, the National Bureau of Investigation (NBI) arrested a RAM member as a co-conspirator in the 1986 salvaging of labor leader Rolando Olalia. The NBI's Manila chief, Salvador Ranin, told the press that the arrest of this suspect, Sergeant Filomento Maligaya, would reopen the case and let them pursue strong evidence against one of RAM's core leaders, Colonel Oscar Legaspi.

For Honasan the timing could not have been less fortunate. Just as he was taking his seat in the Senate, this ten-year-old torture case surfaced with lurid press recollections of the victim's "mutilated" face and "gouged" eyes. Honasan counterattacked. "The NBI should be aware of the ramifications of its actions," he said, "and it should not pursue independent operations against groups currently negotiating with the government." With peace talks in their final stages, he insisted that "all government agencies should be advised to hold the arrest, if any, of Colonel Legaspi, and the immediate release of Sergeant Maligaya."[68]

Just when a breakthrough in the murder seemed near, the same political pressures that had blocked investigation for a decade did so again. Only two days after the case was reopened, NBI director Mariano Mison publicly cleared RAM, saying that the "kidnapping and killing of Olalia . . . were labor related." The next day, the victim's widow, Mrs. Feliciana Olalia, led the left in demanding an investigation of Senator Enrile's role in the murder, calling him the "godfather" of RAM. She also attacked the NBI director for his exoneration of RAM, charging that he was the brother of one of the group's founders, General Salvador Mison. The left appealed to the president to "stop protecting his prodigal sons in the RAM."

Within hours, however, the whole issue collapsed when the NBI's Ranin retracted his accusations and terminated the investigation of Colonel Legaspi on grounds that he was eligible for amnesty. Senator Enrile denied any "ill will" for the victim, and General Almonte urged the media to avoid further accusations.[69] With that, the case was closed.

These revelations raised the pressure on all parties to conclude a peace agreement. RAM's legal ambiguity, once an advantage, had become a liability. Until they received a full amnesty, the rebels would remain vulnerable to investigation for past human rights violations. When Congress convened in August, the Ramos administration, now allied with Senators Enrile and Honasan, did not want the burden of further scandal.[70] After three years of procedural delays, the negotiations were finished in just a few weeks. Under the draft accord, RAM agreed to a "permanent cessation of hostilities" and promised to return its weapons within ninety days. In exchange, the government would reinstate all rebel soldiers and grant "a general and unconditional amnesty for all offenses committed in pursuit of their political beliefs."

Although the president's original 1994 decree had excluded any amnesty for serious human rights violations, the government now conceded this point. Indeed, in the "confidential" final draft of the new decree, government negotiators struck out a clause allowing prosecution on human rights charges. Once Congress approved this decree, RAM would simply submit a list of its five thousand members and all would be granted an automatic amnesty, even for torture and salvaging.[71]

In October 1995, rebel and government representatives met at Camp Aguinaldo for an elaborate signing ceremony. The ritual, marking a formal end to

RAM's revolt, had the air of a historic occasion. "I can guarantee that there will be no more coups," said Senator Honasan. "We give our word of honor," said General Abenina, the RAM negotiator, "all crew-served weapons, light antitank weapons and machine guns will be returned to the government." The president's peace advisor, General Manuel T. Yan, called the signing a "milestone in the peace process."[72]

At a deeper level, this ceremony, involving PMA graduates spanning the institution's entire history, was another benchmark of the military's new prominence in national politics. The government negotiators were led by alumni from the academy's first two decades—from General Yan of Class '41 to General Fotunato Abat ('51) and General Alfredo Filler ('62). On the other side of the table, the rebel coalition covered the academy's next twenty years—from General Abenina ('58) through Lieutenant Colonel Zosimo Paredes, Jr. ('71), and Captain Danilo Lim (PMA '78, USMA '78). As this ceremony and its all-academy cast made clear, peace had been won at the price of conceding the officer corps both control over the military and considerable civil authority.[73]

The next day, Senator Franklin Drilon, a former secretary of justice and advisor to the government panel, criticized the blanket amnesty, asking pointedly, "So now they will give amnesty to the killers of Olalia?"[74] When the ninety-day deadline for the arms surrender expired in January 1996, he demanded "a complete accounting of the firearms" to assure the people that "no further . . . coup d'état can be undertaken by RAM."[75] But Drilon remained a lone voice, marginalized in a peace process controlled by the military. After years of maneuvering, the rebels had finally won impunity for crimes of torture and murder.

Two years later, Senator Drilon's words gained weight when two RAM soldiers came forward with revelations about Olalia's 1986 murder. The case had long been closed and Senator Gringo Honasan, the rebel leader originally implicated, was about to run for vice president, his first step toward the presidency. On 11 January 1998, lawyers for the victim's family presented the Justice Department with a complaint charging twenty RAM members with this eleven-year-old murder. As proof, they attached a twenty-three-page confession from retired sergeant Medardo Barreto, then assigned to the Defense Ministry's Special Operations Group (SOG), a "covert security" unit under Lieutenant Colonel Kapunan. The affidavit contained ninety-eight paragraphs of eyewitness testimony about the murder and a state prosecutor's certification of authenticity.

In Barreto's account, the operation began in early November 1986, when Kapunan assigned him to a seven-man SOG team monitoring Olalia. On the evening of November 13, the team was trailing its target to a union meeting when a RAM officer, Captain Ricardo Dicon, radioed Barreto to proceed with the abduction, assuring him that higher-ups had approved. "It was my understanding," Barreto wrote, "that Captain Dicon's statement meant that the order has the blessings of his superior officers, like Colonels Honasan, Kapunan, and Legaspi." At 7:30 P.M., the RAM team forced Olalia and his driver into their car and took them, handcuffed and blindfolded, to Minister Enrile's safe house in Cubao, Quezon City. When they arrived, a group of RAM leaders—Lieutenant Colonel Tito Legaspi ('71), Major Noe Wong ('75), and Captain Dicon—broke off their meeting to look Olalia over. "Captain Dicon told us informally," the witness explained, "that the order to liquidate Ka Lando [Olalia] and his driver would be given at any time because of the approaching military exercise [coup]." At 1:10 A.M., the squad loaded the victims into two cars and set off for Antipolo with Lieutenant Commander Elpidio Layson ('75) following. Outside Antipolo, the soldiers marched the victims, their mouths stuffed with newspaper, into a grassy field where they were shot and stabbed. In the aftermath, Colonel Honasan, then the ministry's security chief, ordered the SOG sergeants to "clean up" the mess when witnesses identified the license plate of a car used to tail Olalia.[76]

Within a day, this affidavit prompted banner headlines when the ex-sergeant surfaced at the Justice Department to meet the press. As an eyewitness from abduction to execution, Barreto spoke with convincing authority. He had "faithfully" served RAM's leaders Honasan and Kapunan. "I've kept this secret hidden for almost twelve years," he said. "But now I'm ready to face anything, just . . . to secure justice for Ka Lando [Olalia]." He was, moreover, acting out of fear. "Whether I do anything or not," he added, "I might as well be dead. I'm a marked man." To conceal the crime, the RAM leaders had killed six soldiers in their own death squad, starting with Sergeant Dionisio Ramirez in early 1987. After a confessional letter was found under his bed, nine sergeants, acting on Kapunan's orders, strangled him at Enrile's beach house and dumped the body in a deep ravine. Indeed, Barreto himself had heard Kapunan give the kill order and seen his comrades carry out the execution—details that he recalled in a precise, twenty-four step reconstruction of the crime. "At the time," Barreto said, "everyone felt the burden of killing a friend and a colleague." Over the next ten years, five more soldiers connected with the Olalia killing died suspiciously. When a comrade warned he was next, Barreto, convinced Kapunan could kill his own, went public, he said, "out of fear for my own life."[77]

The former RAM leaders dismissed the allegations as politically motivated.

"What else can it be but dirty politics?" Senator Honasan said, charging the revelations were designed to block his campaign for the vice presidency in the upcoming May elections. He denied any role in the coup plot that led to Olalia's death and admitted involvement in only three attempts, including the one in February 1986 that "eventually evolved into people power." Rex Robles, now a navy commodore, implied that the Aquino administration had fabricated the whole November 1986 coup scenario, murder included, to discredit RAM. Noe Wong, now National Police commander for Bulacan, denied "any involvement or knowledge." Kapunan, a rural banker now retired from the military, attacked Barreto as unreliable, saying he had been implicated in graft.[78]

Then, on January 14, Marine Sergeant Eduardo Bueno, another former member of Kapunan's SOG, appeared on ABS-CBN's *TV Patrol* to confess his participation in the Olalia murder. By his count, five former comrades had died under mysterious circumstances. In fact, Bueno admitted to being one of the six sergeants who had murdered the would-be informer Ramirez, holding his hand during the ritualized strangulation. He too was afraid for his life. In the wake of such stunning revelations, the Justice Department moved to provide protection for these key witnesses—a process that was complicated when it was found that six of Olalia's SOG killers were now staffing the Witness Protection Program. In mid-April, after a delay of twelve years, the Justice Department finally filed charges against thirteen RAM members, including Red Kapunan and Tito Legaspi, for the murder of Rolando Olalia. At preliminary hearings, defense lawyers insisted that these crimes were covered by amnesty, a plea that may ultimately block a successful prosecution. Indeed, a year later when the NBI finally arrested the first of these suspects, Captain Ricardo Dicon, an angry Senator Honasan insisted that the Olalia killing was so covered and threatened "terroristic activities" unless the government relented. Significantly, the victim's son, Attorney Rolando Rico Olalia, pointed out the seriousness of the senator's tactical blunder: in claiming amnesty for this crime, something RAM had always avoided, "Honasan made an extrajudicial confession regarding the slayings . . . before the entire citizenry."[79]

Honasan and his RAM comrades are thus an instructive study in the persistent legacy of authoritarian rule. If Barreto's affidavit was accurate, elements of two PMA classes, 1971 and 1975, led by their former barons, had formed a death squad and plotted a brutal execution with military precision. Then, for the next twelve years, they campaigned, without any sign of self-doubt, to win amnesty through legal maneuver and systematic murder. Showing no signs of remorse or restraint, they were working tirelessly to place their leader in the presidency.

Their horrific crimes sparked outrage, but society's collective will could not overcome their political resources. With its perpetrators initially protected by a dictator's denials, and then condoned in a weakened democracy, torture's effect upon the Philippines has persisted long after the actual event, not only in the trauma of the tortured, but in the ambitions of the torturers and the political paralysis of their society.

AMERICAN JUSTICE

As President Ramos moved toward amnesty for his military, their victims were pursuing a massive class-action suit in the U.S. district courts. This juxtaposition of impunity at home, justice abroad was a sign that the trauma of Marcos's terror was becoming imbedded within the Philippines' institutional fabric, constraining its political and creative capacities.

This case began in August 1977, when the dictator's high-strung daughter Imee Marcos, then leader of the youth group *Kabataan Barangay,* appeared before the student body at Mapua Institute of Technology. When a twenty-one-year-old student named Archimedes "Arch" Trajano dared to ask how she became leader of the nation's youth, her bodyguards escorted him to a military camp. Several days later, his mother found the body in a morgue with telltale marks of torture. In 1982, Mrs. Agapita Trajano, still mourning, moved to Honolulu.

Four years later, Marcos fled to Hawaii with President Reagan's promise of safe haven, unaware that he was landing in the only country where he was accountable for his crimes. In 1980, the U.S. Court of Appeals, in *Filartiga v. Peña-Irala,* had found a former Paraguayan police inspector, then residing in the United States, liable for damages since "an act of torture committed by a state official . . . violates established norms of human rights." Not long after Marcos landed in Honolulu, Mrs. Trajano retained activist attorney Sherry Broder, who drew upon this precedent to file a civil case against the former dictator.[80]

Simultaneously, Philadelphia lawyer Robert Swift flew to Manila for the first of many meetings with torture victims and human rights lawyers. By reviewing Marcos's arrest orders and interviewing victims, Swift assembled a class-action suit against the exiled dictator for the torture and murder of 9,541 victims. In San Francisco, renowned litigator Melvin Belli also filed a damage suit on behalf of 23 victims. The U.S. federal courts, through the Multi-District Litigation Panel, consolidated these cases at Honolulu and the litigants then selected Swift as their lead attorney.[81]

Marcos died in 1989, his daughter Imee fled to Morocco, and widow Imelda returned to Manila, but the litigation ground on. By the time the case came to trial in 1992, the Philippine Human Rights Commission had abandoned any attempt at prosecution, making this Honolulu court, by default, the only Filipino forum for justice. The chair of Task Force Detainees, Sister Mariani Dimaranan, testified that her group had documented 5,531 instances of torture under Marcos. Former U.S. ambassador Stephen Bosworth explained that the dictator had exercised "absolute control" over the armed forces and ignored torture. Attorneys for the Marcos estate did not challenge this testimony, trying instead to shift blame to his military subordinates.[82]

In September 1992, the Honolulu jury found Marcos guilty of systematic torture and held his estate liable for damages to all 9,541 victims. Trinidad Herrera-Repuno, an urban poor leader tortured in 1977, declared, "We were given justice for our sufferings." Representative Juan Ponce Enrile, Marcos's former defense minister, denied any culpability, while Senator Wigberto Tañada insisted, "There must be a proper trial in our homeland of these human rights violations." Then, in February 1994, the Honolulu court awarded $1.2 billion in punitive damages—a ruling that attorney Belli hailed as the biggest personal injury verdict in legal history. Lead counsel Swift, indicating future points of contention, added: "We obtained an injunction . . . that froze Marcos assets worldwide, and we know that there are $410 million in assets in Switzerland."[83]

The euphoria was short lived and the litigation soon descended into an ugly brawl over the Marcos millions. In September, after Swift reported that the Ramos administration was trying to split the Swiss deposits with the dictator's family, U.S. Judge Manuel Real extended the injunction to the Philippine government, accusing it of "aiding and abetting the Marcoses." Indeed, only a few days earlier, a Manila regional court had claimed exclusive jurisdiction over the Marcos estate, providing the Ramos government with a pretext for recovering the Swiss deposits.[84]

Within months, the confrontation escalated. In January 1995, the total rose to nearly $2 billion, when the Honolulu court, after reviewing evidence for sample claimants, awarded an additional $766 million in compensatory damages—including $100,000 to claimant Jean C. Tayag, once tortured by then-lieutenant Rodolfo Aguinaldo.[85] But only a week later, President Ramos announced that he would claim the Marcos assets "for the benefit of the sixty-five million Filipinos who are not . . . beneficiaries of the Hawaii rulings." In a biting, personal attack, Manila's *Daily Inquirer* blasted the "moral bankruptcy of the Ramos administration" and reminded him that as constabulary chief under Marcos "it was his men . . . who were applying the water cure to extract confessions and administered electrical shocks to genitals of political detainees." These words were echoed across the political spectrum. The Filipino lawyer for the victims, Romeo Capulong, suggested that Ramos could be charged, a view that Imelda Marcos herself endorsed. "The human rights abuses were mostly done by the then Philippine Constabulary," she said. "And who was the [constabulary] chief then?"[86]

With each passing month, these overlapping, intersecting proceedings in three separate jurisdictions—Philippine, Swiss, and American—twisted the litigation into a lawyers' Gordian knot. In May, U.S. Judge Real ruled that the victims could collect damages from the $450 million in Marcos's Swiss accounts, a decision that brought more litigation and arbitration.[87] Then, in February 1997, a Swiss appeals court ruled in favor of the Marcos family, requiring that the former dictator be found guilty of theft in the Philippines before any funds could be released to the litigants. The *Philippine Daily Inquirer* pointed out that Ramos's "rehabilitation" of so many Marcos cronies now made any such conviction impossible.[88] Indeed, only a few days later, the human rights commissioner, Aurora Reciña, a Ramos appointee and political ally, certified the Marcos family free of any human rights violations. The spectacle of the country's chief human rights enforcer exonerating the dictatorship prompted the victims' group SELDA to charge collusion between the government and the Marcoses to split the Swiss millions.[89]

Whatever its ultimate outcome, the case has, like no other single event, illuminated the compromises that impunity has imposed upon the processes of law and politics in the Philippines. In contrast to the cleansing role of truth commissions elsewhere, the Philippine juxtaposition of American justice with local impunity tainted all who touched it. Seven years after the Honolulu judgment, the victims were factionalized, their lawyers feuding, and the Philippine government colluding with the Marcos heirs to split his stolen billions.[90] For many in Manila, the contrast with the South Korean courts was painfully obvious. In August 1996, when a Seoul judge sentenced two ex-presidents for corruption and brutality, the *Philippine Daily Inquirer* headlined, "Ex-Korea President Sentenced to Death." Directly below, another headline read, "Gov't., Marcoses Share Loot." The next day, President Ramos refused any comparison, saying "that's Korean-style justice," while Senator Honasan agreed, adding, "Maybe we are more forgiving, more humane."[91]

Reflecting on the litigation in a letter to the *Wall Street Journal*, Swift hailed the original Honolulu decision as a "historic precedent" and "a beacon of hope for the victims of abuse and a real credit to the judiciary of the United States."

Under conditions of impunity, the same could not be said of the Philippine courts.[92]

A LEGACY OF TERROR

A decade after Marcos's downfall, an unexpected event compelled the Philippine public to reflect upon the costs of impunity. In June 1996, during the usual morning gridlock along Manila's Katipunan Avenue, six pedestrians approached a Honda Accord bearing license plate "RNA 777," its driver's initials and Marcos's magic number. One pedestrian peered into the driver's side window and nodded, the signal for a burst of pistol fire that shattered the glass. Another opened the door for a final, fatal shot to the temple of Colonel Rolando N. Abadilla, described in the press as "the country's most feared martial-law enforcer and torturer."[93]

Abadilla was not only the creature of Marcos's will, he was its ultimate creation. Born into a merchant family in the town of Banna, Ilocos Norte, Abadilla inherited his mother's distant relationship to Ferdinand Marcos of nearby Batac. After graduating from the PMA in 1965, Abadilla joined the constabulary's elite Special Forces and was picked for "Project Merdeka," Marcos's covert operation to infiltrate terrorists into neighboring Sabah. Along with his superiors, he later faced charges of murdering the operation's Muslim trainees in the notorious Jabidah massacre.

Abadilla later transferred to the constabulary's Metrocom Intelligence and Security Group (MISG), where, under martial law, his aggressive pursuit of anti-Marcos subversives won him command of the unit in 1974. There he developed a close relationship with the constabulary chief, General Ramos, executing his orders with special methods. Asked, after Marcos's flight, if he had not been the "favorite" of General Ver, he replied, "I disagree with that. I was even closer to General Ramos. . . . As a good soldier, I follow legal orders from my superiors. . . . I didn't have much orders from General Ver. Most of them were from General Ramos." Often Abadilla would "go into a frenzy during questioning, lashing with a pistol at the face of a suspect." During his twelve years in command, Abadilla and his band of "MISG boys" evoked what one obituary writer called "countless images of pain—blocks of ice, droplets of ice water, clothes hangers, barbed wire, steel drums, light bulbs, car batteries." Abadilla seemed to revel in this reputation for torture. "With crime groups," he told a reporter who asked about his torture, "I had to be violent with them. Meet force with force. The MISG had to be feared."[94]

After Marcos's downfall, Abadilla mobilized loyalist forces for two abortive coups in 1986–87. When he was found innocent after a court-martial of dubious integrity, the justice secretary urged President Aquino to review the case. But Defense Secretary Fidel Ramos intervened on behalf of his former subordinate, insisting that "there was no valid cause for his further detention." While in prison facing coup charges, Abadilla ran for vice governor of his home province of Ilocos Norte and scored a surprising victory that launched his political career. After Ramos's election to the presidency in 1992, Abadilla joined the Presidential Anti-Crime Commission (PACC), chaired by Vice President Joseph Estrada and led by one of his old MISG men, Panfilo Lacson (PMA '71). Three years later, President Ramos prevailed upon him to run for governor of Ilocos Norte, and Abadilla agreed, saying, "I cannot say no to a former boss who is a good leader." But the ruling Lakas-NUCD party objected to flying the name of a notorious torturer, and Abadilla ran, unsuccessfully, as the party's unofficial "adopted" candidate.[95]

Throughout these years of notoriety, Abadilla flouted the courts. In 1982, the Free Legal Assistance Group (FLAG) filed a civil case against the MISG group for the brutal torture of nine victims. Abadilla simply ignored the summons for six years until President Aquino's Supreme Court finally ordered trial. In 1993, fourteen years after the offense, a Quezon City judge found Abadilla and two former aides, Panfilo Lacson and Rodolfo Aguinaldo, liable for P 550,000 in damages. But the defendants appealed, insuring delays that could continue for another decade or more.[96]

Through such compromises that constitute impunity, the Philippines, beneath the surface of a restored democracy, still suffered the lingering legacy of the Marcos era—collective trauma, pervasive brutality, and an ingrained institutional habit of torture. As defense secretary, Ramos had abolished the constabulary in 1991 without any investigation of its torture under Marcos and merged it with local police, already known as "gross violators of rights," to form the new Philippine National Police (PNP). There torture continued as standard procedure, protected by the police practice of shielding colleagues from investigation.[97] As president, Ramos promoted his constabulary loyalists within the police, producing leadership that, in the words of one observer, "stymied reforms in the PNP." By mid-1997, for example, RAM's strategist Vic Batac, a known torturer, was in line for promotion to chief superintendent along with three of his PMA classmates. Significantly, Ramos's choice to supervise the PNP as interior secretary was Robert Barbers, a career police colonel known as "Dirty Harry Jr." for his alleged involvement in the extrajudicial killings of thirty

Manila drug dealers. Thus, in 1997, the last full year of the Ramos presidency, the military was responsible for only 81 human rights violations, while the PNP accounted for 1,074, nearly 43 percent of the nation's total.[98]

The hidden costs of impunity became visible in the aftermath of Abadilla's killing. In mourning a fallen comrade, the military celebrated him as a hero—challenging society's collective memory of a past marked by salvaging, death, and disappearance. His PMA classmate General Romeo Padiernos hailed Abadilla as "a courageous soldier and a much misunderstood man." At the burial with full military honors, graced by the presence of President Ramos, some twenty of his old "MISG boys," now senior police officers, turned up in dress uniforms to swear revenge. In an emotional eulogy, PNP Chief Superintendent Panfilo Lacson, Abadilla's deputy in the notorious MISG torture unit, swore, "I vow to get your killers all the way up." When a TV reporter asked former Baguio Police Chief Roberto Ortega, another MISG veteran, how he would deal with the killers, he told the camera, "I'll do it my way!" Asked to explain, he replied, "You know the song 'My Way'? I'll do it my way, *markang bungo!*"—a reference to his trademark execution of suspects, both criminal and political. Later asked about the MISG's dubious human rights record in a television interview, Ortega snapped, "What human rights? Only the Reds raise that issue." The public was confronted with the spectacle of a military clique, once synonymous with Marcos's terror, at the apex of national police power.[99]

The Manila press speculated about the implications of a vendetta by these police commanders as a collective shudder went through the body politic. "They will display a cadaver or two and call the whole thing 'justice,'" said an editorial in the newspaper *Today*. "We would not be surprised if the trail leading to this bloody corpse would be strewn with other corpses, hacked and tortured to death in the dead of the night, away from prying eyes, in places were the public cannot have heard the victims' screams."[100]

On June 24, only eleven days after the murder, Interior Secretary Barbers and the PNP director-general called the press to Camp Crame to show off the eight suspects captured by "Task Force Rolly"—all squatting, handcuffed, and visibly battered. Their confessions, officials explained, were corroborated by ballistics tests on the murder weapon. As cameras clicked and zoomed, the police pageant was interrupted as the suspects' relatives cried out, "Get the real killers!" "My brother is not guilty!" "Please don't hurt my husband!" Their lawyers later told the press that the accused were innocent jeepney and tricycle drivers, all plucked randomly from the streets and subjected to nonstop torture—suffocation, electrocution, and beating. "All were found positive for torture," admitted one investigator from the Human Rights Commission. "They were administered electric shock," the wife of an accused told a Senate committee two days later. "They were also repeatedly beaten up." In an editorial on the sordid affair, the *Daily Inquirer* concluded that torture's persistence "reflects a culture in the police that knows no respect for human rights."[101]

Events would soon prove that the malaise had deeper roots. Instead of investigating, the chair of the Human Rights Commission, Aurora Reciña, defended the police. Her own investigators had found clear signs of torture, but she still dismissed the evidence as "not yet conclusive." Echoing the police story, she argued that the bruises could have occurred when the suspects resisted arrest. After an investigator's report finding torture leaked to the press, she ordered his "immediate relief" and canceled public hearings. For the Human Rights Commission, the main constitutional barrier against the abuses of the Marcos era, to ignore such a clear instance of torture seems, from a legal viewpoint, inexplicable.

Viewed politically, however, the decision is revealing of the processes that produce impunity. The commission's founding chair, Jose Diokno, was a renowned human rights activist, but his successor, Reciña, was an otherwise obscure figure who came to office with personal and political impedimenta. As a Ramos appointee, she shared his circumspection on human rights. She was, moreover, the wife of Romeo Reciña (PMA '57)—a retired constabulary general with a problematic human rights record and a longtime Ramos loyalist then well placed in his administration.[102] Thus implicated in the processes of impunity, the commission could not act even when evidence was overwhelming—torture victims on television with visible bruises, reports from its own examiners, and sworn Senate testimony.

The police case finally collapsed when a communist assassination unit, the Alex Boncayo Brigade, claimed the killing. In a message faxed to Manila's top dailies, the communists stated that Abadilla's killing was "part of the Brigade's resolve to render revolutionary justice to the tens of thousands of people who had been victimized by martial rule." When the press voiced doubts about the message's authenticity, the brigade's leader, Sergio Romero, appeared on Sky Cable News to reiterate his responsibility and insist, "We don't know the Tom, Dick, and Harry arrested by the police." Six months later, in January 1997, these claims were confirmed when police announced the arrest of the communist hit-squad leader responsible for the murder.[103]

This incident reveals the distortions that can develop behind a democratic facade when a society fails to confront its authoritarian past. Dictators silence dissent by cultivating a "culture of fear"—fear of arrest, torture, and endless incarceration. Over time, collective trauma deepens to become a continuous undercurrent just beneath the surface of public life. Once the autocrats are gone, their transgressions, like individual trauma, remain imbedded within a society's collective memory and institutional fabric, constraining, perhaps distorting, the political process.[104] Unless addressed, collective trauma, like its individual counterpart, may persist long after the dictator's name has faded from memory.

The Philippines has suffered a subtle political paralysis by seeking a shortcut to democracy without pausing to assess, much less purge, the legacy of the Marcos dictatorship. With the RAM rebels at the palace gates, President Aquino compromised on human rights and halted all investigations of military abuses, past and present. Most importantly, she forged an alliance with Marcos's martial-law enforcer, General Ramos, that assured her survival and his succession to the presidency. Step by step, his administration made a clear break with Aquino's fragile human rights legacy—amnesty for the military rebels, opposition to any compensation for torture victims, political embrace of torturers, and, finally, subversion of the Human Rights Commission. His successor, President Joseph Estrada, completed this general absolution by elevating notorious torturers to police commands, dismissing many of the criminal cases pending against the Marcos cronies, and offering state funerals for both the dictator and his enforcer, General Ver.

Such a policy entailed compromises that, in sum, stretched impunity to its limits. These visible, political bargains have, as the Abadilla aftermath illustrates, slowed the full restoration of democracy and its sine qua non, the rule of law. When a government lacks the will to prosecute the past regime's perpetrators, then, as one legal specialist has argued, "a complete failure of enforcement vitiates the authority of law itself, sapping its power to deter."[105] In the glow of renewed economic growth under Ramos, the Philippines seemed to engage in a collective denial of the thirty-five hundred killed and thirty-five thousand tortured under the Marcos dictatorship. As the economy revived and began to run with Asia's "tigers," there was a tendency to dismiss human rights as a problem of the past. But justice, and the broad social trust it encourages, are, as Robert Putnam reminds us, essential in rebuilding any nation's full fund of "social capital" after the trauma of dictatorship. Societies with "weak civic traditions" that succumbed to dictatorship are now finding as they emerge that "totalitarian rule abused even that limited stock of social capital," wrote Putnam, denying them the "cooperation, trust, reciprocity" needed for sustained growth. A nation cannot develop without high social capital, and social capital cannot grow in a society without justice.[106]

With society's leaders closing ranks to enforce silence about the Philippine past, the voices of dissent have been faint. The activist ex-priest Edicio de la Torre has sensed, since Marcos's fall in 1986, a deep need for reconciliation among both victims and perpetrators—a conviction he illustrates with a parable. When he was arrested in 1974, he said, a young officer "punched me in the stomach and hit me again at the back of the head." Sixteen years later, a uniformed colonel appeared at a public forum on the Philippines in London, proclaiming, "I want to meet Ed de la Torre." Why, the panel asked? "Because I want to give him the pleasure of shouting at me." Why, they asked? "Because," he told the audience, "I was the one who punched him in 1974." De la Torre himself feels that social reconciliation must begin with "public admission of the crimes, including disclosure of the graves and tortures." Then, after confession in the "Catholic tradition," there can there be forgiveness, repentance, and penance. But this process is an impossibility in such a flawed democracy, leading de la Torre to close his parable with a question that borders on despair: "How do you build a society on such shaky ground?" In a similar reflection on Nazi terror, Hannah Arendt shared this belief in the imperative of forgiveness, but she also felt, in words germane to the Philippines, that "men are unable to forgive what they cannot punish."[107]

Freed from judicial review, the torturers of the Marcos era have continued to rise within the police and intelligence bureaucracies, allowing martial law's legacy of military abuse and corruption to persist, unaddressed and largely uncorrected. Although Senator Honasan's campaign for the vice presidency, the usual stepping-stone to the presidency, was blocked by revelations about RAM's death squad, his lesser-known classmates continued their steady rise within the national police. After his inauguration in July 1998, President Joseph Estrada appointed PNP Chief Superintendent Panfilo Lacson to head the powerful Presidential Task Force on Organized Crime. Because Lacson, along with classmate Reynaldo Acop, was still facing charges for the 1995 mass murder of eleven members of the Kuratong Baleleng gang, the appointment sparked protest by human rights groups and the Catholic bishops. Nonetheless, with the president's backing and unlimited operational funds, Lacson soon emerged as the country's most powerful police officer. Within weeks, witnesses recanted and the murder case collapsed. Parallel promotion of three classmates—Ruben Cabagnot, Tiburcio Fusilero, and Acop—to key PNP regional commands made them the most powerful cohort in the police. Significantly, each was notorious for brutal killings in different stages in Class '71's career. Then in December, President Estrada, apparently unaware of the macabre irony, marked the fiftieth anniversary of the U.N. human rights declaration with a palace ceremony honoring his "trusted aide" Lacson, formerly leader of a notorious torture unit and currently facing mass-murder charges, for his "exemplary service to our country." A few weeks later, Lacson and two classmates, notably RAM leader Vic Batac, were promoted to two-star generals in the police hierarchy.

In the midst of his rapid rise, congressional investigators probing massive graft in the AFP's pension fund identified Lacson's PMA classmate, Colonel Oscar Martinez, as a principal in the theft of P 1.75 billion ($44 million). Without pausing to assay the credibility of the charges, both Lacson and Honasan rallied to their class valedictorian, threatening reprisals against any who would harm him. "Remember," said Senator Honasan, "this is the notorious Class '71. There were 106 of us who graduated in that class, and we're solid."[108]

Without a hearing at home, the victims of Marcos's tortures have been forced to seek justice in American courts—a process that both diminishes the Philippine government and exempts it from the burden of reform. Denied an official process of recording and redress akin to those in Argentina or South Africa, the Philippines seemed caught in a long nightmare between remembering and forgetting. "So this is where Philippine society stands 10 years after Marcos left," said the *Philippines Free Press* in its review of the Abadilla murder, "all the way around to where we started. With the military in power in Malacañang, deaths in the streets, absolution in the courts."[109]

THE ACADEMY'S LEGACY

The career of Fidel Ramos reveals much about the ambiguity of President Quezon's military legacy. Historical circumstance provided President Ramos with a unique opportunity to reshape national defense and reform the armed forces. The start of his term in 1992 coincided with extraordinary events—the eruption of Mount Pinatubo and the Philippine Senate's refusal to renew the U.S. bases agreement—which finally, after almost a century, closed the sprawling American military installations at Clark Field and Subic Bay. As the U.S. Navy tugs towed the floating dry docks over the horizon toward Pearl Harbor, the Philippine military assumed, for the first time in its history, full responsibility for the country's defense. Three years later, China occupied a portion of the Spratley Islands claimed by the Philippines. With a clear external threat, Ramos won passage of the AFP Modernization Act of 1995 and an appropriation of thirty billion pesos for air and naval equipment, restoring a sense of military professionalism absent for decades.[110] As he turned the nation's guns outward, Ramos simultaneously eliminated civil unrest by negotiating an end to the Muslim and military revolts, rewarding rebel leaders who laid down their arms with political office. Although the Asian financial crisis would slow and ultimately block any major appropriations before his term ended in mid-1998, Ramos and the Senate had nonetheless defined the future direction of the armed forces.

Such success did not come without a price. Before they would return to the barracks, the Philippine officer corps, like its counterparts elsewhere in the developing world, extracted one concession—impunity for crimes done in service to dictatorship. Many new democracies have resisted these demands, winning concessions to justice that range from confession by the torturers to compensation for the victims. Under Ramos, by contrast, the Philippines practiced an extreme form of impunity by granting the military a blanket amnesty for torture and extrajudicial murder. Rather than punish human rights abuse and purge the offenders, Ramos split the military into a revitalized AFP and a new Philippine National Police. The latter force merged an unreconstructed constabulary and local police to become a refuge for torturers. With the president's support, former military rebels, Colonel Honasan among them, have been elected to provincial and national offices.

Looking at these events from a broader historical perspective, the Ramos administration represents a fulfillment of the political potential inherent in the PMA since its founding in 1936. By training career officers at a state military academy, President Quezon was establishing a corps of professional officers on the American model and thus denying control over the nation's arsenal to established political elites, whether nationalist attorneys from the University of the Philippines or corporate executives from the Ateneo. Ramos, as a graduate of West Point and an honorary member of the PMA's Class of 1951, is a product of this professional military. By preparing the academy's graduates for command, Quezon was trying to foster a tradition of military professionalism. Ramos, in his opposition to nine coups, affirmed these values of apolitical service to the state. By investing the academy with the prestige of his office, Quezon was creating a new elite and opening an alternative path to power. Ramos, in his election and appointments, was a realization of that potential. In building the academy to train a professional officer corps, Quezon showed his wariness of military involvement in partisan politics. Ramos, by his long service to dictatorship and his presidency, gave form to those fears.

But above all, Ramos's ascent makes an important statement about the future role of the country's officer corps. Through his defeat of RAM on the battlefield and in the ballot box, he represented a momentary triumph of the military manager over the heroic commander. Ramos and his coterie of retired officers successfully harnessed "the military's sense of mission as an institution" to realize an ambitious agenda of economic reforms.[111] When human right groups protested Ramos's appointment of so many retired officers as a "silent coup," Public Works Secretary Gregorio Vigilar, a West Point graduate ('53) and former colonel, invoked the military's heroic role. "The same way," he argued, "that military men are almost ready to turn themselves into virtual cannonballs in war, they should not be denied the chance of displaying similar dedication to public service in peace time."[112] The Ramos administration returned the soldiers to the barracks. But it did not resolve the underlying tension between the values of heroism and professionalism embedded within the Philippine officer corps since its founding.

As soon as the PMA graduated its first large "batch" in 1940, the academy's alumni, operating as classes, emerged as an important actor in Philippine politics. As later graduates added both mass and complexity to the officer corps, certain classes, by virtue of character and circumstance, played a critical role in the political process—staffing civilian agencies, shaping defense policy, and promoting or blocking coups.

More broadly, the academy has played an important but little understood role in shaping a modern polity. During the twentieth century, the Philippine state created only two major institutions of higher education: the University of the Philippines, home of secular, nationalist politics; and the PMA, the source of its professional officers and disciplined civil servants. Over time, the military academy has served as a uniquely effective vehicle for social mobility, allowing talented peasants, workers, and petit bourgeois to rise from poverty to command. In establishing this academy, the Philippines fostered a military elite that was a servant of state power, willing to fight and die in its defense. But the PMA also became, over time, the core of a somewhat insular, quasi-hereditary officer corps with a strong sense of patriotism and privilege that has both strengthened and limited the prerogatives of executive power. Their "clannishness and cohesiveness," wrote a retired navy captain, a non-alumnus, "derive from their being products of an exclusive and fraternal character-molding institution, where they undergo . . . nurturing of loyalty and camaraderie manifested in their recognition of one another as 'mistah.'"[113]

Such deep patriotism and tight social bonds can, under stable conditions, support civil supremacy and democracy. But in times of social strife, these same ties can lead military factions to revolt or allow officers to sanction systematic abuse of human rights.

Chapter 10 Reunion

In 1986, fifty years after they entered the Philippine Military Academy, the Class of 1940 began a four-year cycle of reunions that coincided with eight coup attempts by younger alumni. Juxtaposition of reunion and revolt turned their reminiscence into reflection.

Class '40 did not shirk self-congratulation. "What may be considered a most significant achievement," they wrote in their reunion *Golden Book,* "is the fact that after 30 years of service . . . no member of the batch was ever tainted with the breath of scandal; no charges of unexplained or ill-gotten wealth; no charges of human rights violations."[1] Half a century after graduation, they were telling their alma mater that, in the words of the PMA song, the "honor you instill" still "doth guide our will."

To themselves they were saying that, despite the many compromises of active duty, they were still men of honor. To us, observers of their past, they seemed to say that their military socialization had been deep and lasting. They had internalized the rules and lived by them throughout their long careers.

They had remained, in their own eyes, the "same banana" class. In an essay for the *Golden Book,* General Reynaldo Mendoza, the author of the school song and the superintendent whom we met on the first page, argued that the idealism of this exam boycott had inspired them through thirty years of service. Indeed, he concluded, "the 'Same Banana' spirit lives up to this day!" In the tension of that student strike, a forgotten classmate had tossed out those words as a quip, an ephemeral witticism. But the class had, over the decades, recast them into a symbol of integrity.[2] Indeed, the metaphor is instructive in ways that they might not have intended. Clustered around a single stalk and sheathed in near identical skins, this bunch had ripened uniformly with age.

Class '40 was sharply critical of the "solid ranks of gray" that had followed them out of the academy. The first coups—at the Manila Hotel in July 1986 and Camp Aguinaldo in November—were small and inconclusive. But in August 1987, Colonel Gringo Honasan, the former first captain of Class '71, had led officers from nearly twenty academy classes and two thousand rebel troops in an attack on Malacañang Palace. Simultaneously, his classmate, Lieutenant Colonel Red Kapunan, an instructor at the PMA, encouraged an armed revolt by the entire cadet corps. Class '40 found this coup, particularly the academy's participation, abhorrent.

The class denounced "the mutinous acts of the coup plotters" in speeches and letters. At a reunion in October 1987, they heard reports that "no cadet officer has been busted . . . and no dismissals appear to be contemplated." Since these cadets had taken up arms against the state, the class urged, in a collective letter to Chief of Staff Fidel Ramos, that "sterner measures should be undertaken." Specifically, they asked that rebel officers on the academy's staff be court-martialed and all cadets tainted by their ideology be purged.[3]

Several classmates traced the politicization of the officer corps to a decline in the academy's honor system. Writing in the *Golden Book,* Colonel Deogracias Caballero, the moral leader of Class '40, condemned officers who had sold themselves to Marcos and urged a return to the academy's principles—courage, loyalty, integrity. "Courage, as envisioned in the motto, means the will . . . to do what is good, moral and honorable," he wrote. "But it is reported that some First Classmen . . . conspired to arm themselves with their issued firearms . . . and . . . to isolate Baguio City in support of Honasan's rebellion. This courage to commit treason . . . is the height of condemnable acts absolutely unbecoming a soldier." Caballero traced this dishonor to the postwar decline in the academy's standards, particularly the tendency to exempt hazing from the strictures of the honor system. "So, if a hazing occurs," he explained, "the deed is buried in a code of silence. . . . Thus was born the concept that one can do anything provided that he is not caught." Speaking for his class, he said sadly, "We mourn the corruption of the Honor System."[4]

In effect, Caballero was condemning the ascent of personal ties over principle. Rather than offering their highest loyalty to the abstractions of the honor code, younger officers were, in his view, letting personal ties to class and classmates blind them to their duties.

In March Week of 1990, the fiftieth reunion of their graduation, Commodore Ramon Alcaraz, a classmate active in the anti-Marcos opposition, delivered a homily in the chapel before the Corps of Cadets. He celebrated his class, condemned the corruption of its values, and called upon cadets for moral renewal. The whole of Class '90, seated before him on the eve of their graduation, had taken up arms in the 1987 coup, giving a certain poignancy to his words. The old commodore reminded them that his class "was the first to graduate under the four-year curriculum patterned after West Point, introduced by two military greats—MacArthur and Eisenhower," adding proudly that many "academy traditions originated from our class." Instead of wishing Class '90 the usual good luck and Godspeed, he condemned their role in that coup three years before.

> I consider this year's Class of 1990 still fortunate to be graduating tomorrow. . . . You were yearlings during that bloody August 1987 coup attempt led by Gregorio Honasan, whose father, Romeo, was my underclassman. . . . For your questionable behavior during that treasonous coup, all of you should have been summarily dismissed. . . .
>
> As the shadows of my twilight years are lengthening, this is perhaps the last time for me to visit my beloved alma mater. . . . I would like to ask Class 1990 . . . —a last request from the Class of 1940—that after you have taken your oath to uphold and defend the constitution . . . emulate Class 1940. Go forth out there and be a strong moral force in transforming the military into a profession of honor which it used to be.
>
> When we graduated in 1940, the army chief of operations, Brigadier General Vicente Lim, West Point Class 1915 [sic], asked us to be a strong influence in upgrading the Philippine Army. I feel we had completed that mission successfully. It is now your turn to upgrade the military.[5]

These recurring references to honor and morality are striking. Speaking for his class, the commodore seemed to call on the officer corps to return to the academy's original values. Instead of following the violent machismo epitomized by the RAM rebels of Class '71, cadets should emulate his class and embrace a prewar standard of honor based on subordination and discipline.

Even this appeal to principle was not cast in purely abstract terms. From their cadet days, Class '40 had viewed the honor code as the warp in the weave of personal ties that bound them together. After graduation, their adherence to this standard had been strengthened by bonds to class and classmates. For both Commodore Alcaraz and Colonel Caballero, such loyalties should reinforce individual commitment to core values—above all, avoidance of partisan politics and acceptance of civil supremacy. Any deviance demanded severe sanctions. For leading the corps in revolt, instructors should be court-martialed. For joining that coup, the whole of Class '90, like cadets who cheat on an exam, should be expelled from "those solid ranks of gray."

After the 1987 coup, Caballero had mourned and condemned, urging mass dismissal of these dishonorable cadets, Class '90 included. But three years later, faced with the reality of their graduation, Alcaraz encouraged a return to principle by offering them membership in a group larger than their own class, larger than Colonel Honasan's RAM—a long gray line of honorable officers from the long-departed General Lim (USMA '14), to the aging alumni of PMA '40, to themselves, PMA '90.

Class '40's dismay at the corruption of the military and the collapse of its discipline is understandable. Drawn from provinces across the archipelago by the accident of examination, they had forged a strong identity and lifelong bonds at the PMA. More than any other comparable Filipino elite, the officer corps had been created and defined by the nation. No other group had its social role, ideology, and personal values so directly, so fundamentally shaped by the state. Given the strength of these bonds and the depth of their indoctrination, Class '40's anger at RAM's revolt ran deep.

The honor code was a moral touchstone for the cavaliers of Class '40, covering everything from a cadet's exams to a general's protocol for dealing with the president. Implicit within the code was a view of class and corps as autonomous—above and apart from Filipino society. While that perspective reflected the Spartan simplicity of cadet life, it could not encompass the complexity of serving officers subordinated to civil authority. As cadets living in an academy isolated from the wider society, moral choices were clear within a social universe that went no further than the main gate.

Once they reached senior rank, however, serving officers faced pressures from civil authorities who were more and more determined, over the years, to compromise their integrity. When their commander in chief—whether Garcia, Macapagal, or Marcos—decided to play politics with the military, officers had only a limited autonomy for resistance. Their ingrained respect for civil supremacy dictated obedience. Thus, separation of civilian politics from military power was not only the responsibility of the officer corps. Once civilian politicians saw partisan advantage in politicizing the military, commanders faced strong pressures to compromise.

Moreover, the demands of honor and class loyalty were often in conflict. Within the code's narrowly drawn standards for personal integrity, both members of Class '40 who served as chief of staff, Victor Osias and Segundo Velasco, were honorable. Neither had profited personally at the apex of the armed forces. But if we broaden the standards of honor, then there are telling differences. Osias suffered a humiliating dismissal for refusing to comply with Marcos's demands. Velasco, by contrast, was silent while the president formed an illegal, covert-action apparatus within the military. Instead of criticizing Velasco, Class '40 celebrated him. "Gen. 'Gunding' Velasco," the class wrote on the final page of their reunion's *Golden Book*, "was always very proud to state publicly . . . that during his tour, not a single member of the Class . . . ever approached him to ask for any favor that could be misconstrued as 'taking undue advantage' of class relationship."[6] At the apex of the military hierarchy, however, choices were complex in ways that eluded the simple absolutes of the honor code. If we apply the PMA's standards for honor in exams to the requisites of command, Velasco may not have cheated as chief of staff, but he did fail to report those who had.

Thus, their assessments were sometimes uncritical, and they were often ready to forgive a classmate's breach of conduct, whether omission or commission, to preserve their unity. Honored collectively and affirmed individually by Marcos's choice of two classmates as chief of staff, they seemed incapable of distinguishing between one's opposition to his illegal maneuvers and the other's compliance.

In the final analysis, even Class '40 flinched when forced to apply the strict standards of their own honor code and criticize a classmate. In cases of conflict between person and principle, loyalty shaded their moral standards. In their judgment of Velasco, the class overlooked the code's broad demand that a cadet, or chief of staff, report any offense witnessed. The class could not admit Velasco's responsibility for trends within the military that they abhorred. And even if they had seen his dereliction, they still might have found a way to exculpate a beloved classmate. Indeed, their *Golden Book* described Tomas Karingal, who ranked among the most brutal of Marcos's police, as an officer who "served brilliantly," and celebrated Jose Maristela, who spied on classmates for the dictator, as "a person who was not afraid to fight for what he thought was right."[7] In obvious and important ways, Class '40 was held together by some of the same bonds of blind loyalty that led many in Class '71, and the whole of '90, into Colonel Honasan's coup.

With its mix of reunion and graduation, March Week 1990 thus became a dialogue among classes and generations. On the golden anniversary of their commencement, Class '40 criticized the moral choices of '71 to inspire a sense of honor, as they understood it, among the graduates of '90. For observers like us who stand outside the long gray line, there also seems to be, at a deeper level, a debate among generations and even epochs: Class '40 as the heroic past, '71 the troubled present, and '90 the uncertain future. If, as observers, we cast a critical eye upon each of these classes to compare and contrast, we might gain some further insights into the meaning of their military socialization.

COMPARING CLASSES

Looking backward, and leaving aside Class '90 momentarily, the contrast between Class '40 and Class '71 is striking—and perplexing. In real political terms, there is an obvious and important difference between a class unity that prevents coups and one that promotes them. Both classes marched lockstep through an unchanging, four-year progression from degraded plebes to exalted first classmen. Why did two groups of young men, graduating from the same school under similar curricula, turn out so differently?

The answer lies in a confluence of factors internal and external to each class— ranging from individual personalities to global geopolitics. Each class enters the PMA with a particular mix of individual personalities and passes through a four-year experience that is both uniform and unique. Through a regimen of drill, discipline, and indoctrination, the academy seeks to mold every diverse class into a uniform batch for incorporation en bloc into a rigid military hierarchy. But even in the isolation of the mountain campus, unique external elements influence the socialization of each class—the values of entering plebes, political ideas flowing from Manila, and the ruling regime's political agenda. As the class scatters upon graduation and loses its formal coherence, external factors become paramount in shaping their socialization as junior officers. In this sensitive second phase, the regime occupying Malacañang Palace can have a strong, even transformative impact upon the character of each class.

In comparing these two classes, '40 and '71, we cannot overlook the deeper currents of global change concealed beneath the surface of events. During the Philippines' colonial era (1898–1935), the United States built a school system, a national university, and, later, training programs for Filipino officers. To prepare the colony for independence in the 1930s, the U.S. Army encouraged formation of a defense force in its own image led by professional, apolitical officers. In this transitional decade, the Philippine Commonwealth (1935–46), assisted by its American advisors, opened the PMA and produced its first graduates in 1940. For most of the Cold War, the United States was the dominant power in Southeast Asia, backing coups across the region to increase its influence—from Laos in 1960 to Thailand in 1976. But in these postwar decades, a period that coincided with the service of Class '40, Washington still regarded the Philippines as its "showcase of democracy" and did not encourage the political ambitions of its military.

By the early 1970s, however, the United States, anxious over its military bases in the Philippines, supported Marcos's constitutional coup and sanctioned his politicization of the military, Class '71 included. Not only did the Pentagon increase U.S. military aid to the dictator, but there is a possibility that the CIA may have trained his torturers.[8] In the 1980s, Washington's decision to distance itself from Marcos, and then to support his successor Corazon Aquino, contributed to the failure of the six coups led by Class '71.

More broadly, the patterns of Philippine politics coincide with Samuel Huntington's worldwide "waves of democratization." Since independence in 1946, the Philippine state has moved through three constitutional regimes—postwar Republic (1946–72), authoritarian New Society (1972–86), and democratic restoration (1986–present). In his broad overview, Huntington found three waves of democratization sweeping the globe that seem to parallel these Philippine events—a short wave of rising democracy (1943–62), a reverse wave of dictatorship (1958–75), and a third wave of democratization (1974–present).[9] Hence, global currents and geopolitics have played a significant, albeit indirect, role in shaping the career and character of these two classes.

Through a long alliance with the United States, the Philippines opened itself to a global culture that influenced both its identity and politics. In building an army on the American model, the Commonwealth imported a foreign model of masculinity and then, through mass mobilization, drilled Filipino males to its standards. Before Class '40 entered the PMA, many of its future cadets saw the 1934 Hollywood musical *Flirtation Walk,* starring Dick Powell, and found this song-and-dance celebration of military discipline appealing. As cadets, Class '40 acted as extras in two Filipino films inspired by this Hollywood musical, projecting a similar image of disciplined masculinity to audiences across the archipelago. A generation later, Class '71 was inspired by Sergio Leone's Italian

Westerns, starting with *A Fistful of Dollars* in 1966, which starred Clint Eastwood as a man of cool violence. This cinematic celebration of masculine aggression, epitomized in Sylvester Stallone's *Rambo* movies (1982, 1985), resonated within Filipino action films that both reflected and reinforced RAM's coup attempts of the late 1980s.

Juxtaposing two Filipino film posters separated by half a century gives us before-and-after snapshots of this change in masculine ideals. In advertisements for the 1939 Filipino film *Punit na Bandila* (The Torn Flag), leading man Fernando Poe, Sr., stands tall and handsome in the PMA's dress-gray uniform—powerful but disciplined in the manner of a dapper Dick Powell. In 1995, his own son, the action star Fernando Poe, Jr., is an image of insurgent masculinity erupting from Manila's billboards for *Kahit Butas ng Karayom Papasukin Ko* (I Will Pass Through the Eye of a Needle)—gripping an assault rifle, torso dripping with ammunition belts à la Rambo.

State and society also played important roles in shaping the values of each cadet class. Entering the academy between the ages of eighteen and twenty-two, the cadets brought with them the values of their families, schools, and generation. Philippine society of the 1930s trained its young in patriotism and public service as it struggled to build a new nation. By the 1960s, however, radical nationalism inspired a youth movement that questioned the legitimacy of the state. Nationalism began to challenge military professionalism for the cadets' ultimate allegiance.

Reflecting these changes, the academy's institutional ethos experienced a slow shift that contributed to the contrast between these classes, '40 and '71. In the splendid isolation of their new academy, Class '40 imbibed American colonial values of military professionalism. The Commonwealth insured that the PMA's hazing was severe but not brutal. Thus, the class was molded, but not scarred, by a ritual that forged a strong collective identity. By the time Class '71 entered the PMA in the mid-1960s, this rite of passage had degenerated into brutality. The cadets intensified the culture of group bonding over time, but the institution's capacity to inculcate professional ideals seemed to decline.

Under Quezon and Marcos, strong regimes left their imprint on several succeeding classes, making them cohorts marked by distinct sets of shared values. In the Commonwealth era, Class '40 and its underclassmen emerged from the PMA with a strong commitment to military professionalism. After graduation, they joined an army dedicated to national defense and later led a postwar military absorbed in the fight against communist insurgency. By contrast, Class '71 and its successors graduated into Marcos's martial-law regime and were invested with civil powers that inclined them toward political activism. As junior officers, they served Marcos's "New Society" with all their youthful idealism, interrogating dissidents and fighting armed rebels. When they grew disillusioned with the regime's corruption, class leaders formed new political alliances and moved gradually toward revolt.

Among many possible points of comparison, there is one experience that seems particularly significant. Under the Marcos dictatorship, Class '71 became a fist of repression and learned the special skills needed for this mission—torture, intimidation, interrogation, surveillance, penetration, and psychological warfare. As players in Marcos's script of violence, they gradually broke free from the constraints of military discipline. Though trained to serve the state, they gained the will, through these special operations, to become its master. They later preached reform of a corrupted Marcos military, but the RAM boys of Class '71 were its ultimate creatures.

With a heroic self-image rehearsed in safe houses and acted out on EDSA during the people power uprising, the colonels of Class '71 found it impossible to accept the restoration of civil authority under President Aquino. Despite dwindling chances of success, they persisted in their coup attempts until utterly defeated. Transformed by experiences of torture and terror, these colonels were driven by a will to power that seemed to eclipse any military calculus.

In their rebellion against the state, the RAM leaders, almost all of them academy alumni, raise questions about the PMA's success in training junior officers. In its first years, when the state encouraged an apolitical military, the academy used bonding rituals to unify classes like '40 and inspire them to a high standard of professionalism. Class '71 used these same loyalties to sanction torture and rebellion once Marcos made the military a bastion of his authoritarian state.

In her work on the making of torturers, psychologist Janice Gibson notes that ritual humiliation, binding activities, and a sense of elite superiority are conducive, within military organizations, to training torturers.[10] Certainly, the PMA's plebe initiation had all these attributes. In a different political climate, this binding or bonding could, as it did with Class '40, create restraints against such abuse. But once a regime decided to use the military for civil repression, the solidarity of the officer corps could offer only limited resistance. When Marcos, as commander in chief, declared martial law, regular officers had little choice but to obey and enforce his authoritarian rule. Once his regime opened a gulag of safe houses for routine torture, the harsh experience of plebe hazing at the PMA may have allowed the young lieutenants of Class '71 to accept a growing brutalization. Ties to classmates and comrades may have salved doubts and sustained them in a dirty war against a people they were sworn to protect. In sum, this bonding, so successful in defense of the nation against external enemies, can become problematic when officers are charged with domestic repression.

Such systematic torture contributed to the politicization of a generation of regular officers. Through the theatrics of psychological torture, RAM's leaders gained the will to launch a succession of coups that were remarkable for their consistent failure. Instead of focusing on the origins of each coup, we need to reflect upon the implications of their repeated defeat. With few exceptions, military coups d'état in the Third World usually succeed in seizing power. To fail, as RAM did, six times in a row should be something of a record. Why such consistent failure?

Much of the explanation may lie with the incompetence of RAM's leaders. Enamored of their own power, the colonels of Class '71 failed to plan adequately for their coup attempts. Overestimating the power of violence after years of torture and terror, they expected the capital to be awed by the majesty of their blazing rifles, thundering bombs, and rumbling tanks. Even when possessed of superior force in December 1989, Colonel Honasan and his aides were so absorbed in a narcissistic projection of their power that they dallied in a television studio and failed to attack the objectives that would have given them victory.

Not only did RAM's leadership fail, but most officers were not inclined to follow. Simply put, many of these coups did not succeed because the majority of Filipino officers did not believe in seizing state power. A tentative belief in civil supremacy and military professionalism somehow survived the Marcos years among regular officers, most of them PMA alumni. Several of these coups were based on the false presumption that ordinary officers would rally spontaneously once RAM raised its inverted flag of revolt. But the great majority of regular officers had spent the Marcos years leading troops and did not acquire the patrons, power, or habit of torture that might have inclined them toward politics. More than five hundred officers joined RAM's largest coup in 1989, but even then only 30 percent of the officer corps had strong loyalties to either side, rebel or government. Most continued to serve on the line with their troops, avoiding political entanglements.

If the above analysis has merit, then the PMA has both succeeded and failed in its mission of creating an apolitical officer corps. At one level, it produced a minority of active coup plotters who played upon class ties in their bid for power. More broadly, the academy was effective in instilling values of professionalism in a majority of its regular officers.

Before we go further, let us admit the limitations of this exercise. Comparative history, straddling an academic divide like the historical discipline itself, is often more literary analysis, like contextual criticism, than a social science experiment. If we could eliminate all the ambit variables—ethos, era, culture, and circumstance—that distinguish events a decade or a generation apart, and thereby reduce our cases to their structural essentials, it would no longer be history, but something closer to a behavioral experiment. But let us assume that we could, like psychologist Philip Zimbardo, select twenty "average Filipino college males" to role-play two cadet classes, '40 and '71, under the same controlled, laboratory-like conditions in a basement at U.P. or Ateneo. Would our student simulators behave the same when confronted with choices about torture, terror, or revolt? This is a critical question, and the answer must remain tentative and ambiguous. To put it succinctly, yes and no. With its diffuse leadership echelon and quirky individualism ("ask the same question and you will get seventy-nine different answers," one classmate announced at their hangout, the Last Watering Hole), Class '40 was never as easily led as Class '71 under Honasan.

Even so, we must ask: if the individuals who were Class '40 had, somehow, graduated in March 1971, would they have behaved just like the real '71, becoming torturers and launching coups? Would circumstances have overwhelmed their individual and collective moral will? The short answer: maybe. One could make a case that Class '40 member Bartolome Cabangbang, for example, could have emerged as a Honasan-like figure under Marcos, or that Class '40, if plunged into the safe houses, would have emerged, like '71, brutalized and politicized.

While this argument makes some sense at a crude behavioral level, my instinct, after studying these two groups for a decade, somehow recoils from such a simple equation. No doubt Class '40 would have moved in the direction of Class '71, but they might not have gone all the way to become indistinguishable.

So therein lies the strength, and weakness, of this comparative history. Certain similar external factors, such as regime or military mission, play a constant role in forging the character of each class, making them at one level comparable. Yet they remain a particular mix of individuals not easily lifted from the culture, circumstance, and context of their day.

Even with this extended caveat, comparison does seem to reveal some fundamental distinctions in their histories. Class '40 somehow integrated professional values into personal ties in ways that Class '71 did not. When Class '40 moved beyond the academy's all-male formations to family life and individual careers, their sense of solidarity, epitomized in the "same banana" strike, inspired integrity. As cadets, they had learned, through invention and indoctrination, a military masculinity that balanced subordination with self-assertion, group loyalty with individual integrity. In the years after graduation, the group itself may have dissolved into memory, but its ideals and loyalties intertwined in ways that helped to restrain an entire generation of officers from corruption and coups. Despite many individual failings, Class '40, bound together by code and camaraderie, maintained a reasonable standard of professionalism throughout their decades on active duty.

As it turned out, the slender ties of class, corps, and honor were the threads of the Philippine Republic's constitutional fabric. Wispy though they might have been, these ties made the Philippine military an honorable service for over a quarter-century. Once this fabric began to fray, regular officers could turn their guns on the people to become instruments of a martial-law dictatorship or commanders of a coup d'état.

A LASTING LEGACY

While this contrast between Classes '40 and '71 offers insight about the country's past, a comparison of '40 and '90 allows us a glimpse into its future. Through this latter comparison, we can see how the interaction between ruling regime and each academy class, so important in the military's history for the past sixty years, might continue to influence the Philippine armed forces well into the twenty-first century.

There are striking similarities in the cadet experiences of these two classes that seem to defy the half-century separating them. Both moved through the same four-year progression from neophyte to cadet commander. Both emerged with the same core values—honor, patriotism, and, above all, class unity. If we juxtapose Class '40's "kaydet" days, as reconstructed in Chapter 2, with the class history in the *Sword '90* yearbook, we can see strong parallels in both the details and larger design of these narratives.

Like Class '40 before them, the entering plebes of '90 felt the "sure shock" of beast barracks, where they were "treated as the lowest mammalians." But they emerged from this suffering with a strong sense of "the solidity that this class had . . . achieved." Like their elders, Class '90 internalized the values of the honor system so thoroughly that, they wrote, it "became precious to us." At the end of their first year, these plebes were moved by the recognition ceremony, "when rare handshakes took place." As yearlings, they assumed their first responsibilities and learned "respect for the chain of command." In their third or "cow year," the class supervised the "molding of the new cadets," by "barking at the plebes for their laxities . . . and instilling in them the . . . traditions of the Cadet Corps." As seniors, they reveled in "being the newly-installed Immaculates of the Corps." They received their diplomas during March Week with "morale as high as the stars" and a feeling that "graduation is . . . but a beginning of real struggle."[11]

Even the extraordinary events in the history of these classes seem, paradoxically, similar. Periodically, the cadet corps has reacted to a perceived moral crisis with protests, sometimes a spectacular strike. Class '40 protested unfair exams, '71 staff corruption, and '90 government hypocrisy. Threatened with dismissal, both '40 and '90 persisted in the face of official pressure and emerged with their solidarity affirmed. Just as Class '40, in their cow year, had continued in their "same banana" strike until the chief of staff promised redress, so '90, as yearlings, maintained their armed protests until the undersecretary of defense met their demands. Both classes emerged from their respective three-day strikes with their solidarity and sense of integrity affirmed.[12]

But similarity can conceal difference. Class '40's "same banana" strike reinforced their commitment to the chain of command, while the abortive protest of '71 later inspired revolt against this same hierarchy. The legacy of Class '90's involvement in the 1987 coup will not become evident for another quarter-century. As each class graduates into ever-changing historical circumstance, it draws upon its academy experience and subsequent military socialization to make political choices with serious import for its nation's future.

Historical context, particularly this interaction between class and regime, might offer some hints about the future role of these younger officers. There are some obvious, telling differences in the external forces that shaped the cadet experience of Classes '40 and '90. Entering the PMA just after the Commonwealth's inauguration with independence on the horizon, Class '40 shared in the country's common purpose and sense of promise. They internalized the academy's values as model officers for a new national army. Their "same banana" strike was a protest against an abuse of authority—an affirmation of regulations, not a rebellion. Throughout their careers, they insisted on going "by the book" in matters large and small: no corruption, no coups, chain of command, civil supremacy.

By contrast, the cadet days of Class '90 coincided with four years of unprecedented ferment. Arriving at the academy only weeks after Marcos's downfall, they witnessed a fatal bombing on the parade ground, bitter political divisions, and several coup attempts. With the nation in flux and only their ideals constant, the members of this class, many still teenagers, were pressed to make some serious political choices. When RAM launched its first overt coup in August 1987, they followed the rest of the corps in armed revolt. But two years later when Colonel Honasan attempted his largest coup, Class '90, then only weeks from graduation, "preferred to support the constitution" and insured that the PMA would remain aloof. Even so, their yearbook praised RAM's fight for "the Filipino people" and condemned the political leaders who "only knew to promise and instead make their pockets healthy through public money."[13] The meaning that Class '90 draws from these experiences and the ways these lessons might be confirmed or challenged by their service as junior officers will have a lasting influence on them, on the armed forces, and, ultimately, on the nation.

Each regime thus leaves an invisible legacy in the values of the junior officers it has trained and commissioned. At the peak of their power, strong presidents such as Quezon, Marcos, and Ramos can have a profound impact upon the military. But they and their regimes fade or fall after little more than a decade. Long after a president's name is remembered only in street signs and schoolbooks, an entire cohort of officers will continue to bear the imprint of his or her decisions. Presidents come and go, but the officer corps, with its academy and classes, remains.

In the final phase of their thirty-year careers, each class carries the lessons of these early, formative experiences as it rises to command of the armed forces. Quezon's influence on cadets at the PMA and University of the Philippines during the 1930s had a strong and surprisingly direct role in shaping the armed forces of the 1960s. Similarly, Marcos's impact on PMA classes of the 1970s and Ramos's on those of the 1990s will continue to influence the character of the Philippine military and their nation well into the twenty-first century.

Notes

ABBREVIATIONS

BMC	*Baguio Midland Courier*
BT	*Bulletin Today* (Manila)
DG	*Daily Globe* (Manila)
DM	*Daily Mirror* (Manila)
FEER	*Far Eastern Economic Review* (Hong Kong)
MB	*Manila Bulletin*
MC	*Manila Chronicle*
MDB	*Manila Daily Bulletin*
MS	*Manila Standard*
MT	*Manila Times*
NARA	U.S. National Archives and Records Administration
NYT	*New York Times*
PDE	*Philippine Daily Express* (Manila)
PDI	*Philippine Daily Inquirer* (Manila)
PN	*Philippine Newsday* (Manila)
PS	*Philippine Star* (Manila)
PFP	*Philippines Free Press* (Manila)

SSD   *Sun Star Daily* (Cebu City)
TT   *Tribune* (Manila)

PREFACE

1. *Assembly* 52, no. 5 (May 1994), 157–58.
2. Letter from Lieutenant General Dennis P. McAuliffe (ret.) to Colonel William E. Burr II (ret.), 5 November 1993.
3. Letter from Lieutenant General David E. Ott (ret.) to Colonel William E. Burr II (ret.), 21 November 1993.
4. Letter from Brigadier General Douglas Kinnard (ret.) to Colonel William E. Burr II (ret.), 21 November 1993.

CHAPTER 1: CLASS AND CORPS

1. Interview with General Reynaldo Mendoza (ret.), Camp Aguinaldo, 13 August 1996; Major Rogelio S. Lumabas, "Whither the MAP," *The Cavalier* 7, no. 1 (November–December 1967), 5–6.
2. Kenneth W. Kemp and Charles Hudlin, "Civil Supremacy over the Military: Its Nature and Limits," *Armed Forces & Society* 19, no. 1 (Fall 1992), 8; Eric A. Nordlinger, *Soldiers in Politics: Military Coups and Governments* (Englewood Cliffs, N.J.: Prentice-Hall, 1977), 6; Samuel P. Huntington, *The Third Wave: Democratization in the Late Twentieth Century* (Norman: University of Oklahoma Press, 1991), 21; Claude E. Welch, Jr., and Arthur K. Smith, *Military Role and Rule: Perspectives on Civil-Military Relations* (North Scituate, Mass.: Duxbury Press, 1974), ix; Republic of the Philippines, *The Final Report of the Fact-Finding Commission (Pursuant to R.A. No. 6832)* (Manila: Bookmark, 1990), 97.
3. The literature on civil-military relations has relatively little close analysis of military takeovers. Since the 1950s, writing on civil-military relations, derived largely from Western experience, minimizes the issue, while studies of developing nations, where coups are central, often do not engage their dynamics in detail. For example, only 4 percent of the 1,325 titles in a bibliography of military sociology deals with coups. (See Kurt Lang, *Military Institutions and the Sociology of War: A Review of the Literature with Annotated Bibliography* [Beverly Hills: Sage, 1972], 238–42.) As a partial corrective, there is large specialist literature on coups, which is, however, often descriptive or narrowly taxonomic. For work on civil-military relations, see Samuel P. Huntington, *The Soldier and the State: The Theory and Politics of Civil-Military Relations* (Cambridge: Harvard University Press, 1957); Morris Janowitz, *The Professional Soldier: A Social and Political Portrait* (New York: Free Press, 1960); and Morris Janowitz and Stephen D. Wesbrook, eds., *The Political Education of Soldiers* (Beverly Hills: Sage, 1983).

    For general studies of the military in developing nations, see John J. Johnson, *The Role of the Military in Underdeveloped Countries* (Princeton: Princeton University Press, 1962); René Lemarchand, "Civil-Military Relations in Former Belgian Africa: The Military as Contextual Elite," in Steffan W. Schmidt and Gerald A. Dorfman, eds., *Soldiers in Politics* (Los Altos: Geron-X, 1974), 69–96; Claude E. Welch, Jr., "Civilian Control of the Military: Myth and Reality," in Claude E. Welch, Jr., ed., *Civilian Control of the Military: Theory and Cases from Developing Countries* (Albany: State University of New York Press, 1976), 1–41; Sheldon W. Simon, ed., *The Military and Security in the Third World: Domestic and International Impacts* (Boulder: Westview, 1978).

For literature on coups, often descriptive or taxonomic, see, Khhalid B. Sayeed, "The Role of the Military in Pakistan," in Jacques Van Doorn, ed., *Armed Forces and Society: Sociological Essays* (The Hague: Mouton, 1968), 274–96; Amos Perlmutter, "The Praetorian State and the Praetorian Army: Toward a Taxonomy of Civil-Military Relations in Developing Politics," *Comparative Politics* 1 (1969), 382–404; Martin Needler, "The Causality of the Latin American Coup d'État: Some Numbers, Some Speculations," in Schmidt, *Soldiers in Politics,* 145–59; Amos Perlmutter, *The Military and Politics in Modern Times: On Professionals, Praetorians, and Revolutionary Soldiers* (New Haven: Yale University Press, 1977), 89–114; José Z. Garcia, "Military Factions and Military Intervention in Latin America," in Simon, *Military and Security in the Third World,* 47–75.

4. S. E. Finer, *The Man on Horseback: The Role of the Military in Politics* (London: Pall Mall, 1962), 5–6.
5. Rigoberto D. Tiglao, "Rebellion from the Barracks: The Military as Political Force," in Philippine Center for Investigative Journalism, *Kudeta: The Challenge to Philippine Democracy* (Manila: Philippine Center for Investigative Journalism, 1990), 15; Philippines, *Final Report,* appendix J.
6. *PDI,* 12/4/89; Philippine Center, *Kudeta,* 209–15.
7. *PDI,* 12/15/89, 12/16/89; Ruben Fulgueras Ciron, "Civil-Military Relations in the Philippines: Perceptions of PMA-Trained Officers" (Ph.D. diss., University of the Philippines, 1993), 82, 167.
8. Jose M. Mendoza, ed., *Batch '36 Golden Book* (Manila: PMA Class '40 Association Inc., 1986), 348.
9. Interview with Colonel Pedro Bersola (ret.), Quezon City, 25 August 1996.
10. *Singapore Business Times,* 9–10 July 1994.
11. Cadet Corps, *The Academy Scribe* (Manila: The Academy Scribe Organization, 1988), 682–89; Ciron, "Civil-Military Relations," 15, 55.
12. Michelle Zimbalist Rosaldo, "Woman, Culture, and Society: A Theoretical Overview," in Michelle Zimbalist Rosaldo and Louise Lamphere, eds., *Woman, Culture and Society* (Stanford: Stanford University Press, 1974), 17–41; Henrietta L. Moore, *Feminism and Anthropology* (Minneapolis: University of Minnesota Press, 1988), chapters 1, 2; Micaela di Leonardo, "Introduction: Gender, Culture, and Political Economy," in Micaela di Leonardo, ed., *Gender at the Crossroads of Knowledge: A Feminist Anthropology in the Postmodern Era* (Berkeley: University of California Press, 1991), 1–48.
13. Michael Roper and John Tosh, "Introduction," in Michael Roper and John Tosh, eds., *Manful Assertions: Masculinities in Britain Since 1800* (London: Routledge, 1991), 2; Harry Brod, "Introduction: Themes and Theses of Men's Studies," in Harry Brod, ed., *The Making of Masculinities: The New Men's Studies* (Boston: Allen & Unwin, 1987), 2; David H. J. Morgan, *Discovering Men* (London: Routledge, 1992), 41–42.
14. Wendy Chapkis, "Sexuality and Militarism," in Eva Isaksson, ed., *Women and the Military System* (New York: St. Martin's, 1988), 106–13; Cynthia Enloe, "Beyond Steve Canyon

and Rambo: Feminist Histories of Militarized Masculinity," in John R. Gillis, ed., *The Militarization of the Western World* (New Brunswick, N.J.: Rutgers University Press, 1989), 121–22; Carol Cohn, "Sex and Death in the Rational World of Defense Intellectuals," *Signs: Journal of Women in Culture and Society* 12, no. 4 (1987), 689. In the 1960s, anthropologist Lionel Tiger argued that males, influenced by their prehistoric role as hunters, formed bonds that influenced their contemporary behavior. His emphasis on biology as the cause of this behavior limits the relevance of his work to this study. See Lionel Tiger, *Men in Groups* (New York: Random House, 1969), xiii–xx.

15. Cynthia Enloe, "It Takes Two," in Saundra Pollock Sturdevant and Brenda Stoltzfus, *Let the Good Times Roll: Prostitution and the U.S. Military in Asia* (New York: The New Press, 1992), 23–24.

16. After reading an early draft of this chapter, Cynthia Enloe wrote that this study details a group that "stands out for their professionalized political restraint. So many of the other . . . studies that consciously explore identities, focus on groups of soldiers who have stood out because of their brutality." (Letter to author, 28 April 1995.)

17. David H. J. Morgan, "Theater of War: Combat, the Military, and Masculinities," in Harry Brod and Michael Kaufman, eds., *Theorizing Masculinities* (Thousand Oaks, Calif.: Sage, 1994), 169–70.

18. Edward A. Shils and Morris Janowitz, "Cohesion and Disintegration in the Wehrmacht in World War II," *Public Opinion Quarterly* 12 (Summer 1948), 280–88; John P. Lovell, "The Professional Socialization of the West Point Cadet," in Morris Janowitz, ed., *The New Military: Changing Patterns of Organization* (New York: Russell Sage Foundation, 1964), 142–45; Janice Gibson, "Training People to Inflict Pain: State Terror and Social Learning," *Journal of Humanistic Psychology* 31, no. 2 (1991), 72–81.

19. Correlli E. Barnett, "The Education of Military Elites," *Journal of Contemporary History* 2, no. 3 (1967), 22–23.

20. Ibid., 23; Sanford M. Dornbush, "The Military Academy as an Assimilating Institution," *Social Forces* 33, no. 4 (1955), 318; Gary L. Wamsley, "Contrasting Institutions of Air Force Socialization: Happenstance or Bellwether?" *American Journal of Sociology* 78, no. 2 (1972), 405–9; Joel Newman, "An Exploratory Study of Elite Philippine Academy Cadets: Social Origins: Family Structure, Function, and Symbolism: Individual Psychodynamics: Occupational Selection: Intergenerational Mobility: and Early Career Success" (Ph.D. diss., University of Chicago, 1989), 541–42.

21. Huntington, *Soldier and the State,* 19–58 (quote is from p. 58); Janowitz, *Professional Soldier,* 233–79.

22. Janowitz, *Professional Soldier,* 277–79 (discussion of Weber is found in a note on p. 278).

23. Alfred W. McCoy, "'An Anarchy of Families': The Historiography of State and Family in the Philippines," in Alfred W. McCoy, ed., *An Anarchy of Families: State and Family in the Philippines* (Quezon City: Ateneo de Manila University Press, 1994), 1–32.

24. For an eloquent example of such heroic military history, see Carlos Quirino, *Filipinos at War* (Quezon City: Vera-Reyes, 1981).

25. Robert R. Reed, *Colonial Manila: The Context of Hispanic Urbanism and Process of Morphogenesis* (Berkeley: University of California Press, 1978), 50; Antonio de Morga, *Sucesos de las Islas Filipinas* (Cambridge: Cambridge University Press, 1971), 282–83; C.L.F.F.R.

Sainte-Croix, *Voyage commercial et politique aux Indes orientales, aux isles Philippines* (Paris: Clament Frères, Aux Archives du Droit Français, 1810), 188–89; William Henry Scott, "The Spanish Occupation of the Cordillera in the 19th Century," in Alfred W. McCoy and Edilberto de Jesus, eds., *Philippine Social History: Global Trade and Local Transformations* (Quezon City: Ateneo de Manila University Press, 1982), 39–56; James Warren, "Slavery and the Impact of External Trade: The Sulu Sultanate in the Nineteenth Century," in McCoy and de Jesus, *Philippine Social History,* 415–46.

26. Theodore Grossman, "The Guardia Civil and Its Influence on Philippine Society," *Archiviana* (December 1982), 4–7; Ruby R. Paredes, "The Partido Federal, 1900–1907: Political Collaboration in Colonial Manila" (Ph.D. diss., University of Michigan, 1990), 51–56.

27. Isagani R. Medina, *Cavite before the Revolution (1571–1986)* (Quezon City: College of Social Sciences and Philosophy, University of the Philippines, 1994), 63–105, 214; Warren, "Slavery and the Impact of External Trade," 433; Bruce Cruikshank, *Samar: 1768–1898* (Manila: Historical Conservation Society, 1985), 84–92.

28. Grossman, "Guardia Civil," 3–4; Q. A. Salazar, Z. Yulo, and A. Navarro, *Talaarawan 1996 Handog sa Sentenaryo Himagsikang 1896* (Quezon City: Miranda Bookstore, 1995), entry for May 23.

29. Carlos Quirino, *The Young Aguinaldo: From Kawit to Biyak-na-Bato* (Manila: Aguinaldo Centennial Year, 1969), 85–91.

30. Glenn Anthony May, *A Past Recovered* (Quezon City: New Day, 1987), 162; Glenn Anthony May, *Battle for Batangas: A Philippine Province at War* (New Haven: Yale University Press, 1991), 60; Santiago Alvarez, *Recalling the Revolution: Memoirs of a Filipino General* (Madison: Center for Southeast Asian Studies, University of Wisconsin, 1992), 34–35, 177–80.

31. Quirino, *Young Aguinaldo,* 71–72. For a careful study of how a similar elite revolutionary army formed in nearby Batangas Province, see May, *Battle for Batangas,* 76–79; and May, *Past Recovered,* 60–62.

32. May, *Past Recovered,* 60–63, 156–57, 159–60; Glenn Anthony May, "The Philippine Levee of 1899: Conscription, Nationalism, and National Amnesia" (ms, Seminar on Force in History, Institute for Advanced Study, Princeton, N.J., September 1966), 8, 13; Alvarez, *Recalling the Revolution,* 58–59, 67–69; Quirino, *Young Aguinaldo,* 71–72; Luis Camara Dery, *The Army of the First Philippine Republic and Other Historical Essays* (Manila: De La Salle University Press, 1995), 21–24; Milagros C. Guerrero, "The Provincial and Municipal Elites of Luzon during the Revolution, 1898–1902," in McCoy and de Jesus, *Philippine Social History,* 167–79; Vicente L. Rafael, "Nationalism, Imagery, and the Filipino Intelligentsia in the Nineteenth Century," *Critical Inquiry* 16, no. 3 (1990), 605–7.

33. May, *The Battle for Batangas,* 96–98; May, *A Past Recovered,* 152–54; Dery, *The Army of the First Philippine Republic,* 40.

34. Edward M. Coffman, "The Philippine Scouts, 1899–1942: A Historical Vignette," *Acta No. 3: Teheran 6/16 VII 1976* (Bucharest: International Commission of Military History, 1978), 73; James R. Woolard, "The Philippine Scouts: The Development of America's Colonial Army" (Ph.D. diss., Ohio State University, 1975), 3–13; Edward M. Coffman "Batson of the Philippine Scouts," *Parameters* 7, no. 3 (1977), 70–71; John A. Larkin, *Sugar and the Origins of Modern Philippine Society* (Berkeley: University of California Press,

1993), 29; Charles H. Franklin, Warrant Officer, U.S. Army, "History of the Philippine Scouts, 1899–1934" (Fort Humphreys, D.C., May 1935), RG 407, AG314.73 (5–1–35), NARA, 1–2.
35. Mrinalini Sinha, *Colonial Masculinity: The 'Manly Englishman' and the 'Effeminate Bengali' in the Late Nineteenth Century* (Manchester: Manchester University Press, 1995), 69–95 (quote is from page 81); Douglas M. Peers, "'The Habitual Nobility of Being': British Officers and the Social Construction of the Bengal Army in the Early Nineteenth Century," *Modern Asian Studies* 25, no. 3 (1991), 545–69; Nadzan Haron, "Colonial Defense and British Approach to the Problems in Malaya, 1874–1918," *Modern Asian Studies* 24, no. 2 (1990), 275–95; Myron Echenberg, *Colonial Conscripts: The* Tirailleurs Sénégalais *in French West Africa, 1857–1960* (Portsmouth, N.H.: Heinemann, 1991), 28–29; Cynthia H. Enloe, *Ethnic Soldiers: State Security in Divided Societies* (Athens: University of Georgia Press, 1980), 26–27.
36. Woolard, "Philippine Scouts," 13, 225; Franklin, "History of the Philippine Scouts," 11, 14–17.
37. Franklin, "History of the Philippine Scouts," 22, 29; Woolard, "Philippine Scouts," 168–69, 226–29, 259; Coffman, "Philippine Scouts, 1899–1942," 76; letter from Judge-Advocate General Geo. B. Davis to the Adjutant General, 11 June 1908; Memorandum for Records, (signed) W.L.P., 16 November 1926, Bureau of Insular Affairs, War Department, RG 350 (Bureau of Insular Affairs), E-5, Box, 766, #11685, NARA.
38. George Y. Coats, "The Philippine Constabulary, 1901–1917" (Ph.D. diss., Ohio State University, 1968), 4. The constabulary restricted Filipino access to firearms by limiting gun licenses and aggressively pursuing all unregistered arms. In Ilocos Norte Province, for example, the constabulary allowed only twenty-four gun licenses from 1901 to 1919. (See Capt. L. E. Quintero, "Firearm Problems," *Khaki and Red* 10, no. 6 [June 1930], 9.)
39. Vic Hurley, *Jungle Patrol: The Story of the Philippine Constabulary* (New York: E. P. Dutton, 1938), 298–99; Lieutenant Colonel Harold Hanne Elarth, *The Story of the Philippine Constabulary* (Los Angeles: Philippine Constabulary Officers Association, 1949), 14–15.
40. Captain Alfonso A. Calderon, "The Philippine Constabulary 1901–1951: Half-Century of Service," in Capt. Alfonso A. Calderon and Marciano C. Sicat, eds., *Golden Book Philippine Constabulary* (Manila: Philippine Constabulary, 1951), 28–29; Cadet Corps, *Academy Scribe,* 22, 29; Jaime J. Bugarin, "Six Decades of Noble Soldiering," in Philippine Constabulary, *1960 Yearbook Philippine Constabulary* (Manila, 8 August 1960), 20–21; Registry Committee, Association of General and Flag Officers, *General and Flag Officers of the Philippines (1896–1983)* (Quezon City: Association of General and Flag Officers, 1983), 106.
41. Anne-Marie Hilsdon, *Madonnas & Martyrs: Militarism and Violence in the Philippines* (Quezon City: Ateneo de Manila University Press, 1995), 48, 51, 89; Colonel Sinforoso L. Duque, *Soldier Heroes: A Handbook on the Winners of the Major Medals Awarded by the Philippine Constabulary and Armed Forces Since 1902* (Manila: National Media Production Center, 1981), vii.
42. Ma. Luisa Camagay, *Working Women of Manila in the 19th Century* (Quezon City: University of the Philippines Press, 1995), 119; Cristina Blanc Szanton, "Collision of Cultures: Historical Reformulations of Gender in the Lowland Visayas, Philippines," in Jane Mon-

nig Atkinson and Shelly Errington, eds., *Power and Difference: Gender in Island Southeast Asia* (Stanford: Stanford University Press, 1990), 344–83; Shelly Errington, "Recasting Sex, Gender, and Power: A Theoretical and Regional Overview," in Atkinson and Errington, *Power and Difference,* 1–58; Ann Laura Stoler, "Carnal Knowledge and Imperial Power: Gender, Race, and Morality in Colonial Asia," in di Leonardo, ed., *Gender at the Crossroads of Knowledge,* 51–101.

43. Cadet Corps, *Academy Scribe,* 19–23, 526; H.J. Res. 123, 31 January 1908, RG-350, E-5, Box 647, No. 11,685, NARA; Woolard, "Philippine Scouts," 229–30; Coffman, "Philippine Scouts, 1899–1942," 75; Franklin, "History of the Philippine Scouts," Table F-4; letter from Assistant Chief, Bureau of Insular Affairs, To: The Adjutant General, U.S. Army, 10 December 1912, RG-350, E-5, Box 647, No. 11,685, NARA; Adjutant General, To: Chief, Bureau of Insular Affairs, 15 February 1923, RG-350, E-5, Box 647, No. 11,685, NARA; Richard Bruce Meixsel, "An Army for Independence? The American Roots of the Philippine Army" (Ph.D. diss., Ohio State University, 1993), 137; Clarence E. Endy, Jr., "The Gentlemen from the Philippines," *Bulletin of the American Historical Collection* 11, no. 3 (1983), 10–11.

44. Endy, "Gentlemen from the Philippines," 10–11; Ernesto O. Rodriguez, *Commodore Alcaraz: First Victim of President Marcos* (New York: Vantage, 1986), 120; Meixsel, "Army for Independence," 282–84.

45. Charles Burke Elliott, *The Philippines to the End of the Commission Government: A Study in Tropical Democracy* (Indianapolis: Bobbs-Merrill, 1917), 176; Jose G. Syjuco, *Military Education in the Philippines* (Quezon City: New Day, 1977), 15.

46. Woolard, "Philippine Scouts," 170–84, 196.

47. Joseph Ralston Hayden, *The Philippines: A Study in National Development* (New York: Macmillan, 1955), 734–35.

48. J. L. Panis, "A Filipino Citizen Army in the Making," *The Philippine Republic* 2, no. 2 (February 1925), 14–15; Rafael Palma, memo to U.S. Secretary of War, 22 March 1924, RG-350, E-5, Box 1298, No. 28003, NARA; Captain Ambrosio P. Peña, *After Action Report,* vol. 1, *The Story of the First Regular Division* (Manila: Bureau of Printing, 1953), 1–2. In 1912, the Philippine Constabulary started a smaller, less formal program of military training at the University of the Philippines, but it failed to lay the foundations for a full military science department (See Syjuco, *Military Education in the Philippines,* 47.)

49. Ex-Cadet Colonel Jesus C. Perlas, "The Department in the Making," *The Philippinensian* (Manila: Senior Students of the University of the Philippines, 1930), 407; "The Corps in Review 1936–1937," *The Philippinensian 1937* (Manila: The Senior Class, University of the Philippines, 1937), 256, 276; Third Lieutenant Tomas C. Benitez, "The U.P. Cadet Corps," *The 1938 Philippinensian* (Manila: The 1938 Senior Class of the University of the Philippines, 1938), 294–96; Colonel Julio Martinez, "We March On," *1938 Philippinensian,* 313.

50. "Our Prize Cadets," *Ateneo Monthly* 1, no. 1 (July 1922), 9–10; *Ateneo de Manila Annual 1922* (Manila), 94–96; Teodoro Evangelista, "Cadet Corps and Colonel Rhodes," *Ateneo Monthly* 1, no. 4 (October 1922), 123–25; Ignacio Lim, "Military," *Ateneo Monthly* (March 1924), 394–95.

51. *Ateneo Monthly* (March 1927), 495–99; *Ateneo Monthly* (March 1928), 579–81; Ignacio

Salazar, "Ateneo Cadet Corps," *Ateneo Monthly* (March 1925), 482–86; Manuel Colayco, "The Corps," *Ateneo Monthly* (Commencement 1926), 527–31; *The 1937 Ateneo Aegis* (Manila: Ateneo de Manila, 1937), 178; Cadet Major Mariano A. Yenko, Jr., "Military Department," *Aegis 1939* (Manila: The Senior Classes of the Ateneo de Manila, 1939), 194; interview with Captain Eugenio Lara (ret.), Anaheim Hills, Calif., 2 January 1994.

52. J. L. Panis, "Filipino Citizen Army," 14–15; Cadet Captain Pacifico M. Stuart del Rosario, "The Department of Military Science and Tactics," *The 1931 Philippinensian* (Manila: The Class of 1931 of the University of the Philippines, 1931), 359; Meixsel, "Army for Independence," 163; Vicente Lim, *To Inspire and to Lead: The Letters of Vicente Lim, 1938–1942* (Manila: privately printed, 1980), 63.

53. Interview with Lara; Meixsel, "Army for Independence," 288–90; Syjuco, *Military Education in the Philippines*, 47.

54. Florencio F. Magsino and Rogelio S. Lumabas, *Men of PMA*, vol. 1 (Manila: by the authors, 1978), 69–70; Lieutenant Colonel Robert M. Carswell, "Philippine National Defense," *Coast Artillery Journal* 84, no. 2 (March–April 1941), 122; Sergio Osmeña, *National Defense and Philippine Democracy: Address Delivered by Honorable Sergio Osmeña at the Commencement Exercises of the Philippine Military Academy, Baguio, March 15, 1940* (Manila: Bureau of Printing, 1940), 7; Robert H. Ferrell, ed., *The Eisenhower Diaries* (New York: W. W. Norton, 1981), 8–10. For contemporary reports of the plan, see *NYT*, 11/20/34, 11/25/34, 5/30/36, 6/20/36.

55. Ferrell, *Eisenhower Diaries*, 10, 20; Lim, *To Inspire and to Lead*, 89.

56. Ricardo Trota Jose, *The Philippine Army, 1935–1942* (Quezon City: Ateneo de Manila University Press, 1992), 227.

57. May, *Past Recovered*, 60–63.

58. Maurice Garnier, "Technology, Organizational Culture, and Recruitment in the British Military Academy," in George A. Kourvetaris and Betty A. Dobratz, eds., *World Perspectives in the Sociology of the Military* (New Brunswick, N.J.: Transaction, 1977), 81–91; George A. Kourvetaris and Betty A. Dobratz, "Social Recruitment and Political Orientations of the Officer Corps in Comparative Perspective," in Kourvetaris, and Dobratz, *World Perspectives*, 93–117.

59. *Philippinensian 1937*, 232–43.

60. Janowitz, *Professional Soldier*, 277.

61. Harold W. Maynard, "A Comparison of Military Elite Role Perceptions in Indonesia and the Philippines" (Ph.D. diss., American University, 1976), 489–92.

62. Jennifer Morrison Taw, *Thailand and the Philippines: Case Studies in U.S. IMET Training and Its Role in Internal Defense and Development* (Santa Monica: RAND, 1994), 42; U.S. Agency for International Development, *U.S. Overseas Loans and Grants and Assistance from Internal Organizations* (Washington, D.C.: Office of Planning and Budgeting, Bureau for Program and Policy Coordination, CONG-R-0105, 1990), 81; Stephen Rosskamm Shalom, *The United States and the Philippines: A Study of Neocolonialism* (Philadelphia: Institute for the Study of Human Issues, 1981), 108; Felipe Miranda and Ruben F. Ciron, "Development and the Military in the Philippines: Military Perceptions in a Time of Continuing Crisis," in J. Soedjati Djiwandono and Yong Mun Cheong, eds.,

Soldiers and Stability in Southeast Asia (Singapore: Institute for Southeast Asian Studies, 1988), 169.
63. Nick Cullather, *Illusions of Influence: The Political Economy of United States-Philippines Relations, 1942–1960* (Stanford: Stanford University Press, 1994), 90; Brigadier General Edward G. Lansdale, "Lessons Learned: The Philippines 1946–1953" (Washington, D.C.: U.S. Department of State, Foreign Service Institute, typescript, 26 September 1963), 3.
64. Maynard, "Comparison of Military Elite," 489–99; Donald L. Berlin, "Prelude to Martial Law: An Examination of Pre-1972 Philippine Civil-Military Relations" (Ph.D. diss., University of South Carolina, 1982), 185–99. Among the alumni of the University of the Philippines ROTC program close to Marcos were Isamel Lapus (1935), Fred Ruiz Castro (Law 1936), Eduardo Garcia (Law 1939), Romeo Espino (Agriculture 1937), Fabian Ver (student in Law, 1937–41), and Carmelo Barbero (Arts, ca. 1936).
65. Miranda, "Development and the Military," 169; Berlin, "Prelude to Martial Law," 202–03.
66. Raymond Bonner, *Waltzing with a Dictator: The Marcoses and the Making of American Policy* (New York: Times Books, 1987), 50–51, 74–75; Miranda, "Development and the Military," 175; Berlin, "Prelude to Martial Law," 205–08.
67. Berlin, "Prelude to Martial Law," 200–01.
68. The three members of Class '51 were General Ignacio Paz, chief of the Intelligence Service of the AFP (ISAFP); General Thomas Diaz, Philippine Constabulary Zone I commander; and Colonel Alfredo Montoya, commander of the Philippine Constabulary's Metropolitan Command (Metrocom). Significantly, two of these men served in Jakarta as assistant armed forces attaché during the period when the Indonesian army was taking power—Paz in 1965–67 and Montoya in 1968. See Registry Committee, *General and Flag Officers*, 123, 267, 307; Carolina G. Hernandez, "The Philippines," in Zakaria Haji Ahmad and Harold Crouch, eds., *Military-Civilian Relations in Southeast Asia* (Singapore: Oxford University Press, 1985), 184–85; Bonner, *Waltzing with a Dictator*, 3–4, 99–100.
69. Francisco T. San Miguel, ed., *The Silver Sword of PMA Class 1951* (Manila: Philippine Military Academy Class of 1951, 1976); Philippine Military Academy, *The CCAFP 76 Magilas Sword* (Fort Del Pilar, Baguio City, 1976).
70. Felipe B. Miranda, "At the Crossroads of Politicization," in Lorna Kalaw-Tirol, ed., *Duet for EDSA: 1996, Looking Back, Looking Forward* (Manila: Foundation for Worldwide People Power, Inc., 1995), 72–78, 82–85.
71. *PDI*, 1/28/96, 3/8/96. In April 1997, a Manila newspaper published a list of 108 retired AFP officers serving in the Ramos administration, and reported that of these 95 percent were PMA alumni. (See *PDI*, 4/15/97, 4/16/97, 4/21/97.)
72. *PDI*, 8/13/97, 12/9/97, 12/16/97, 1/30/98; *NYT*, 9/22/97; *The Economist*, 12 April 1997, 13 September 1997; *MB*, 11/8/97.

CHAPTER 2: KAYDET DAYS

1. Jose M. Mendoza, ed., *Batch '36 Golden Book* (Manila: PMA Class '40 Association Inc., 1986), 16. In 1985, as the fiftieth reunion of the class's admission to PMA approached, an

editorial committee, chaired by Colonel Jose Mendoza (ret.), compiled biographies for all 120 of the original plebes and wrote a long collective biography. Aside from extended interviews with thirty-two members of the class of '40, this volume is the single most important source for this chapter. Those classmates interviewed were Manuel Acosta (18 March 1995), Ramon Alcaraz (30 December 1993, 30 December 1994, 4 October 1997), Felix Apolinario (16 March 1995, 13 August 1996), Pedro Baban (11 February 1989), Uldarico Baclagon (8 November 1988, 16 March 1995), Pedro Bartolome (24 August 1996), Pedro Bersola (12 March 1995, 25 August 1996), Reynaldo Bocalbos (15 March 1995, 24 August 1996, 11 February 1999), Deogracias Caballero (12 October 1988, 15 October 1988, 19 March 1995, 24 August 1996), Felicissimo Castillo (23 January 1989), Ciceron de la Cruz (12 March 1995, 29 August 1996), Francisco del Castillo (18 March 1995, 12 August 1996), Justiniano Cortez (19 March 1995), Jose Esguerra (11 February 1989), Licurgo Estrada (22 November 1988, 24 January 1989), Horacio Farolan (14 March 1995), Ricardo Foronda (27 January 1989, 31 August 1996), Amos Francia (21 January 1989), Lucendro Galang (19 March 1995), Ramon Gelvezon (18 November 1988), Francisco Jimenez (4 January 1995), Jose Mendoza (22 October 1988, 26 October 1988, 1 February 1989), Reynaldo Mendoza (27 October 1988, 14 March 1995, 13 August 1996), Conrado Nano (14 March 1995), Edmundo Navarro (22 January 1989, 7 February 1989), Victor Osias (23 November 1988, 3 February 1989, 8 February 1989), David Pelayo (19 November 1988), Salvador Piccio (5 November 1988), Eduardo Soliman (20 January 1989), Hospicio Tuazon (21 January 1989), Segundo Velasco (22 November 1988), Pedro Yap (14 August 1996).

2. Vicente Lim, *To Inspire and to Lead: The Letters of Vicente Lim, 1938–1942* (Manila: privately printed, 1980), 99–100.

3. Florencio F. Magsino and Rogelio S. Lumabas, *Men of PMA,* vol. 1 (Manila: by the authors, 1978), 69–70; Richard Bruce Meixsel, "An Army for Independence? The American Roots of the Philippine Army" (Ph.D. diss., Ohio State University, 1993), 2, 297–302; Robert H. Ferrell, ed., *The Eisenhower Diaries* (New York: W. W. Norton, 1981), 8–10; Lieutenant Colonel Robert M. Carswell, "Philippine National Defense," *Coast Artillery Journal* 84, no. 2 (March–April 1941), 122; Sergio Osmeña, *National Defense and Phillipine Democracy: Address Delivered by Honorable Sergio Osmeña at the Commencement Exercises of the Philippine Military Academy, Baguio, March 15, 1940* (Manila: Bureau of Printing, 1940), 7; *NYT,* 11/20/34, 11/25/35, 5/30/36, 6/20/36.

4. Cadet Corps, *The Academy Scribe* (Manila: The Academy Scribe Organization, 1988), 50; Osmeña, *National Defense,* 7–8, 10.

5. Commonwealth of the Philippines, *Report of the President of the Philippines on the Activities of the Philippine Army for the Period January 1 to December 31, 1936* (Manila: Bureau of Printing, 1937), quoted in David Y. Nanney, "Philippine Military Academy" (ms, 1938), 6, Julius Ruiz Papers, Filipino American Historical Society, Seattle, Wash.; Ronald G. Bauer, "Military Professional Socialization in a Developing Country" (Ph.D. diss., University of Michigan 1973), 1–2.

6. Mendoza, *Golden Book,* i; Douglas MacArthur, *Reminiscences* (New York: McGraw-Hill, 1964), 104; Lieutenant Conrado B. Rigor, "The Philippine Military Academy," *Coast Artillery Journal* 84, no. 4 (July–August 1941), 326.

7. John P. Lovell, "The Professional Socialization of the West Point Cadet," in Morris

Janowitz, ed., *The New Military: Changing Patterns of Organization* (New York: Russell Sage Foundation, 1964), 127–39; Sanford M. Dornbusch, "The Military Academy as an Assimilating Institution," *Social Forces* 33, no. 4 (May 1955), 316–21; Gary L. Wamsley, "Contrasting Institutions of Air Force Socialization: Happenstance or Bellwether?" *American Journal of Sociology* 78, no. 2 (September 1972), 399–417.

8. Correlli Barnett, "The Education of Military Elites," *Journal of Contemporary History* 2, no. 3 (1967), 22–23.

9. Charles W. Larned, "West Point and Higher Education," *Army and Navy Life* 8 (June 1906), 18; Benjamin G. Rader, "The Recapitulation Theory of Play: Motor Behaviour, Moral Reflexes, and Manly Attitudes in Urban America, 1880–1920," in J. A. Mangan and James Walvin, eds., *Manliness and Morality: Middle-Class Masculinity in Britain and America, 1800–1940* (New York: St. Martin's, 1987), 126–27; Morris Janowitz, *The Professional Soldier: A Social and Political Portrait* (Glencoe: The Free Press, 1960), 128; Lovell, "Professional Socialization," 139–42; D. Clayton James, *The Years of MacArthur*, vol. 1, *1880–1941* (Boston: Houghton Mifflin, 1970), 276–77.

10. Nanney, "Philippine Military Academy," 7.

11. Frederick J. Manning, "Morale, Cohesion, and Esprit de Corps," in Reuven Gal and A. David Mangelsdorff, *Handbook of Military Psychology* (Chichester: John Wiley & Sons, 1991), 456–57; Janowitz, *The Professional Soldier*, 138.

12. Interview with Colonel Felicissimo Castillo (ret.), Quezon City, 23 January 1989.

13. Mendoza, *Golden Book*, 349.

14. Carolina G. Hernandez, "The Extent of Civilian Control of the Military in the Philippines: 1946–1976" (Ph.D. diss., State University of New York at Buffalo, 1979), 169–72.

15. Republic of the Philippines, *The Final Report of the Fact-Finding Commission (Pursuant to R.A. No. 6832)* (Manila: Bookmark, October 1990), 70–74.

16. Hernandez, "The Extent of Civilian Control of the Military," 178; Bauer, "Military Professional Socialization"; Donald L. Berlin, "Prelude to Martial Law: An Examination of Pre-1972 Philippine Civil-Military Relations" (Ph.D. diss., University of South Carolina, 1982); Harold W. Maynard, "A Comparison of Military Elite Role Perceptions in Indonesia and the Philippines" (Ph.D. diss., American University, 1976), 423; Joel Newman, "An Exploratory Study of Elite Philippine Academy Cadets: Social Origins: Family Structure, Function, and Symbolism: Individual Psychodynamics: Occupational Selection: Intergenerational Mobility: and Early Career Success" (Ph.D. diss., University of Chicago, 1989), 541–42; Viberto Selochan, "Professionalization and Politicization of the Armed Forces of the Philippines" (Ph.D. diss., Australian National University, 1990).

17. Mendoza, *Golden Book*, 110–11.

18. Michelle Zimbalist Rosaldo, "Woman, Culture, and Society: A Theoretical Overview," in Michelle Zimbalist Rosaldo and Louise Lamphere, eds., *Woman, Culture, and Society* (Stanford: Stanford University Press, 1974), 25; David Gilmore, *Manhood in the Making: Cultural Concepts of Masculinity* (New Haven: Yale University Press, 1990), 14; Mark C. Carnes, "Middle-Class Men and the Solace of Fraternal Ritual," in Mark C. Carnes and Clyde Griffen, eds., *Meanings for Manhood: Constructions of Masculinity in Victorian America* (Chicago: University of Chicago Press, 1990), 38–40; Mark C. Carnes, *Secret Ritual and Manhood in Victorian America* (New Haven: Yale University Press, 1989), 3, 14.

19. Roger M. Keesing, "Introduction," in Gilbert H. Herdt, ed., *Rituals of Manhood: Male Initiation in Papua New Guinea* (Berkeley: University of California Press, 1982), 32–34; Gilbert H. Herdt, "Fetish and Fantasy in Sambia Initiation," in Herdt, *Rituals of Manhood*, 57–61.
20. Thomas M. Keifer, *The Tausug: Violence and Law in a Philippine Moslem Society* (New York: Holt, Rinehart and Winston, 1972); Renato Rosaldo, *Ilongot Headhunting, 1883–1974: A Study in Society and History* (Stanford: Stanford University Press, 1980), 139–40.
21. Arnold van Gennep, *The Rites of Passage* (Chicago: University of Chicago Press, 1960), vii, 11.
22. Pierre Clastres, *Society Against the State: Essays in Political Anthropology* (New York: Zone Books, 1987), 177–88; Gilmore, *Manhood in the Making*, 9–29; R. Brian Ferguson and Neil L. Whitehead, "The Violent Edge of Empire," in R. Brian Ferguson and Neil L. Whitehead, eds., *War in the Tribal Zone* (Santa Fe: School of American Research Press, 1992), 2, 18–25.
23. J. A. Mangan, "Social Darwinism and Upper-Class Education in Late Victorian and Edwardian England," in Mangan and Walvin, *Manliness and Morality*, 150–53; J. A. Mangan, *Athleticism in the Victorian and Edwardian Public School* (Cambridge: Cambridge University Press, 1981), 22–41; J. A. Mangan, *The Games Ethic and Imperialism: Aspects of the Diffusion of an Ideal* (New York: Viking, 1986), 33–36; Michael Rosenthal, *The Character Factory: Baden-Powell and the Origins of the Boy Scout Movement* (London: Collins, 1986), 1–6.
24. George Mosse, *Fallen Soldiers: Reshaping the Memory of the World Wars* (New York: Oxford University Press, 1990), 15, 72; Samuel Stouffer et al., *The American Soldier*, vol. 2, *Combat and Its Aftermath* (Princeton: Princeton University Press, 1949), 131–32.
25. George Mosse, *Nationalism and Sexuality: Middle-Class Morality and Sexual Norms in Modern Europe* (Madison: University of Wisconsin Press, 1985), 31–32, 100; Jessica Benjamin and Anson Rabinbach, "Foreword," in Klaus Theweleit, *Male Fantasies*, vol. 2, *Male Bodies: Psychoanalyzing the White Terror* (Minneapolis: University of Minnesota Press, 1989), xvi; David B. Ralston, *Importing the European Army: The Introduction of European Military Techniques and Institutions into the Extra European World, 1600–1914* (Chicago: University of Chicago Press, 1990), 179.
26. Cadet Lieutenant Colonel Emilio Viardo, "Character Building in the U.P. Corps of Cadets," *The 1931 Philippinensian* (Manila: The Class of 1931 of the University of the Philippines, 1931), 381.
27. Cadet Sergeant Fred Ruiz Castro, "Training in the Corps of Cadets," *The 1932 Philippinensian* (Manila: The Class of 1932 of the University of the Philippines, 1932), 355; Antonio Quirino, "Our Sponsors," *The Philippinensian* (Manila: The Senior Students of the University of the Philippines, 1930), 427.
28. Doris G. Nyuda, *The Beauty Book: A History of Philippine Beauty from 1908–1980* (Manila: Mr. & Mrs. Publishing Co., 1980) entries for 1920, 1922, 1931.
29. *TT*, 2/16/35, 2/21/35, 2/22/35, 3/3/35.
30. *TT*, 2/17/35, 2/19/35, 3/3/35, 1/24/37, 2/5/37; Joseph Ralston Hayden, *The Philippines: A Study in National Development* (New York: Macmillan, 1955), 203–04.
31. *TT*, 2/15/36, 2/16/36, 2/25/36, 2/26/36; *The Sunday Tribune Magazine*, 1 March 1936.

32. *TT,* 1/30/37, 2/9/37, 2/11/37, 2/2/37, 2/3/37, 2/15/38, 2/16/38; Nyuda, *The Beauty Book,* entry for 1937.
33. *The Philippinensian 1937* (Manila: Senior Class of the University of the Philippines, 1937), 237, 257–59, 268–69; Ferdinand Marcos, "Capas Memoirs," in Manuel E. Buenafe, *The Voice of the Veteran: An Anthology of the Best in Song and Story by the Defenders of Freedom* (Manila: Republic Promotion, 1946), 18–20.
34. *TT,* 3/25/36; *The Philippines Herald,* 3/25/36; Meixsel, "Army for Independence," 295–96.
35. Commonwealth of the Philippines, Army Headquarters, Bulletin No. 17, 17 April 1936, Record Group 350, Bureau of Insular Affairs, Entry 5, Box 235, NARA; Meixsel, "Army for Independence," 301.
36. *TT,* 2/10/39.
37. Ignacio M. Sarmiento, "Nice Work If You Can Get It!" *PFP,* 24 June 1939, 8–10; letter from Jorge B. Vargas to Samuel Goldwyn, 17 July 1939, Series VII, Box No. 345, Subject File: Moving Picture/Music & Musicians, Manuel Quezon Papers, Philippine National Library; message from Samuel Goldwyn to President Manuel Quezon, 23 August 1939, Series V, Box No. 110, Subject File: General Correspondence 1939, August 13–31, Manuel Quezon Papers, Philippine National Library.
38. Commonwealth of the Philippines, Army Headquarters, Bulletin No. 119, 14 August 1937, Department of Manuscripts, Cornell University Library.
39. Janowitz, *The Professional Soldier,* 138; Erving Goffman, *Asylums: Essays on the Social Situation of Mental Patients and Other Inmates* (Garden City: Anchor, 1961).
40. *The Sword of Nineteen Hundred and Thirty-Eight* (Baguio: Cadet Corps of the Army of the Philippines, Philippine Military Academy, 1938), 46–48, 104.
41. Commonwealth of the Philippines, Philippine Army, Philippine Military Academy, *Bulletin of Information 1938* (Manila: Bureau of Printing, 1938), 16–19, RG-350, E-5, Box 1276, No. 28003, NARA.
42. Rosaldo, *Ilongot Headhunting,* 35–37.
43. Brian Fegan, "Mang Dionisio Macapagal: A Peasant Rebel Matures" (paper presented at the Association for Asian Studies Annual Meeting, Washington, D.C., 7 April 1995). See also Cristina Blanc Szanton, "Collision of Cultures: Historical Reformulation of Gender in the Lowland Visayas, Philippines," in Jane Monnig Atkinson and Shelly Errington, eds., *Power and Difference: Gender in Island Southeast Asia* (Stanford: Stanford University Press, 1990), 350.
44. Cadet Sergeant Macario Peralta [Jr.], "Activities of the Corps during the Year," *The 1932 Philippinensian,* 358.
45. *The 1934 Philippinensian* (Manila: Senior Class of the University of the Philippines, 1934), 396.
46. Interview with Captain Eugenio Lara (ret.), Anaheim Hills, California, 2 January 1994; *Khaki and Red* 10, no. 12 (December 1930), 26.
47. Ricardo Trota Jose, *The Philippine Army, 1935–1942* (Quezon City: Ateneo de Manila University Press, 1992), 61, 238; Dale O. Smith, *Cradle of Valor: The Intimate Letters of a Plebe at West Point* (Chapel Hill: Algonquin, 1988), 17–18; interview with Lara; Ferrell, *The Eisenhower Diaries,* 15; Cadet Corps, *Academy Scribe,* 50.

48. Jose G. Syjuco, *Military Education in the Philippines* (Quezon City: New Day, 1977), 15–16; Cadet Corps, *Academy Scribe,* 50; Rigor, "Philippine Military Academy," 325.
49. *Khaki and Red* 10, no. 4 (April 1930), 25–26; Syjuco, *Military Education,* 10–11, 15–16; Cadet Corps, *Academy Scribe,* annex J-4; *Nineteen Hundred and Thirty-Eight,* 36–39; PMA, *Bulletin of Information 1938,* 14–40.
50. Cadet Corps, *Academy Scribe,* 53.
51. Mendoza, *Golden Book,* 15, 180, 264; Liberato Picar, "The Scrapbook of Liberato Picar" (ms, 1987, provided by Mrs. Beatriz Apostol vda. de Picar), 1–2.
52. Letter to the author from Colonel Deogracias Caballero, 22 December 1995; interview with General Horacio Farolan, Villamor Air Base, 14 March 1995; Mendoza, *Golden Book,* 15. Among the thirty-two classmates interviewed, eight were valedictorians, one was salutatorian, and another eleven were honor students.
53. Cadet Corps, *Academy Scribe,* 247, 349–53; *The Corps* 2, no. 10 (Baguio, Mid-Summer 1939), 16–17; Ernesto O. Rodriguez, *Commodore Alcaraz: First Victim of President Marcos* (New York: Vantage, 1986), 19, 125.
54. van Gennep, *Rites of Passage,* 10–11, 106–07; Mendoza *Golden Book.* 17–19; Paul T. Bartone and Faris P. Kirkland, "Optimal Leadership in Small Army Units," in Gal and Maglesdorff, *Handbook of Military Psychology,* 396–98; *Nineteen Hundred and Thirty-Eight,* 193.
55. *Nineteen Hundred and Thirty-Eight,* 114; Mendoza, *Golden Book,* 17–19; interview with Lara; interview with Captain Pedro Yap (ret.), Cebu City, 14 August 1996.
56. Rodriguez, *Commodore Alcaraz,* 122; interview with Commodore Ramon A. Alcaraz (ret.), Orange, Calif., 30 December 1993; *Nineteen Hundred and Thirty-Eight.* 51.
57. Interview with General Ramon Gelvezon (ret.), Pasig, Manila, 18 November 1988.
58. Mendoza, *Golden Book,* 17, 19; *Nineteen Hundred and Thirty-Eight,* 193–96.
59. Noli R. Reyes, "365 Days Make a Man," *The Sunday Tribune Magazine,* 7 November 1937, 20–21, 36; interview with Commodore Felix Apolinario (ret.), Quezon City, 16 March 1995.
60. Mendoza, *Golden Book,* 23, 312; *Nineteen Hundred and Thirty-Eight,* 200; interview with Colonel Hospicio Tuazon (ret.), Quezon City, 21 January 1989.
61. Interviews with Gelvezon; interview with Colonel Jose Mendoza (ret.), Quezon City, 22 October 1988.
62. Edmundo Navarro, *Beds of Nails* (Manila: by the author, 1988), 9–13.
63. Cadet Corps. *Academy Scribe,* 59.
64. Colonel Ciceron de la Cruz, "Memoirs" (ms, 1993), 15; interview with Colonel Ciceron de la Cruz (ret.), Quezon City, 12 March 1995.
65. Mendoza, *Golden Book,* 37; interview with Colonel David Pelayo (ret.), Quezon City, 19 November 1988; Jose M. Crisol, 3/c, "The 'Blessed' Hand-Shake," *The Corps* 2, no. 10 (Baguio, Mid-Summer 1939), 8.
66. Interview with Colonel Pedro Bartolome (ret.), Quezon City, 25 August 1986. A review of military psychology by a researcher at Walter Reed Army Institute noted that the "importance of group solidarity for effective military performance . . . has been reflected in the elevation of close-order drill to near sacramental status: that is, physical unity was the explicit goal." (See Manning, "Morale, Cohesion," 456.)

67. Interview with General Reynaldo Mendoza (ret.), Camp Aguinaldo, Quezon City, 14 March 1995; interview with Ciceron de la Cruz; interview with Colonel Pedro Bersola (ret.), Quezon City, 12 March 1995.
68. Mendoza, *Golden Book*, 26; Lim, *To Inspire and to Lead*, 33.
69. Interview with Jose Mendoza.
70. Interview with Farolan.
71. Mendoza, *Golden Book*, 41.
72. Cadet Corps. *Academy Scribe*, 59; Antonio M. Gonzalez, ed., *Philippine Military Academy: The Golden Sword Class of 1942* (Manila: Class of 1942, 1991), 48; interview with Colonel Francisco del Castillo (ret.), Quezon City, 18 March 1995; interview with General Victor Osias (ret.), Makati, Manila, 22 November 1988; Crisol, "The 'Blessed' Hand-Shake."
73. Interviews with Alcaraz; Osias.
74. Interview with General Salvador Piccio (ret.), Quezon City, 5 November 1988; interview with Major Edmundo Navarro (ret.), Malate, Manila, 22 January 1989; Lim, *To Inspire and to Lead*, 105.
75. Mendoza, *Golden Book*, 41–42, 164, 234; interview with Colonel Francisco Jimenez (ret.), North Hollywood, Calif., 4 January 1995; interview with Colonel Eduardo Soliman (ret.), Quezon City, 20 January 1989; *Nineteen Hundred and Thirty-Eight*, 207.
76. *Nineteen Hundred and Thirty-Eight*, 146–51; interview with Bartolome.
77. *Nineteen Hundred and Thirty-Eight*, 146–51.
78. Interview with Navarro.
79. Interview with Osias.
80. In interviews between 1988 and 1996, I asked seventeen classmates to recall their class rank and eleven gave it exactly. Another five gave more general answers ("middle") that may have sprung from modesty rather than forgetfulness.
81. Interview with Colonel Pedro Bersola (ret.), Quezon City, 12 March 1995.
82. *Nineteen Hundred and Thirty-Eight*, 201; interview with Alcaraz. Among the thirty-two classmates interviewed, eight cited their admiration for the colonial constabulary as a significant factor in their decision to take the PMA examination.
83. Mendoza, *Golden Book*, 128; Cadet Corps, *Academy Scribe*, 675. Among the thirty-two classmates interviewed, nine offered extensive, often laudatory recollections of these West Point alumni on the PMA's staff—notably, Captain Jose S. Esguerra (Damortis, La Union, 11 February 1989); Colonel Felipe Fetalvero (Manila, 28 January 1989); and Colonel Soliman.
84. First Class of the Philippine Military Academy, *The Sword of 1940* (Manila: The Sword, 1940), 226–29; Smith *Cradle of Valor*, 256–57; Mendoza, *Golden Book*, 23.
85. Interview with Washington Sagun's PMA roommate, Ramon A. Alcaraz; *Sword of 1940*, 165; Mendoza, *Golden Book*, 23, 44, 126–27, 298.
86. TT, 10/11/39, 2/10/39; Gonzalez, *Golden Sword*, 138; interview with Colonel Manuel Acosta (ret.), Quezon City, 18 March 1995; *Nineteen Hundred and Thirty-Eight*, 119, 124.
87. Commonwealth of the Philippines, Army Headquarters, Bulletin No. 119; PMA, *Bulletin of Information 1938*, 35–36, 70; *Sword of 1940*, 136; Mendoza, *Golden Book*, 348.
88. Interview with Colonel Deogracias Caballero (ret.), Quezon City, 12 October 1988.

89. Letter to author from Colonel Deogracias Caballero, 22 December 1995.
90. Letter from Caballero; interview with Navarro.
91. Mendoza, *Golden Book*, 47, 292.
92. Reynaldo A. Mendoza, "The 'Same Banana,'" *Batch '36 Golden Book*, 50, 114; interview with Alcaraz; interview with Ricardo Foronda, Quezon City, 31 August 1996; interview with Bartolome.
93. Mendoza, "Same Banana," 51–52, 116–17; Cadet Corps, *Academy Scribe*, 57; Rodriguez, *Commodore Alcaraz*, 122–23; interview with Alcaraz.
94. Interview with Caballero; letter from Caballero.
95. Among all the various accounts of this incident, Colonel de la Cruz provided the most detailed and convincing recollection of these details (see Ciceron de la Cruz, "Memoirs," 66.)
96. Interviews with Jose Mendoza; Bersola.
97. Mendoza, "Same Banana," 115; interview with Colonel Reynaldo Bocalbos (ret.), Quezon City, 15 March 1995; interviews with Jimenez; Ciceron de la Cruz; Jose Mendoza.
98. Mendoza, "Same Banana," 115–17; Mendoza, *Golden Book*, 52; Navarro, *Bed of Nails*, 18; Rodriguez, *Commodore Alcaraz*, 122–23.
99. Interviews with Caballero; Jose Mendoza.
100. Mendoza, "Same Banana," 116; *TT*, 3/9/39; Rodriguez, *Commodore Alcaraz*, 123; interview with Alcaraz; Navarro, *Bed of Nails*, 18–19.
101. Mendoza, *Batch '36 Golden Book*, 52.
102. Rigor, "Philippine Military Academy," 326; *Nineteen Hundred and Thirty-Eight*, 107.
103. Letter from Caballero.
104. Interview with Colonel Lucendro Galang (ret.), Cawit, Cavite, 19 March 1995; interview with Colonel Francisco del Castillo (ret.), Quezon City, 18 March 1995; interview with Colonel Conrado Nano (ret.), Quezon City, 14 March 1995; interview with Colonel Eduardo Soliman (ret.), Quezon City, 20 January 1989; interviews with Osias; Foronda.
105. *Sword of 1940*, 161.
106. Mendoza, *Golden Book*, 112; Deogracias F. Caballero, "The PMA Motto and the Erosion of Its Values," in Mendoza, *Golden Book*, 127.
107. Interview with Reynaldo Mendoza; Mendoza, *Golden Book*, 23.
108. Cadet R. Mendoza and Cadet Q. Evangelista, "P.M.A. Oh! Hail to Thee," *Sword of 1940*, 225.
109. Klaus Theweleit, *Male Fantasies*, vol. 1: *Women Floods Bodies History* (Minneapolis: University of Minnesota Press, 1987).
110. *Nineteen Hundred and Thirty-Eight*, 158, 161; letter to author from Major Edmundo Navarro (ret.), the photo editor of *The Corps*. 18 March 1994.
111. *Nineteen Hundred and Thirty-Eight*, 108.
112. *Sword of 1940*, 49, 108.
113. Mendoza, *Golden Book*, 61.
114. Conrado B. Rigor, "The Academy Comes of Age," *Sunday Tribune Magazine*, 10 March 1940, 23–25; *Sunday Tribune Magazine*, 17 March 1940.
115. *TT*, 3/14/40, 3/15/40; *Sunday Tribune Magazine*, 17 March 1940; *The News (Behind the News)*, 17 March 1940.

116. Rigor, "Academy Comes of Age."
117. Mendoza, *Golden Book*, 61–62, 258; First Class, *Sword of 1940*, 111.
118. Mendoza, *Golden Book*, 62.
119. Osmeña, *National Defense*, 14–15.
120. Interview with Caballero; Mendoza, *Golden Book*, 62; *TT*, 3/16/40; telephone interview with Commodore Ramon Alcaraz, Orange, Calif., 4 October 1997.
121. In August 1996, some fifty-eight or fifty-nine years after the photograph was taken, I showed nine members of Class '40 a photograph of the cadet First Battalion and asked them to identify the faces. Those interviewed were Felix Apolonario, Pedro Bersola, Reynaldo Bocalbos, Deogracias Caballero, Francisco del Castillo, Ciceron de la Cruz, Ricardo Foronda, Reynaldo Mendoza, and Pedro Yap.
122. Interviews with Piccio; Jose Mendoza.

CHAPTER 3: BAPTISM BY FIRE

1. Letter to author from Commodore Ramon A. Alcaraz (ret.), 25 July 1993; Ernesto O. Rodriguez, *Commodore Alcaraz: First Victim of President Marcos* (New York: Vantage, 1986), 124–25.
2. Rodriguez, *Commodore Alcaraz*, 125.
3. Interview with Commodore Ramon Alcaraz, Orange, Calif., 30 December 1993.
4. Jose M. Mendoza, ed., *Batch '36 Golden Book* (Manila: PMA Class '40 Association Inc., 1986), 132–326.
5. Rodriguez, *Commodore Alcaraz*, 126–27; *Philippines* 1, no. 8 (November 1941).
6. Commodore Ramon A. Alcaraz (ret.), "The Saga of the Q-Boats" (ms, n.d.), 3–6; *Life*, 9 February 1942, 42.
7. Alcaraz, "Saga of the Q-Boats," 5; interview with Alcaraz; Rodriguez, *Commodore Alcaraz*, 128.
8. Rodriquez, *Commodore Alcaraz*, 133–35; interview with Alcaraz.
9. Letters to author from Alcaraz, 28 February 1994, 16 March 1994.
10. *Philippines* 3, no. 4 (15 June 1943), 5; letter from Alcaraz, 28 February 1994; Rodriguez, *Commodore Alcaraz*, 137–39; Mendoza, *Batch '36 Golden Book*, 138; Alcaraz, "The Saga of the Q-Boats," 8–10; Uldarico Baclagon, *They Served with Honor: Filipino War Heroes of World War II* (Quezon City: DM, 1965), 92.
11. Mendoza, *Batch '36 Golden Book*, 138; Rodriguez, *Commodore Alcaraz*, 137–59; Alcaraz, "Saga of the Q-Boats," 1–19; letter from Alcaraz, 28 February 1994.
12. Alcaraz, "Saga of the Q-Boats," 3.
13. Interview with Captain Pedro Yap (ret.), Cebu City, 14 August 1996.
14. Interview with Colonel Hospicio Tuazon (ret.), Quezon City, 21 January 1989; interview with Colonel Pedro Bersola (ret.), Quezon City, 12 March 1995; interview with Colonel David Pelayo (ret.), Quezon City, 19 November 1988. See also Mendoza, *Batch '36 Golden Book*, 232.
15. Interview with Pelayo.
16. Captain Ambrosio P. Peña, *After Action Report*, vol. 1: *The Story of the First Regular Division* (Manila: Bureau of Printing, 1953), 10, 19, 21–22; Rico T. Jose, "Beach Defense:

The First Regular Division at Mauban, December 23–26, 1941," *Bulletin of the American Historical Collection* 17, no. 2 (1989), 8–9.

17. Peña, *After Action Report,* 18–19, 21–23, 131–32.
18. Jose, "Beach Defense," 8–9, Peña, *After Action Report,* 28–30; interview with Pelayo.
19. Interview with Pelayo; Peña, *After Action Report,* 38–41, 44.
20. Jose, "Beach Defense," 12–13; Celedonio A. Ancheta, ed., *The Wainwright Papers,* vol. 1 (Manila: New Day, 1980), 99.
21. Peña, *After Action Report,* 44–45.
22. *The Sword of Nineteen Hundred and Thirty-Eight* (Baguio: Cadet Corps of the Army of the Philippines, Philippine Military Academy, 1938), 69; Jose, "Beach Defense," 12–13; interview with Pelayo.
23. Baclagon, *They Served,* 11; interview with Pelayo.
24. Interview with Pelayo; Peña, *After Action Report,* 48–49; Jose, "Beach Defense," 15–16.
25. Interview with Pelayo; Jose, "Beach Defense," 16–17.
26. Jose, "Beach Defense," 17; Uldarico Baclagon, *Philippine Campaigns* (Manila: Graphic House, 1952), 162–63.
27. Jose, "Beach Defense," 17; Peña, *After Action Report,* 38–49, 49–50; interview with Pelayo.
28. Interview with Pelayo; Jose, "Beach Defense," 18; Peña, *After Action Report,* 50.
29. Mendoza, *Batch '36 Golden Book,* 334; letter from Alcaraz, 5 March 1996.
30. Peña, *After Action Report,* 52–53, 107–08, 132; interview with Pelayo; Mendoza, *Batch '36 Golden Book,* 334.
31. Jose "Beach Defense," 20–23.
32. *Reports of General MacArthur: Japanese Operations in the Southwest Pacific Area,* vol. 2, pt. 1 (Washington, D.C.: U.S. Government Printing Office, 1966), 95; Louis Morton, *United States Army in World War II: The War in the Pacific: The Fall of the Philippines* (Washington, D.C.: Office of the Chief of Military History, Department of the Army, 1953), 142; Baclagon, *They Served,* 11–13.
33. Jose Mendoza, "Experiences" (ms, ca. 1985), 63–140.
34. Mendoza, *Batch '36 Golden Book,* 246.
35. Ibid., 80.
36. Interview with Colonel Jose Mendoza (ret.), Manila, 1 February 1989; interview with Colonel Deogracias Caballero (ret.), Manila, 15 October 1988; Donald Blackburn, interview, Senior Officers Oral History Program, U.S. Military History Institute, Carlisle, Pa. 172–73.
37. Interview with Commodore Felix Apolinario (ret.), Quezon City, 16 March 1995; interview with General Horacio Farolan (ret.), Villamor Air Base, 14 March 1995; interview with General Reynaldo Mendoza (ret.), Camp Aguinaldo, 14 March 1995.
38. Gavan Daws, *Prisoners of the Japanese: POWs of World War II in the Pacific* (New York: William Morrow, 1994), 73–87; Donald Knox, *Death March: The Survivors of Bataan* (San Diego: Harcourt Brace and Jovanovich, 1981), 153–55.
39. Interview with Ricardo Foronda, Quezon City, 27 January 1989; interview with Colonel Francisco Jimenez (ret.), North Hollywood, Calif., 4 January 1995; interview with

Colonel Manuel Acosta (ret.) Quezon City, 18 March 1995; interviews with Reynaldo Mendoza; Pelayo; Bersola; Farolan.
40. Interview with General Salvador Piccio (ret.), Quezon City, 5 November 1988; interview with Colonel Conrado Nano (ret.), Quezon City, 14 March 1995.
41. Interview with Colonel Ciceron de la Cruz (ret.), Quezon City, 12 March 1995; interviews with Acosta Tuazon, 13 March 1995, 21 January 1989; Mendoza, *Batch '36 Golden Book*, 300.
42. Mendoza, *Batch '36 Golden Book*, 79, 82, 194, 332; interview, Tuazon, 21 January 1989.
43. Mendoza, *Batch '36 Golden Book*, 68, 150, 172, 202; "Heroes of the Philippine Army Air Corps," *Philippines* 4, no. 2 (March 1944), 12; Colonel Jesus A. Villamor, *They Never Surrendered: A True Story of Resistance in World War II* (Quezon City: Vera-Reyes, 1982), 30–40.
44. Mendoza, *Batch '36 Golden Book*, 78, 170.
45. Bartolome Cabangbang, "General Information on Corregidor, Bataan, Concentration Camps and Guerrillas," RG-16, Philippine Project, Select Messages: Cabangbang, Bartolome C., MacArthur Archives, Norfolk, Va.
46. Cabangbang, "General Information"; Bartolome Cabangbang, "The Bombardment of Corregidor," RG-16, Philippine Project, Select Messages: Cabangbang, Bartolome C., MacArthur Archives.
47. Interview with Alcaraz; interview with General Salvador Piccio (ret.), Quezon City, 5 November 1988.
48. Mendoza, *Batch '36 Golden Book*, 82, 296.
49. Cabangbang, "General Information."
50. Villamor, *They Never Surrendered*, 115; Uldarico Baclagon, *Filipinos in the Allied Intelligence Bureau* (Makati: St. Paul, 1989), 74; Florencio F. Magsino and Rogelio S. Lumabas, *Men of PMA*, vol. 1 (Manila: by the authors, 1978), 133; Carlos A. Mendigo, "President Carlos P. Garcia and the Resistance Movement," *Philippines Free Press*, 5 July 1958, 18, 42.
51. Villamor, *They Never Surrendered*, 98, 114–15; Baclagon, *Filipinos in the Allied Intelligence Bureau*, 52; Mendoza, *Batch '36 Golden Book*, 160, 172; letter from: Villamor, To: General MacArthur, NR 154, 11 October 1943, RG-16, Philippine Project, Select Messages: Villamor—January–October 1943, MacArthur Archives; letter to: General MacArthur, From: Sparks, NR 170, 22 October 1943, RG-16, Philippine Project, Select Messages: Villamor—January–October 1943, MacArthur Archives.
52. Villamor, *They Never Surrendered*, 207–08.
53. Subject: Trained Personnel, memo to: Colonel C. Whitney, 26 July 1944, Headquarters 5217th Reconnaissance Battalion (Provisional), United States Army Forces in the Far East, RG-16: Whitney, Personnel Files-Special, Personnel 5217th Recon. Bn. (Prov.), MacArthur Archives; memo from S. J. C. to Chief of Staff, 6 August 1944, General Headquarters, Southwest Pacific Area, RG-16, Philippine Project, Guerrilla Records, General Files: Philippine Sub-Division, June–August 1944, MacArthur Archives.
54. Interview with Pelayo.
55. Alejo Santos, "A Brief History of the Bulacan Military Area," *The Journal of History* 8, no. 2 (June 1960), 14–15; interview with Alcaraz.

56. Message from: Cabangbang, To: General MacArthur, NR 581, 18 December 1944, RG-16, Philippine Project, Select Messages: Cabangbang, Bartolome C., June 1944–January 1945, MacArthur Archives; message from: Cabangbang, To: General MacArthur, NR 624, 25 December 1944, RG-16, Philippine Project, Messages: Cabangbang, September 1944–March 1945, MacArthur Archives; message from: Cabangbang, To: General MacArthur, NR 831, 17 January 1945, RG-16, Philippine Project, Messages: Cabangbang, September 1944–March 1945, MacArthur Archives; Baclagon, *They Served with Honor,* 206–07.
57. B. C. Cabangbang to Commander-in-Chief Hukbalahap, 17 January 1945, RG-16, Philippine Project—Hukbalahaps, 1943–1945, MacArthur Archives; Message from: Cabangbang, To: General MacArthur, NR 846, 18 January 1945, RG-16, Philippine Project—Hukbalahaps, 1943–1945, MacArthur Archives; Message from Cabangbang, To General MacArthur, NR 857, 19 January 1945 RG-16, Philippine Project—Hukbalahaps, 1943–1945, MacArthur Archives.
58. Mendoza, *Batch '36 Golden Book,* 170; C. W. to Chief of Staff, 28 August 1945, RG-16, Philippine Subdivision—Administration, June–August 1945, MacArthur Archives; Baclagon, *They Served with Honor,* 206–07; message from: MacArthur, To: Cabangbang, NR 8, 2 December 1944, RG-16, Philippine Project, Select Messages: Cabangbang, Bartolome C., June 1944–January 1945, MacArthur Archives; memo from C. W. to Chief of Staff, 28 August 1945, MacArthur Archives; message from: Cabangbang, To: MacArthur, NR 366, 6 December 1944, Philippine Project, Select Messages: Cabangbang, Bartolome C., June 1944–January 1945, MacArthur Archives; Capt. B. C. Cabangbang, General Order No. 1, 1 January 1945, RG-16, Philippine Sub-Division-Administration, January 1945, MacArthur Archives. Cabangbang's order identified four of those promoted as former PMA cadets—Santos Dizon, Pablo Ignacio, Benjamin Jacinto (PMA '45), and Jorge Siacungco ('45). Of these, only two are listed on the roster of the PMA Alumni Association.
59. Magsino and Lumabas, *Men of PMA,* 72.
60. Interview with General Victor Osias (ret.), Manila, 8 February 1989.
61. Colonel Uldarico S. Baclagon, *Last 130 Days of the USAFFE* (Makati: Astra Ink Corp., 1982), 104.
62. Interview with Osias.
63. Ibid.
64. Ibid.
65. Interviews with Colonel Licurgo Estrada (ret.), Quezon City, 22 November 1988, 24 January 1989; Mendoza, *Batch '36 Golden Book,* 272, 316.
66. Magsino and Lumabas, *Men of PMA,* 116–17, 121, 128, 131–33.

CHAPTER 4: CAREER SOLDIERS

1. Jose M. Mendoza, ed., *Batch '36 Golden Book* (Manila: PMA Class '40 Association Inc., 1986), 104–05.
2. Ronald G. Bauer, "Military Professional Socialization in a Developing Country" (Ph.D. diss., University of Michigan, 1973), 27.
3. Harold W. Maynard, "A Comparison of Military Elite Role Perceptions in Indonesia and the Philippines" (Ph.D. diss., American University, 1976), 456–61; Amos Perlmutter, *The Military and Politics in Modern Times: On Professionals, Praetorians, and Revolutionary Soldiers* (New Haven: Yale University Press, 1977), 115–19, 132–33.

4. Mendoza, *Golden Book,* 349; Milton W. Meyer, *Diplomatic History of the Philippine Republic* (Honolulu: University of Hawaii Press, 1965), 18–19; interview with Colonel Francisco Jimenez (ret.), North Hollywood, Calif., 4 January 1995; interview with General Horacio Farolan (ret.), Villamor Air Base, Manila, 14 March 1995.
5. Mendoza, *Golden Book,* 87–88; interview with Colonel Jose Mendoza (ret.), Quezon City, 22 October 1988.
6. Interview with General Salvador Piccio (ret.), Quezon City, 5 November 1988; Mendoza, *Golden Book,* 84.
7. Mendoza, *Golden Book,* 88; *MT,* 2/12/48, 2/22/48, 2/26/48, 2/28/48, 2/29/48.
8. Interview with Jose Mendoza; *MT,* 2/22/48, 2/26/48, 2/28/48, 2/29/48, 3/13/48, 5/11/48, 5/17/48, 7/18/49; Mendoza, *Golden Book,* 88–92.
9. Reynaldo A. Mendoza, "The 'Same Banana,'" in Mendoza, *Golden Book,* 117.
10. Ernesto O. Rodriguez, *Commodore Alcaraz: First Victim of President Marcos* (New York: Vantage, 1986), 163–64, 250.
11. First Class of the Philippine Military Academy, *The Sword of 1940* (Manila: The Sword, 1940), 156; Mendoza, *Golden Book,* 98, 156; interview with General Reynaldo Mendoza (ret.), Quezon City, 13 March 1995.
12. Mendoza, "Same Banana," 117; interview with Ricardo Foronda, Quezon City, 27 January 1989; Mendoza, *Golden Book,* 208.
13. *Sword of 1940,* 106; Mendoza, *Golden Book,* 88, 99; interview with Foronda; interview with Captain Pedro Yap (ret.), Cebu City, 14 August 1996.
14. Uldarico Baclagon, *Lessons From the Huk Campaign in the Philippines* (Manila: M. Colcol, 1960), 3, 36, 248; U.S. Department of the Army, Office of the Assistant Chief of Staff, G-2, "Intelligence Research Project: The Philippine Constabulary" (Project 7557, 15 December 1952), U.S. Military History Institute, Carlisle Barracks, Pa.
15. Baclagon, *Lessons From the Huk Campaign,* 6; Major Lawrence M. Greenberg, *The Hukbalahap Insurrection: A Case Study of a Successful Anti-Insurgency Operation in the Philippines—1946–1955* (Washington, D.C.: U.S. Army Center of Military History, 1987), 82–85, 112–17; Colonel Napoleon D. Valeriano and Lieutenant Colonel Charles T. R. Bohannan, *Counter-Guerrilla Operations: The Philippine Experience* (New York: Frederick A. Praeger, 1962), 122–24, 207–08; *MT,* 7/27/50; OIR Report No. 5209, 27 September 1950, 39, *Declassified Documents Reference System* (Washington, D.C.: Carrolton, 1979), 220-B; Donn Hart, "Magsaysay: Philippine Candidate," *Far Eastern Survey* 22, no. 6 (May 1953), 66–67; A. H. Peterson, G. C. Reinhardt, and E. E. Conger, eds., *Symposium on the Role of Airpower in Counterinsurgency and Unconventional Warfare: The Philippine Huk Campaign* (Santa Monica, Calif.: The Rand Corporation, July 1963), 18.
16. Interview with Colonel Conrado Nano (ret.), Quezon City, 14 March 1995; Peterson, Reinhardt, and Conger, *Symposium on the Role of Airpower,* 18; Mendoza, *Golden Book,* 92–94, 97–98; Joint United States Military Advisory Group to the Republic of the Philippines, Monthly Summary of Activities for August 1953, 8 September 1953, Record Group 218, Joint Chiefs of Staff, NARA; Joint United States Military Advisory Group to the Republic of the Philippines, Annual Report, 1 January–31 December 1953, 1 February 1954, U.S. Military Historical Institute, Carlisle Barracks, Pa.; Benedict J. Kerkvliet, *The Huk Rebellion: A Study of Peasant Revolt in the Philippines* (Berkeley: University of California Press, 1977), 245.

17. Interview with Colonel Conrado Nano (ret.), Quezon City, 14 March 1995.
18. *MT,* 7/11/53, 11/4/53; Jose Veloso Abueva, *Ramon Magsaysay: A Political Biography* (Manila: Solidaridad, 1971), 221, 225; Colonel Ed Lansdale of JUSMAG, "Memorandum Prepared for the Ambassador in the Philippines (Spruance)," in John Glennon, ed., *Foreign Relations of the United States, 1952–54,* vol. 12, *East Asia and the Pacific,* pt. 2 (Washington, D.C.: U.S. Government Printing Office, 1987), 548, 551; Nick Cullather, *Illusions of Influence: The Political Economy of United States–Philippines Relations, 1942–1960* (Stanford: Stanford University Press, 1994), 116, 217.
19. Interview with Reynaldo Mendoza.
20. Interview with Colonel Deogracias Caballero (ret.), Camp Aguinaldo, Quezon City, 18 March 1995; File: Cabal, General Manuel, *MT* Clipping Files, Lopez Memorial Museum, Pasig, Metro-Manila; *MT,* 1/29/59.
21. Mendoza, *Golden Book,* 98.
22. Ciceron de la Cruz, "Memoirs" (ms, 1993), 73–74; *DM,* 1/26/61; *MT,* 1/27/61, 2/7/61.
23. Interview with Major Edmundo Navarro (ret.), Malate, Manila, 22 January 1989.
24. Interview with Jose Mendoza.
25. Mendoza, *Golden Book,* 100.
26. Ibid., 54, 333; *MT,* 8/3/61.
27. *DM,* 8/1/61, 8/2/61, 8/21/61.
28. *DM,* 3/29/60, 4/2/60, 8/12/60, 10/7/60, 10/21/61, 11/3/61, 2/10/62; *MT,* 3/3/60, 3/24/60, 8/10/61, 9/28/61, 1/7/62, 4/12/62; interview with Commodore Ramon A. Alcaraz (ret.), Orange, Calif., 30 December 1993.
29. Reynaldo Mendoza, "Retaliation," *Cavalier* (n.d.), (clipping provided by its author); interview with Colonel Pedro C. Bersola (ret.), Quezon City, 12 March 1995.
30. Mendoza, *Golden Book,* 100; Association of General and Flag Officers, *General and Flag Officers of the Philippines (1896–1977)* (Quezon City: Association of General and Flag Officers, 1977).
31. Interview with Colonel Lucendro Galang (ret.), Cawit, Cavite, 19 March 1995; interview with Alcaraz; Ernesto O. Rodriguez, *Commodore Alcaraz: First Victim of President Marcos* (New York: Vantage, 1986), 76; Mendoza, *Golden Book,* 302, 341.
32. Interview with Jimenez.
33. Francisco T. San Miguel, ed., *The Silver Sword of PMA Class 1951* (Manila: Philippine Military Academy Class of 1951, 1976).
34. Mendoza, *Golden Book,* 50, 218, 340; letter to author from Colonel Deogracias Caballero, 22 December 1995.
35. Interview with General Ramon G. Gelvezon (ret.), Manila, 18 November 1988; *Sword of 1940,* 87.
36. Mendoza, *Golden Book,* 218; interview with Gelvezon.
37. Mendoza, *Golden Book,* 340–41; interview with Navarro.
38. *DM,* 7/23/55; *MT,* 8/24/55, 6/18/56, 8/24/58, 10/25/58, 11/2/58, 12/3/58, 2/3/59, 3/17/59.
39. Colonel Ciceron de la Cruz, "Memoirs," 21.
40. Mendoza, *Golden Book,* 340–41; interview with Gelvezon.
41. Rodriguez, *Commodore,* 226; interview with Gelvezon.

42. Interview with Gelvezon; Mendoza, *Golden Book*, 340–41.
43. Interview with Gelvezon; Mendoza, *Golden Book*, 104, 106.
44. Interview with General Pedro Baban (ret.), La Trinidad, 11 February 1989.
45. Mendoza, *Golden Book*, 100, 152, 335–36.
46. Interviews with Reynaldo Mendoza; Farolan; Association of General and Flag Officers, *General and Flag Officers of the Philippines*, 48, 107.
47. Interview with Jimenez; Curriculum Vitae of Colonel Francisco M. Jimenez 0–1670 FA (GSC): C. Résumé While in USA, 1973–1992, 30 June 1992.
48. Interview with General Victor Osias (ret.), Manila, 8 February 1989.
49. Ibid.
50. Ibid.
51. Ibid.
52. Interview with General Segundo Velasco (ret.), Manila, 22 November 1988; *Sword of 1940*, 124; appendix A in Mendoza, *Golden Book*, 351.
53. Mendoza, *Golden Book*, 320.
54. Interview with Colonel Francisco del Castillo (ret.), Quezon City, 18 March 1995.
55. Baclagon, *Lessons from the Huk Campaign*, 84.
56. Letter from Colonel Francisco Jimenez to Commodore Ramon A. Alcaraz, 7 February 1996; letter from Commodore Ramon A. Alcaraz to the author, 5 March 1996.
57. Mendoza, *Golden Book*, 320.
58. Interview with del Castillo; letter from Alcaraz.
59. Interview with Colonel Manuel Acosta (ret.), Quezon City, 18 March 1995; interview with Velasco.
60. *MT*, 10/8/67.
61. Interview with Alcaraz.
62. *MT*, 3/21/68; Ernesto O. Rodriguez, *Working with Heroes and Exiles* (New York: Vantage, 1989), 167.
63. *MT*, 3/21/68.
64. *MT*, 3/22/68.
65. Ibid.
66. *MT*, 3/29/68; *PFP*, 4/6/68, 66.
67. *DM*, 4/4/68; *PFP*, 4/16/68, 67, 68.
68. *PFP*, 4/6/68, 66; *DM*, 2/19/71.
69. *PFP*, 4/6/68, 66; interview with Velasco; *MT*, 5/26/68.
70. "Speech Delivered by General Segundo Velasco, Outgoing AFP Chief of Staff, during Turnover Ceremonies at Fort Aguinaldo on 28 May 1968," *MT* Clipping Files, Lopez Memorial Museum, Pasig, Metro-Manila.
71. Letter from Jimenez to Alcaraz.
72. *DM*, 1/25/71.
73. *MT*, 8/22/71; interview with a senior AFP officer with experience in demolition and sabotage.
74. Press Release, "Lt. Col. Tomas B. Karingal," 28 January 1960, File: Col. Karingal, *MT* Clipping Files, Lopez Memorial Museum, Pasig, Metro-Manila.
75. Mendoza, *Golden Book*, 206, 238, 282, 316.

76. Interview with Colonel David Pelayo (ret.), Quezon City, 19 November 1988; Mendoza, *Golden Book,* 206, 238, 316; Press Release, "Lt. Col. Tomas B. Karingal."
77. Press Release, "Lt. Col. Tomas B. Karingal"; *MT,* 7/23/64; Mendoza, *Golden Book,* 87, 99, 238.
78. Commission on Elections, "Resolution," Case No. 445, 3 November 1963; Letter from Tomas Benitez to Joe Bautista, 24 October 1963; Statement, Emmanuel Pelaez, Office of the Vice President (n.d.), File: Col. Karingal, *MT* Clipping Files, Lopez Memorial Museum, Pasig, Metro-Manila; interview with Alcaraz.
79. Mendoza, *Golden Book,* 238; *MT,* 7/11/64, 7/23/64, 7/25/64, 7/31/64, 4/14/65.
80. *Ang Bayan* 16, no. 4 (June 1984), 10.
81. Mendoza, *Golden Book,* 238.
82. Interviews with Alcaraz; Mendoza, *Golden Book,* 238; *Metro Times,* 2/17/90; *MB,* 2/23/90; *PS,* 2/28/90; *DG,* 3/4/90; *Malaya,* 3/8/90; *"Ako ang Batas": Gen. Karingal,* Director Francis Jun Posadas, Seiko Films, 1990.
83. Nicasio Tumulak and Uldarico Baclagon, *The Fortunes of the Poor* (Manila: privately printed, 23 October 1964), 1; *Sword of 1940,* 67; Uldarico Baclagon, "Bartolome Cabangbang," in Mendoza, *Golden Book,* 170.
84. B. C. Cabangbang to Courtney Whitney, 17 June 1945, RG-16, Philippine Project, Select Messages: Cabangbang, Bartolome C., June 1944–January 1945, MacArthur Archives; Baclagon, "Bartolome Cabangbang," 170; *MT,* 8/31/57; Tumulak and Baclagon, *Fortunes of the Poor.*
85. Tumulak and Baclagon, *Fortunes of the Poor.*
86. Interview with Colonel Uldarico Baclagon (ret.), Manila, 8 November 1988; interview with Farolan.
87. Tumulak and Baclagon, *Fortunes of the Poor;* Alfred W. McCoy, *The Politics of Heroin: CIA Complicity in the Global Drug Trade* (New York: Lawrence Hill, 1991), 162–78; William Corson, *Armies of Ignorance: The Rise of the American Intelligence Empire* (New York: Dial, 1977), 320–22.
88. Interview with Baclagon.
89. Ibid.; Baclagon, "Bartolome Cabangbang," 170; *MT,* 8/31/57, 1/28/59, 2/21/59, 9/8/59; Resil Mojares, *The Man Who Would Be President: Serging Osmeña and Philippine Politics* (Cebu City: Maria Cacao, 1986), 88, 90, 94.
90. *MT,* 1/30/55.
91. *PFP,* 12/27/58, 63; *MT,* 12/12/58.
92. Cadet Corps, *The Academy Scribe* (Manila: The Academy Scribe Organization, 1988), 77–78; Antonio M. Gonzalez, ed., *Philippine Military Academy: The Golden Sword Class of 1942* (Manila: Class of 1942, 1991), 322; Florencio F. Magsino and Rogelio S. Lumabas, *Men of PMA,* vol. 1 (Manila: by the authors, 1978), 119–20; Mendoza, *Golden Book,* 84, 86. Among the members of PMA's Class of 1944 who served in the Hunters ROTC were Eleuterio Adevoso (commander), Pacifico Jose (G-1), Frisco San Juan (G-2), and Juanito Ferrer (See Baclagon, *Philippine Campaigns,* 248–49).
93. Proculo L. Mojica, *Terry's Hunters (The True Story of the Hunters ROTC Guerrillas)* (Manila: Benipayo, 1965), 616–20; Viberto Selochan, "Professionalization and Politicization of the Armed Forces of the Philippines" (Ph.D. diss., Australian National Uni-

versity, 1990), 117–19, 125–27; Donald L. Berlin, "Prelude to Martial Law: An Examination of Pre-1972 Philippine Civil-Military Relations" (Ph.D. diss., University of South Carolina, 1982), 142–43.
94. Selochan, "Professionalization," 125–26; Cullather, *Illusions of Influence,* 171.
95. *PFP,* 11/1/58, 63.
96. *PFP,* 11/22/58, 62.
97. *MT,* 1/23/59.
98. Interview with Piccio; *PFP,* 22 November 1958, 62; *MT,* 1/23/59.
99. *MT,* 10/31/58, 11/4/58; Berlin, "Prelude to Martial Law," 150–51.
100. *MT,* 11/2/58, 11/5/58, 11/7/58.
101. *PFP,* 12/6/58, 6, 63; Berlin, "Prelude to Martial Law," 149–51.
102. *MT,* 11/4/58, 11/8/58, 11/15/58, 11/16/58, 11/21/58, 11/29/58, 11/20/58.
103. *MT,* 7/19/59; Mojica, *Terry's Hunters,* 620.
104. Berlin, "Prelude to Martial Law," 151–55; *MT,* 12/6/58, 12/9/58, 12/12/58, 12/26/58.
105. Berlin, "Prelude to Martial Law," 141–59.
106. *MT,* 2/18/59, 3/10/59.
107. Mendoza, *Golden Book,* 138.
108. *MT,* 4/19/63; *DM,* 4/22/63, 1/19/67; Tumulak and Baclagon, *Fortunes of the Poor.*
109. Tumulak and Baclagon, *Fortunes of the Poor; DM,* 8/31/65; *MT,* 1/8/66.
110. Interview with del Castillo; Mendoza, *Golden Book,* 132, 170, 172.
111. Mendoza, *Golden Book,* 338; interview with Baclagon; Filemon Rodriguez, *The Marcos Regime: Rape of the Nation* (New York: Vantage, 1985), 68.
112. *Morning Times* (Cebu), 4/15/78; *The Freeman* (Cebu), 4/15/78.
113. *Morning Times,* 5/14/78; *Philippines Herald* (Chicago), 5/16/78; *Vistas* (Cebu), 4/11/82, 8–9.
114. Mendoza, *Golden Book,* 172; Rodriguez, *Working with Heroes,* 78–79; interview with Alcaraz.
115. Baclagon, "Bartolome Cabangbang," 170, 338; interview with Baclagon; *Malaya,* 9/15/85.
116. *FEER,* 5/1/81, 16; 6/26/81, 16.
117. David Wurfel, *Filipino Politics: Development and Decay* (Ithaca, N.Y.: Cornell University Press, 1988), 252; *FEER,* 6/5/81, 32.
118. *FEER,* 6/19/81, 13; Rodriguez, *Working with Heroes,* 200; interview with Alcaraz, 30 December 1994.
119. Interviews with Alcaraz, 30 December 1993, 30 December 1994; Baclagon, "Bartolome Cabangbang," 170, 338; *Malaya,* 9/14/85.
120. *Malaya,* 9/14/85; interview with Baclagon.

CHAPTER 5: MYTH OF THE MAHARLIKA

1. Ernesto O. Rodriguez, *Commodore Alcaraz: First Victim of President Marcos* (New York: Vantage, 1986), 111–19; interviews with Commodore Ramon Alcaraz (ret.), 30 December 1993, 30 December 1994.
2. Rodriguez, *Commodore,* 122–23.

3. Ibid., 147–59.
4. Ibid., 161–65, 168–69.
5. Lieutenant Isidro Espela, PN, "Philippine Navy," Registry Committee, Association of General and Flag Officers, *General and Flag Officers of the Philippines (1896–1977)* (Manila: Association of General and Flag Officers, 1977), 9; Rodriguez, *Commodore,* 162–67; interviews with Alcaraz.
6. Rodriguez, *Commodore,* 166–67, 191; interview with Commodore Felix Apolinario (ret.), Quezon City, 16 March 1995.
7. Interviews with Alcaraz.
8. Interview with Apolinario; Rodriguez, *Commodore,* 250.
9. Rodriguez, *Commodore,* 196–97, 202–13.
10. Ibid., 183, 239–45, 246–47; *MT,* 10/9/54, 5/27/55, 5/28/55; *DM,* 6/29/54, 7/10/54, 8/18/54, 9/20/54, 5/26/55, 6/27/55, 7/12/56, 12/3/65; *The People of the Philippines, plaintiff, vs. Major General Calixto Duque et al., defendants,* Violation of AW #97 & #95, Armed Forces of the Philippines, "Memorandum in Support of Motion for Dismissal," by Ferdinand E. Marcos, Counsel for Major General Calixto Duque, 13 September 1954, File: Army, *MT* Clipping Files, Lopez Memorial Museum, Pasig, Metro-Manila.
11. Daniel F. Doeppers, *Manila, 1900–1941: Social Change in a Late Colonial Metropolis* (New Haven: Yale University Southeast Asia Studies, 1984), 20–22, 126–27; Robert R. Reed, "The Tobacco Industry," in Robert E. Huke, ed., *Shadows on the Land: An Economic Geography of the Philippines* (Manila: Bookmark, 1963), 353–56; Frank H. Golay, *The Philippines: Public Policy and National Economic Development* (Ithaca: Cornell University Press, 1961), 163–66, 336; U.S. Department of Commerce, Bureau of International Commerce, *Philippines: A Market for U.S. Products* (Washington, D.C.: U.S. Government Printing Office, 1965), 8.
12. Barbara Harvey, "Tradition, Islam and Rebellion: South Sulawesi, 1905–1965" (Ph.D. diss., Cornell University, 1974), 264, 322; Kathryn M. Robinson, *Stepchildren of Progress: The Political Economy of Development in an Indonesia Mining Town* (Albany: State University of New York Press, 1986), 85–86; Herbert Feith, *The Decline of Constitutional Democracy in Indonesia* (Ithaca: Cornell University Press, 1962), 488–94; A. V. H. Hartendorp, *History of Industry and Trade of the Philippines: The Magsaysay Administration,* vol. 2 (Manila: Philippine Education Company, 1961), 273–74; Thomas M. McKenna, "The Defiant Periphery: Routes of Iranun Resistance in the Philippines," *Social Analysis* 35 (April 1994), 21.
13. *DM,* 3/12/63; *MT,* 10/1/62, 2/23/63, 3/12/63, 7/9/63, 8/12/63.
14. Interview with Apolinario.
15. Interviews with Alcaraz; Rodriguez, *Commodore,* 248–49.
16. *MT,* 9/3/66, 12/9/66, 1/21/67; interviews with Alcaraz; *DM,* 1/20/67; John Sidel, "Walking in the Shadow of the Big Man: Justiniano Montano and Failed Dynasty Building in Cavite, 1935–1972," in Alfred W. McCoy, ed., *An Anarchy of Families: State and Family in the Philippines* (Quezon City: Ateneo de Manila University Press, 1994), 140.
17. Armed Forces of the Philippines, "Report on Smuggling of BSC into the Philippines for the Years 1962, 63, 64 and 65," 30 November 1965, *MT* Clipping Files, Lopez Memorial Museum, Pasig, Metro-Manila; *MT,* 12/11/63, 12/11/65, 12/17/65.
18. Interview with Colonel Francisco Jimenez (ret.), North Hollywood, Calif., 4 January

1995; phone interview with Jimenez, San Jose, Calif., 26 February 1996; Curriculum Vitae of Colonel Francisco M. Jimenez 0-1670 FA (GSC), n.d.
19. Rodriguez, *Commodore,* 249–50.
20. Interview with Apolinario.
21. Ibid.
22. *MT,* 3/10/64, 3/12/64, 3/16/64, 3/18/64.
23. Rodriguez, *Commodore,* 250; *MT,* 11/29/65.
24. Rodriguez, *Commodore,* 1–4; interviews with Alcaraz; *MT,* 12/1/65.
25. Interviews with Alcaraz; Rodriguez, *Commodore,* 3–4, 30–32; *MT,* 12/2/65, 12/4/65.
26. Interviews with Alcaraz; Sidel, "Walking in the Shadow of the Big Man," 140–44.
27. Rodriguez, *Commodore,* 4–5, 9–10; interviews with Alcaraz.
28. Rodriguez, *Commodore,* 4–5, 9–15, 16; *MDB,* 1/21/66; *Evening News* (Manila), 1/21/66, 1/22/66.
29. Interviews with Alcaraz; Rodriguez, *Commodore,* 4–5, 19–20.
30. Interviews with Alcaraz.
31. Rodriguez, *Commodore,* 41–51.
32. Interview with Apolinario.
33. Rodriguez, *Commodore,* 53–57, 253; Ernesto O. Rodriguez, *Working with Heroes and Exiles* (New York: Vantage, 1989), 185; interviews with Alcaraz.
34. Interviews with Alcaraz; Rodriguez, *Working with Heroes,* 12; Rodriguez, *Commodore,* 57, 59–63, 100.
35. Interviews with Alcaraz; *PDI,* 11/29/98.
36. Rodriguez, *Commodore,* 97–108.
37. Ibid., 103–04; F. E. Marcos, *Notes on the New Society of the Philippines* (1973), 25.
38. *Philippines Sunday Express,* 7/20/75.
39. Gaston Z. Ortigas and Sylvia L. Mayuga, *A Revolutionary Odyssey: The Life and Times of Gaston Z. Ortigas* (Manila: Anvil Publishing, 1994), 98; interviews with Alcaraz.
40. Rodriguez, *Working with Heroes,* 22–27.
41. Ibid., 124–25; Ortigas and Mayuga, *Revolutionary Odyssey,* 107–09.
42. Rodriguez, *Working with Heroes,* 161–62, 172; interviews with Alcaraz.
43. Ortigas and Mayuga, *Revolutionary Odyssey,* 109–10, 112–13; Kapatid, *Pintig 2: Anthology of Prose and Poems from Philippine Prisons* (Manila: Kapatid, 1985), 169–72.
44. Rodriguez, *Working with Heroes,* 176–84, 209; Ortigas and Mayuga, *Revolutionary Odyssey,* 112–13.
45. Rodriguez, *Working with Heroes,* 78–79, 86, 92.
46. Ibid., 81–82.
47. Ibid., 21, 29–30, 107; interviews with Alcaraz.
48. Interview with Colonel Jose Mendoza (ret.), 22 October 1988; interview with General Reynaldo Mendoza (ret.), Camp Aguinaldo, Quezon City, 14 March 1995; interview with Colonel Ciceron de la Cruz (ret.), Quezon City, 12 March 1995. Colonel Ciceron de la Cruz, "Memoirs" (ms, 1995), 59.
49. Rodriguez, *Working with Heroes,* 214–23; interviews with Alcaraz.
50. Arturo C. Aruiza, *Ferdinand E. Marcos: Malacañang to Makiki* (Quezon City: ACAruiza Enterprises, 1991), 434; Rodriguez, *Working with Heroes,* 120–21, 278–79, 286–89, 308–09; *PDI,* 3/2/90.

51. *Los Angeles Times,* 2/26/86; *BT,* 3/3/86; *Orange County Register,* 3/5/86.
52. Interview with Colonel Uldarico Baclagon (ret.), Quezon City, 8 November 1988; interviews with Alcaraz.
53. Jose M. Mendoza, ed., *Batch '36 Golden Book* (Manila: PMA Class '40 Association Inc., 1986), 154; interview with Baclagon.
54. First Class of the Philippine Military Academy, *The Sword of 1940* (Manila: The Sword, 1940), 59, 230; interview with Baclagon; Mendoza, *Golden Book,* 154, 351.
55. Interview with Baclagon, 16 March 1995; Colonel Eduardo Dimacali, "Introduction," in Colonel Uldarico S. Baclagon, *Last 130 Days of the USAFFE* (Makati: Astra Ink Corp., 1982), vi–vii.
56. *Sword of 1940,* 59.
57. Interviews with Baclagon.
58. Ibid.
59. Uldarico S. Baclagon, *They Chose to Fight: The Story of the Resistance Movement in Negros and Siquijor Islands* (Quezon City: Capitol, 1962), 2–9; Uldarico Baclagon, *Philippine Campaigns* (Manila: Graphic House, 1952), 259; interview with Baclagon.
60. Baclagon, *They Chose to Fight,* 13, 15; interviews with Baclagon.
61. Baclagon, *They Chose to Fight,* 30–31, 49; *Sword of 1940,* 30.
62. Baclagon, *Campaigns,* 264; Baclagon, *They Chose to Fight,* 50–51.
63. Baclagon, *Campaigns,* 264–65, 373–74; Mendoza, *Golden Book,* 166.
64. Interviews with Baclagon.
65. Ibid.
66. Baclagon, *Campaigns,* 264–65, 372–73.
67. Mendoza, *Golden Book,* 154; Uldarico Baclagon, *They Served with Honor: Filipino War Heroes of World War II* (Quezon City: DM Press, 1965), 252–54.
68. Interview with Baclagon; Baclagon, *Last 130 Days,* vi; Mendoza, *Golden Book,* 88.
69. Baclagon, *Campaigns,* vii.
70. Teodoro A. Agoncillo, "Philippine Historiography in the Age of Kalaw," *Solidarity* 5, no. 99 (1984), 3–16; Baclagon, *Campaigns,* 35.
71. Ibid., 41–43, 53–54, 110, 124.
72. Ibid., 201–03, 361–64, 374–76, 381–84; Teodoro A. Agoncillo, *The Fateful Years: Japan's Adventure in the Philippines, 1941–45,* vol. 2 (Quezon City: R. Garcia, 1965), 645–777.
73. Interviews with Baclagon; Mendoza, *Golden Book,* 154.
74. Uldarico Baclagon, *Lessons From the Huk Campaign in the Philippines* (Manila: M. Colcol, 1960), 198, 230–32.
75. Interviews with Baclagon; Mendoza, *Golden Book,* 336.
76. Interviews with Baclagon.
77. Mendoza, *Golden Book,* 337.
78. "Ang Mga Maharlika," Box 298, Philippine Archives, File 60, RG 407, NARA.
79. Major Ferdinand E. Marcos, letter to: Adjutant General, Philippine Army, Subject: Complete Roster, Submission of, 18 August 1945, "Ang Mga Maharlika," Box 298, Philippine Archives, File 60, RG 407, NARA.
80. Major Harry McKenzie, letter to: Lieutenant Colonel James W. Davis, Guerrilla Coordinator, 16 September 1945, "Ang Mga Maharlika," Box 298, Philippine Archives, File 60, RG 407, NARA.

81. Lieutenant William D. MacMillan, "Report on the 'Ang Mga Maharlika,'" n.d., "Ang Mga Maharlika," Box 298, Philippine Archives, File 60, RG 407; Thomas J. Brown, letter to: Major Ferdinand Marcos, 7 June 1947, "Ang Mga Maharlika," Box 298, Philippine Archives, File 60, RG 407; Captain E. R. Curtis, letter to: Lieutenant Colonel W. M. Hanes, Subject: Ferdinand E. Marcos, 24 March 1948, "Ang Mga Maharlika," Box 298, Philippine Archives, File 60, RG 407, NARA.
82. Jose M. Crisol and Uldarico S. Baclagon, *Valor: World War II Saga of Ferdinand E. Marcos* (Quezon City: Development Academy of the Philippines, 1983), 21–22; Republic of the Philippines, Senate of the Philippines, *Official Directory 1960–61* (Manila: Bureau of Printing, 1960), 20; interview with de la Cruz; *MDB*, 23 December 1963.
83. Hartzell Spence, *Marcos of the Philippines* (Cleveland: World Publishing, 1969), 7–10, 123; Kerima Polotan, *Imelda Romualdez Marcos: A Biography* (Cleveland: World Publishing, 1969), 126–27.
84. *Sunday Times Magazine*, 10 April 1966, 9 April 1967; *MDB*, 9 May 1966; *Variety*, 2 April 1967; *MT*, 4/8/67; Col. Manuel A. Acosta, "Our Military Memorials," *Self-Reliance in Freedom: Speeches and Writings on Philippine Defense and National Growth* (Manila: Philippine Education Promotion, 1976), 231–36; *MT*, 4/9/69, 4/10/69; *BT*, 2/22/81.
85. Baclagon, *They Served with Honor*, 38, 166.
86. Alfred W. McCoy, "In Search of the Lost Eden—The Historiography of the Philippine Peasant," *Asian Studies Association of Australia Review* 6, no. 2 (November 1982), 70–71.
87. *BT*, 3/11/77.
88. Ferdinand E. Marcos, "Every Nation Needs a Bataan," in *Self-Reliance in Freedom*, 4–17.
89. Interviews with Baclagon; Uldarico S. Baclagon, *Military History of the Philippines* (Manila: St. Mary's, 1975), 296, 305–06.
90. *BT*, 9/12/80.
91. Remedios F. Ramos, E. Arsenio Manuel, Florentino H. Hornedo, Norma G. Tiangco, *Si Malakas at Si Maganda* (Manila: Jorge Y. Ramos, 1980), 5–9. I am grateful to Vina Lanzona for this translation.
92. Ferdinand E. Marcos, *Tadhana: The History of the Filipino People*, vol. 2, *The Formation of the National Community (1565–1896)* (Manila: by the author, 1976), i, ii, vi–vii.
93. Republic of the Philippines, Batasang Pambansa, First Regular Session, Parliamentary Bill No. 195, introduced by Assemblymen Bautista and Ilarde, 14 August 1978; *BT*, 12/22/82; Bonifacio Gillego, "The Other Version of FM's War Exploits," *We Forum*, November 1982.
94. Bonifacio Gillego, "Marcos: The Hero of Kiangan Who Never Was," *Philippine News* (Chicago), 22–28 September 1982; Ruben A. Tan, *The Philippine Congress, 1987–1992* (Metro-Manila: Ruben A. Tan, 1988), 166.
95. William C. Rempel, *Delusions of a Dictator: The Mind of Marcos as Revealed in His Secret Diaries* (Boston: Little Brown, 1993), 194; Gillego, "FM's War Exploits," *Philippine News*, 22–28 September 1982; *BT*, 12/8/82, 12/9/82.
96. *BT*, 1/22/83, 1/23/83, 1/27/83, 12/27/84; *Washington Post*, 12/18/83.
97. Crisol and Baclagon, *Valor*, 9, 13–15, 28–29; interview with Baclagon.
98. Interviews with Baclagon; Alcaraz.
99. *Los Angeles Times*, 1/24/86; *Mr & Ms* (Manila), 10–16 January 1986.

100. *NYT,* 1/23/86; Raymond Bonner, *Waltzing with a Dictator: The Marcoses and the Making of American Policy* (New York: Times, 1987), 393–95; *PDI,* 1/27/86; *Business Day,* 1/27/86; Gemma Nemenzo Almendral, "The Fall of the Regime," in Aurora Javate-De Dios et al., eds., *Dictatorship and Revolution: Roots of People's Power* (Manila: Conspectus, 1988), 197–98.
101. *NYT,* 1/24/86, 1/26/86; *Washington Post,* 2/6/86.
102. *Malaya,* 1/26/86; *BT,* 1/26/86.
103. *PDI,* 2/1/86.
104. Aruiza, *Malacañang to Makiki,* 243, 285.
105. Interview with Fe Navarro, widow of Eduardo Navarro, Malate, Metro-Manila, 15 March 1995.
106. Interviews with Baclagon; Alcaraz; D. H. Soriano and Isidro L. Retizos, *Who's Who* (Makati: Who's Who Publishers, 1981), 194; Colonel Sinforoso L. Duque, *Soldier Heroes: A Handbook on the Winners of the Major Medals Awarded by the Philippine Constabulary and Armed Forces Since 1902* (Manila: National Media Production Center, 1981), 137; Aruiza, *Malacañang to Makiki,* 444.

CHAPTER 6: TORTURE

1. Marcelo Suarez-Orozco, "A Grammar of Terror: Psychocultural Responses to State Terrorism in Dirty War and Post-Dirty War Argentina," in Carolyn Nordstrom and JoAnn Martin, eds., *The Paths to Domination, Resistance, and Terror* (Berkeley: University of California Press, 1992), 230, 237.
2. *NYT,* 7/9/96; William Rees-Mogg, "The Torture Industry," in Rehabilitation and Research Centre for Torture Victims, *Annual Report 1995* (Copenhagen: Rehabilitation and Research Centre for Torture Victims, 1996), 5–6; Inge Genefke, "Some Steps towards a World with Less Torture," in Rehabilitation and Research Centre for Torture Victims, *Annual Report 1995,* 15–16; Rehabilitation and Research Centre for Torture Victims, *Annual Report 1995,* 21–23, 32–34; Keith Carmichael et al., "The Need for REDRESS," *Torture* 6, no. 1 (1966), 7; Erik Holst, "International Efforts on the Rehabilitation of Torture Victims," in June C. Pagaduan Lopez and Elisabeth Protacio Marcelino, eds., *Torture Survivors and Caregivers: Proceedings of the International Workshop on Therapy and Research Issues* (Quezon City: University of the Philippines Press, 1995), 8–14, 190–91; Helena Cook, "The Role of Amnesty International in the Fight Against Torture," in Antonio Cassese, ed., *The International Fight Against Torture* (Baden-Baden: Nomos Verlagsgesellschaft, 1991), 172–86.
3. Metin Basoglu et al., "Psychological Effects of Torture: A Comparison of Tortured with Nontortured Political Activists in Turkey," *The American Journal of Psychiatry* 151, no. 1 (1994), 76–81; Finn Somnier, Peter Vesti, Marianne Kastrup, Inge Kemp Genefke, "Psycho-Social Consequences of Torture: Current Knowledge and Evidence," in Metin Basoglu, ed., *Torture and Its Consequences: Current Treatment Approaches* (Cambridge: Cambridge University Press, 1992), 56–71; Janusz Heitzman and Krzysztof Ruthkowski, "Mental Disorders in Persecuted and Tortured Victims of the Totalitarian System in Poland," *Torture* 6, no. 1 (1996), 19–22; Dimocritos Sarantidis et al., "Long-Term Effects of Torture

of Victims during the Period of Dictatorship in Greece," *Torture* 6, no. 1 (1996), 16–18; Amnesty International, *Report on Torture* (London: Duckworth, 1975), 44.

4. Amnesty International, *Torture in the Eighties* (London: Martin Robinson, 1984), 19; Carlos Madariaga, M.D., "Torture Prevention as a Public Health Problem," *Torture* 6, no. 4 (1996), 86–89; Diana Kordon, Lucila Edelman, et al., "Torture in Argentina," in Basoglu, *Torture*, 433–51.

5. June C. Pagaduan Lopez, "The History of the Study of Psycho-Social Trauma (PST)," in Lopez, *Torture Survivors*, 82–83; Ellen Sherwood, "The Power Relationship Between Captor and Captive," *Psychiatric Annals* 16, no. 11 (November 1986), 653–55.

6. Within this growing literature on victims of torture, those sources consulted for this chapter included F. E. Somnier and I. K. Genefke, "Psychotherapy for Victims of Torture," *British Journal of Psychiatry* 149 (1986), 323–29; Derrick Silove et al., "Psychosocial Needs of Torture Survivors," *Australian and New Zealand Journal of Psychiatry* 25, no. 4 (December 1991), 481–90; Stuart Turner and Caroline Gorst-Unsworth, "Psychological Sequelae of Torture: A Descriptive Model," *British Journal of Psychiatry* 157 (1990), 475–80.

7. Bruno Bettelheim, *The Informed Heart: Autonomy in a Mass Age* (New York: The Free Press, 1960), 118.

8. Daniel Jonah Goldhagen, *Hitler's Willing Executioners: Ordinary Germans and the Holocaust* (New York: Vintage, 1997), 5–9; Alex P. Schmid, "A Dutch Think Tank on Torture," *Torture* 2, no. 1 (1992), 2, 9–10; Duncan Forrest, *A Glimpse of Hell: Reports on Torture Worldwide* (New York: Amnesty International and New York University Press, 1996), 87; Federico Allodi, "Somoza's National Guard: A Study of Human Rights Abuses, Psychological Health and Moral Development," in Ronald D. Crelinsten and Alex P. Schmid, eds., *The Politics of Pain: Torturers and Their Masters* (Boulder: Westview, 1995), 117.

9. In its first report on torture published in 1973, Amnesty International, reflecting the earlier psychology literature, concluded that torturers are "grossly abnormal personalities." (See Amnesty, *Report on Torture*, 63–68.) Literature since the 1960s emphasizes the "normal" nature of torturers and thus represents, in sum, a revision of Adorno's earlier focus on psychological abnormality as a key to understanding fascism, prejudice, and political violence. (See T. W. Adorno et al., *The Authoritarian Personality* [New York: Harper & Brothers, 1950], 961–76.) In this gradual shift from psychological to situational factors, Henry Milgram's obedience experiments, described below, "stimulated a remarkable tide of reaction." (See Arthur G. Miller, *The Obedience Experiments: A Case Study of Controversy in Social Science* [New York: Praeger, 1986], 1–15, 67–69, 252–55.)

In support of these first two points, a 1993 survey of ten countries with systematic torture found that the military "sadist" was not "prevalent" since effective torture requires a controlled, detached demeanor. (See Ronald D. Crelinsten, "The World of the Torturer," *Torture* 3, no. 1 [1993], 5–10.) My third point is supported by psychologist Wolfgang Heinz, who found, through interviews with Argentine officers about their "dirty war," that the elite counter-guerrilla units carried out torture by stigmatizing victims as foreign agents, thus stripping them of their "quality of being human." (See Wolfgang F. Heinz, "Torture and the Armed Forces: The Process of Learning to Accept Torture by Military Officers in Argentina and Uruguay," *Torture* 2, no. 2 [1992], 55–56.) The last point is influenced by a study of the Nazi concentration camps which argued that their horror was the product of "normal people placed in deranged and degrading circumstances." (See John P. Sabini and Maury Silver, "Destroying the Innocent with a Clear Conscience: A Sociopsychology of the Holocaust," in Joel E. Dimsdale, ed., *Survivors, Victims, and Perpetrators: Essays on the Nazi Holocaust* [Washington: Hemisphere, 1980], 356.)

10. Janice T. Gibson, "Training People to Inflict Pain: State Terror and Social Learning," *Journal of Humanistic Psychology* 31, no. 2 (Spring 1991), 72–73, 80–81; Janice T. Gibson and Mika Haritos-Fatouros, "The Education of a Torturer," *Psychology Today* 20, no. 11 (November 1986), 56; Amnesty International, *Torture in Greece: The First Torturers' Trial, 1975* (London: Amnesty International, 1977), 28, 35–42.

11. June Pagaduan Lopez, M.D. et al., "Human Rights Violations: What Perpetuates the Perpetrator" (Quezon City: Psychosocial Trauma Program, Center for Integrative and Development Studies, University of the Philippines, 1995), 2–7, 9, 18; Lan Mercado Carreon, "Interviews with Perpetrators," *PST Quarterly* 1, no. 2 (July–September 1996), 8; Mika Haritos-Fatouros, "The Official Torturer: A Learning Model for Obedience to the Authority of Violence," *Journal of Applied Social Psychology* 18, no. 13 (October 1988), 1107–20.

12. Stanley Milgram, *Obedience to Authority: An Experimental View* (New York: Harper & Row, 1974), 1–43; Stanley Milgram, "Group Pressure and Action Against a Person," *Journal of Abnormal and Social Psychology* 9, no. 2 (1964), 137–43; Miller, *Obedience Experiments*; Christopher R. Browning, *Ordinary Men: Reserve Police Battalion 101 and the Final Solution in Poland* (New York: Harper Perennial, 1992), 171–77.

13. Philip G. Zimbardo, "On the Ethics of Intervention in Human Psychological Research: With Special Reference to the Stanford Prison Experiment," *Cognition* 2, no. 2 (1973), 243–44: Philip G. Zimbardo and colleagues, "The Mind Is a Formidable Jailer: A Pirandellian Prison," *New York Times Magazine*, 8 April 1973, 38–60; Browning, *Ordinary Men*, 167–69. The guards wore masculine uniforms; the prisoners dresslike gowns, perhaps symbolizing their emasculation. (See Sabini, "Destroying the Innocent," 348).

14. Sofia Salimovich, Elizabeth Lira, and Eugenia Weinstein, "Victims of Fear," in Juan E. Corradi, Patricia Weiss Fagen, and Manuel Antonio Garreton, eds., *Fear at the Edge: State Terror and Resistance in Latin America* (Berkeley; University of California Press, 1992), 76–78; Jean Franco, "Gender, Death and Resistance: Facing the Ethical Vacuum," in Corradi, *Fear at the Edge*, 105–07.

15. Dr. Diana R. Kordon et al., "Torture in Argentina," in Diana R. Kordon et al., eds., *Psychological Effects of Political Repression* (Buenos Aires: Sudamericana/Planeta, 1988), 102; Rees-Mogg, "The Torture Industry," 5–6.

16. Sherwood, "Power Relationship," 653–55; Bruno Bettelheim, "Individual and Mass Behavior in Extreme Situations," in Harold Prohansky and Bernard Seidenberg, eds., *Basic Studies in Social Psychology* (New York: Holt, Rinehart and Winston, 1965), 630, 637; Bettelheim, *Informed Heart,* 109–10; Klaus Theweleit, *Male Fantasies,* vol. 2, *Male Bodies: Psychonalyzing the White Terror* (Minneapolis: University of Minnesota Press, 1989), 304–05; Ernest Federn, "The Terror as a System: The Concentration Camp," *The Psychiatric Quarterly Supplement* 22, pt. 1 (1948), 57.

17. Otto Doerr-Zegers, Lawrence Hartmann, Elizabeth Lira, and Eugenia Weinstein, "Tor-

ture: Psychiatric Sequeli and Phenomenology," *Psychiatry* 55, no. 2 (May 1992), 178–79; Lawrence Hartmann et al., "Psychopathology of Torture Victims," *Torture* 3, no. 2 (1993), 36–38.
18. Doerr-Zegers, "Torture," 179–83. Similarly, a group of six Argentine psychotherapists found from treating survivors that "torture targets the individual's identity," which they defined as "a complex of representations of self . . . that produces the feeling of oneness and allows one to maintain internal coherence in time." (Kordon, "Torture in Argentina," in Basoglu, *Torture,* 433–51.)
19. José A. Saporta, Jr., and Bessel A. van der Kolk, "Psychobiological Consequences of Severe Trauma," in Basoglu, *Torture,* 151–57.
20. Amnesty, *Torture in Greece,* 32; Marguerite Feitlowitz, *A Lexicon of Terror: Argentina and the Legacies of Torture* (New York: Oxford University Press, 1998), 10–11; Elaine Scarry, *The Body in Pain: The Making and Unmaking of the World* (New York: Oxford University Press, 1985), 28, 47, 53, 54, 56.
21. Scarry, *Body in Pain,* 41–42; Patricia Weiss Fagen, "Repression and State Security," in Corradi, *Fear at the Edge,* 62; Juan Rial, "Makers and Guardians of Fear: Controlled Terror in Uruguay," in Corradi, *Fear at the Edge,* 98; Ximena Bunster-Burotto, "Surviving Beyond Fear: Women and Torture in Latin America," in June Nash and Helen Safa, eds., *Women and Change in Latin America* (South Hadley, Mass.: Bergin and Garvey, 1985), 297–99, 314.
22. Christopher Simpson, *Science of Coercion: Communication Research and Psychological Warfare 1945–1960* (New York: Oxford University Press, 1994), 4–5, 72–73, 114–15; *Baltimore Sun,* 1/27/97; *Washington Post,* 1/28/97; *NYT,* 1/29/97.
23. "Torturer in U.S. for Training," *Tanod* (Manila) 1, no. 3 (September 1978), 3; Task Force Detainees, Association of Major Religious Superiors in the Philippines, *Pumipiglas: Political Detention and Military Atrocities in the Philippines* (Manila: TFDP, 1980), 106–07.
24. The Philippine military apparently coined this neologism to describe its torture operations in the mid-1970s and it was soon taken up by the country's human rights groups. (See Task Force Detainees, *Political Detainees of the Philippines Book 3* [Manila: Association of Major Religious Superiors in the Philippines, March 1978], 41–43.)
25. Michel Foucault, *Discipline and Punish: The Birth of the Prison* (New York: Vintage, 1979), 49; E. P. Thompson, "Patrician Society, Plebeian Culture," *Journal of Social History* 7, no. 4 (1974), 389–90.
26. Frank Graziano, *Divine Violence: Spectacle, Psychosexuality, and Radical Christianity in the Argentine "Dirty War"* (Boulder: Westview, 1992), 203; Amnesty, *Torture in Greece,* 28, 35–42; Scarry, *Body in Pain,* 56–57; Michael Taussig, *Shamanism, Colonialism, and the Wild Man: A Study in Terror and Healing* (Chicago: University of Chicago Press, 1987), 83, 133; Feitlowitz, *Lexicon of Terror,* 3–4.
27. Republic of the Philippines, *The Final Report of the Fact-Finding Commission (Pursuant to R.A. No. 6832)* (Manila: Bookmark, 1990), 42, 44.
28. Ray Bonner, *Waltzing with a Dictator: The Marcoses and the Making of the American Policy* (New York: Times, 1987), 126–27; Senator Benigno "Ninoy" Aquino, Jr., *A Garrison State in the Make and Other Speeches* (Makati: Benigno S. Aquino, Jr., Foundation, 1985),

345–51; Filemon C. Rodriguez, *The Marcos Regime: Rape of the Nation* (New York: Vantage, 1985), 85–86; Ferdinand E. Marcos, *Notes on the New Society of the Philippines* (Manila: Marcos Foundation, 1973), 2–3.
29. Marcos, *New Society*; Amnesty International, *Report of an Amnesty International Mission to the Republic of the Philippines, 11–28 November 1981* (London: Amnesty International, 1982), 56–57.
30. *NYT*, 2/23/86; Bonner, *Waltzing with a Dictator*, 125, 468. Not only was this retired officer one of my most reliable sources, but he provided me with corroboration that, in a Philippine context, was convincing. (Interview with a member of the Class of 1940, Metro-Manila, August 1996.)
31. Philippines, *Final Report*, 43–50; Rigoberto D. Tiglao, "The Consolidation of the Dictatorship," in Aurora Javate-De Dios et al., eds., *Dictatorship and Revolution: Roots of People's Power* (Manila: Conspectus, 1988), 53.
32. Fagen, "Repression and State Security," 49–55, 58–60; Neil J. Kritz, ed., *Transitional Justice: How Emerging Democracies Reckon with Former Regimes*, vol. 3, *Laws, Rulings, and Reports* (Washington, D.C.: U.S. Institute for Peace, 1995), 146–47; Joan Dassin, ed., *Torture in Brazil* (New York: Vintage, 1986), 204–05, 235–38; *NYT*, 11/10/86; Kritz, *Transitional Justice*, vol. 2, 431; Lawrence Weschler, *A Miracle, A Universe: Settling Accounts with Torturers* (New York: Pantheon, 1990), 53.
33. Philippines, *Final Report*, 52; Tiglao, "Consolidation of the Dictatorship," 55.
34. Sanford M. Dornbusch, "The Military Academy as an Assimilating Institution," *Social Forces* 33, no. 4 (May 1955), 316–21.
35. Quotations are from all of the yearbooks found in the PMA's library during an August 1996 visit: *The 1953 Sword, The 1972 Sword, The CCAFP 76 Magilas Sword, The Sword '77, The Sword of Class 1978, The Sword 1987, The 89 Makatao Swordbook, Sword '90, The Sword '91*.
36. Harold W. Maynard, "A Comparison of Military Elite Role Perceptions in Indonesia and the Philippines" (Ph.D. diss., American University, 1976), 456; Carolina G. Hernandez, "The Extent of Civilian Control of the Military in the Philippines: 1946–1976" (Ph.D. diss., State University of New York at Buffalo, 1979), 252.
37. Maynard, "Comparison of Military Elite," 423; Felipe Miranda, *The Politicization of the Military* (Quezon City: University of the Philippines Press, 1992), 13–14; Philippines, *Final Report*, 72–74.
38. Philippines, *Final Report*, 70.
39. Ibid., appendix J.
40. Jose G. Ayap, *The 1971 Sword* (Baguio City: Cadet Corps, Armed Forces of the Philippines, 1971).
41. Ruben Fulgueras Ciron, "Civil-Military Relations in the Philippines: Perceptions of PMA-Trained Officers" (Ph.D. diss., University of the Philippines, 1993), 15, 55, 152.
42. Cadet Corps, *The Academy Scribe* (Baguio City: The Academy Scribe Organization, 1989), 142.
43. Benjamin N. Muego, "Civilian Rule in the Philippines," in Constantine P. Danopoulos, ed., *Civilian Rule in the Developing World: Democracy on the March?* (Boulder: Westview, 1992), 215–17; Cadet Corps, *The Academy Scribe*, 227, 242, 391; Dante Simbulan, "A Study

of the Socio-Economic Elite in Philippine Politics, 1946–1963" (Ph.D. diss., Australian National University, 1965); interview with Jose "Pete" Lacaba, Canberra, 14 February 1988.
44. Rigoberto D. Tiglao, "Rebellion from the Barracks: The Military as Political Force," in Philippine Center for Investigative Journalism, *Kudeta: The Challenge to Philippine Democracy* (Manila: Philippine Center for Investigative Journalism, 1990), 15; Viberto Selochan, "Professionalization and Politicization of the Armed Forces of the Philippines" (Ph.D. diss., Australian National University, 1990), 67; Jo-Ann Q. Maglipon, *Primed: Selected Stories 1972–1992* (Pasig: Anvil, 1993), 227; Colonel A. P. Aguirre, *A People's Revolution of Our Time* (Quezon City: Colonel Alexander P. Aguirre, 1986), v–vi.
45. Ayap, *1971 Sword*; interviews with Lieutenant Colonel Victor Corpus, Quezon City, 1 April 1987, 5 April 1987.
46. *Mr. & Ms.*, March 21–27, 1986, 17; Maglipon, *Primed*, 227.
47. Rigoberto D. Tiglao, "Rebellion from the Barracks: The Military as Political Force," in Philippine Center for Investigative Journalism, *Kudeta*, 15.
48. Ayap, *1971 Sword*; interviews with Corpus.
49. Ayap, *1971 Sword*; PDI, 10/7/94.
50. *PC Journal*, July 1986. Despite centuries of Spanish rule, the term "gringo," common in Mexico, was never used in the Philippines. During the late 1960s, the term first appeared in the Philippines when Sergio Leone's Italian Westerns were shown.
51. Cadet Corps, *Academy Scribe*, 95, 101; Selochan, "Professionalization," 66; interview with Colonel Deogracias Cabellero (ret.), Quezon City, 12 October 1988; interview with General Reynaldo Mendoza (ret.), Camp Aguinaldo, Quezon City, 14 March 1995; interview with Colonel Francisco del Castillo (ret.), Quezon City, 18 March 1995.
52. Cadet Corps, *Academy Scribe*, 126–27, 130.
53. *MT*, 7/10/56, 7/16/56.
54. *MT*, 7/11/56, 7/17/57, 6/29/61.
55. Interview with General Salvador Piccio (ret.), Quezon City, 5 November 1988.
56. Interview with Castillo; interview with Colonel Francisco Jimenez (ret.), North Hollywood, Calif., 4 January 1995.
57. Major Rogelio S. Lumabas, "Whither the MAP," *The Cavalier* 7, no. 1 (November–December 1967), 5.
58. Maglipon, *Primed*, 226–27.
59. Ayap, *1971 Sword*.
60. Maglipon, *Primed*, 226–27.
61. *BMC*, 4/6/69, 4/20/69, 3/1/70; *MT*, 4/19/69, 4/20/69, 4/22/69, 4/26/69.
62. Ronald G. Bauer, "Military Professional Socialization in a Developing Country" (Ph.D. Diss., University of Michigan, 1973), 40–42; *DM*, 4/5/71.
63. Lopez, "Human Rights Violations," 9, 17–18.
64. Selochan, "Professionalization," 64–65; interview with Jose Duran, Quezon City, 19 March 1995.
65. Lopez, "Human Rights Violations," 13; Lan Mercado Carreon, "A Torturer's Tale," 9–10.
66. While Honasan's 1995 campaign literature claimed three Gold Cross medals, the AFP register showed only one for "leading his men in a fire fight against numerically superior

rebels . . . [t]hough hit in the right thigh" at Lebak Cotabato in June 1973. (See Colonel Sinforoso L. Duque, *Soldier Heroes: A Handbook on the Winners of the Major Medals Awarded by the Philippine Constabulary and Armed Forces Since 1902* [Manila: National Media Production Center, 1981], 91; "Tungkol Kay Gringo B. Honasan," *Para Sa Bansa Gringo Honasan Sa Senado* [Quezon City: Benny J. Brizuela, 1995.])

67. Selochan, "Professionalization," 188; Sheila Coronel, "RAM: From Reform to Revolution," Philippine Center for Investigative Journalism, *Kudeta*, 60; *Mr. & Ms.*, 27 February–5 March 1987, 6; Fortunato U. Abat, *The Day We Nearly Lost Mindanao: The CEMCOM Story* (San Juan: FUA 1993), 105; *PC Journal*, July 1986.

68. Philippines, *Final Report*, appendix J, Table VI-6, 442–43.

69. Amnesty International, *Report of Amnesty International Mission to the Republic of the Philippines, 22 November–5 December 1975* (London: Amnesty International, 1976), 13, 57, 72–73, 85.

70. Amnesty, *Philippines 1981*, 2, 8; *Philippine Graphic*, 18 January 1993; *NYT*, 11/10/86; Amnesty International, *Philippines: The Killing Goes On* (New York: Amnesty International, 1992), 14.

71. Hernandez, "Extent of Civilian Control," 216; Selochan, "Professionalization," 57, 68, 216.

72. Suarez-Orozco, "A Grammar of Terror," 235–36; Martin Edward Anderson, *Dossier Secreto: Argentina's Desaparecidos and the Myth of the "Dirty War"* (Boulder: Westview, 1993), 2, 5, 205–19; *Nunca Mas: The Report of the Argentine National Commission on the Disappeared* (New York: Farrar Straus Giroux, 1986), 209, 233; Feitlowitz, *Lexicon of Terror*, 8, 25, 45, 165–68, 172–74.

73. Task Force Detainees of the Philippines, *Pumipiglas: Political Detention and Military Atrocities in the Philippines 1981–1982* (Quezon City: Task Force Detainees of the Philippines, Association of Major Religious Superiors in the Philippines, 1986), 44–45; Association of Major Religious Superiors in the Philippines, *Political Detainees in the Philippines Book Two* (Manila: AMRSP, 31 March 1977), 1; Amnesty International, *Human Rights Violations in the Philippines* (New York: Amnesty International USA, 1982), 1.

74. Lawyers Committee for International Human Rights, *The Philippines: A Country in Crisis* (New York: Lawyers Committee, 1983), 32–49; Amnesty, *Philippines 1981*, 62; Ma. Serena I. Diokno, "Unity and Struggle," in De Dios, *Dictatorship and Revolution*, 146–47; Rev. La Verne D. Mercado and Sister Mariani Dimaranan, *Philippines: Testimonies on Human Rights Violations* (Manila: World Council of Churches, 1986), 89–135; Richard J. Kessler, *Rebellion and Repression in the Philippines* (New Haven: Yale University Press, 1989), 137.

75. In a lead editorial on the "summary killings and disappearances" of the Marcos era, the *Philippine Daily Inquirer* (6/29/96) commented, "we call this 'salvaging,' demonstrating our talent to reinvent the English language."

76. Maynard, "Comparison of Military Elite," 461.

77. Gemma Nemenzo Almendral, "The Fall of the Regime," in De Dios, *Dictatorship and Revolution*, 200; Tiglao, "Consolidation of Dictatorship," 54.

78. Amnesty, *Philippines 1981*, 21–23; *PFP*, 29 June 1996; Task Force Detainees, *Pumipiglas* (1980), 64; Biodata, Panfilo Lacson, Library, *PDI*.

79. *Amnesty, Human Rights Violations,* 12–17.
80. Amnesty, *Philippines 1981,* 1–12, 22–23; Satur C. Ocampo, "Leaving the Pain Behind," *PST Quarterly* 1, no. 2 (July–September 1996), 13; Amnesty International, *Philippines: Unlawful Killings by Military and Paramilitary Forces* (New York: Amnesty International, 1988), 5–6.
81. Miranda, *The Politicization of the Military,* 12.
82. Maynard, "Comparison of Military Elite," 462.
83. Leoncio Co, University of the Philippines, Interdisciplinary Forum on Political Detainees, 16 April 1986.
84. Alfred W. McCoy, *Priests on Trial* (New York: Penguin, 1984), 212–15.
85. Father Edgardo Kangleon, "A Moment of Uncertainty" (ms, 8 December 1982), 7, enclosed in letter to Dear Papa/Mama/Rey, 30 September 1983 (copy furnished by Father Niall O'Brien, St. Columban's Mission Society, Bacolod City). An excerpt of this letter was published in Promotion of Church People's Rights, *That We May Remember* (Quezon City: PCPR, May 1989), 168–73.
86. Kangleon, "Moment of Uncertainty," 11–12.
87. Ibid., 13–16.
88. Promotion of Church People's Rights, *That We May Remember,* 172–73.
89. Mercado, *Philippines,* 52.
90. Lawyers Committee for Human Rights, *"Salvaging" Democracy: Human Rights in the Philippines* (New York: Lawyers Committee for Human Rights, 1985), 56–57.
91. *PDI,* 7/23/93; Task Force Detainees, *Pumipiglas* (1980), 103.
92. Ibid., 103–07.
93. Amnesty, *Report of 1975,* 21, 28.
94. Telephone interview with Luis Jalandoni, Utrecht, Netherlands, 18 February 1988; Task Force Detainees, *Pumipiglas* (1986), 59–64.
95. Eliseo C. Tellez, Jr., Proof of Claim Form for Torture Victims, 8 December 1992, Samahan ng mga Ex-Detainee Laban sa Detensyon at para sa Amnestia (SELDA), Manila. Other affidavits and reports citing Lieutenant Aguinaldo's abuse of genitals were filed by Alfonso Abrazado (n.d.), Roberto Verzola y Sevilla (n.d.), Monico Atienza y Montenegro (n.d.), and Oliver G. Teves (13 July 1993).
96. According to one of Aguinaldo's prisoners, Satur Ocampo, "late in the night he would ask for these women detainees to be brought to his office. He just talked to them, how lonely he is and all that. But he would not touch them." (Interview, Quezon City, 27 August 1996.)
97. *PDI,* 7/23/93. Aguinaldo's victims included journalists Satur Ocampo, Pete Lacaba, Julius Fortuna, and Oliver Teves; U.P. professors Roger Posadas and Temario Rivera; photographer Anacleto Ocampo; artist Nestor Buyayong; and activists Allan Jazmines and Nilo Tayag.
98. Dr. Temario C. Rivera, "Details of Torture Inflicted on Temario C. Rivera" (ms, courtesy of the author, August 1996.)
99. Association of Major Religious Superiors in the Philippines, *Political Detainees in the Philippines, Book Two* (Manila: AMRSP, 31 March 1977), 8.
100. Ibid., 15.

101. Interview with Satur Ocampo, Quezon City, 27 August 1996; Ocampo, "Leaving the Pain Behind," 13.
102. Task Force Detainees, *Pumipiglas* (1980), 106–07.
103. Coronel, "RAM," 65. An early report about Batac stated: "Other officers of the 5th CSU who have been implicated in torture accounts of political detainees and against whom no public investigation has been made: Capt. (now Major) Cecilio Penilla, Capt. Virgilio Saldajeno, Lt. Rodolfo Aguinaldo, Lt. Victor Batac, Lt. Robert Delfin, Lt. Cesar Alvarez." (See Task Force Detainees, *Political Detainees of the Philippines, Book Three,* 5.) Other details of Batac's treatment of prisoners are found in interviews with Jose Ma. Sison, Alan Jazmines, and "Gene," University of the Philippines, Interdisciplinary Forum on Political Detainees, 16 April 1986.
104. Danilo P. Vizmanos, Proof of Claim Form for Torture Victims, *In Re. Estate of Ferdinand E. Marcos Human Rights Litigation,* MDL No. 840—Class Action (5 May 1993); *Honolulu Star-Bulletin,* 9/11/92; *BT,* 8/15/76.
105. AMRSP, *Political Detainees in the Philippines, Book Two,* 8–9; *Daily Cardinal* (Madison, Wisc.), 10/16/86.
106. *Daily Cardinal,* 10/17/86; *Capital Times* (Madison, Wisc.), 10/17/86.
107. Interview with Maria Elena Ang, Sydney, Australia, 9 May 1989.
108. *The Australian* (Sydney), 4/26–27/86.
109. Interview with Randall Echanis, University of the Philippines, Interdisciplinary Forum on Political Detainees, n.d.
110. For references to Lieutenants Bibit, Batac, and Aguinaldo working together as torturers, see Domingo Luneta (22 November 1992), Marcelio M. Talam, Jr. (4 November 1992), Ma. Paz Castronuevo Talam (3 November 1992), Eliseo C. Tellez, Jr. (8 December 1992), Oliver G. Teves (13 July 1993), Proof of Claim Form for Torture Victims, SELDA, Manila; and Monica Atienza y Montenegro (date of arrest: 4 October 1974; Present Confinement: 5th Constabulary Security Unit, Camp Crame), SELDA. For references to Rolando Abadilla and his two MISG comrades, Roberto Ortega and Panfilo Lacson, see Romeo I. Chan (11 March 1993), Damaso de la Cruz (27 October 1992), Proof of Claim Form for Torture Victims, SELDA, Manila.
111. *Malaya,* 8/29/87.
112. Coronel, "RAM," 60; interview with Sheila Coronel, Manila, 5 January 1988; Neni Sta. Romana-Cruz, "Reformists Night Out: In Uniform But Into Fun," *Mr. & Ms.,* March 21–27, 1986, 19–20; Maglipon, *Primed,* 228.
113. *Mr. & Ms.,* March 21–27, 1986; interview with Navy Captain Rex Robles, Manila, 25 July 1986; interview with RAM leader, Manila, July 1986.
114. Interview with Robles.

CHAPTER 7: MUTINY

1. Secretary Jose T. Almonte, "The Emergence of East Asian Democracies" (paper presented at Kennedy School of Government, Harvard University, 26 September 1996), 2.
2. Captain Danilo Vizmanos (PN, ret.), "Can a Military Man Be Civilianized?" *Today* (Manila), 30 April 1997.

3. Mark R. Thompson, *The Anti-Marcos Struggle: Personalistic Rule and Democratic Transition in the Philippines* (New Haven: Yale University Press, 1995), i, 155–56.
4. Interview with General Rafael Ileto, Manila, 2 July 1986.
5. Rigoberto D. Tiglao, "The Consolidation of the Dictatorship," in Aurora Javate-De Dios et al., eds., *Dictatorship and Revolution: Roots of People's Power* (Manila: Conspectus, 1988), 50, 404; Arturo C. Aruiza, *Ferdinand E. Marcos: Malacañang to Makiki* (Quezon City: ACAruiza Enterprises, 1991), 43. By 1975, the ROTC regulars in service commands were General Rafael Zagala (army), General Jose Rancudo (air force), and Admiral Hilario Ruiz (navy). (See Harold W. Maynard, "A Comparison of Military Elite Role Perceptions in Indonesia and the Philippines" [Ph.D. diss., American University, 1976], 489–92).
6. Aruiza, *Malacañang to Makiki,* 83–84.
7. *DM,* 25 May 1963, 15 March 1971; Ray Bonner, *Waltzing with a Dictator: The Marcoses and the Making of American Policy* (New York: Times, 1987), 347; Aruiza, *Malacañang to Makiki,* 52–53.
8. Ferdinand E. Marcos, "Presidential Security Command: An Elite Unit of the AFP," in *Self-Reliance in Freedom: Contemporary Speeches and Writings on Philippine Defense and National Growth* (Manila: Philippine Educational Promotion, 1977), 71–75; Republic of the Philippines, *The Final Report of the Fact-Finding Commission (Pursuant to R.A. No. 6832)* (Manila: Bookmark, October 1990), 46; Angela Stuart Santiago, *Duet for EDSA: 1986 Chronology of a Revolution* (Manila: Foundation for Worldwide People Power, 1995), 11; Aruiza, *Malacañang to Makiki,* 42–43.
9. Santiago, *Duet for EDSA: Chronology,* 11; Cecilio T. Arillo, *Breakaway: The Inside Story of the Four-Day Revolution in the Philippines, February 22–25, 1986* (Mandaluyong: CTA & Associates, May 1986), 139; Aruiza, *Malacañang to Makiki,* 109–11, 438.
10. HPC [Headquarters Philippine Constabulary] Historical Committee, *Ferdinand E. Marcos: 77 Days in Eastern Pangasinan* (Manila: Office of Media Affairs, 1981), ix–x; Major Narciso Ramos, "Identification," 17 February 1945, Special Roster, Guerrilla Unit: Ang Mga Maharlika (Ilocos Norte Regiment, Sarrat Unit), 1 June 1945, Box 298, Philippine Archives, File 60, RG 407, NARA.
11. Belinda A. Aquino, "Political Violence in the Philippines: Aftermath of the Aquino Assassination," in *Southeast Asian Affairs* (Singapore: Institute of Southeast Asian Studies, 1984), 274–75.
12. Colonel A. Aguirre, *A People's Revolution of Our Time: Philippines, February 22–25, 1986* (Quezon City: Pan-Service Master Consultants, 1986), 27.
13. Belinda Aquino, "The Philippines in 1987: Beating Back the Challenge of August," *Southeast Asian Affairs 1988* (Singapore: Institute of Southeast Asian Studies, 1988), 203–04.
14. Santiago, *Duet for EDSA: Chronology,* 11; Arillo, *Breakaway,* 138, 143–44; Aruiza, *Malacañang to Makiki,* 83–90.
15. Ma. Serena I. Diokno, "Unity and Struggle," in De Dios, *Dictatorship and Revolution,* 168–69, 653.
16. Arillo, *Breakaway,* 142.
17. Marilies von Brevern, *The Turning Point: Twenty-Six Accounts of February Events in the Philippines* (Manila: Lyceum Press, 1986), 35–37; Ma. Serena I. Diokno, "Unity and Struggle," 162–63; Gemma Nemenzo Almendral, "The Fall of the Regime," in De Dios, *Dic-*

tatorship and Revolution, 185, 209–10, 666; Santiago, Duet for EDSA: Chronology, 13; Aruiza, Malacañang to Makiki, 88–89; NYT, 2/4/86.

18. Criselda Yabes, *The Boys from the Barracks: The Philippine Military After EDSA* (Manila: Anvil, 1991), 112–15.
19. Aruiza, *Malacañang to Makiki*, 166, 239; interview with Colonel Antonio Sotelo, Villamor Air Base, Manila, 21 July 1986; interview with Colonel Braulio Balbas, Fort Bonifacio, Manila, 25 July 1986.
20. Arillo, *Breakaway*, 131–32; Aruiza, *Malacañang to Makiki*, 86; Maynard "Comparison of Military Elite," 492–97.
21. Ronald G. Bauer, "Military Professional Socialization in a Developing Country" (Ph.D. diss., University of Michigan, 1973), 27; Maynard, "Comparison of Military Elite," 423.
22. Aruiza, *Malacañang to Makiki*, 41–43, 66, 86; Philippines, *Final Report*, 56.
23. Maynard, "Comparison of Military Elite," 489, 491.
24. Philippines, *Final Report*, 72–74.
25. Viberto Selochan, "Professionalization and Politicization of the Armed Forces of the Philippines" (Ph.D. diss., Australian National University, 1990), 69.
26. Amos Perlmutter, *The Military and Politics in Modern Times: On Professionals, Praetorians, and Revolutionary Soldiers* (New Haven: Yale University Press, 1977), 12, 17, 103; von Brevern, *The Turning Point*, 96; Aruiza, *Malacañang to Makiki*, 47.
27. Colonel Hector M. Tarrazona, *After EDSA*, vol. 1 (Pasay City: Hector M. Tarrazona, 1989), 55; *Para Sa Bansa Gringo Honasan Sa Senado* (Quezon City: Benny J. Brizuela, 1995 [campaign leaflet]); Aruiza, *Malacañang to Makiki*, 47–50.
28. Ministry of National Defense, Transcript of Press Conference, 22 February 1986, 1–2; Arillo, *Breakaway*, 137, 166.
29. Arillo, *Breakaway*, 135–37, 141.
30. Yabes, *Boys from the Barracks*, 161; Santiago, *Duet for EDSA: Chronology*, 11.
31. Interview with Juan Ponce Enrile by the *National Times* (Sydney), September 1986.
32. Selochan, "Professionalization," 184; Tarrazona, *After EDSA*, 21, 25; "Preliminary Statement of Aspirations," *Crossroads to Reform* (Manila: Education Committee, Reform the AFP Movement, January 1986), 9–10.
33. *Crossroads to Reform*, 13–18.
34. Tarrazona, *After EDSA*, 21; Lieutenant Gregorio Catapang, "Le Mouvement de Réforme de l'Armée (RAM) et les Forces Armée de Philippines (AFP)," *Les Temps Moderne* 44, no. 508 (November 1988), 82–83; *Asia Week*, 24 May 1985; *Mr. & Ms.*, May 31–June 6, 1985, 6; Santiago, *Duet for EDSA: Chronology*, 11.
35. *Asia Week*, 14 June 1985, 17; Tarrazona, *After EDSA*, 23; interview with Captain Rex Robles, Manila, 25 July 1986; Aruiza, *Malacañang to Makiki*, 45; Lawyers Committee for Human Rights, *"Salvaging" Democracy*, 145.
36. *We Belong* 1, no. 3, (18 July 1985), 4–5, 9–10; Lawyers Committee for Human Rights, *"Salvaging" Democracy*, 167; *Metro Manila Times*, 5/10/85; *Mr. & Ms.*, May 31–June 6, 1985, 7.
37. "Regarding Aim No. (ii): A Fundamental Area of Corruption," *We Belong* 1, no. 3 (18 July 1985), 5.
38. Lawyers Committee for Human Rights, *"Salvaging" Democracy: Human Rights in the Philippines* (New York: The Lawyers Committe for Human Rights, 1985), 143–44; *PDE*, 7/6/85; *BT*, 7/6/85.

39. Philippines, *Final Report*, 119–20.
40. Interview with Colonel Eduardo "Red" Kapunan, Manila, 6 July 1986.
41. Interview with Colonel Gregorio "Gringo" Honasan, 24 July 1986.
42. Philippines, *Final Report*, 124–25; Almendral, "Fall of the Regime," 199; Tarrazona, *After EDSA*, 37–38; interview with Colonel Hector Tarrazona, Manila, 8 July 1986; interview with Captain Felix Turingan, Manila, 28 August 1986; interview with Kapunan; De Dios, *Dictatorship and Revolution*, 702–05.
43. Interview with Robles.
44. Interviews with Kapunan; Honasan
45. Interview with Lieutenant Colonel Marcelino "Jake" Malajacan, Manila, 29 August 1986; Rigoberto D. Tiglao, "The Consolidation of the Dictatorship," in De Dios, *Dictatorship and Revolution*, 53.
46. Interview with General Artemio Tadiar, Fort Bonifacio, 31 August 1986; Association of General and Flag Officers, *General and Flag Officers of the Philippines (1896-1990)* (Quezon City: Association of General and Flag Officers, 1991), 136, 450, 496; letter from Brigadier General Antonio Palafox to General Fidel V. Ramos, Chief of Staff, 27 February 1986.
47. Philippines, *Final Report*, 125; interviews with Kapunan, Honasan.
48. Interview with Ileto.
49. De Dios, *Dictatorship and Revolution*, 742.
50. Letter from Alejandro Melchor to Her Excellency President Corazon C. Aquino, 25 April 1986; interview with Felix Turingan, Manila, 28 August 1986. Newspaper columnist Amando Doronilla recalled meeting Jose Almonte in Cambodia in the 1960s, where he told him "that he had infiltrated the Vietcong and Ho Chi Minh trail on an undercover mission" (*PDI*, 12/22/95). According to his citation for the AFP's Distinguished Service Star, then captain Almonte, on orders from President Marcos, had "infiltrated" the Central Office for the Liberation of South Vietnam (COSVN). (See Colonel Sinforoso L. Duque, *Soldier Heroes: A Handbook on the Winners of the Major Medals Awarded by the Philippine Constabulary and Armed Forces Since 1902* [Manila: National Media Production Center, 1981], 41–42.)
51. Interviews with Turingan; Honasan.
52. Interview with Tadiar.
53. Interview with Captain Ricardo Morales, Quezon City, 28 August 1986.
54. Interview with Honasan.
55. Ibid.
56. Interview with Tadiar; Aruiza, *Malacañang to Makiki*, 37–40.
57. Interview with Kapunan.
58. Ibid.
59. Interview with Tadiar; Santiago, *Duet for EDSA: Chronology*, 13–14; Aruiza, *Malacañang to Makiki*, 40.
60. Interview with Turingan.
61. Interview with Malajacan.
62. While General Tadiar said that the Fifth Battalion was at the palace by Saturday noon, the Davide Commission reported that it was there at 4:00 A.M. (Interview with Tadiar; Philippines, *Final Report*, 127.) Interview with Kapunan; Arillo, *Breakaway*, 5–6.

63. Interview with Honasan; Lewis M. Simons, *Worth Dying For* (New York: William Morrow, 1987), 272.
64. Interviews with Honasan; Morales. Aruiza, *Malacañang to Makiki*, 40–41.
65. Aruiza, *Malacañang to Makiki*, 33.
66. Aguirre, *People's Revolution*, 17.
67. Interview with Turingan; Santiago, *Duet for EDSA: Chronology*, 19; Almendral, "Fall of the Regime," 209; von Brevern, *The Turning Point*, 36; Arillo, *Breakaway*, 12, 146–47; Alfred W. McCoy, Marian Wilkinson, and Gwen Robinson, "Coup!" *Veritas*, Special Edition (October 1986), 8; Aguirre, *People's Revolution*, 21; Aruiza, *Malacañang to Makiki*, 44.
68. Transcript, Press Conference, Ministry of National Defense, 22 February 1986; Simons, *Worth Dying For*, 276.
69. Arillo, *Breakaway*, 19.
70. Santiago, *Duet for EDSA: Chronology*, 36.
71. General Prospero Olivas, "Narration of Activities of CG PCM/DIR, MPF, 22–28 Feb. 86."
72. Santiago, *Duet for EDSA: Chronology*, 45.
73. Interview with Colonel Hector Tarrazona, Manila, 8 July 1986; Aguirre, *People's Revolution* 20; Santiago, *Duet for EDSA: Chronology*, 68; Major General Alexander Aguirre AFP (ret.), *A People's Revolution of Our Time: Philippines, February 22–25, 1986* (Quezon City: Fastprint, 1996), 32; Margarita R. Cojuangco et al., *Konstable: The Story of the Philippine Constabulary, 1901–1991* (Manila: ABoCan, 1991), 163; interview with General Romeo M. Reciña, Davao City, 17 July 1987; Corps of Cadets, *The Sword* (Baguio: Cadet Corps, Armed Forces of the Philippines, 1957).
74. Interview with Jaime Cardinal Sin, Manila, 21 July 1986.
75. Olivas, "Narration of Activities"; Aguirre, *People's Revolution*, 27–28; Aruiza, *Malacañang to Makiki*, 11, 127.
76. Interview with Jesus "Chito" V. Ayala, Davao City, 17 July 1987; *PDI*, 24 February 1986.
77. Transcript, Press Conference, Maharlika Public Affairs, 22 February 1986; Arillo, *Breakaway*, 34–37.
78. Interview with Kapunan.
79. Quijano de Manila, *Quartet of the Tiger Moon: Scenes from the People-Power Apocalypse* (Manila: Book Stop, 1986), 35.
80. Aruiza, *Malacañang to Makiki*, 5, 12, 104–107, 123–25, 179–80, 352–53.
81. Memorandum to Chief of Staff, NAFP, from Brigadier General Isidoro de Guzman, Subject: Command Actions During the Crisis, 28 February 1986.
82. Interview with General Artemio Tadiar, Fort Bonifacio, 31 August 1986; *MT*, 3/3/60, 3/24/60; *DM*, 3/29/60, 4/2/60, 8/12/60, 10/7/60.
83. Interview with General Artemio Tadiar, Fort Bonifacio, 31 August 1986; Brigadier General Artemio A. Tadiar, Jr., "Narrative Report of the Events Covering the Period 22–26 February 1986," 7; *FEER*, 1 August 1985, 10.
84. Santiago, *Duet for EDSA: Chronology*, 70.
85. Interview with Honasan; Aguirre, *People's Revolution*, x.

86. Interview with Tadiar; Aguirre, *People's Revolution*, 36; *Mr. & Ms.*, 21–27 March 1986; Santiago, *Duet for EDSA: Chronology*, 72–78.
87. Lieutenant Ferdinand S. Golez, aide to General Fabian Ver, "The Philippine Revolution of 1986: A Personal Account of Events of Malacañang on 20–26 February 1986" (n.d.), 6; Memorandum to Chief of Staff (New) Armed Forces of the Philippines, from General Felix Brawner, Subject: After Operations Report (The Final Hour: 22–27 February 1986), 5 March 1986, 4; interview with Malajacan.
88. Interview with General Antonio Sotelo, Villamor Air Base, 21 July 1986; Santiago, *Duet for EDSA: Chronology*, 107.
89. Interview with Commondore Tagumpay Jardiniano, Manila, 27 August 1986.
90. Santiago, *Duet for EDSA: Chronology*, 102–03.
91. Interview with Balbas; Tarrazona, *After EDSA*, 51; Aguirre, *People's Revolution*, 30–31, 36.
92. Interviews with Balbas; Tadiar. Memorandum to Chief of Staff, NAFP, from Colonel Braulio B. Balbas, Jr., Subject: Participation Report, 1 March 1986, 3–6.
93. Aruiza, *Malacañang to Makiki*, 123, 180; interview with Balbas; Balbas to Chief of Staff, NAFP, 3–6. Originally a member of Class '60 at the PMA, Oscar Florendo was turned back for hazing and graduated with the Class of 1962.
94. Interview with Turingan.
95. The Class '71 officers in the attack were Lieutenant Colonels Eduardo S. Matillano, Francisco Zuiba, and Teodorico Viduya (see Philippines, *Final Report*, 132); Jo-Ann Q. Maglipon, *Primed: Selected Stories, 1972–1992* (Pasig: Anvil, 1993), 227.
96. Arillo, *Breakaway*, 87.
97. Santiago, *Duet for EDSA: Chronology*, 112–13, 137.
98. Aruiza, *Malacañang to Makiki*, 127–28.
99. Brawner to Chief of Staff, (New) Armed Forces of the Philippines, 6; Arillo, *Breakaway*, 87; Cojuangco, *Konstable*, 163; interview with General Reciña.
100. Tadiar, "Narrative Report," 7; interview with Ileto.
101. Interview with Secretary of Finance Jimmy Ongpin, Manila, 30 August 1986.
102. Ibid.; Monina Allarey Mercado, ed., *People Power* (Manila: Reuter Foundation, 1986), 233–34.
103. Santiago, *Duet for EDSA: Chronology*, 160; *MT*, 3/1/86.
104. McCoy et al., "Coup!"; Aruiza, *Malacañang to Makiki*, 140–69, 186–87, 193. These two officers were Lieutenant Colonels Ricardo de Leon and Prospero Ocampo.
105. Interview with Honasan; von Brevern, *The Turning Point*, 97.
106. Interview with Jaime Cardinal Sin, Manila, 21 July 97.
107. Virgilio T. J. Suerte Felipe, *Cardinal Sin and the February Revolution* (Manila: TJ Publications, 1987), 18.
108. Antonio B. Lambino, S.J., "Theological Reflection on the Filipino Exodus: August 21, 1983 to February 25, 1986," in Pedro S. de Achutegui, S.J., ed., *The "Miracle" of the Philippine Revolution: Interdisciplinary Reflections* (Manila: Loyala School of Theology, 1986), 14.
109. Aruiza, *Malacañang to Makiki*, 142–43.
110. Interview with Sister Cres Lucero, Quezon City, 3 July 1986.

CHAPTER 8: COUP D'ETAT

1. Emilio F. Mignone, "Beyond Fear: Forms of Justice and Compensation," in Juan E. Corradi, Patricia Weiss Fagen, and Manuel Antonio Garreton, eds., *Fear at the Edge: State Terror and Resistance in Latin America* (Berkeley: University of California Press, 1992), 250–63; Frank Graziano, *Divine Violence: Spectacle, Psychosexuality, and Radical Christianity in the Argentine "Dirty War"* (Boulder: Westview, 1992), 54–59, 222–25; Neil J. Kritz, ed., *Transitional Justice: How Emerging Democracies Reckon with Former Regimes*, vol. 2, *Country Studies* (Washington, D.C.: U.S. Institute for Peace, 1955), 323–81; Aldo Rico, *El Desafío Argentino* (Buenos Aires: Ediciones del Bicentenario, 1995), 79–80.
2. Lawyers Committee for Human Rights, *Impunity: Prosecutions of Human Rights Violations in the Philippines* (New York: Lawyers Committee for Human Rights, 1991), 18–22; José Zalaquett, "Confronting Human Rights Violations Committed by Former Governments: Principles Applicable and Political Constraints," in Kritz, *Transitional Justice*, vol. 1, *General Considerations*, 18–19.
3. Graziano, *Divine Violence*, 20.
4. Maria Serena I. Diokno, "Peace and Human Rights: The Past Lives On," in Lorna Kalaw-Tirol, ed., *Duet for EDSA: 1996 Looking Back, Looking Forward* (Manila: Foundation for Worldwide People Power, Inc., 1995), 92; Lawyers Committee for Human Rights, *Impunity*, 18–22, 57–58; International Commission of Jurists, *The Failed Promise: Human Rights in the Philippines Since the Revolution of 1986* (Geneva: International Commission of Jurists, 1991), 144–47; Belinda A. Aquino, "The Human Rights Debacle in the Philippines," in Naomi Roht-Arriaza, ed., *Impunity and Human Rights in International Law and Practice* (New York: Oxford University Press, 1995), 231–36.
5. Jo-Ann Q. Maglipon, *Primed: Selected Stories 1972–1992* (Manila: Anvil, 1993), 228; Lieutenant Gregorio Catapang (PMA '81), "Le Mouvement de Réforme de l'Armée (RAM) et les Forces Armée de Philippines (AFP)," *"Les Temps Moderne* 44, no. 508 (November 1988), 96.
6. Ruben Fulgueras Ciron, "Civil-Military Relations in the Philippines: Perceptions of PMA-Trained Officers" (Ph.D. diss., University of the Philippines, 1993), 140, 178–79, 195.
7. Republic of the Philippines, *The Final Report of the Fact-Finding Commission (Pursuant to R.A. No. 6832)* (Manila: Bookmark, October 1990), 476.
8. Gary Hawes, "Aquino and Her Administration: A View from the Countryside," *Pacific Affairs* 62, no. 1 (1989), 14.
9. Alfred W. McCoy, Marian Wilkinson, and Gwen Robinson, "Coup!" *Veritas*, Special Edition (October 1986), 4; Philippines, *Final Report*, 478; Ciron, "Civil-Military Relations," 66–67, 139.
10. Colonel Hector M. Tarrazona, *After EDSA*, vol. 1 (Pasay City: by the author, 1989), 31; Philippines, *Final Report*, 477.
11. *Mr. & Ms.*, 28 February–6 March 1986, 3; *Mr. & Ms.*, 21–27 March 1986, 19–20; *PC Journal*, July 1986.
12. Bryan Johnson, *Four Days of Courage* (New York: Free Press, 1987), 33.
13. Lewis M. Simons, *Worth Dying For* (New York: William Morrow, 1987), 264.

14. Cynthia Enloe, "Beyond Steve Canyon and Rambo: Feminist Histories of Militarized Masculinity," in John R. Gillis, ed., *The Militarization of the Western World* (New Brunswick: Rutgers University Press, 1989), 123; Emmanuel A. Reyes, *Notes on Philippine Cinema* (Manila: De La Salle University Press, 1989), 52.
15. Interviews with General Rafael Ileto (ret.), Manila, 7 January 1988, 21 January 1988.
16. Interview with Captain Felix Turingan, Manila, 28 August 1986.
17. *Mr. & Ms.*, 21–27 March 1986, 17; Johnson, *Four Days of Courage*, 39.
18. *PC Journal*, July 1986.
19. Interview with Colonel Eduardo "Red" Kapunan, Manila, 6 July 1986.
20. Task Force Detainees, *Political Detainees of the Philippines Book Three* (Manila: Association of Major Orders of Religious Superiors, March 1978), 5–6; Sheila Coronel, "RAM: From Reform to Revolution," in Philippine Center for Investigative Journalism, *Kudeta: The Challenge to Philippine Democracy* (Manila: Philippine Center for Investigative Journalism, 1990), 71–73; Philippines, *Final Report*, 141, 479; Criselda Yabes, *The Boys from the Barracks: The Philippine Military after EDSA* (Pasig: Anvil, 1991), 19.
21. Francisco Nemenzo, "From Autocracy to Elite Democracy," in Aurora Javate-De Dios et al., eds., *Dictatorship and Revolution: Roots of People's Power* (Manila: Conspectus, 1988), 225–29; Salvador H. Laurel, *Neither Trumpets nor Drums: Summing Up the Cory Government* (Manila: by the author, 1992), 58–59; *PFP*, 26 July 1986.
22. Philippines, *Final Report*, 135; Arturo C. Aruiza, *Ferdinand E. Marcos: Malacañang to Makiki* (Quezon City: ACAruiza Enterprises, 1991), 231.
23. Coronel, "RAM," 71–72; Philippines, *Final Report*, 141, 479.
24. Philippines, *Final Report*, 118.
25. Coronel, "RAM," 72–73.
26. Cecilio T. Arillo, *Breakaway: The Inside Story of the Four-Day Revolution in the Philippines, February 22–25, 1986* (Mandaluyong: CTA & Associates, May 1986), 130; interviews with Captain Rex Robles, Manila, 7 July 1986, 21 August 1986.
27. *NYT*, 2/26/88; *Sydney Morning Herald*, 2/26/88.
28. Yabes, *Boys from the Barracks*, 44; Coronel, "RAM," 72–73.
29. Belinda A. Aquino, "EDSA as Vision and Liberation," in Kalaw-Tirol, *Duet for EDSA*, 26; Coronel, "RAM," 73; *PDE*, 11/13/86.
30. *National Midweek*, 26 November 1986, 8–9; *Observer* (Manila), 11/10/86; *MC*, 10/28/86.
31. Manuel F. Martinez, *Aquino vs. Marcos: The Grand Collision* (Hong Kong: by the author, 1987), 277–79.
32. Interview with Kapunan, 21 July 1986.
33. Gemma N. Almendral, *EXTRA: The Lonely Crusade of Victor Corpus* (Manila: Special promotional pamphlet for film *The Victor Corpus Story*, n.d.), 5; *National Midweek* (Manila), 19 November 1986, 6–7; Victor N. Corpus, *Silent War* (Quezon City: VNC Enterprises, 1989), 11–13.
34. *National Midweek*, 19 November 1986, 45–46; interview with Jose Lacaba, Canberra, 14 February 1988.
35. Interviews with Lieutenant Colonel Victor Corpus, Quezon City, 1 April 1987 and 5 April 1987; Nemenzo, "From Autocracy to Elite Democracy," 264.
36. Philippines, *Final Report*, 149.

37. *Midday,* 11/7/86; *MT,* 11/7/86; *DE,* 11/7/86; *PDI,* 11/7/86; *Malaya,* 11/9/86; Almendral, *EXTRA,* 8; interviews with Corpus.
38. *Mr. & Ms.*, 14–20 November 1986, 6–9, 12–13; Philippines, *Final Report,* 149–51.
39. Philippines, *Final Report,* 151; *Observer,* 11/24/86; *PDI,* 8/10/95; Jun Cruz Reyes, *Mga Daluyong, Mga Unos Sa Panahon ni Rolondo Olalia* (Manila: Kalikasan, 1989), 207–09.
40. Reyes, *Mga Daluyong, Mga Unos,* 208–09.
41. *National Midweek,* 9 March 1988, 42; *PDI,* 2/20/88, 2/21/88, 2/22/88, 2/25/88, 2/27/88, 12/9/89, 8/5/95, 8/6/95, 8/10/95; *Sydney Morning Herald,* 2/26/88.
42. *PDI,* 8/6/95; Rigoberto D. Tiglao, "Restoration and Global Incorporation," in Kalaw-Tirol, *Duet for EDSA,* 123.
43. Philippines, *Final Report,* 203.
44. *MT,* 11/25/86; *PDI,* 11/24/86.
45. Coronel, "RAM," 76–77; Nemenzo, "From Autocracy," 264.
46. Interview with Lieutenant Colonel Ricardo Alqueza, Provincial Commander, Camp Addurru, Cagayan Province, 10 January 1988; Philippines, *Final Report,* 152–53. In Cebu City, Lieutenant Colonel Tiburcio Fusilero, RAM's regional leader, mobilized the right for a rally. In Butuan City, Colonel Ruben Cabagnot, a member of Class '71, led an antigovernment protest. (See *PDI,* 12/11/89; *National Midweek,* 10 December 1986, 8–9.)
47. Coronel, "RAM," 76–77.
48. Ibid.; Philippines *Final Report,* 154–55. These battalion commanders were Lieutenant Colonel Jake Malajacan (PMA '71); Lieutenant Colonel Rafael Galvez ('71); and Lieutenant Colonel Saulito Aromin ('74).
49. *Malaya,* 11/24/86; *PDI,* 11/24/86; Coronel, "RAM," 77.
50. Philippines, *Final Report,* 154–55; *Malaya,* 11/24/86; *PDI,* 11/24/86; Mark R. Thompson, "The 'Little Left' in the Politics of the Philippines," *Field Staff Report No. 17* (Universities Field Staff International, April 1990), 5; Lawyers Committee for Human Rights, *Impunity,* 9, 18–22.
51. Interviews with Ileto; Philippines, *Final Report,* 155–57; *FEER,* 4 February 1988, 13–14.
52. Coronel, "RAM," 78–79.
53. Belinda Aquino, "The Philippines in 1987," *Southeast Asian Affairs 1988* (Singapore: Institute of Southeast Asian Affairs, 1988), 197; Aurora Javate-de Dios, "Intervention and Militarism," in de Dios, *Dictatorship and Revolution,* 295; Felipe B. Miranda, "At The Crossroads of Politicization," in Kalaw-Tirol, *Duet for EDSA,* 79; Ciron, "Civil-Military Relations," 66–67, 158.
54. Report of General Fidel Ramos to President Corazon Aquino, 1 September 1987.
55. Felipe Miranda, *The Politicization of the Military* (Quezon City: University of the Philippines Press, 1992), 6.
56. Philippines, *Final Report,* 176.
57. *FEER,* 1 August 1985; Yabes, *Boys from the Barracks,* 69–70, 131, 217–28.
58. Yabes, *Boys from the Barracks,* 75; Philippines, *Final Report,* 181.
59. Philippines, *Final Report,* 180–82.
60. Report of Ramos to President, 1 September 1987.
61. Philippines, *Final Report,* 182; *MC,* 8/30/87.

62. *PDI*, 8/29/87.
63. *PDI*, 6/15/90; notes from the original Channel 13 broadcast, Manila, 28 August 1987; Philippines, *Final Report*, 187–88.
64. de Dios, "Intervention and Militarism," in de Dios, *Dictatorship and Revolution*, 303; Philippines, *Final Report*, 182, 185; interview with National Security advisor Emmanuel Soriano, Manila, 6 January 1988.
65. *MC*, 8/29/87; author's notes, EDSA, just south of Ortigas Avenue, 2:30 P.M., 28 August 1987; Philippines, *Final Report*, 186; de Dios, "Intervention and Militarism," 310.
66. de Dios, "Intervention and Militarism," 303; *Time Australia*, 7 September 1987, 19.
67. Philippines, *Final Report*, 200.
68. Social Weather Stations, Inc., "Social Weather Stations Survey of Public Opinion on the December 1, 1989 Coup Attempt," 2–3; *PDI*, 9/6/87; *The Independent* (Manila), 14 September 1987, 2.
69. *PDI*, 9/17/87.
70. Coronel, "RAM," 81. Colonel Oscar "Tito" Legaspi led the attack on Villamor Air Base, while Lieutenant Colonel Eduardo Matillano commanded rebel forces in an attack on Channel 4. At Baguio Lieutenant Colonel Red Kapunan inspired the PMA mutiny; in Pampanga Lieutenant Colonel Reynaldo Berroya held a general hostage; at Basa Air Base Major Francisco Baula readied fighter cover; in Albay Lieutenant Colonel Batac tried to occupy the airport; and in Cebu Lieutenant Colonel Tiburcio Fusilero acted as the de facto rebel deputy. (See Coronel, "RAM," 82–83, 85.)
71. Report of Ramos to President, 1 September 1987.
72. Selochan, "Professionalization," 70; Cadet Corps, Philippine Military Academy, Armed Forces of the Philippines, *The Sword 1987* (Fort Del Pilar, Baguio City: by author, 1987); Cadet Corps, *The Academy Scribe* (Manila: The Academy Scribe Organization, 1988), 714; Yabes, *Boys from the Barracks*, 71.
73. Cadet Corps, *Academy Scribe*, 181–83, 709; Selochan, Professionalization," 71–72; Philippines, *Final Report*, 198; Yabes, *Boys from the Barracks*, 71.
74. Report of Ramos to President, 1 September 1987; Philippines, *Final Report*, 199.
75. Bigkis-Lahi Class '90, *Sword '90* (Fort Del Pilar, Baguio City: Cadet Corps, Philippine Military Academy, Armed Forces of the Philippines, 1990), 228–29.
76. *MC*, 10/8/87, 11/14/87.
77. *MC*, 11/12/87; Yabes, *Boys from the Barracks*, 102.
78. *MC*, 12/18/87.
79. Philippines, *Final Report*, 84; *MC*, 4/22/88; Yabes, *Boys from the Barracks*, 107; *PDI*, 4/2/97.
80. *PDI*, 6/13/90; Glenda Gloria, "YOU: The Soldier as Nationalist," in Philippine Center for Investigative Journalism, *Kudeta*, 133–37; Yabes, *Boys from the Barracks*, 39–41; *MT*, 5/22/91; de Dios, *Dictatorship and Revolution*, 705.
81. Yabes, *Boys from the Barracks*, 127, 206, 210; *PDI*, 5/5/90; Gloria, "YOU," 133–37.
82. Philippines, *Final Report*, 98; Yabes, *Boys from the Barracks*, 140.
83. *PDI*, 11/20/89.
84. de Dios, "Intervention and Militarism," 313–14; Miranda, "At the Crossroads of Politi-

cization," 82; Ciron, "Civil-Military Relations," 179; Amnesty International, *Philippines: The Killing Goes On* (New York: Amnesty International, 1992), 17–19.

85. Miranda, "At the Crossroads of Politicization," 79–80; Diokno, "Peace and Human Rights," 94, 96–97; Lawyers Committee for Human Rights, *Impunity,* 20–21, 59–63; Amnesty International, *Philippines: The Killing Goes On,* 72–80; Aquino, "Human Rights Debacle in the Philippines," 239; Ciron, "Civil-Military Relations," 195; International Commission of Jurists, *Failed Promise,* 238–39; Human Rights Watch/Asia, *Bad Blood: Militia Abuses in Mindanao, the Philippines* (New York: Human Rights Watch, 1992), 6–7.

86. *PDI,* 11/22/88, 1/23/88, 2/11/88; Yabes, *Boys from the Barracks,* 115; *FEER,* 4 February 1988, 13–14, 10 March 1988, 20–21.

87. While Ileto used this father-son image, Ramos himself called Ileto "my mentor, elder brother, friend and superior." (Interview with Ileto; *FEER,* 4 February 1988, 12–13; *PDI,* 1/23/88, 1/25/88.)

88. Miranda, "At the Crossroads of Politicization," 70–71; Laurel, *Neither Trumpets nor Drums,* 144–45; U.P. School of Economics, "A Time for Hard Decisions," *Kasarinlan* 5, nos. 1–2 (1989), 39–40.

89. Randolf S. David, "The December First Coup Attempt: Persistent Questions and Tentative Responses," *Kasarinlan* 5, nos. 1–2 (1989), 4; Aruiza, *Malacañang to Makiki,* 421–25; Miranda, "At the Crossroads of Politicization," 68–70.

90. Philippines, *Final Report,* 483.

91. Ibid., 222, 234, 233–38, 252, 444, 447.

92. Ibid., 233–39.

93. Ibid., 489, 491; Yabes, "Seven Days in December," *Kudeta,* 91.

94. Philippines, *Final Report,* 206, 447.

95. Rigoberto D. Tiglao, "Rebellion from the Barracks: The Military as Political Force," Philippine Center for Investigative Journalism, *Kudeta,* 15; Philippine Center for Investigative Journalism, *Kudeta,* 209–15.

96. *PDI,* 12/4/89.

97. Philippines, *Final Report,* 261–64, 291, 292–93.

98. Ibid., 303–08; *PDI,* 12/14/89; Yabes, "Seven Days in December," 94.

99. Philippines, *Final Report,* 269–71.

100. Ibid., 340–43; Yabes, *Boys from the Barracks,* 171.

101. Yabes, *Boys from the Barracks,* 167; Philippines, *Final Report,* 496–97.

102. Philippines, *Final Report,* 347–48; *MC,* 2/13/90.

103. Philippines, *Final Report,* 229, 313.

104. Yabes, *Boys from the Barracks.* 175.

105. Philippines, *Final Report,* 226, 270, 325.

106. Yabes, "Seven Days in December," 99; Republic of the Philippines, The Fact Finding Commission (Pursuant to Republic Act No. 6832), *Interim Report,* no. 2, *The Role of the Philippine Air Force in the Destruction of Rebel Air Assets in Sangley Point* (Manila, 12 May 1990), 12–14; *PDI,* 12/11/89.

107. Philippines, *Final Report,* 495; Yabes, "Seven Days in September," 99; Memorandum

from Defense Secretary Fidel Ramos to President Corazon Aquino, 1 December 1989; interview with U.S. Central Intelligence Agency official, National Intelligence Estimates, 22 November 1996; Philippines, *Interim Report*, no. 2, 6–7.
108. Philippines, *Final Report*, 495–96; Yabes, "Seven Days in December," 99.
109. Philippines, *Final Report*, 232, 322, 334–36.
110. Yabes, "Seven Days in December," 100; Philippines, *Final Report*, 344–45.
111. Philippines, *Final Report*, 225, 297, 348; Yabes, "Seven Days in December," 101.
112. Philippines, *Final Report*, 273–74, 276.
113. Ibid., 503–07; Philippines, *Interim Report*, no. 2, 15–16; Laurel, *Neither Trumpets nor Drums*, 128–29; Yabes, "Seven Days in December," 105.
114. *MC*, 12/4/89, 12/10/89; *PDI*, 12/4/89; Philippines, *Final Report*, 228, 319.
115. Philippines, *Final Report*, 319–20; *PDI*, 12/4/89, 12/12/89.
116. *MC*, 12/4/89; Yabes, "Seven Days in December," 111–12.
117. *MC*, 12/5/89, 12/10/89; *PDI*, 12/5/89.
118. Yabes, "Seven Days in December," 120; Yabes, *Boys from the Barracks*, 191; Philippines, *Final Report*, 280–82.
119. Yabes, "Seven Days in December," 125.
120. *SSD*, 12/2/89, 12/3/89.
121. *PDI*, 12/4/89, 12/6/89, 12/7/89, 12/8/89, 12/10/89, 12/11/89.
122. *PDI*, 10/10/89, 12/12/89; *SSD*, 12/10/89.
123. *PDI*, 12/10/89, 12/11/89.
124. *PDI*, 2/1/93.
125. Ciron, "Civil-Military Relations," 97, 137, 176, 195.
126. Philippines, *Final Report*, 376–78; Amando Doronilla, "Foreign Policy: Redefining Themes and Emphasis," in Kalaw-Tirol, ed., *Duet for EDSA*, 57; Eduardo C. Tadem, "The Aquino Government Under Siege," *Kasarinlan* 5, nos. 1–2 (1989), 29–33; *MC*, 12/17/89; *NYT*, 12/20/89.
127. *PDI*, 12/15/89.
128. *MT*, 2/23/90; Coronel, "RAM," 53; Joel Rocamora, "EDSA on My Mind," in Kalaw-Tirol, *Duet for EDSA*, 225–26.
129. *PDI*, 1/14/90; *MC*, 1/18/90.
130. *PDI*, 2/7/91; *DG*, 11/24/90.
131. Yabes, *Boys from the Barracks*, 222.
132. Ibid., 225.
133. *PDI*, 12/15/89, 12/16/89.
134. Human Rights Watch/Asia, *Bad Blood*, 16; Yabes, *Boys from the Barracks*, 224.
135. *PDI*, 10/5/90.
136. *MS*, 10/8/90; Yabes, *Boys from the Barracks*, 226; *MT*, 10/5/90; *PDI*, 10/7/90.
137. Yabes, *Boys from the Barracks*, 229–30.
138. Miranda, "At the Crossroads of Politicization," 69–70.
139. *DG*, 4/9/90; YOU, "The Story Behind the Extrication of Lieutenant Colonel Billy Bibit," Philippine Center for Investigative Journalism, *Kudeta*, 189–91; Gloria, "YOU," 136–37; *PDI*, 4/10/90.

140. *MC*, 3/10/90; *PDI*, 2/15/94.
141. *PDI*, 12/1/90; interview with Satur C. Ocampo, Manila, 27 August 1996; interview with Rick Reyes, Aix-en-Provence, 28 April 1997.

CHAPTER 9: IMPUNITY

1. June Pagaduan Lopez, M.D., "Mainstreaming Is Prevention," *PST Quarterly* 1, no. 1 (April–June 1996), 2.
2. "Impunity," Rehabilitation and Research Centre for Torture Victims, *Annual Report 1995* (Copenhagen: Rehabilitation and Research Centre for Torture Victims, 1996), 6.
3. Ernesto San Julian, "SERSOC: Uruguay Organizes a Rehabilitation Centre," *Torture* 2, no. 1 (1992), 16–17; Elisabeth Marcelino, "The Philippine Experience: Rehabilitation of Survivors of Torture and Political Violence Under A Continuing Stress Situation," *Torture* 2, no. 1 (1992), 19–21; Leslie London, M.D., "Conference Review: The VII International Symposium, 'Caring for Survivors of Torture: Challenges for the Medical and Health Professions,'" *Torture* 6, no. 3 (1996), 66; Christian Pross, "Social Isolation of Survivors of Persecution in a Post-Totalitarian Society: The Therapist's Dilemma," *Torture* 6, no. 1 (1996), 5–6; Lucila Edelman and Diana Kordon, "Psychosocial Effects of Impunity," in Rehabilitation and Research Centre for Torture Victims, *Annual Report 1995*, 9–10; Simona Ruy-Pérez, "Speech," *Torture* 6, no. 4 (1996), 98–99; Simona Ruy-Pérez, "Experiences of Human Rights Organizations with Transitions to Democratic Rule in the Southern Cone," in June C. Pagaduan Lopez and Elisabeth Protacio Marcelino, eds., *Torture Survivors and Caregivers: Proceedings of the International Workshop on Therapy and Research Issues* (Quezon City: University of the Philippines Press, 1995), 15–22.
4. London, "Conference Review," 69.
5. Lucila Edelman and Diana Kordon, "Incidence of Social Belonging, Personal Identity, and Historical Memory in Different Approaches to Psychological Therapy," *Torture* 6, no. 1 (1996), 4–5; Edelman, "Psychosocial Effects," 9.
6. Dr. Diana R. Kordon and Dr. Lucila I. Edelman, "Psychological Effects of Political Repression," pt. 2, in Diana R. Kordon et al., eds., *Psychological Effects of Political Repression* (Buenos Aires: Sudamericana/Planeta, 1988), 174; Dr. Diana R. Kordon and Dr. Lucila I. Edelman, "Violation of Human Rights: Text or Context in Couple and Family Analysis," in Kordon et al., *Psychological Effects of Political Repression*, 92.
7. Leslie London, "Dealing with the Pain of Apartheid's Past: South Africa's Truth and Reconciliation Commission," in Rehabilitation and Research Centre for Torture Victims, *Annual Report 1995*, 7–8.
8. Jorge Correa, "Dealing with Past Human Rights Violations: The Chilean Case After Dictatorship," in Neil J. Kritz, ed., *Transitional Justice: How Emerging Democracies Reckon with Former Regimes*, vol. 2, *Country Studies* (Washington, D.C.: U.S. Institute for Peace, 1995), 478–94; Kritz, *Transitional Justice*, vol. 3, *Laws, Rulings, and Reports*, 685–95; Dimocritos Sarantidis et al., "Long-Term Effects of Torture of Victims During the Period of Dictatorship in Greece," *Torture* 6, no. 1 (1996), 16–18; Dimistris Pantazis, "Greece: Living Memories of Wars," in Lopez, *Torture Survivors*, 64–65; Lawrence Weschler, *A Miracle, A Universe: Settling Accounts with Torturers* (New York: Pantheon,

1990), 236; Luc Huyse, "Justice After Transitions: On the Choices Successor Elites Make in Dealing with the Past," in Kritz, ed., *Transitional Justice*, vol. 1, *General Considerations*, 115.
9. Weschler, *A Miracle, A Universe*, 191.
10. Maria Serena I. Diokno, "Peace and Human Rights: The Past Lives On," in Lorna Kalaw-Tirol, ed., *Duet for EDSA: 1996 Looking Back, Looking Forward* (Manila: Foundation for Worldwide People Power, Inc., 1995), 92; Belinda A. Aquino, "The Human Rights Debacle in the Philippines," in Naomi Roht-Arriaza, ed., *Impunity and Human Rights in International Law and Practice* (New York: Oxford University Press, 1995), 236–39.
11. *PDI*, 2/7/91; *MS*, 2/7/91; *MT*, 2/8/91, 4/3/91; *MC*, 4/12/91.
12. *PN*, 9/18/91; *DG*, 11/3/91; *PDI*, 9/13/91, 11/2/91, 11/6/91.
13. *PDI*, 11/5/91; *DG*, 11/8/91; *PDI*, 11/21/91.
14. Republic of the Philippines, Office of the President, Executive Order No. 19, Fidel V. Ramos, 1 September 1992; *MB*, 11/13/92; *PS*, 11/14/92; Rebulosyonaryong Alyansang Makabansa (RAM), "Talking Points," National Unification Commission, Seventh Regular Meeting, 29 October 1992, in "Press Statements on National Unification Commission and Peace Process" (Quezon City: National Unification Commission, 1993).
15. *Philippine Graphic*, 4 January 1993, 8.
16. *PS*, 12/24/92; Republic of the Philippines, National Unification Commission, Press Statement, 23 December 1992; *DG*, 12/27/92; *PDI*, 12/25/92, 12/29/92; *Philippine Graphic*, 18 January 1993, 5.
17. *MC*, 3/14/93.
18. *PDI*, 7/23/93.
19. *MB*, 2/7/93; Republic of the Philippines, National Unification Commission, "Principles for Characterization of Offenses for Confidence-Building and Amnesty," 11 March 1993; *Malaya*, 5/10/93; *PDI*, 5/28/93, 7/2/93; *PS*, 7/16/93.
20. *PDI*, 1/15/90, 10/21/90; interview with Lieutenant Colonel Ricardo Alqueza, Constabulary Provincial Commander, Camp Addurru, Cagayan Province, 10 January 1988.
21. *PDI*, 3/5/90; Sheila S. Coronel, *Coups, Cults, and Cannibals: Chronicles of a Troubled Decade (1982–1992)* (Manila: Anvil, 1993), 210.
22. Interview with Alqueza; *Countryside Standard Magazine*, January 1988, 15–16; Coronel, *Coups, Cults, and Cannibals*, 209.
23. *The Australian*, 2/2/88.
24. *MC*, 12/5/89; *PDI*, 1/8/90, 1/11/90, 1/15/90, 3/15/90; *FEER*, 1 February 1990.
25. Criselda Yabes, *The Boys from the Barracks: The Philippine Military After EDSA* (Manila: Anvil, 1991), 213, 222; *MC*, 3/6/90, 3/7/90, 3/10/90; *DG*, 3/10/90, 6/16/90; *MT*, 3/10/90; *PDI*, 3/1/90, 3/3/90, 3/4/90, 3/5/90, 3/6/90, 6/13/90, 5/25/92.
26. *MC*, 7/15/92; *PDI*, 7/10/92, 9/10/92.
27. Interview with Satur C. Ocampo, Manila, 27 August 1996; Satur C. Ocampo, "Leaving the Pain Behind," *PST Quarterly* 1, no. 2 (July–September 1996), 11–13.
28. *MB*, 8/16/93, 8/19/93.
29. *MB*, 8/16/93.
30. *MB*, 9/22/94, 9/24/94; *Malaya*, 9/22/94.
31. *Colonel Billy Bibit—RAM!* director William G. Mayo, producer William Leary, screenplay by Eddie G. Mayo and Henry Nadong, Viva Films, 1994.

32. Felipe B. Miranda, "At the Crossroads of Politicization," in Kalaw-Tirol, *Duet for EDSA*, 77.
33. Joe Taruc, "Liberty in Action," Radio D-ZRH (Manila), 3 February 1994, 8:45 to 9:45 A.M.
34. Interview with Professor Francisco Nemenzo, Honolulu, Hawaii, 15 April 1996; *Malaya*, 2/10/94; *PDI*, 2/9/94; *MT*, 3/5/94.
35. *PDI*, 2/23/94.
36. *MB*, 2/23/94; *MT*, 2/23/94; *PDI*, 2/23/94.
37. *PDI*, 2/25/94.
38. *PDI*, 2/25/94.
39. *MB*, 3/27/94.
40. *MT*, 3/31/94, 4/13/94.
41. Republic of the Philippines, Senate, *Senate Journal*, 24 May 1994, 12–13; 31 May 1994, 65, 71.
42. Ibid., 24 May 1994, 20; *PDI*, 3/27/94, 3/30/94, 11/29/94; *MB*, 11/19/94.
43. *MT*, 4/13/94; *MS*, 3/17/94, 4/13/94; *Singapore Business Times*, 7/9–10/94; *MB*, 3/30/94, 5/8/94, 8/12/94; *PDI*, 2/26/94, 8/12/94, 5/7/96, 3/9/97.
44. *PDI*, 12/8/94, 12/9/94.
45. *PDI*, 1/4/95, 2/22/95.
46. *PDI*, 5/11/95.
47. *PDI*, 2/5/95.
48. *Malaya*, 3/11/94; *PDI*, 2/15/95.
49. *PDI*, 12/30/94, 1/28/96; Coronel, *Coups, Cults, and Cannibals*, 87.
50. Randy David, "Public Forum," TV Channel 5 (Manila), 20 February 1995, 10:30 to 11:30 P.M.
51. "Gringo Honasan Independent" (Quezon City: paid for by Benny J. Brizuela, printed by Book Media Press, 1995).
52. *MC*, 3/13/95, 3/14/95, 3/15/95; *Malaya*, 3/14/95.
53. *PDI*, 5/16/95.
54. *PDI*, 2/24/95, 4/11/95, 4/20/95, 4/22/95.
55. *PDI*, 5/5/95, 5/7/95, 6/18/95, 6/23/95; interview with Nemenzo; Joel Rocamora, "EDSA on My Mind," in Kalaw-Tirol, *Duet for EDSA*, 228.
56. *PDI*, 5/8/95.
57. *PDI*, 5/8/95, 5/9/95, 5/10/95.
58. *PDI*, 5/15/95, 5/16/95.
59. *PDI*, 5/7/95, 5/8/95, 5/10/95, 5/12/95, 5/15/95, 5/16/95, 5/18/95, 5/21/95.
60. *PDI*, 5/17/95.
61. *PDI*, 5/16/95, 5/17/95.
62. *PDI*, 5/25/95.
63. *PDI*, 5/26/95, 5/27/95, 6/7/95.
64. *PDI*, 5/14/95, 5/19/95.
65. *PDI*, 6/14/95, 6/29/95.
66. Interview with Professor Temario C. Rivera, Quezon City, 11 December 1995; *PDI*, 5/22/96.
67. Rocamora, "EDSA on My Mind," 227.

68. *PDI*, 8/7/95, 8/10/95; *MB*, 8/5/95.
69. *PDI*, 8/6/95, 8/7/95, 8/8/95. After a reinvestigation, the prosecution moved to dismiss charges against Sergeant Filomeno Maligaya and he was released in January 1996. In August 1995, Judge Mauricio Rivera denied a motion by Colonel Oscar Legaspi's lawyers to quash the case against him and he remained at large as of September 1996. (See *PDI*, 9/16/96.)
70. *Philippine Graphic*, 28 August 1995, 14; *PDI*, 9/3/95.
71. *MB*, 10/14/95; General Agreement for Peace between the Government of the Republic of the Philippines and the Rebolusyonaryong Alyansang Makabansa-Soldiers of the Filipino People-Young Officers Union, 13 October 1995; "Confidential" Proclamation, Malacañang Palace, Granting Amnesty to Members and Supporters ... (RAM-SFP-YOU), Fidel V. Ramos (n.d.). A year later, in October 1996, a Quezon City judge dismissed rebellion charges against loyalist leaders Jose Ma. Zumel, Reynaldo Cabauatan, and Rafael Recto for their participation in the December 1989 coup. In his decision, the judge found that the government's grant of amnesty in June 1995 "obliterated" all charges "as if no act or crime has been committed by the accused." (See *PDI*, 10/13/96.)
72. *PDI*, 10/14/95.
73. *PDI*, 10/15/95.
74. In a "confidential" draft of the proclamation granting amnesty to RAM that government peace negotiators carried during the negotiations, the insertions and deletions of key clauses, including one conceding the blanket amnesty, carried the notation "Based on Sen. Drilon's Recommendation." See "Confidential" Proclamation, Malacañang Palace, Granting Amnesty to Members and Supporters ... (RAM-SPF-YOU), Fidel V. Ramos; *PDI*, 10/15/95.
75. *PDI*, 1/14/96; *MT*, 1/15/96. A year after the surrender, Senator Franklin Drilon claimed that RAM had returned only twenty-six of the sixty light antitank weapons that it held. (See *PDI*, 10/5/96.)
76. *PDI*, 1/12/98, 1/13/98, 1/15/98; Republika ng Pilipinas, Lungsod ng Maynila, Sinumpaang Salaysay, Medardo D. Barreto, 12 January 1998.
77. *PDI*, 1/13/98.
78. *PDI*, 1/13/98, 1/14/98, 1/17/98.
79. *PDI*, 1/15/98, 1/17/98, 1/24/98, 3/21/98, 3/24/98, 4/14/98, 2/3/99, 2/9/99, 2/11/99; *Today*, 2/5/99, 2/6/99, 2/10/99; *MT*, 2/5/99; Republika ng Philipinas, Lungsod ng Maynila, Sinumpaang Salaysay, Eduardo E. Bueno, 23 January 1998.
80. Dr. Belinda A. Aquino, "Hidden Wealth vs. Rights," *PDI*, 12/23–25/92; *PDI*, 1/27/95; *NYT*, 3/3/95; Keith Carmichael et al., "The Need for REDRESS," *Torture* 6, no. 1 (1996), 8–9; Richard B. Lillich, "Damages for Gross Violations of International Human Rights," *Torture* 6, no. 3 (1996), 56–57; Robert F. Drinan, S.J., and Teresa T. Kuo, "Putting the World's Oppressors on Trial: The Torture Victim Protection Act," *Human Rights Quarterly* 15, no. 3 (1993), 605–24.
81. Aquino, "Hidden Wealth vs. Rights," *PDI*, 12/23–25/92; Aquino, "The Human Rights Debacle in the Philippines," 239–41; *PDI*, 1/27/95.
82. *Honolulu Star-Bulletin*, 9/11/92; Aquino, "Hidden Wealth vs. Rights"; *PDI*, 9/23/92, 9/25/92, 9/26/92.

83. *PDI*, 9/26/92, 9/29/92; *Honolulu Star-Bulletin*, 2/24/94; letter from Robert Swift et al., *Wall Street Journal*, 1/9/96; *Honolulu Advertiser*, 2/24/94.
84. *Honolulu Star-Bulletin*, 9/13/94; *Today*, 9/14/94; *PDI*, 9/14/94, 9/15/94.
85. *In RE: Estate of Ferdinand E. Marcos Human Rights Litigation*, U.S. District Court for the District of Hawaii, MDL No. 840, "Verdict—Class Action" (18 January 1995), 2.
86. *Honolulu Star-Bulletin*, 1/25/95; *PDI*, 1/27/95; *DG*, 9/26/92; *Today* (Manila), 4/3/96.
87. *NYT*, 10/28/95, 12/10/95.
88. *PDI*, 2/27/97, 2/28/97, 3/2/97, 1/16/98.
89. *PDI*, 4/17/97, 4/20/97.
90. Letter from Dan Vizmanos, chairman, SELDA, *MC*, 9/29/95; *Today*, 10/22/95; letter from Loretta Ann P. Rosales, chairperson, Claimants 1081 Inc., *Today*, 10/24/95; *PDI*, 1/15/96; *Hong Kong Standard*, 1/16/96.
91. *PDI*, 8/27/96, 8/28/96.
92. *Wall Street Journal*, 1/9/96.
93. *PFP*, 29 June 1996; *Today*, 6/14/96, 6/15/96. For documentation of torture by Rolando Abadilla and the MISG, see Task Force Detainees, *Pumipiglas: Political Detention and Military Atrocities in the Philippines* (Manila: Task Force Detainees Philippines, 1981), 64–65; Task Force Detainees of the Philippines, *Pumipiglas: Political Detention and Military Atrocities in the Philippines 1981–1982* (Manila: Task Force Detainees Philippines, Association of Major Religious Superiors of the Philippines, 1986), 53–64.
94. *Philippine Graphic*, 1 July 1996; *PFP*, 29 June 1996; *PDI*, 1/30/88; *Today*, 6/15/96.
95. Joker Arroyo, "Military Justice Is No Justice at All," *PDI*, 6/20/96; *Philippine Graphic*, 1 July 1996; *PFP*, 29 June 1996.
96. Satur C. Ocampo, "The Boys from MISG," *Philippine News and Features*, 22 June 1996; *Philippine Graphic*, 1 July 1996.
97. Human Rights Watch/Asia, *Bad Blood: Militia Abuses in Mindanao, the Philippines* (New York: Human Rights Watch, 1992), 8–9; Lawyers Committee for Human Rights, *Impunity: Prosecutions of Human Rights Violations in the Philippines* (New York: Lawyer's Committee for Human Rights, September 1991), 5–6, 23–28. The quotation is from the chair of the Human Rights Commission, Mary Concepcion Bautista, as cited in Rod B. Gutang, *Pulisya: The Inside Story of the Demilitarization of Law Enforcement in the Philippines* (Quezon City: by the author, 1991), 26–28, 127–32.
98. *PDI*, 6/3/97, 3/10/97; *Asia Week*, 5 April 1996; *MT*, 12/10/98.
99. *PFP*, 29 June 1996; *Today*, 6/18/96, 6/24/96; *Philippine Graphic*, 1 July 1996.
100. *PFP*, 29 June 1996; *Today*, 6/16/96.
101. *PFP*, 6 July 1996; *PDI*, 6/27/96, 6/29/96, 7/4/96.
102. *PDI*, 6/27/96, 7/3/96, 7/4/96. As a constabulary officer, Romeo Reciña developed close ties to his chief Fidel Ramos during the Marcos dictatorship. On the first day of the 1986 revolt, Reciña was commander of Regional Unified Command (RUC) 11 in Davao City when Ramos called asking for his support. Siding with the rebels, Reciña contacted all constabulary commanders in Region 11. "I called General Ramos and told him that all the constabulary were behind their commander and that greatly boosted his morale," Reciña recalled. Four months later, he was promoted to general upon recommendation by Chief of Staff Ramos. As regional commander in 1987, General Reciña supervised the

operations of the vigilante groups *Alsa Masa* and *Nakasaka,* both notorious for human rights violations. Although he was in command when their extralegal violence reached its peak, attracting national and international concern, Reciña dismissed all allegations about their abuses as "leftist." After his retirement, President Ramos appointed Reciña the administrator of PHIVIDEC, a major quasi-government investment firm. (Interview with General Romeo M. Reciña, Davao City, 17 July 1987; Lawyers Committee for Human Rights, *Vigilantes in the Philippines: A Threat to Democratic Rule* [New York: Lawyers Committee for Human Rights, 1988], 23–28, 141–42; Association of General and Flag Officers, *General and Flag Officers of the Philippines (1896–1990)* [Quezon City: Association of General and Flag Officers, 1991], 511; *PDI,* 4/15/97.)

103. *PDI,* 6/20/96, 6/21/96, 7/2/96, 7/5/96, 7/24/96, 1/28/97.
104. Juan E. Corradi et al., "Fear: A Cultural and Political Construct," in Juan E. Corradi, Patricia Weiss Fagen, and Manuel Antonio Garreton, eds., *Fear at the Edge: State Terror and Resistance in Latin America* (Berkeley: University of California Press, 1992), 1–3; Manuel Antonio Garreton, "Fear in Military Regimes," in Corradi et al., *Fear at the Edge,* 23–25.
105. Diane F. Orentlicher, "Settling Accounts: The Duty to Prosecute Human Rights Violations of a Prior Regime," in Kritz, *Transitional Justice,* vol. 1, 377.
106. Robert D. Putnam, *Making Democracy Work: Civic Traditions in Modern Italy* (Princeton: Princeton University Press, 1993), 177–78, 181–85.
107. Edicio de la Torre, "The Valuing Process Among Caregivers," in Lopez, *Torture Survivors.,* 117–31, 221; Hannah Arendt, *The Human Condition* (Chicago: University of Chicago Press, 1958), 212–23, 236–43.
108. *MC,* 7/25/97, 7/28/97; *PDI,* 7/5/98, 7/9/98, 8/25/98, 8/26/98, 9/12/98, 11/24/98, 11/27/98, 12/5/98, 12/12/98, 12/13/98, 1/8/99; *Today,* 9/3/98, 9/12/98, 11/7/98, 11/9/98, 12/6/98, 12/12/98, 12/13/98, 1/8/99; *MT,* 11/26/98, 11/27/98/, 12/5/98, 12/6/98, 12/12/98, 12/13/98.
109. *PFP,* 29 June 1996.
110. Miranda, "At the Crossroads of Politicization," 72–76, 83–96; *PDI,* 7/22/96, 12/22/96, 8/28/98; *Today,* 11/6/98, 11/8/98; *MT,* 11/8/98.
111. Rocamora, "EDSA on My Mind," 226.
112. *PDI,* 4/17/97, 4/20/97.
113. Captain Danilo Vizmanos (PN, ret.), "Can a Military Man Be Civilized?" *Today,* 4/30/97.

CHAPTER 10: REUNION

1. Jose M. Mendoza, ed., *Batch '36 Golden Book* (Manila: PMA Class '40 Association Inc., 1986), 350.
2. Reynaldo A. Mendoza, "The Same Banana," in Mendoza, *Golden Book,* 116–17.
3. Mendoza, *Golden Book,* 350; PMA Class '40, "PMA Motto," in Mendoza, *Golden Book,* 125.
4. Deogracias F. Caballero, "The PMA Motto and the Erosion of Its Values," in Mendoza, *Golden Book,* 126–27.

5. "Homily" by Commodore Ramon A. Alcaraz (ret.), Class of 1940, presented at the Alumni Memorial Mass, PMA Chapel, Ft. Del Pilar, Baguio City, 17 February 1990 (copy provided by the author). For press reports on the speech, see *Sunday Times,* 2/18/90; *PDI,* 3/2/90.
6. Mendoza, *Golden Book,* 350.
7. Ibid., 238, 333.
8. *Baltimore Sun,* 1/27/97; *Washington Post,* 1/28/97; *NYT,* 1/29/97.
9. Jose V. Abueva, "A Global Perspective: High Point in a 'Decade of Democracy,'" in Lorna Kalaw-Tirol, ed., *Duet for EDSA: 1996 Looking Back, Looking Forward* (Manila: Foundation for Worldwide People Power, Inc., 1995), 14–19; Samuel P. Huntington, *The Third Wave: Democratization in the Late Twentieth Century* (Norman: University of Oklahoma Press, 1991).
10. Janice Gibson, "Training People to Inflict Pain: State Terror and Social Learning," *Journal of Humanistic Psychology* 31, no. 2 (Spring 1991), 72–81.
11. Cadet Corps, Armed Forces of the Philippines, *Sword '90* (Fort Del Pilar, Baguio City: Bigkis-Lahi Class '90, 1990), 220–29.
12. Ibid., 229.
13. Ibid., 240.

Index

Abadia, Lisandro C. (PMA '62), 302, 303
Abadilla, Rolando N. (PMA '65): biography, 330–33; CIA training, 190; in Manila Hotel coup, 267–68; MISG commander, 206; murder of, 330; in politics, 331; as top torturer, 190, 217, 242
Abat, Fortunato U. (PMA '51), 324
Abcede, Salvador, 162, 165
Abdul Latif. *See* Martelino, Eduardo
Abenina, Edgardo M. (PMA '58), 286, 290, 293, 324
Abra (Q-boat), 76–77
Acenas, Alberto K. (PMA '40), 96
Acop, Ruben (PMA '71), 335
Acosta, Manuel A. (PMA '40): as cadet, 62, 69; curator Dambana ng Kagitingan, 170; in Death March, 92; and loyalty of Segundo Velasco, 126; retirement of, 106
Acosta, Melchor A. (PMA '40), 109
Acosta, Melchor I., Jr. (PMA '71), 276
Adevoso, Eleuterio ("Terry") (PMA '44), 135, 154–55, 157
Afable, Silvestre, 235
AFP Modernization Act of 1995, 336–37
Agoncillo, Teodoro, 165
Aguinaldo, Emilio, 15–16
Aguinaldo, Rodolfo E. (PMA '72): biography, 304–8; in coups, 252, 267, 273, 290, impunity, 307–8; in politics, 304–5; as torturer, 190, 206, 207–14, 304
Aguinaldo: The True-to-Life Story of Gov. Rodolfo Aguinaldo of Cagayan (film, 1993), 308–10
Agusan (Q-boat), 76
Ako ang Batas—General Karingal (film, 1990), 132
Alano, Heracleo J. (PMA '40), 76–77, 96, 137
Albano, Jerry L. (PMA '71), 251

409

Albert, Carlos ("Charlie") J. (USNA '37), 99, 131, 137
Alcaraz, Ramon A. (PMA '40): biography, 144–59; on honor code, 125, 341; retirement, 113; "same banana" leader, 65; service, 76–78, 95, 97, 108; and Uldarico Baclagon, 175–76
Alejo, Benedicto A. (PMA '41), 88
Alex Boncayao Brigade, 132, 333
Alfonsin, Raul, 260
All Stars, The (band), 100
Allen, Theodore, 255
Allied Intelligence Bureau. *See* U.S. Army, Allied Intelligence Bureau
Almonte, Jose T. (PMA '56): on coups, 223, 238, 313; on military academy, 8, 32; as presidential advisor, 32; RAM patron, 315, 321
Amateur Frolic (stage play, PMA), 68
Amhara (tribe, Ethiopia), 41
Amnesty International: on the military, 282; 1975 report on torture, 204; 1981 finding on torture, 204–5; 1984 report, 185; on Rodolfo Aguinaldo, 212
Andaya, Tito, 309–10
Anderson, Bernard, 97
Andrada, Jose B. (USNA '30), 76–77, 145–46
Ang, Maria Elena, 215–16
Anti-Smuggling Action Center (ASAC), 166
Apolinario, Felix M. (PMA '40), 53, 146, 148, 150, 153
April 6 Liberation Movement (1980), 156
Aquino, Benigno ("Ninoy") A.: assassination of, 30, 131, 228; on Operation Merdeka, 128, 156, 192, 270
Aquino, Benigno III ("Noynoy"), 276
Aquino, Corazon ("Cory") C.: administration of, 31, 261–62, 269, 282–83; during EDSA revolt, 246, 249, 253–54; on Ferdinand Marcos's medals, 176; on impunity, 260, 301, 334
Aquino, Tony, 99

Araneta Cup (drill award), 71, 145
Aranzaso, Alberto S. (PMA '40), 92–93
Arayat, Mount, 124
Arellano, Alfonso (PCA '29), 135–37
Arendt, Hannah, 335
Argao, Delfin E. (PMA '40), 95
Argentina, 187, 192, 260
Arias, Jose, 200
Aristide, Jean-Bertrand, 224
Armed Defenders of Democracy (magazine), 232
Armed Forces of the Philippines (AFP): budget of, 28, 31, 193; during Ramos administration, 337; founding, 4, 26, 105; officer corps of, 10, 29–30; politicization of, 5, 14, 40, 205
Armed Forces Officer Personnel Act of 1948, 105
Armies: colonial, 14–19, 104; national, 19, 23–26; tribal, 18, 42
Army, Philippine Revolutionary, 15–17
Aromin, Saulito R. (PMA '74), 237, 239, 243, 247
Arrest Search and Seizure Order (ASSO), 207
Articles of War, 314
Arucan, Miguel, 202
Aruiza, Arturo C. (PMA '67), 230–31, 243, 247
Arula, Jibun, 127
Ateneo de Manila University, 22, 337, 349
Atienza, Danilo, 289
August 1987 Coup, 275–80
Ausejo, Crisostomo, 200
Authoritarian rule, 5, 29, 32, 191–93, 206
Authoritarianism: era of, 183–84; in the Philippines, 19, 30, 171–72, 184; terror in, 183–84; use of military, 29–30, 191–93, 206

Baban, Pedro (PMA '40), 52, 54, 96, 100, 120–21
Bacalla, Alexander I. (PMA '62), 158

Baclagon, Uldarico S. J. (PMA '40): biography, 159–80; as military historian, 83, 124, 144; and Ramon Alcaraz, 175–76; resignation of, 106; as war hero, 163
Baden-Powell, Lord, 43
Baguio City, 3, 24, 60–61, 100
Balao, Eulogio (PCA '31), 110–11, 116, 125
Balbanero, Pedro R. (PMA '57), 227, 236, 245
Balbas, Braulio B., Jr. (PMA '60), 229, 251
Balgos, Mariano, 131, 132
Balmaceda, Arthur G. (PMA '71), 252
Bangkok (Thailand), 40, 147, 223
Barbero, Carmelo, 147, 153
Barbers, Robert, 331–33
Barreto, Medardo, 324–25
Barreto, Renato L. (PMA '38), 53, 54
Barrios, Pacifico B. (PMA '40), 106
Bartolome, Pedro J. (PMA '40), 55, 58
Basa, Cesar, 99
Basa Air Base (Cavite), 236
Basiao, Ruben S. (PMA '88), 279
Bataan Peninsula, 26, 77, 99, 172
Batac, Gonzalo L. (PMA '43), 283
Batac, Vicente Mauro G. (PMA '86), 296
Batac, Victor G. ("Vic") (PMA '71): in coups, 268, 273, 284, 286–87; promotion, 331, 336; RAM leader, 195, 230, 281; torturer, 214–17, 304
Batang Matapang (comic book), 176
Batangas Province, 148, 150
Batasan Pambansa, 138
Battalion Combat Teams (BCT), 107, 109, 125, 165
Battle for San Carlos, 163
Battle of Bataan, 75, 78, 86, 130, 171
Battle of Bessang Pass, 120, 169, 171
Battle of Langemarck, 43
Baula, Francisco P., Jr. (PMA '73), 252
Bayron, Sofio E. (PMA '40), 96
Beijing (China), 223
Belli, Melvin, 327–28
Benedicto, Roberto S., 25
Benguet Auto Line, 35

Berberabe, Carlos, 149
Berberabe, Pablo, 149
Bersola, Pedro C. (PMA '40), 7, 79, 91
Bettelheim, Bruno, 186
Biazon, Rodolfo G. (PMA '61), 246, 292, 314, 321
Bibit, Billy C. (PMA '72): in coup, 284; in film, 308; in politics, 315, 320; release of, 296; and Satur Ocampo, 307; as torturer, 215, 217
Bicol region, 130, 229
Black propaganda. *See* Psychological warfare
Blando, Marcelo C. (PMA '60), 286, 289, 292
Blue seal smuggling, 147–53, 167. *See also* Smuggling
Blue seals (American cigarettes), 28
Bocalan, Lino, 148–50
Bohol Island, 96, 101, 133
Bombing: City Hall, 129; Commission on Elections, 156; Esso and Caltex, 129; Philippine International Convention Center, 156; Philippine Village Hotel, 156; Plaza Miranda, 129, 269; and RAM, 269; Rustan's Department Store, 156; Sulo Hotel, 156; Supreme Court, 129; terror, 29, 275
Bonifacio, Andres, 15–16
Bornales, Abenir D. (PMA '40), 162
Borromeo, Emilio O. (PMA '42), 136
Bosworth, Stephen, 243, 328
Boy Scouts, 43
Brawner, Felix A., Jr. (PMA '57), 236, 253
Brawner, Franklin A. (PMA '71), 286
Brazil, 192
Brillantes, Ricardo T. (PMA '72), 239, 243
Broder, Sherry, 327
Bueno, Eduardo, 326
Bulacan Military Area (BMA), 97
Bureau of Constabulary, 95, 104, 130, 145
Burke, Edmund, 5
Burma, 5, 134, 136
Burnham, Daniel K., 61

Cabagnot, Ruben V. (PMA '71), 202, 335
Cabal, Manuel F. (PCA '33): corruption, 111–12; on coup, 108; resignation, 112
Caballero, Deogracias F. (PMA '40): on coups, 108, 340; in Death March, 86, 89; on honor code, 60, 63–68; on initiation, 199–200; service, 136, 167
Cabangbang, Bartolome C. (PMA '40): biography, 133–41; in politics, 125, 129; service, 126, 166; in World War II, 92–98
Cadet Prayer (PMA), 160
Cagayan, 100, 232, 305
Cagayan Province, 90, 304–5
Calajate, Domingo H. (PMA '60), 286, 303
Caldoza, Urbano B. (PMA '40), 92, 96, 134, 138–40
Camp Aguinaldo (formerly Camp Murphy, Quezon City), 235, 277–78, 289–90
Camp Allen (Baguio City), 50
Camp Aquino (Tarlac), 236
Camp Capinpin (Rizal), 236
Camp Crame (Quezon City), 235, 245, 253–54
Camp Major General Tomas B. Karingal (Quezon City), 132
Camp Murphy (Quezon City), 79, 81, 99, 136
Camp O'Donnell (Tarlac), 90–92
Camp Olivas (Pampanga), 213
Campaign for the Abolition of Torture, 185
Campo, Isagani V. (PCA '32), 137
Cancio, Manuel C. (PMA '41), 92
Canieso, Rodolfo A. (PMA '56), 272, 274, 295
Cannu, Ramon, 236
Capas (concentration camp), 75, 78, 86, 88, 94–95
Capulong, Romeo, 329
Carpio, J. Antonio, 272
Carruncho, Eric, 263

Carter, Jimmy, 205
Castillo, Felicisimo S. (USMA '40), 40
Castillo, Lazaro, 215
Castro, Fred Ruiz, 25, 44
Catholic Bishops Conference, 335
Causin, Florencio C. (PMA '40), 72
Cavite Province, 14–15, 148–49, 152
Cebu City, 96, 139
Central Intelligence Agency (CIA), 134, 238
Cepeda, Emmanuel S. (USMA '33), 64–65
Channel 4, 252, 253, 287, 289
Chile, 188, 192, 300
China, 336
Christmas 1989 Coup, 4, 31, 284–95
Ciron, Aida, 249
Ciron, Ruben T. (PMA '68), 231, 249, 262, 294
Citizens League of Quezon City, 131
Civil Affairs Office, AFP, 128
Civil Aviation Administration (CAA), 133
Civilian supremacy, 15, 25, 28, 38, 63, 71; Class of 1940 and, 342–43; and Fidel Ramos, 273; and Filipino officers, 348; under martial law, 194; RAM leaders and, 258
Clark Field (Pampanga), 27, 336
Class of 1937, 55
Class of 1938, 52–53, 55
Class of 1940: academic achievement, 52, 60; advanced training, 104, 131, 190; fiftieth, 119, 120, 132; as first classmen, 59–63; graduation, 70–73; on honor code, 66–68, 101–2, 109, 114, 141, 341; initiation of, 52–55, 57–58; and masculinity, 62, 68–70; promotion, 104–9, 113–14; retirement, 106, 112, 113–14, 122, reunion, 36, 67, 72, 104; in "same banana," 63–65, 349; service, 74, 91, 163; silver, 114; social background, 51–52; socialization of, 6, 36, 73, 141, 179; on solidarity, 55–56, 60, 92, 101, 109; as upper-classmen, 57–59

Class of 1942, 57, 80
Class of 1943, 80, 283
Class of 1944, 80, 135, 137, 199
Class of 1945, 80, 199
Class of 1951, 29–30, 114, 199, 200
Class of 1953, 194
Class of 1957, 237
Class of 1971: in coups, 4, 8–9, 30, 195, 204; impunity, 302–4; initiation of, 201–2; as interrogators, 8, 154, 207; in Marcos dictatorship, 347; RAM movement, 198, 230–34; service, 8, 203–4, 207, 347; socialization of, 197–99; in torture, 206–7
Class of 1972, 194
Class of 1975, 326
Class of 1976, 30, 194
Class of 1977, 194
Class of 1987, 279
Class of 1990, 341, 350–52
Class of 1991, 194
Cleofe, Senen A. (PMA '38), 54
Club Filipino (San Juan), 254
Co, Leoncio, 207
Coast Artillery, 74, 85–92
Cojuangco, Eduardo ("Danding"), 284, 293
Cojuangco, Enrique, 284
Colonel Billy Bibit-RAM (film, 1994), 311–12
Colonialism: in Africa, 18; armies and, 14–19; in Asia, 17, 345; ideology and, 27; and masculinity, 17–18; in Philippines, 114–19, 344–45; United States, 20–21, 26, 344–45
Commendador, Jose B. (PMA '59), 286, 293
Commission on Appointments, 109–10
Commission on Elections (Comelec), 131, 139, 156
Commission on Human Rights, 216, 261, 278, 282, 328
Committee on Human Rights. *See* Commission on Human Rights

Communist Party of the Philippines, 270
Communist Revolt, 107, 117, 131
Conscription: colonial, 15–16; Commonwealth, 23, 37; Europe, 42–43; film and, 47; gender representation of, 11, 19; peasants, 115; Philippines, 16, 26, 43–44, 46, 79; World War II, 80–81, 162
Constabulary Academy, 20, 24, 135. *See also* Philippine Constabulary Academy
Constabulary Investigative Service (CIS), 112, 116
Constabulary Revenue-Customs (CRC), 116
Constabulary School. *See* Constabulary Academy
Constitutional Commission, 268
Constitutional Referendum (1987), 31, 268, 275
Coronel, Sheila, 218–19
Corps of Sponsors (U.P.), 44
Corps, The (PMA magazine), 160
Corpus, Victor N. (PMA '67), 197–98, 270–71
Corregidor Island (Bataan), 78, 93, 127
Corruption: in Armed Forces of the Philippines, 336; Fabian Ver and, 120–21, 126; under Ferdinand Marcos, 29, 152–54, 347
Counter Intelligence Command, AFP, 302
Coups d'état: against Corazon Aquino, 261; against Ferdinand Marcos, 154–55, 222–24, 234–36; of August 1987, 275–80; of Christmas 1989, 4–5, 31, 284–93; EDSA revolt, 240–45; God Save the Queen, 268–73; Magsaysay inspired, 108, 136; Manila Hotel, 266–68; Mindanao enclave, 295–96; in November 1986, 273–75. *See also individual coups*
Crame, Rafael, 19
Crisol, Jose M. (PMA '42), 57, 129, 167, 175, 177

Crisologo, Floro, 125, 138, 225
Crisologo, Juan, 225
Crispin Tagamolila Command, NPA, 302
Cuenco, Mariano J., 135

Dambana ng Kagitingan (shrine), 170
Danao City (Cebu), 118
David, Randy, 316–17
Davide Commission: August 1987 coup, 278; Christmas 1989 coup, 284; on coups, 195, 262, 268, 273; on hazing, 41; on Rodolfo Aguinaldo, 305
Davis, Bette, 62
De la Cruz, Ciceron P. (PMA '40), 91, 109, 116, 158
De la Paz, Dr. Remberto ("Bobby"), 233, 303
De la Torre, Edicio, 335
De Villa, Renato S. (PMA '57), on coups, 245, 271; service, 32, 229, 283, 303
Death March, 26, 75, 86–89, 103, 168
Del Castillo, Francisco D. P. (PMA '40), 124–26, 167
Del Rosario, Cesario, 129
Del Rosario, Rosa, 62
Dicon, Ricardo, 325
Dictatorship. *See* Authoritarian rule
Dimaranan, Sister Mariani, 328
Diokno, Jose W., 261, 333
Diokno, Ma. Socorro, 314
Dioneda, Leovic R. (PMA '78), 320
Dizon, Victor, 121
Doerr-Zegers, Otto, 187–88
Domingo, Emilio A. (PMA '44), 136
Doromal, Edgardo M. (PMA '74), 239–40
Doronilla, Amando, 320
Drilon, Franklin, 314, 324
Dupaya, Teresa, 305
Durano, Ramon, 118
Durian, Jesus I. (PMA '60), 249
Durian, Vangie, 249
Duvalier, François, 224

Eastwood, Clint, 198, 263, 346
Echanis, Randall, 266
EDSA Revolution, 4, 10, 32; day one, 30–31, 242–47; day two, 247–50; day three, 250–54; day four, 254–55; international impact, 224; outcome, 255–58
Eisenhower, Dwight, 23, 49–50
Elarth, Harold H., 18, 22
Elections: male-electorate, 19; 1953 presidential, 4, 108; 1957, 135; 1965, 118–19, 125, 138, 225; 1969, 138, 154; 1978, 129, 138, 204; 1981 presidential, 140; 1986 snap, 120, 158, 176, 239; 1992 presidential, 35, 302–3; 1998, 32
Elite families, Filipino, 13, 25, 72, 197
Engeniero, Ismael, 96
Enloe, Cynthia, 11
Enrile, Cristina, 246
Enrile, Juan Ponce: ambush, 192; coup patron, 30, 193, 267, 293; Defense Minister, 29, 225, 254, 262; martial-law, 229; in politics, 315, 320–21; resignation, 274; on torture, 328
Epifanio de los Santos Avenue (EDSA), 235, 247
Escalante (Negros), 228
Escandor, Johnny, 217, 266, 303
Escobar, Nolasco M. (PMA '40), 92
Esguerra, Arsenio, 216
Eslao, Nelson G. (PMA '71), 280
Espino, Amado, Jr., 213
Espino, Romeo, 25, 127–29, 225
Estrada, Eva. *See* Kalaw, Eva Estrada
Estrada, Joseph, 318, 331, 335
Estrada, Licurgo E. (PMA '40), 54, 64, 71, 100–1
Evangelista, Quirico C. (PMA '40), 67

Fajardo, Tirso G. (USMA '34), 61, 70, 109
Farolan, Horacio N. (PMA '40), 91–92, 104, 121
February 1987 Plebiscite. *See* Constitutional Referendum (1987)
Federal Party, 139–40

Federation of Women's Clubs, 45
Felix, Pedro L. (PMA '38), 52–53
Fellers, Bonner E., 39
Ferdinand E. Marcos: 77 Days in Eastern Pangasinan (memoir), 226
Fernandez, Proceso C. (PMA '59), 288
Ferrer, Jaime, 135
Fetalvero, Felipe J. (PMA '40), 54
Field Artillery, AFP, 74
5th Constabulary Security Unit (CSU), 206, 211
5th Marine Landing Team, 242
Figueroa, Hernani F. (PMA '66): in coups, 231, 273; interrogator, 208–11, 304: RAM chairman, 233
Filart, Alfredo S. (PMA '40), 81, 130
Filipinization policy, 19
Filipino Heroes (Uldarico S. Baclagon), 174
Filler, Alfredo L. (PMA '62), 324
Finer, S. E., 5
1st Scout Rangers, 236, 275
Fistful of Dollars A. (film, 1966), 198, 346
Flirtation Walk (film, 1934), 62, 345
Florendo, Oscar E. (PMA '62), 200, 252, 306
Flores, Manuel T. (PCA '34), 112, 249
For Every Tear a Victory (Hartzell Spence), 170
Foronda, Ricardo A. (PMA '40), 90, 106
Fort, Guy, 161
Fort Benning (Georgia), 104, 163
Fort Bonifacio (Makati City), 212–13, 235–36, 247, 287
Fort Leavenworth (Kansas). *See* U.S. Command and General Staff College
Fort Magsaysay (Nueva Ecija), 274
Fort Sam Houston (Texas), 131
Fort Sill (Oklahoma), 104
Fort Wint (Grande Island), 85
49th Infantry Battalion, 237, 239
Foucault, Michel, 191
Foundation for Clean Elections, 321
14th Infantry Battalion, 169, 174, 242, 276

Francisco, Carlos V., 171
Francisco, Guillermo, 130
Francisco, Jose V. (USNA '31), 146
Free Legal Assistance Group (FLAG), 314, 331
Fusilero, Tiburcio N. (PMA '71), 219, 286, 293, 335

Gador, Tirso H. (PMA '66), 231, 239, 241
Galang, Lucendro L. (PMA '40), 54, 112–13
Galicia, Gilberto, 272
Galido, Alejandro A. (PMA '58), 284
Galvez, Rafael C. (PMA '71), 286–87, 292
Garcia, Carlos, 96, 131, 135–37, 342
Garcia, Eddie, 132
Garrett, Norbert, 241
Gasmin, Segundo, 149–50
Gause, Damon, 93
Gelvezon, Ramon G. (PMA '40), 53, 72, 114–20, 137
Gender: under colonialism, 19; feminine, 19–20, 44–46, 68–70; film representation of, 12, 47, 345–46; masculine, 43–46, 115, 345; military and, 11–12, 44, 58–59, 69; theory of, 11–14
Genefke, Inge, 185
Gibson, Janice, 186, 347
Gidaya, Ernesto S. (PMA '51), 30
Gillego, Bonifacio, 155, 157–58, 174
God Save the Queen Coup (1986), 268–73
Gojo, Romelino D. R. (PMA '70), 284–87, 290, 292
Goldwyn, Samuel, 47
Gomez, Tomas, 294
Good, the Bad and the Ugly, The (film, 1966), 263
Goyena, Lucita, 47
Guardia Civil, 15
Guardian Brotherhood (society), 267
Guevara, Santiago G. (USMA '23), 80
Gun smuggling, 232

Habeas Corpus, 29
Habib, Philip, 243
Haiti, 224
Hazing, 35, 39, 112, 119–201
Hernandez, Jun, 312
Heroism, 144, 178–79
Higaonon (tribe), 296, 320
Historiography: Filipino, 165; and masculinity, 165; military, 12–13, 165, 177, 179; national, 165; U.S. colonial, 165
Hitler, Adolf, 43
Home Defense Force, 129
Honasan, Gregorio ("Gringo") B. (PMA '71): alliances, 280–82; biography, 198–99; in coups, 263, 267, 272, 274, 276, 284; EDSA revolt, 254, 256–57; on human rights violations, 216–17, 266; myth of, 263–66; at PMA, 4, 197–98, 218; in politics, 312, 315, 320; RAM leader, 7, 30, 183, 223, 230, 286; service, 203, 218, 231; surrender, 302
Honasan, Romeo G. (PMA '43), 108, 283
Hornedo, Florentino, 172
Howitzer, The (yearbook), 20
Huk revolt. *See* Hukbalahap guerrillas
Hukbalahap guerrillas, 27, 98, 107
Human Resource Exploitation Training Manual (Central Intelligence Agency), 190
Human rights: abuse of, 192–93, 282, 300–1; under Marcos regime, 192–93; in Philippines, 204–7, 208–17, 261, 326–27; under Ramos presidency, 332; UN conventions, 185; United States, 190, 327–30
Human Rights Commission. *See* Commission on Human Rights
100th Nite Show (stage play, PMA), 70
Hunters ROTC guerrillas, 135, 137
Huntington, Samuel, 12

Ibaloi (tribe), 52, 120
Ideal Theatre (Manila), 62
Iginuhit ng Tadhana (film, 1965), 170
Ileto, Rafael M. (USMA '43), 71, 238, 274, 283
Iloilo Province, 114
Ilongot (tribe), 42
Impunity: in Aquino administration, 260, 301; in Estrada administration, 301; Latin America, 260; legacy of, 299–300; Philippines, 260, 301–2; in Ramos administration, 301, 303, 314, 329
Infantry, AFP, 79–85
Initiation: Ethiopia, 41; Germany, 43; Kenya, 41; male, 41–43; military, 42; military academy, 199; Papua New-Guinea, 42; Philippines, 199; United Kingdom, 42; United States, 41–43
Integrated National Police (INP), 228
Intelligence Service of the AFP (ISAFP), 227
International Symposium on Torture: VI (Buenos Aires, 1993), 300; VII (Capetown, 1995), 300

Jabidah massacre (1968), 127, 330
Jalandoni, Fr. Luis, 212
Janowitz, Morris, 12
Japanese invasion, 37, 75
Jardiniano, Tagumpay R. (PMA '57), 236, 245, 250
Javier, Jose M. (PMA '40), 79, 82–83, 85
Jimenez, Francisco M. (PMA '40), 103, 122, 129, 150, 200
Jimenez, Nicanor, 175, 178
John Paul II, Pope, 170
Johnson, Bryan, 264, 266

Kabataang Barangay (youth organization), 327
Kabataang Makabayan (KM), 197
Kahit Butas ng Karayom Papasukin Ko (film, 1995), 346
Kakarong Regiment, 97
Kalaw, Eva Estrada, 46
Kamalayan '86, 234

Kamlon, Hadji, 131–32
Kangleon, Fr. Edgardo, 208–11, 303
Kapunan, Eduardo ("Red") E., Jr. (PMA '71): impunity, 302; in politics, 315, 320; as RAM leader, 199, 230, 266–68, 286–87, 340; service, 204, 231, 274; torturer, 217–19, 261
Karingal, Tomas B., 129–33
Katipunan (society), 15–16
Keesing, Roger, 42
Kempeitai, 100, 130
Kilusang Bagong Lipunan (KBL), 139, 176, 238
Kiss of the Spider Woman (film, 1985), 192
Kuratong Baleleng Gang, 335

Lacaba, Jose ("Pete"), 270
Lacson, Panfilo ("Ping") M. (PMA '71), 206, 217, 267, 331–32, 335–36
Lakas-NUCD Party, 331
Lambino, Antonio B., S.J., 257
Lamon Bay (Quezon), 81–82
Langemarck, Battle of. *See* Battle of Langemarck
Lansdale, Edward, 108
Lapid, Lito, 308, 310, 315
Lapus, Patrocinio D. (PMA '41), 92
Lara, Eugenio C. (PMA '38), 49
Latin America, 5, 187
Laurel, Jose, 133, 136
Laurel, Salvador, 235, 280, 291, 293
Lawrence of Arabia (film, 1962), 128
Laxalt, Paul, 254
Layson, Elpidio S. (USNA '75, PMA '74), 272, 325
Lazatin, Rafael, 117
Lee, Robert S. (PMA '81), 277
Legaspi, Oscar ("Tito") E. (PMA '71): in coups, 238, 272, 284, 286, 288; at PMA, 199–201; in Rolando Olalia's murder, 322–24, 325–26; surrender, 292
Leone, Sergio, 198, 345
Lessons from the Huk Campaign (Uldarico S. Baclagon), 165

Libarnes, Benjamin N. (PMA '67), 213
Liberal Party, 118, 271
Libingan ng mga Bayani (Makati City), 313
Light A Fire Movement (1980), 156
Lim, Alfredo, 133, 246
Lim, Danilo (PMA '78, USMA '78), 285, 292, 324
Lim, Pilar H., 45
Lim, Roberto (USNA '42), 57
Lim, Vicente P. (USMA '14), 20, 23, 36–37, 71–72
Lopez, Eugenio, Jr., 155
Lopez, Eugenio, Sr., 156
Lopez, June, 186, 202
Love Pirates of Hawaii. See 100th Nite Show (stage play, PMA)
Lovely, Victor, Jr., 157
Loyalty Status Review Board, 104
Lubao (Pampanga), 87
Lucero, Fr. Pedrito, 211
Lucero, Sr. Cres, 258
Lumen, Francisco (PMA '40), 85–86, 88
Luzon (Q-boat), 76–78

Macapagal, Diosdado, 112–13, 151, 157, 170, 342
MacArthur, Douglas, 38
Maceda, Ernesto, 151
Mactan Air Base (Cebu), 293
Madaling Araw (film, 1938), 62
Madame Panseni (Japanese spy), 130
Madarang, Casimiro M., Jr., 139
Maglipon, Jo-Ann, 219
Magluyan, Juan B. (PMA '37), 147
Magsaysay, Ramon: death, 138; in the military, 26–28, 107–8, 116; in politics, 105, 135
Maharlika, Ang Mga (Guerrilla unit), 168–69
Maharlika, Ang Mga—Its History in Brief (memoir), 168
Maharlika Highway, 173

Malacañang Palace (Manila), 235, 241, 287, 289
Malajacan, Teodoro ("Jake") Q., Jr. (PMA '71), 239, 242–43, 297, 314
Malay, Bobby, 297
Male initiation. *See* Initiation
Maligalig, Proceso ("Boy") L. (PMA '69), 286, 312, 320
Maligaya, Filomeno, 322–23
Manglapus, Raul, 135, 155–56
Manila, 14, 17, 98
Manila Bay, 77–79, 148, 281
Manila Carnival (festival), 22, 44–45
Manila Hotel Coup (1986), 266–68, 340
Manobo (tribe), 296
Manriquez, Romulo A. (PMA '36), 158, 174
Manuel, E. Arsenio, 172
Marcos, Ferdinand ("Bongbong"), Jr., 241, 319, 321
Marcos, Ferdinand E.: authoritarian rule under, 7–8, 154–55, 224–25; and Class of 1971, 203; corruption, 29, 151–52; as defense secretary, 191; exile, 255; failing health, 176, 226–27; military abuses, 192–93; torture, 204–7; use of military, 193, 342; war record, 158, 167–70, 176–78
Marcos, Imee, 327
Marcos, Imelda R., 132, 172, 226–28
Marcos loyalists, 120, 255, 267, 273, 280
Maristela, Jose C. (PMA '41), 111–12, 155, 157
Martelino, Eduardo, 128–29
Martelino, Pastor (USMA '20), 61, 64
Martial Law, 5, 114, 129, 192–93
Martial races, 18
Martinez, Oscar O. (PMA '71), 336
Masai (Kenya), 41
Masculinity: under colonialism, 19; hazing and, 41–42; in military academy, 11–12, 38–40, 47–48, 62, 68–70; and modern military, 12–13, 41, 44, 218; representation in films, 12, 47, 62, 345–46; traditional, 48; tribal, 42
Massera, Emilio, 205
Mata, Ernesto S. (PMA '37), 113, 152–53, 161–62
Mauban (Quezon Province), 81–85
Mayo, Job T. (PMA '40), 81, 136
Mayor, Arcadio (PMA '41), 84
McKenzie, Harry, 170
Melchor, Alejandro, 225, 238
Melchor, Alejandro, Jr. (USNA '52), 63
Mendoza, Godofredo F. (PMA '38), 82–83, 85
Mendoza, Jose M. (PMA '40): class historian, 65, 73; as plebe, 54; promotion, 109–11; service, 85–89
Mendoza, Reynaldo A. (PMA '40): academy training, 199–200; on coup, 108; on honor code, 341; on initiation, 54, 56, 201; PMA superintendent, 3, 199; promotion, 125; and "same banana," 340; service, 85, 90
Mercado, Rogaciano, 153
Metrocom Intelligence and Security Group, Constabulary (MISG), 131, 206, 330
Metropolitan Command (Metrocom), 236
Milgram, Stanley, 186
Military, Philippine: abuse, 28; history, 12, 14, 159; during Martial Law, 29, 30; officers in, 23–24; professionalism in, 5, 7; promotions in, 28; women in, 20; after World War II, 28, 104–7
Military Assistance Agreement, 27
Military conscription. *See* Conscription
Military history, 9–10, 164–66
Military junta, 155, 183, 254, 275
Military reform, 232–34
Military socialization. *See* Socialization
Mindanao, 8, 101, 203, 295
Mindanao enclave coup, 295–96
Miranda, Felipe, 207
Mison, Mariano, 323
Mison, Salvador M. (PMA '55), 323

Missing (film, 1982), 192
Misuari, Nur, 127
Mitsubishi "Zero" fighters, 92
Molina, Pedro Quezon (PMA '37), 49
Montano, Delfin, 127
Montemayor, Cesar U. (PMA '40), 41, 60
Morales, Ricardo C. (PMA '77), 239, 243, 246, 269
Moran, Manuel, 112
Morton, Luis, 85
Mosse, George, 43
Mount Pinatubo (Zambales). *See* Pinatubo, Mount (Zambales)
Mount Samat (Bataan). *See* Samat, Mount (Bataan)
Movement for Advancement of Nationalism (MAN), 153
Movement for Free Philippines (MFP), 155
My Kaydet Girl (song), 61–62, 70
Myth, 144, 158–59, 167–73, 263–66

Nacionalista Party, 138, 166, 291
Nagtahan Bridge (Manila), 276–77
Nano, Conrado V. E. (PMA '40), 66, 91, 107
National Amnesty Commission, 314
National Bureau of Investigation (NBI), 154, 269, 322
National Defense Act of 1935, 21, 23, 26, 37–38
National Defense Intelligence Office, 206
National Heroes Day, 46
National Intelligence and Security Agency (NISA), 206, 215, 226–27
National Press Club, 233
National Unification Commission (NUC), 303
Nationalism: Filipino, 19, 20, 22, 62, 346; Filipino historiography and, 164, 174; gender representation of, 20; masculinity and, 346; and society, 346; and state, 346
Navarro, Edmundo G. (PMA '40), 54, 58, 65, 109, 178

Negrito (tribe), 305
Negros Island, 96, 101
Never Again Movement (NAM), 319
New People's Army (NPA), 132, 192, 197, 262
New Republic (1981), 139
New Society, 29, 174
Nguyen Cao Ky, 123
Nichols Field (Pasay City), 99
Nieva, Antonio, 313
Nituda, Victor, 172
Noble, Alexander O. (PMA '69): amnesty, 314; in coups, 286, 288–89, 295–96; in politics, 315, 320
Nosce, Ramon Q. (PMA '40), 54
Notes on New Society (Ferdinand E. Marcos), 155
November 1986 Coup, 273–75
Nuval, Santiago C. (PMA '38), 147, 149–50

Ocampo, Satur, 213–14, 297, 306–7
Ochoco, Brillante C. (PMA '55), 227, 250, 253
Ochosa, Jose Reynaldo B. (PMA '72), 278, 286
Off-Shore Patrol (OSP). *See* Philippine Army, Off-Shore Patrol (OSP)
Officers: American, 19; colonial, 17, 19; Filipino, 19–23, 104; as politicians, 31, 312–22; politicization, 29–31, 40, 340; recruitment, 24; training, 27, 37–43
Ohta, Shoji, 85
Olalia, Feliciana, 323
Olalia, Rolando, 269, 272, 303, 322, 326
Olalia, Rolando Rico, 326
Olbes, Ramon A. (PMA '40), 72, 76, 106
Olivares, Flaviano, 118
Olivas, Prospero A. (PMA '53), 240, 245–46
Oliveros, Domingo, 286
Ongpin, Jaime, 254
Operation Dagdag-Bawas, 321
Operation Merdeka, 128, 330

Ople, Blas, 257–58, 291
Orias, Epimaco V. (PMA '40), 92, 101
Oropesa, Cirilo, 236
Ortega, Jose, 82–83
Ortega, Roberto, 217, 323
Osias, Camilo, 99, 122
Osias, Victor M. (PMA '40): biography, 122–24; on patronage, 121; at PMA, 52, 57, 59, 68–69; service, 98–110, 343
Osmeña, Sergio, 72, 98
Osmeña, Sergio ("Serging"), Jr., 118, 135, 154, 157
Osmeña, Sergio III, 155
Otis, Elwell S., 18
O'Toole, Peter, 128

P-26 pursuit planes, 92
Padiernos, Romeo A. (PMA '65), 332
Padilla, Robin, 311, 315
Pagulayan, Leonie, 278
Paje, Alan, 297
Palafox, Antonio, 237
Palencia, Alfonso P. (PMA '38), 118
Palma, Rafael, 21
Palo, Marco, 212
Pampanga River, 78
Pangasinan Province, 98, 169–70
Papua-New Guinea, 42
Paredes, Allen T. (PMA '88), 280
Paredes, Brigido T. (PMA '60), 271
Paredes, Zosimo M., Jr. (PMA '71), 315, 320, 324
Parker, George, 81
Patronage politics, 7, 30, 103: and Bartolome Cabangbang, 134; and Jose Mendoza, 110–14; and Pedro Baban, 121; and Segundo Velasco, 142
Pattugalan, Roland I. (PMA '57), 227, 236, 243, 250
Pelayo, David A. (PMA '40), 55, 79–85, 97, 155
People Power. *See* EDSA Revolution
People's Reform Party (PRP), 315

Peralta, Macario, Jr., 48, 105, 112–13, 149–51
Perez, Antonio P. (PMA '40), 71
Philippine Air Force, 74–75, 92–94
Philippine Army: formation of officers, 23–24; founding, 19, 46; at Manila Carnival, 45; mobilization in 1945, 104–7; training, 21–24; after World War II, 26–28
Philippine Army, Off-Shore Patrol (OSP), 74, 76–79, 145–46
Philippine Army Air Corps (PAAC), 74–75, 92–94
Philippine Campaigns (Uldarico S. Baclagon), 159, 164, 172
Philippine Civic Action Group (PHILCAG), 29
Philippine Commonwealth, 23–26, 44, 49, 345–46
Philippine Constabulary (P.C.), 17–78, 107, 329, 331
Philippine Constabulary Academy (PCA), 21, 35, 49–50
Philippine Consulate (Los Angeles), 158
Philippine Marines, 236, 277
Philippine Medical Action Group, 205
Philippine Military Academy (PMA): alumni, 5–6, 103, 224, 338; cadets, 80, 196; curriculum, 24, 38, 48, 50, 197; disciplinary code, 66; family ties, 9, 195; founding, 20, 24, 35, 37–38; indoctrination, 193–94; initiation, 40–41, 48, 194, 200, 347; recruitment in, 8–9; women in, 19–20
Philippine National Guard (PNG), 21
Philippine National Police (PNP), 132, 331, 337
Philippine Navy Patrol, 146–47
Philippine Red Cross, 95
Philippine Scouts, 17–18, 20, 45, 99
Philippine-American Friendship Group, 139
Philippine-American War, 20, 62, 164
Philippinensian (yearbook), 46, 48

Philippines, Congress—House, National Defense Committee, 105, 135, 153
Philippines, Congress—Senate, 4, 105
Philippines Legislature, 19–20
Picar, Liberato R. (PMA '40), 51, 77
Piccio, Salvador T. (PMA '40), 91, 95–96, 104, 126, 200
Piccio, Vicente, 253
Pimentel, Aquilino ("Nene"), 316–17, 321
Pinatubo, Mount (Zambales), 336
Pizarro, Danilo E. (PMA '63), 288
Planting Rice (song), 100
Platt, Nicholas, 289
Plaza Miranda bombing. *See* Bombing; Terror
P.M.A. Forever (song), 48
P.M.A., Oh! Hail to Thee (song), 3, 67, 250
Poe, Fernando, Jr., 346
Poe, Fernando, Sr., 47, 346
Politico-military crisis of 1958, 135
Powell, Dick, 62, 345
Power, Tyrone, 62
Preliminary Statement of Aspirations (pamphlet), 232
President Quezon's Own Guerrillas II, 130
Presidential Agency on Reforms and Government Operations (PARGO), 126, 138
Presidential Anti-Crime Commission (PACC), 331
Presidential Commitment Order (PCO), 207
Presidential Security Command (PSC), 236, 238
Presidential Security Guard, 183
Presidential Security Unit, 123, 154, 226
Presidential Task Force on Organized Crime, 335
Proclamation Number One, 238
Proclamation *1081*. *See* Martial Law
Proclamation *347* and *348*, 314
Project Mactan (terror campaign), 155–56
Promotion, 232

Propaganda, black. *See* Psychological warfare
Psinakis, Steve, 156
Psychological warfare, 184–91
Public Forum (TV show), 316
Punit na Bandila (film, 1939), 47, 62, 346
Purugganan, Abraham (PMA '78): capture of, 302; in coups, 284, 286, 289, 291–92; on Gregorio Honasan, 275
Pusyong Bisaya, 138–39
Putnam, Robert, 334–35

Q-111 (torpedo boat). See *Luzon* (Q-boat)
Q-112 (torpedo boat). See *Abra* (Q-boat)
Quezon, Manuel: and Class of 1940, 71; on hazing, 49–50; on masculinity, 47; and the military, 7, 336–37; officers training, 20–24, 37–40, 58, 352; role of women, 45
Quirino, Elpidio, 105, 108

Radio Veritas, 246
Ramas, Josephus: collector, 112; in EDSA revolt, 248, 251; service, 227, 236, 240
Rambo movies (1982, 1985), 346
Ramirez, Dionisio, 325
Ramos, Fidel V. (USMA '50): defection, 31, 243; and Fabian Ver, 228; on impunity, 303; military ties, 229; in politics, 31, 301, 303; at Rolando Abadilla's funeral, 332; service, 141, 175, 225, 240, 254; on torture, 31
Ramos, Honorato, 84
Ramos, Leticia. *See* Shahani, Leticia Ramos
Ramos, Narciso, 227, 229
Rangoon (Burma), 40, 223
Ranin, Salvador, 273, 322–23
Raphael, Victor, 240
Raval, Vicente, 110
Razon, Avelino I., Jr. (PMA '74), 243
Reagan, Ronald, 157, 174
Reagan administration, 228
Real, Manuel, 328–29

Real Glory, The (film, 1939), 47
Rebolusyonaryong Alyansang Makabayan (RAM). *See* Reform the Armed Forces Movement
Reciña, Aurora N., 329, 333
Reciña, Romeo M. (PMA '57), 253, 333
Reform the Armed Forces Movement (RAM): during Aquino administration, 262; coup plan, 30–31, 183, 234–37; impunity, 314, 323; leaders, 7, 184, 221, 304; during Martial Law, 18; origin, 193, 230–344; and politics, 223, 312–22; on torture, 233, 266
Regional Unified Command (RUC), 228
Rehabilitation and Research Centre for Torture Victims (RCT), 185
Renan, Ernest, 301
Republic Act 207, 105
Republic Act 291. *See* Armed Forces Officer Personnel Act of 1948
Repuno, Trinidad Herrera, 328
Reserve Officers' Training Corps (ROTC), 21–22, 24, 46
Retirement Law, 122
Revised Penal Code, 314
Reyes, Noli, 63
Reyes, Rick, 297
Rico, Aldo, 260
Rivera, Temario, 213, 321–22
Rizal, Jose, 15
Robles, Rex C. (PMA '65): and black propaganda, 204, 269; in coups, 231, 238, 268; as RAM spokesman, 219–21, 233; in Rolando Olalia's murder, 273, 326
Rodriguez, Jose (PMA '40), 95–96
Romero, Rufo (USMA '31), 52, 61
Romero, Sergio, 333
Romualdez, Alfredo, 153
Romulo, Carlos, 45
Roosevelt, Franklin, 23, 26
Rosaldo, Michelle, 41
Rosales, Carmen, 62
Roxas, Manuel, 104–5, 146, 169

Sabah, 127
Sabite, Feliciano, 306
Sagun, Washington M. (PMA '40), 61, 92
Saint-Jean, Iberico, 205
Salvaging, 191, 205–6, 272. *See also* Torture
Samar Island, 209, 211
Samat, Mount (Bataan), 170–72
San Juan, Frisco F. (PMA '44), 135
Sanchez, Luisito G. (PMA '67), 277, 286
Sandhurst (military academy), 39
Sangley Point (Luzon), 236, 287–88
Santiago, Miriam Defensor, 315, 318
Santos, Alejo, 97–98, 125, 137, 140, 151
Santos, Alfredo, 80–82
Santos, Arsenio C., Jr. (PMA '72), 239
Santos, Luis, 306
Scarry, Elaine, 189, 191
Sebastian, Faustino R. (PMA '40), 152
Segovia, Epifanio E. (PMA '40), 162
Segundo, Fidel J. (USMA '17), 59, 80–82
SELDA, 205, 329
Shahani, Leticia Ramos, 229
Simbulan, Dante C. (PMA '52), 197
Simons, Lewis, 264
Sin, Cardinal Jaime, 235, 246, 257
Singson, Fidel V. C. (PMA '57), 236
Sison, Jose Ma., 153, 197, 214, 308
Sistoza, Pedro G. (PMA '60), 252
Smuggling: Batangas, 48; Berberabe brothers, 149; Cavite, 148; copra, 148; corruption from, 148; Felix Apolinario, 148–49; Ferdinand Marcos, 151–53; Indonesia, 148; Lino Bocalan, 148, 152; Philippine Constabulary and, 149; Philippines, 148; at the PMA, 197; politics of, 147, 153; PX goods, 134, 137, 147; Ramon Alcaraz and, 147; Sandakan, 148–49; Sulawesi, 148; Sulu Sea, 148; tobacco (blue seals), 147–53
Socialization: female, 44–46; male, 41–42, 68–70; military, 32, 37, 41, 44, 79–80; in military academy, 36, 38, 184, 193–95, 346; ties in, 40

Soldiers. *See* Officers; Warrior
Soldiers of the Filipino People (SFP), 280
Soliman, Eduardo P. (PMA '40), 54
Sotelo, Antonio, 229, 250
Southeast Asia Treaty Organization (SEATO), 165
Spanish-American War, 16–17
Spanish military, 15–16
Special Action Force, PC, 232
Special Air Service (SAS), 231
Spratley Islands (South China Sea), 336
St. Cyr (military academy), 39
State: armies, 14–19, 23–26, 42–43; coup against, 223, 260–61; defense against coups, 297–98; ideology and, 12–13; use of male initiation, 38–39, 41–43; modern, 42
Statehood USA Movement, 133, 138, 140
Stotsenberg Educational Training Center, 95
Subic Bay (Zambales), 27, 85, 147, 336
Sueo Oe, 83
Sulawesi (Indonesia), 148
Sulo Hotel (Quezon City), 156
Sulu Sultanate, 14–15
Supreme Court, 174, 306
Swift, Robert, 327–29
Switzerland, 134, 328
Sword, The (PMA yearbook): Class of 1938, 50, 53, 58, 60; Class of 1940, 61, 67, 69–70, 115; Class of 1971, 195–96, 199, 201; Class of 1990, 350; others, 194

Tabuena, Luis, 284
Tadhana (Ferdinand E. Marcos), 173
Tadiar, Artemio A., Jr. (PMA '59): assault at Camp Aguinaldo, 248–49, 251–53; in EDSA revolt, 227, 239, 241, 243
Tañada, Lorenzo, 153
Tañada, Wigberto, 314, 328
Tarrazona, Hector M. (PMA '68), 231
Taruc, Luis, 98, 109, 124–25

Task Force Detainees (TFD), 211, 328
Tatad, Francisco ("Kit"), 140, 176
Tausug (tribe), 128
Tayag, Jean Cacayorin, 212, 328
Tayag, Nilo, 281
Teachers Camp (Baguio City), 50, 61
Tecson, Arsenio L. (PMA '70), 284, 286, 288, 290
Teodosio, Emmanuel R. (PMA '72), 277
Terror, 155–56, 191–93
Thailand, 136
Thompson, E.P., 191
Thornycroft, Ltd. (England), 76
Tirona, Tomas C. (PMA '40), 100–1
Tolentino, Arturo, 267
Tolentino, Ben ("Toots"), 86–87
Torres, Ruben, 322
Torres High School (Manila), 95
Torture: academic research, 185–88; CIA research, 190; legacy of, 187; litigation, 208; methods, 189–90; perpetrators, 186, 191, 217–21; psychological, 185, 188–90; UN conventions, 185; victims, 185, 187–88
Towards a Filipino Ideology (Ferdinand E. Marcos), 233
Trajano, Agapita, 327
Trajano, Archimedes, 327
Trans-Asiatic Airlines, 134
Trauma: psychological, 185–91; social, 300, 333–35
Trono, Pedro, 115, 118
Truth and Reconciliation Commission (South Africa), 300
Tuazon, Hospicio B. (PMA '40), 53, 79, 91–92
Tugung, Ulbert Ulama, 273
Turingan, Felix ("Boy") L. (PMA '65): in Christmas 1989 coup, 286, 288, 290; in coups, 231, 238, 273, 281, 284; on Gringo Honasan, 264–65; Mindanao enclave, 295

United Democratic Opposition (Unido), 140
United Nations Convention Against Torture, 185, 261
United Nations Declaration on Human Rights, 185
U.S. Air Force (Clark Field), 121, 133, 255, 290, 336
U.S. Army, 16–18, 21–22, 75, 100, 163
U.S. Army, Allied Intelligence Bureau, 96, 98
U.S. Army, Guerrilla Affairs Division, 169–70
U.S. Army, Historical Division, 163
United States Army Forces in the Far East (USAFFE), 26, 77, 161–62
U.S. Coast Guard Academy, 12
U.S. Command and General Staff College (Fort Leavenworth), 190
U.S. Congress, 20, 103
U.S. Defense Intelligence Agency (DIA), 238, 240
U.S. Department of Commerce, 148
U.S. District Courts, 327
U.S. Embassy, 29, 228, 241, 289
U.S. Infantry, 18
U.S. Library of Congress, 163
U.S. Marines, 145
U.S. Military Academy (USMA), 17–18, 20, 38–40
U.S. Military Mission, 23
U.S. Mutual Defense Treaty, 145
U.S. National Archives, 175, 177, 180
U.S. Naval War College, 147, 163
U.S. Veterans Administration (Manila), 109
U.S. War Department, 18, 21, 23
University of Santo Tomas, 80, 115
University of the Philippines (U.P.), 21, 24–25, 58–59, 337–38, 349
USAFIP–Northern Luzon, 169

Valdes, Basilio, 65, 71–72
Valdez, Simeon, 168
Valencia, Fausto C. (PMA '41), 79, 82
Valenzona, Benedicto, 100, 130
Valeriano, Napoleon E. (PMA '37), 101, 108
Valeroso, Diosdado T. (PMA '82), 281, 296
Valor (Jose M. Crisol and Uldarico S. Baclagon), 159, 175, 178
van Gennep, Arnold, 42
Vanguard Fraternity (U.P.), 22, 123, 227
Vargas, Jesus (PCA '29), 100, 135–37
Velasco, Segundo P. (PMA '40), 123–24, 126–29, 343
Velasquez, Jaime C. (USMA '31), 61, 160–63
Ver, Fabian: biography, 224–30; corruption, 119–20, 126; in EDSA revolt, 242–55; exile, 255; service, 123, 206, 223, 225
Ver, Irwin P. (PMA '70), 226, 230, 240, 247, 252
Ver, Rexor P., 226
Ver, Wyrlo P., 226, 252
Vergara, Elba, 189
Veritas (magazine), 262
Veterans Bank, 129
Veterans, Filipino, 103–5
Victor Corpus Story, The (film, 1986), 270
Viduya, Teodorico E. (PMA '71), 252
Vigilar, Gregorio R. (USMA '53), 338
Villamor, Jesus, 92–93, 96–97, 133
Villamor Air Base (Pasay City), 235, 237, 239, 252–53, 287–88
Villareal, Luis, 272
Villegas, Juan, 213
Vizmanos, Danilo, 214–15

Wakaoji, Nobuyuki, 273
Warfare: colonial, 16, 22, 27; modern, 242–43; tribal, 16, 42, 96
Warrior: colonial, 15, 17, 123–26, 168; heroic, 47, 168; military manager, 13; myths about, 168; national, 19, 26–28; nationalism and, 20, 22, 62, 346

Washington, D.C., 158, 174
We Belong (newsletter), 233
Weber, Max, 13
Weinberger, Caspar, 174
West Point. *See* U.S. Military Academy
West Point of the Philippines, The (film, 1937), 62
Women, Filipino, 20, 44, 320
Women's rights, Filipino, 45, 319
Wong, Noe A. (PMA '75), 239, 325–26
Wood, Leonard, 21
World Conference on Human Rights, 185
World War I, 19, 21
World War II, 7, 12, 20, 80, 115; Class of 1940, 52, 75; Uldarico Baclagon on, 164
Writ of Habeas Corpus. *See* Habeas Corpus

Yabes, Criselda, 280–81
Yamashita treasure, 177
Yan, Manuel T. (PMA '41), 57, 71, 127, 324
Yap, Pedro M. (PMA '40), 106
Yorac, Haydee, 304
Young Officers Union (YOU), 281, 285, 294

Zablan Field (Camp Murphy), 85, 93, 98
Zamboanga Province, 14
Zamora, Fe, 266
Zimbardo, Philip, 187, 349
Zumel, Jose Ma. D. L. (PMA '59): in coups, 267, 284, 286, 295; PMA superintendent, 30; surrender of, 303